Springer Series in Chemic

Editors: V. I. Goldanskii R. Gomer

1 **Atomic Spectra and Radiative Transitions** By I. I. Sobelman
2 **Surface Crystallography by LEED** Theory, Computation and Structural Results. By M. A. Van Hove, S. Y. Tong
3 **Advances in Laser Chemistry** Editor: A. H. Zewail
4 **Picosecond Phenomena** Editors: C. V. Shank, E. P. Ippen, S. L. Shapiro
5 **Laser Spectroscopy** Basic Concepts and Instrumentation By W. Demtröder 2nd Printing
6 **Laser-Induced Processes in Molecules** Physics and Chemistry Editors: K. L. Kompa, S. D. Smith
7 **Excitation of Atoms and Broadening of Spectral Lines** By I. I. Sobelman, L. A. Vainshtein, E. A. Yukov
8 **Spin Exchange** Principles and Applications in Chemistry and Biology By Yu. N. Molin, K. M. Salikhov, K. I. Zamaraev
9 **Secondary Ion Mass Spectrometry SIMS II** Editors: A. Benninghoven, C. A. Evans, Jr., R. A. Powell, R. Shimizu, H. A. Storms
10 **Lasers and Chemical Change** By A. Ben-Shaul, Y. Haas, K. L. Kompa, R. D. Levine
11 **Liquid Crystals of One- and Two-Dimensional Order** Editors: W. Helfrich, G. Heppke
12 **Gasdynamic Laser** By S. A. Losev
13 **Atomic Many-Body Theory** By I. Lindgren, J. Morrison
14 **Picosecond Phenomena II** Editors: R. M. Hochstrasser, W. Kaiser, C. V. Shank
15 **Vibrational Spectroscopy of Adsorbates** Editor: R. F. Willis
16 **Spectroscopy of Molecular Excitions** By V. L. Broude, E. I. Rashba, E. F. Sheka
17 **Inelastic Particle-Surface Collisions** Editors: E. Taglauer, W. Heiland
18 **Modelling of Chemical Reaction Systems** Editors: K. H. Ebert, P. Deuflhard, W. Jäger
19 **Secondary Ion Mass Spectrometry SIMS III** Editors: A. Benninghoven, J. Giber, J. László, M. Riedel, H. W. Werner
20 **Chemistry and Physics of Solid Surfaces IV** Editors: R. Vanselow, R. Howe
21 **Dynamics of Gas-Surface Interaction** Editors: G. Benedek, U. Valbusa
22 **Nonlinear Laser Chemistry** Multiple-Photon Excitation By V. S. Letokhov
23 **Picosecond Phenomena III** Editors: K. B. Eisenthal, R. M. Hochstrasser, W. Kaiser, A. Laubereau
24 **Desorption Induced by Electronic Transitions DIET I** Editors: N. H. Tolk, M. M. Traum, J. C. Tully, T. E. Madey
25 **Ion Formation from Organic Solids** Editor: A. Benninghoven
26 **Semiclassical Theories of Molecular Scattering** By B. C. Eu
27 **EXAFS and Near Edge Structures** Editors: A. Bianconi, L. Incoccia, S. Stipcich
28 **Atoms in Strong Light Fields** By N. B. Delone, V. P. Krainov
29 **Gas Flow in Nozzles** By U. Pirumov, G. Roslyakov
30 **Theory of Slow Atomic Collisions** By E. E. Nikitin, S. Ya. Umanskii
31 **Reference Data on Atoms, Molecules, and Ions** By A. A. Radzig, B. M. Smirnov
32 **Adsorption Processes on Semiconductor and Dielectric Surfaces I** By V. F. Kiselev, O. V. Krylov
33 **Surface Studies with Lasers** Editors: F.R. Aussenegg, A. Leitner, M.E. Lippitsch
34 **Inert Gases** Potentials, Dynamics, and Energy Transfer in Doped Crystals. Editor: M. L. Klein
35 **Chemistry and Physics of Solid Surfaces V** Editors: R. Vanselow, R. Howe
36 **Secondary Ion Mass Spectrometry, SIMS IV** Editors: A. Benninghoven, J. Okano, R. Shimizu, H. W. Werner
37 **X-Ray Spectra and Chemical Binding** By A. Meisel, G. Leonhardt, R. Szargan
38 **Ultrafast Phenomena IV** By D. H. Auston, K. B. Eisenthal
39 **Laser Processing and Diagnostics** Editor: D. Bäuerle

CHPH

49

Springer Series in Chemical Physics

Edited by Vitalii I. Goldanskii

Springer Series in Chemical Physics

Editors: Vitalii I. Goldanskii   Fritz P. Schäfer   J. Peter Toennies

Managing Editor: H. K. V. Lotsch

Volume 40 **High-Resolution Spectroscopy of Transient Molecules**
By E. Hirota

Volume 41 **High Resolution Spectral Atlas of Nitrogen Dioxide 559–597 nm**
By K. Uehara and H. Sasada

Volume 42 **Antennas and Reaction Centers of Photosynthetic Bacteria**
Structure, Interactions, and Dynamics
Editor: M. E. Michel-Beyerle

Volume 43 **The Atom-Atom Potential Method.** Applications to Organic Molecular Solids
By A. J. Pertsin and A. I. Kitaigorodsky

Volume 44 **Secondary Ion Mass Spectrometry SIMS V**
Editors: A. Benninghoven, R. J. Colton, D. S. Simons, and H. W. Werner

Volume 45 **Thermotropic Liquid Crystals, Fundamentals**
By G. Vertogen and W. H. de Jeu

Volume 46 **Ultrafast Phenomena V**
Editors: G. R. Fleming and A. E. Siegman

Volume 47 **Complex Chemical Reaction Systems**
Editors: J. Warnatz and W. Jäger

Volume 48 **Ultrafast Phenomena VI**
Editors: T. Yajima, K. Yoshihara, C. B. Harris, and S. Shionoya

Volume 49 **Vibronic Interactions in Molecules and Crystals**
By I. B. Bersuker and V. Z. Polinger

Volume 50 **Molecular and Laser Spectroscopy**
By Zu-Geng Wang and Hui-Rong Xia

Volumes 1–39 are listed on the back inside cover

I. B. Bersuker V. Z. Polinger

Vibronic Interactions in Molecules and Crystals

With 86 Figures

Springer-Verlag Berlin Heidelberg New York
London Paris Tokyo Hong Kong

Professor Dr. Isaac B. Bersuker
Dr. Sci. Victor Z. Polinger
Department of Quantum Chemistry, Institute of Chemistry
Academy of Sciences, MoSSR, Grosul Street 3
SU-277028 Kishinev, USSR

Series Editors

Professor Dr. Fritz P. Schäfer
Max-Planck-Institut für
Biophysikalische Chemie
D-3400 Göttingen-Nikolausberg, FRG

Professor Vitalii I. Goldanskii
Institute of Chemical Physics
Academy of Sciences, Kosygin Street 3
Moscow V-334, USSR

Professor Dr. J. Peter Toennies
Max-Planck-Institut für Strömungsforschung
Böttingerstrasse 6–8
D-3400 Göttingen, FRG

Managing Editor: Dr. Helmut K. V. Lotsch
Springer-Verlag, Tiergartenstrasse 17
D-6900 Heidelberg, Fed. Rep. of Germany

This edition is an updated version of the original Russian edition:
Vibronnuie Vsaimodeistviya v Molekulakh i Kristallakh
© Nauka, Moscow 1983

ISBN13:978-3-642-83481-3 e-ISBN-13:978-3-642-83479-0
DOI: 10.1007/978-3-642-83479-0

Library of Congress Cataloging-in-Publication Data. Bersuker, I. B. (Isaak Borisovich) [Vibronnye vzaimodeĭstviia v molekulakh i kristallakh. English] Vibronic interactions in molecules and crystals / I. B. Bersuker, V. Z. Polinger. p. cm.–(Springer series in chemical physics ; v. 49) Translation of: Vibronnye vzaimodeĭstviia v molekulakh i kristallakh. Bibliography: p. Includes index. ISBN 0-387-19259-X (U.S.) 1. Jahn-Teller effect. 2. Molecules. 3. Crystals. I. Polinger, V. Z. (Viktor Zigfridovich) II. Title. III. Series. QD461.B46313 1989 539'.6–dc 19 88-16100

Softcover reprint of the hardcover 1st edition 1989

Typesetting: ASCO Trade Typesetting Ltd., Hong Kong
2154/3150-543210 – Printed on acid-free paper

Preface

Vibronic interaction effects constitute a new field of investigation in the physics and chemistry of molecules and crystals that combines all the phenomena and laws originating from the mixing of different electronic states by nuclear displacements. This field is based on a new concept which goes beyond the separate descriptions of electronic and nuclear motions in the adiabatic approximation. Publications on this topic often appear under the title of the Jahn-Teller effect, although the area of application of the new approach is much wider: the term *vibronic interaction* seems to be more appropriate to the field as a whole.

The present understanding of the subject was reached only recently, during the last quarter of a century. As a result of intensive development of the theory and experiment, it was shown that the nonadiabatic mixing of close-in-energy electronic states under nuclear displacements and the back influence of the modified electronic structure on the nuclear dynamics result in a series of new effects in the properties of molecules and crystals. The applications of the theory of vibronic interactions cover the full range of spectroscopy [including visible, ultraviolet, infrared, Raman, EPR, NMR, nuclear quadrupole resonance (NQR), nuclear gamma resonance (NGR), photoelectron and x-ray spectroscopy], polarizability and magnetic susceptibility, scattering phenomena, ideal and impurity crystal physics and chemistry (including structural as well as ferroelectric phase transitions), stereochemistry and instability of molecular (including biological) systems, mechanisms of chemical reactions and catalysis.

The most interesting achievements of the theory of vibronic interactions are related to structural phase transitions in crystals (or, in a wider sense, to structural phase transformations in condensed media), the physics of impurity centers in crystals, and spectroscopy of polyatomic systems. In particular, structural aspects of the high-temperature superconductivity discovered recently have, beyond doubt, a vibronic nature.

This volume emerged from investigations on the theory of vibronic interactions carried out in collaboration with co-workers and colleagues at the Moldavian Academy of Sciences over a period of more than 25 years. It is the first book on this topic to be published for more than 15 years, during which time new ideas have been developed and some problems of primary importance have been solved. The authors hope that all main results on the theory of vibronic interactions in molecules and crystals obtained until now are well represented in this book, with the exception of some chemical and biological applications and some other special problems. Compared with the Russian book on which this edition is based, this volume is more complete, in particular due to the addition of a survey of Green's function approaches, the solutions of multimode and multicenter problems, evaluations of additional interesting cases of optical and photoelectron spectra, polarizabilities and birefringence, Peierls transitions, and a general model for structural phase transitions in condensed media.

This book is mainly addressed to physicists – theoreticians and experimentalists (researchers, professors, graduate students, etc.) – specialists in the field of the structure and properties of molecules and crystals. However, this book may also be useful for specialists in related fields, in particular for chemists, biologists, and materials scientists; many new properties of materials have proved to be of vibronic origin.

I. B. Bersuker

Kishinev, January 1989 *V. Z. Polinger*

Acknowledgements

We are grateful to our co-workers and colleagues. Without our shared research and continuous personal contact the general presentation of the subject given in this book would not have been possible. With special pleasure we acknowledge fruitful discussions with B. G. Vekhter, M. D. Kaplan, I. Ya. Ogurtsov, Yu. E. Perlin, Yu. B. Rosenfeld, D. I. Khomskii and B. S. Tsukerblat.

We also acknowledge the discussions and cooperation with C. A. Bates, L. S. Cederbaum, R. Englman, F. S. Ham, V. V. Hizhnyakov, N. N. Kristoffel, K.-A. Müller, M. C. M. O'Brian, D. Reinen, H. Thomas and M. Wagner.

Contents

1. Introduction

As is well known, the fundamental laws of electronic and nuclear motions determining the structure and properties of molecules and crystals were revealed in the 1930s immediately after the discovery of quantum mechanics. Because of mathematical difficulties encountered in the quantum-mechanical description of polyatomic systems, several essential approximations are usually employed in the solution of the appropriate Schrödinger equation, and among them the adiabatic approximation is the most important [1.1, 2]. This approximation is based on the difference in the masses (and hence velocities) of electrons and nuclei; due to this difference, for every position of the nuclei at any instant a stationary distribution of the electrons is attained. Without the adiabatic approximation the notion of spatial structure (nuclear configuration) becomes uncertain.

Almost immediately after the introduction of the adiabatic approximation the most striking deviations from this approximation were also understood. In 1934 Landau, in a discussion with *Teller* [1.3], pointed out that if the electronic state of a certain symmetric nuclear configuration is degenerate, then this configuration is unstable with respect to nuclear displacements which remove the degeneracy. Presumably Landau's arguments were based on ideas similar to that used in the proof of the von Neumann-Wigner theorem about crossing electronic terms [1.4] (see also [1.5]), which had already been published at that time.

The Landau statement was afterwards verified by E. Teller and H. Jahn and more rigorously formulated as what is now known as the Jahn-Teller theorem [1.6]. Actually, not only degenerate but also relatively close-in-energy (quasi-degenerate or pseudodegenerate) electronic terms cannot be treated by the adiabatic approximation.

The real content of the Jahn-Teller theorem is as follows: if the adiabatic potential of the system, which is a formal solution of the electronic part of the Schrödinger equation, has several crossing sheets, then at least one of these sheets has no extremum at the crossing point. It is clear that knowledge of this feature of the adiabatic potential is not enough in order to specify the nuclear behaviour. The latter can be determined only after the solution of the nuclear part of the Schrödinger equation. Therefore, the Jahn-Teller theorem itself, strictly speaking, does not directly contain predictions concerning observable magnitudes.

The neglect of this important circumstance resulted in a peculiar situation when for a long period of time all attempts to obtain experimental evidence of the instability predicted by the Jahn-Teller theorem (of the so-called Jahn-Teller effect) were unsuccessful or unconvincing. In 1939 *Van Vleck* wrote [1.7] that "it is a great merit of the Jahn-Teller effect that it disappears when not needed", and in 1960 *Low* in his book [1.8] added that "it is a property of the Jahn-Teller effect that whenever one tries to find it, it eludes measurement". These declarations reflect the situation when the theory of vibronic interactions was not sufficiently developed, and there was no proper understanding of the observable effects that should be expected as a consequence of the Jahn-Teller theorem. At present the situation is completely changed, and in addition to a deep understanding of vibronic coupling phenomena, achieved by means of theoretical investigations, a series of new effects and laws based on the theory of vibronic interactions have been predicted and observed experimentally, and new trends in the physics and chemistry of molecules and crystals have been initiated in this area [1.9–20].

In the 1950s the development of the ESR technique made possible observations of fine structure effects in polyatomic systems. Even the first experiments on the ESR spectra of bivalent copper compounds in twofold-degenerate electronic states revealed an unusual fine and hyperfine structure of the spectrum and their rather complicated temperature dependence. These results could be explained only by taking account of the nuclear dynamics due to the Jahn-Teller effect. Work on ESR spectra with Jahn-Teller effects has now grown into an independent research area that is presented in separate chapters in monographs and handbooks [1.9–13, 18, 20].

In the late 1950s and early 1960s the investigation of electronic and vibrational spectra taking account of vibronic coupling began. In particular, the expected fine structure of electronic transition bands was calculated, and the tunneling of the energy levels in Jahn-Teller systems with strong vibronic coupling was predicted. The experimental confirmation of these effects stimulated fast development of the area as a whole. Meanwhile, it was shown that besides the proper Jahn-Teller effect in systems with electronic degeneracy, similar effects are expected in the case of mixing of close-in-energy electronic states (pseudo-Jahn-Teller effect), which together with the Renner effect (instability of linear molecules) form the so-called vibronic effects. At present the correct interpretation of the observed shapes of broad bands of optical absorption and luminescence, in particular, the number of maxima, the halfwidth of component bands, their asymmetry, intensity and temperature changes, cannot be carried out without considering the vibronic effects. The same is true for the interpretation of the fine structure of spectra of molecules in the gas phase or in matrices, as well as of impurity crystals at low temperatures [1.9, 11, 13, 16, 17, 20].

In the early 1960s it became clear that the interaction of dynamically unstable Jahn-Teller centers in crystals at a certain temperature may result in an ordering of the local distortions manifested as a phase transition (cooperative Jahn-Teller effect). Simultaneously it was shown that in the case of dipolarly unstable centers in crystals (i.e. when the vibronic effects result in distortions with dipole moment formation) such an ordering may lead to the spontaneous polarization of the crystal and ferroelectric phase transitions. The experimental confirmation of these predictions resulted in the development of a new trend in the physics of structural phase transitions and ferroelectricity, which is still continuing today [1.11, 13, 14, 19].

Besides the examples listed above, the vibronic effects have been shown to be important in infrared, Raman, NMR, nuclear quadrupole resonance (NQR), nuclear gamma resonance (NGR), photoelectron and x-ray spectroscopy, dipole (multipole) moments, polarizabilities and magnetic succeptibilities, crystal chemistry, stereochemistry and instability of molecules, mechanisms of chemical reactions, and catalysis [1.11, 16, 28].

All the aforesaid demonstrates the general significance of the vibronic approach to the solution of various problems of the modern physics and chemistry of molecules and crystals. This allows one to speak about the formation of a new concept in the physics of polyatomic systems—the concept of vibronic interactions. The distinguishing feature of this concept is that it considers the non-separability of electronic and nuclear motions, i.e. it goes beyond the adiabatic approximation. Papers on this area of specialization are in most cases published under the historically established title of the "Jahn-Teller effect".

Concerning the subjects of the vibronic interaction theory, i.e. polyatomic systems to be treated by the vibronic approach, a priori there are no exceptions so far, provided these systems contain more than two atoms. Indeed, in any such systems there are either electronic degenerate (ground or excited) states, or pseudodegenerate ones (the energy gap between the mixing states may be very large). Effects formally similar to those mentioned above may be present in other physical systems, e.g. in elementary particles, such as in the meson-nucleon interaction in quantum field theory [1.21], in the α-cluster description of light nuclei [1.22], in heavy nuclei (giant resonances) [1.23], in the analysis of resonant interaction of light with matter [1.24, 25], etc. [1.26].

As is clear from the title, this book is devoted to vibronic effects only in molecules and in crystals. Its main goal is to give a systematic presentation of the theoretical background to the concept of vibronic interactions, the main ideas, methods and applications of this approach. Great attention is paid to the explanation of the physical meaning of the results at the expense of a more-detailed description of the experimental data. The latter are given only as illustrations of the main conclusions of the theory. Certainly, the list of references cannot pretend to completeness. All the works published before the end of 1979 can be found in the Bibliographic Review [1.27].

The distribution of the material in the chapters is as follows. In the following three chapters the theory of vibronic interactions is presented. In Chap. 2, starting with the Born-Oppenheimer approximation, the vibronic Hamiltonian which rigorously considers the coupling between electronic and nuclear motions is introduced. The discussion of the Jahn-Teller theorem is also given there. In Chap. 3 the adiabatic potentials in cases of electronic degeneracy or pseudodegeneracy are evaluated and analysed. These adiabatic potentials are used in

Chap. 4 to solve the vibronic equations and to obtain the energy spectrum and wave functions of polyatomic systems. The second half of the book (Chaps. 5 and 6) is devoted to the applications of the theory of vibronic interactions to problems of spectroscopy and structural phase transitions. In Chap. 5 the shape of electronic transition bands, their fine structure, zero-phonon lines, infrared absorption, Raman and ESR spectra are presented. The theory of the cooperative Jahn-Teller effect and structural phase transitions in crystals, including the vibronic theory of ferroelectricity, is given in Chap. 6.

2. Vibronic Interactions and the Jahn-Teller Theorem

In this chapter the first principles which form the basis of the concept of vibronic interactions, the concept of mixing electronic states by nuclear displacements, are discussed. After analysis of the traditional adiabatic approximation and its limitations (Sec. 2.1), the so-called vibronic Hamiltonian which takes into account the nonadiabatic interaction of electrons and nuclei is deduced (Sect. 2.2). The proof and discussion of the Jahn-Teller theorem are presented in Sect. 2.3 by considering the shape of the adiabatic potential surface in the neighborhood of the point of electronic degeneracy where the deviations from the adiabatic approximation are most marked. This theorem served as a starting point for the development of the field as a whole, and therefore the physical consequences of the nonadiabatic vibronic mixing of electronic states are often called the Jahn-Teller effects.

2.1 Adiabatic Approximation

2.1.1 Proper Adiabatic Approximation

The polyatomic system considered as a set of interacting electrons and nuclei is a quantum system, and in most cases it can be described by means of wave functions Ψ, which are solutions of the appropriate Schrödinger equation $H\Psi = E\Psi$. In the general case of a large number of particles linked by the Coulomb interaction, this equation cannot be solved analytically and its numerical solution by means of computers is inaccessible for practical reasons in most cases. Therefore the necessity of models and approximate solutions (based on these models) arises.

The first and most important approximation is the adiabatic one [2.1, 2]. This is based on the fact that the nuclear masses are much larger than the electronic ones (for the hydrogen atom the ration of these masses is $5:10^4$) and therefore, on average, the nuclei of the polyatomic system move much more slowly than the electrons. The latter are thus able to follow the nuclear displacements, i.e. their distribution in space is determined by the instantaneous nuclear configuration. This allows us to assume, as a first approximation, that the nuclei are fixed, and to consider their displacements by means of perturbation theory in higher-order approximations. Owing to the uniformity and isotropy of space, the translational and rotational degrees of freedom of an isolated polyatomic system can be described by cyclic coordinates, and, generally speaking, they can be separated. Note, however, that the separation of the rotational degrees of freedom is not trivial [2.3]. Considering this problem solved, we can write the Hamiltonian of the remaining degrees of freedom of the polyatomic system in the form

$$H(\mathrm{r},Q) = H(\mathrm{r}) + V(\mathrm{r},Q) + T(Q) \ . \tag{2.1.1}$$

Here $H(\mathrm{r})$ includes the kinetic energy of the electrons and all the interelectronic interactions, r is the coordinate of the electronic degrees of freedom, including space and spin ones, Q is the configurational coordinate of the vibrational degrees of freedom, $V(\mathrm{r},Q)$ is the operator considering the energy of both electron-nuclei and nuclei-nuclei Coulomb interactions, and $T(Q)$ is the operator of the kinetic energy of the nuclei. The relativistic interactions of the electron spin with the nuclear motions are neglected.

As mentioned before, the kinetic energy T of the nuclei may be regarded as a small perturbation. Let us assume that the Schrödinger equation for the electronic motion in the field of arbitrary fixed positions of the nuclei is solved in zero order of the perturbation theory:

$$[H(\mathrm{r}) + V(\mathrm{r},Q)]\varphi_n(\mathrm{r},Q) = \varepsilon_n(Q)\varphi_n(\mathrm{r},Q) \ . \tag{2.1.2}$$

The eigenvalues $\varepsilon_n(Q)$ and eigenfunctions $\varphi_n(\mathrm{r},Q)$ have the nuclear coordinates Q as parameters, the latter forming a complete functional orthonormalized set. Now we look for the exact eigenfunctions of the exact Hamiltonian (2.1.1) in the form of the expansion

$$\Psi(\mathrm{r},Q) = \sum_n \chi_n(Q)\varphi_n(\mathrm{r},Q) \ , \tag{2.1.3}$$

where the coefficients $\chi_n(Q)$ are obviously functions of Q. In order to find them, we substitute (2.1.3) into the Schrödinger equation with the Hamiltonian (2.1.1), multiply the latter from the left by $\varphi_m^*(\mathrm{r},Q)$, and integrate it with respect to r. Taking into account that $T(Q) = -\hbar^2\sum_i(2M_i)^{-1}\partial^2/\partial Q_i^2$, where i labels the nuclear degrees of freedom (except the separated translational and rotational ones) and M_i is the reduced mass of the appropriate symmetrized coordinate Q_i, we obtain

$$\sum_n [T_{mn}(Q) + U_{mn}(Q)]\chi_n(Q) = E\chi_m(Q) \ . \tag{2.1.4}$$

Here

$$\begin{gathered} U_{mn}(Q) = \varepsilon_n(Q)\delta_{mn} \ , \qquad T_{mn}(Q) = T(Q)\delta_{mn} + \Lambda_{mn}(Q) \ , \\ \Lambda_{mn}(Q) = -\hbar^2\sum_i \frac{1}{M_i}\left[A_{mn}^{(i)}(Q)\frac{\partial}{\partial Q_i} + \frac{1}{2}B_{mn}^{(i)}(Q)\right] \ , \end{gathered} \tag{2.1.5}$$

$$A_{mn}^{(i)} = \int \varphi_m^* \frac{\partial \phi_n}{\partial Q_i}\mathrm{d}\mathrm{r} \ , \qquad B_{mn}^{(i)} = \int \varphi_m^* \frac{\partial^2 \phi_n}{\partial Q_i^2}\mathrm{d}\mathrm{r} \ . \tag{2.1.6}$$

The system of coupled differential equations (2.1.4) may be regarded as an eigenvalue problem for the matrix Hamiltonian $\|H_{mn}\| = \|T_{mn} + U_{mn}\|$, the matrix elements of which remain operators in the Q space. (The double vertical lines denote a matrix, e.g. $\|H_{mn}\|$ is a matrix $\hat{H}$ with matrix elements H_{mn}.) It should be emphasized that (2.1.4) is obtained from the Schrödinger equation with the Hamiltonian (2.1.1) by means of indentity transformations, and in this sense it is an exact system.

As is well known, the perturbation theory for singlet states is different from that for degenerate states. In the former case it leads to the traditional adiabatic approximation, whereas for degenerate states it results in the vibronic Hamiltonian which allows for the nonadiabatic mixing of electronic states having the same or close energies.

Consider first the nondegenerate electronic state $\varphi_n(\mathrm{r}, Q)$. Following the rules of perturbation theory, we consider in the first order just the diagonal elements of the matrix Hamiltonian $\|H_{mn}\|$. The resulting Hamiltonian for the $\chi_n(Q)$ functions

$$H_n = T(Q) + \Lambda_{nn} + \varepsilon_n(Q) \tag{2.1.7}$$

is equivalent to neglecting all the terms in (2.1.3) except the nth one:

$$\Psi(\mathrm{r}, Q) = \chi_n(Q)\varphi_n(\mathrm{r}, Q) \ . \tag{2.1.8}$$

This result is not trivial. Indeed, considering the significant interaction of electrons with nuclei, the behavior of the nuclear subsystem has to be described, generally speaking, by means of the density matrix obeying the quantum Liouville equation. The neglect of the off-diagonal matrix elements $T_{nm} = \Lambda_{nm}$, $m \neq n$ permits the change from mixed states of the nuclear subsystem to pure ones.

This result has a clear-cut physical meaning. The distribution of electrons is described by the factor $\varphi_n(\mathrm{r}, Q)$, which depends on the nuclear coordinates. This means that the motion of the nuclei leads only to the deformation of the electron distribution and not to transitions between different electronic states. In other words, a stationary distribution of electrons is obtained for each instantaneous position of the nuclei, i.e. the electrons follow the nuclear motion adiabatically. The distribution of the nuclei is described by the wave function $\chi_n(Q)$ determined by the Schrödinger equation with the Hamiltonian (2.1.7), where its potential energy includes $\varepsilon_n(Q)$, the mean field of all the electrons of the system. This is the *adiabatic approximation.* The operators Λ_{mn}, $m \neq n$, not taken into account in this approximation, lead to transitions between the states φ_n and φ_m. They are called *operators of nonadiabaticity*, while the magnitudes $\varepsilon_n(Q)$ are called *adiabatic potentials.*

The stationary states $\varphi_n(\mathrm{r}, Q)$ may be chosen real. In this case Λ_{nn} does not include differential operators, since, as can be easily shown, all the $A_{nn}^{(i)}$ values are equal to zero. For the most interesting lowest nuclear states localized near the minima of the potential $\Lambda_{nn}(Q) + \varepsilon_n(Q)$ in the Hamiltonian (2.1.7), the dependence of $\Lambda_{nn}(Q)$ on Q, as a rule, can be neglected in comparison with the relatively strong function $\varepsilon_n(Q)$. Just the latter determines the existence and the position of the extrema of the potential $\Lambda_{nn}(Q) + \varepsilon_n(Q)$. Therefore, the adiabatic potentials $\varepsilon_n(Q)$ often allow true conclusions to be drawn about the nature of the nuclear motions without detailed solution of the Schrödinger equation.

For example, if $\varepsilon_n(Q)$ has a relatively deep minimum, one can state a priori that the lowest states near the bottom of the adiabatic potential correspond to small vibrations near the minimum point. This statement can be easily proved by means of the harmonic approximation. Let us assume that in the neighborhood of the absolute minimum Q_0 the potential energy $\Lambda_{nn}(Q) + \varepsilon_n(Q)$ is a smooth function that may be expanded in a power series. Consider small vibrations near this point, which is equivalent to the restriction to quadratic terms in the potential energy

$$\Lambda_{nn} + \varepsilon_n \approx E_n^{(0)} + \tfrac{1}{2} \sum_{ij} W_{ij}(Q_i - Q_i^{(0)})(Q_j - Q_j^{(0)}) \ , \tag{2.1.9}$$

where $E_n^{(0)} = \Lambda_{nn}(Q_0) + \varepsilon_n(Q_0)$, and $W_{ij} = \partial^2[\varepsilon_n(Q_0) + \Lambda_{nn}(Q_0)]/\partial Q_i \partial Q_j$. The Schrödinger equation with such a quadratic potential can easily be solved by means of transforming to normal coordinates, resulting in the well-known oscillator solution.

The adiabatic approximation is justified when the solution of the exact system of equations (2.1.4) differs a little from the approximate solution (2.1.8) given by the Hamiltonian (2.1.7). This is satisfied when the criterion of perturbation theory is fulfilled. This means that the first-order correction to the zeroth-order wave function (2.1.8) has to be small. Taking the oscillator solutions of the Schrödinger equation with the quadratic potential (2.1.9) as the $\chi_n(Q)$ functions, and as the perturbation operator, the approximate expressions for $T_{nm} = \Lambda_{nm}$, $n \neq m$, where the smooth functions $A_{nm}^{(i)}(Q)$ and $B_{nm}^{(i)}(Q)$ are replaced by their values at the point Q_0, we obtain as a simple criterion for the applicability of the adiabatic approximation [2.2]

$$\frac{\hbar\omega}{|E_n^{(0)} - E_m^{(0)}|} \ll 1 \ . \tag{2.1.10}$$

Here ω is the largest frequency of the small nuclear vibrations near the point Q_0. (For a more detailed discussion of the criteria of applicability of the adiabatic approximation see [2.4].)

The criterion (2.1.10) refers to the case of stable polyatomic systems when the matrix of second derivatives W_{ij} in (2.1.9) is a positive definite matrix and the approximation (2.1.9) leads to a discrete vibrational spectrum. In the case of a continuous spectrum (unstable configurations) the criterion has another form and can be reduced to the requirement that the so-called Massey parameter, the ratio $|E - \varepsilon_n(Q)|/|\varepsilon_m(Q) - \varepsilon_n(Q)|$, is small [2.5].

Some estimates of orders of magnitude which demonstrate the extent of approximation of the results obtained can be presented. According to the uncertainty relation, we have $\bar{p} \sim \hbar/d$, where $\bar{p} = m\bar{v}$ is the mean momentum of the electrons, m and $\bar{v}$ being their mass and velocity, respectively, and d is the linear dimension of the electron distribution. From this we find $\bar{v} \sim \hbar/(md)$.

For the nuclei which vibrate near the point Q_0, considering the virial theorem we have $M\bar{V}^2/2 \sim M\omega^2\bar{Q}^2/2 \sim \hbar\omega$, from which it follows that $\bar{V} \sim (\hbar\omega/M)^{1/2}$ and $\bar{Q} \sim (\hbar/M\omega)^{1/2}$. Here M is the mass of the nuclei and $\bar{V}$ and $\bar{Q}$ are the mean square values of the velocities and coordinates, respectively, the latter having the equilibrium value Q_0 as their zero. In order to estimate the frequency ω one can determine the force constant $M\omega^2$ by considering the Coulomb interaction of the frozen electronic distribution with the nucleus displaced from its center: $M\omega^2 \sim e^2/d^3 \sim \hbar^2/md^4$, where e is the charge of the electron. Here the virial theorem is also employed, according to which the potential energy of the electrons ($\sim e^2/d$) is of the same order of magnitude as its kinetic energy ($\sim \hbar^2/md^2$). It follows that $\omega \sim (\hbar/md^2)(m/M)^{1/2}$, and taking into account the above estimates, we have

$$\bar{V} \sim \bar{v}\left(\frac{m}{M}\right)^{3/4}, \qquad \bar{Q} \sim d\left(\frac{m}{M}\right)^{1/4}. \tag{2.1.11}$$

It is also of interest to estimate the value of $\overline{\Delta E}$, the characteristic average energy gap in the spectrum of the electronic subsystem. In order to do this, the uncertainty relation between the energy and time, $\overline{\Delta E} \sim \hbar/\overline{\Delta t}$, can be used, where $\overline{\Delta t} \sim d/\bar{v} \sim md^2/\hbar$ is the period of the finite motion of the electrons. It follows that $\overline{\Delta E} \sim \hbar^2/md^2$, and considering the above estimate of ω, we get $\hbar\omega/\overline{\Delta E} \sim (m/M)^{1/2}$. Thus, as expected, the criterion of validity of the adiabatic approximation (2.1.10) is determined by the difference in the masses of electrons and nuclei.

It should be emphasized that the above estimates are valid only for the corresponding mean values and may not be valid for some concrete values of the magnitudes under consideration. For instance, the estimate $\overline{\Delta E} \sim \hbar^2/md^2 \sim 1$–$10$ eV gives the order of magnitude of the mean spacing in the energy spectrum of the electronic subsystem, but the electronic spectrum may contain degenerate ($\Delta E = 0$) or quasi-degenerate (with small ΔE values) states for which the above estimations are incorrect. Indeed, using the uncertainty relation for energy one can find that $\bar{v} \sim \Delta E\, d/\hbar$, from which it follows that in the case of degenerate or quasi-degenerate electronic states the idea of fast electrons loses its physical meaning and the adiabatic approximation is invalid. These cases and related special effects which are beyond the framework of the simple adiabatic theory will be discussed in the following sections.

2.1.2 Born-Oppenheimer Approximation

Note that, as follows from the estimate (2.1.11) the harmonic approximation (2.1.9) used in the previous section is justified by the smallness of the value $(m/M)^{1/4}$, i.e., by the difference in the masses of electrons and nuclei, as is the adiabatic approximation. This means that in fact the value $\kappa = (m/M)^{1/4}$ is the small parameter of the theory. Consequently, one can perform the substitution $Q = Q_0 + \kappa q$, where the new coordinates $q = \{q_1, q_2, \ldots\}$ do not contain small parameters, separate explicitly the small parameter κ in the initial Hamiltonian, and carry out perturbation theory calculations with respect to this parameter [2.1, 3]. In order to obtain the Hamiltonian of the zeroth-order approximation from the exact one (2.1.1), it is necessary not only to exclude the kinetic energy of the nuclei, but also to assume that $Q = Q_0$ (since in the initial approximation $\kappa q = Q - Q_0 = 0$).

Thus, as an initial basis in this case the states $\psi_n(\mathrm{r}) = \phi_n(\mathrm{r}, Q_0)$, the eigenfunctions of the Hamiltonian

$$H(\mathrm{r}, Q_0) = H(\mathrm{r}) + V(\mathrm{r}, Q_0) \ , \tag{2.1.12}$$

are used, and instead of the expansion (2.1.3) we have

$$\Psi(\mathrm{r}, Q) = \sum_n \chi_n(Q)\psi_n(\mathrm{r}) \ . \tag{2.1.13}$$

The physical background for such an approach can be understood if one takes into account that in the ground vibrational state the nuclear configuration is described by the distribution $|\chi_n(q)|^2 = (\kappa\sqrt{\pi})^{-1}\exp[-(Q-Q_0)^2/\kappa]$, which for small κ values to a good approximation has the properties of a δ-function (this can be easily proved by taking the limit $\kappa \to 0$). This allows the solution (2.1.8) to take the form[1]

$$\Psi(\mathrm{r}, Q) \simeq \chi_n(Q)\varphi_n(\mathrm{r}, Q_0) = \chi_n(Q)\psi_n(\mathrm{r}) \ . \tag{2.1.14}$$

In other words, in the limit of very heavy nuclei the nuclei are localized at the minima Q_0 of the potential surface, and it is sensible to consider just the electronic states φ_n corresponding to these fixed nuclear configurations.

Now substitute the wave function (2.1.13) into the Schrödinger equation with the Hamiltonian (2.1.1), multiply it from the left by $\psi_m^*(\mathrm{r})$, and integrate with respect to r. As a result one obtains a system of an infinite number of coupled differential equations of the form (2.1.4), where $T_{mn}(Q) = T(Q)\delta_{mn}$ and

$$U_{mn}(Q) = \int \psi_m^*(\mathrm{r})[H(r) + V(\mathrm{r}, Q)]\psi_n(\mathrm{r})\,dr \ . \tag{2.1.15}$$

Thus, the matrix of the operator of kinetic energy of the nuclei in the new basis is diagonal, whereas the matrix of the remaining part of the Hamiltonian is nondiagonal.

In this case, too, the system (2.1.4) is obtained by exact solution of the Schrödinger equation with the Hamiltonian (2.1.1) by means of identity transformations, and in this sense it is an exact system.

Further consideration of the perturbation theory depends on whether the electronic state $\psi_n(\mathrm{r})$ under consideration is a singlet, or whether it is one of the degenerate or quasi-degenerate states.

---

[1] The identity $f(Q)\delta(Q - Q_0) = f(Q_0)\delta(Q - Q_0)$ is valid provided $f(Q)$ is not singular.

Consider first the case of a nondegenerate state where the singlet term $E_n^{(0)} = \varepsilon_n(Q_0)$ is separated from the other electronic terms by large energy gaps. In first order perturbation theory one has to take into account only the diagonal elements of the matrix Hamiltonian $\|H_{nm}\| = \|T_{nm} + U_{nm}\|$, from which we obtain

$$H_n = T(Q) + U_{nn}(Q) \tag{2.1.16}$$

as the Hamiltonian for the determination of the $\chi_n(Q)$ function. This is equivalent to neglecting all the terms in (2.1.13) except the nth one, i.e., to looking for the solution in the form of (2.1.14).

As follows from (2.1.16), the matrix elements $U_{nn}(Q)$ in this approximation play the role of the potential energy of the nuclei. Therefore from now on $\|U_{nm}\|$ will be called the matrix of potential energy of the nuclei. The unitary transformation of the basis which leads to the diagonal form of $\|U_{nm}\|$ makes the matrix of the kinetic energy operator nondiagonal, thus returning us to the initial formulation of the problem (to the basis states $\varphi_n(\mathrm{r}, Q)$). The eigenvalues of the matrix $\|U_{nm}\|$ coincide with the adiabatic potentials $\varepsilon_n(Q)$ determined in (2.1.2).

The approach based on the expansion (2.1.13) leading to the approximate wave function of the form (2.1.14), as distinguished from the proper adiabatic approximation considered above, is called the *crude adiabatic approximation*, or the *Born-Oppenheimer approximation*[2], after the authors who first suggested it in 1927 [2.1]. Both approaches are linked by the unitary transformation of the electronic basis, and from this point of view they are equivalent. But this equivalency is valid only for exact solutions. The convergency of approximate solutions in these two cases is different.

In particular, the Born-Oppenheimer approximation necessarily involves the expansion of the potential energy in a power series with respect to nuclear displacements from the point Q_0, and this, as mentioned, leads to the convergency determined by the small parameter $\kappa = (m/M)^{1/4}$, and not by $\kappa^3 = (m/M)^{3/4}$ as in the case of the adiabatic approximation.

Moreover, in some cases the weak convergency or even invalidity of the Born-Oppenheimer approximation is caused not by the large contribution of the operator of nonadiabaticity, but by the significant anharmonicity of the potential surface $\varepsilon_n(Q)$. If this anharmonicity results in a large amplitude $Q - Q_0$ of nuclear vibrations, then the transformation to new coordinates $q = (Q - Q_0)/\kappa$ with consequent expansion of the Hamiltonian in a power series with respect to κ has no physical sense.

The applicability of the adiabatic approximation combined with the invalidity of the *Born-Oppenheimer* approximation may be important in the case of the so-called nonrigid molecules [2.3], provided the energy gap to the nearest upper sheet of the adiabatic potential is large enough. The potential energy $\varepsilon_n(Q)$ in these cases has several minima, which are comparable in energy and are divided by rather low barriers. It should be emphasized once more that the adiabatic

[2] Note that there is a discrepancy in the names of different versions of the adiabatic approximation (see the review article [2.6]).

approximation may be valid for some nonrigid molecules, and the main difficulties of the Born-Oppenheimer approach are caused here in fact by the complicated nuclear dynamics in significantly anharmonic potentials $\varepsilon_n(Q)$. On the other hand, the Born-Oppenheimer approach is convenient because it allows the use of symmetry considerations, since the electronic states $\psi_n(\mathrm{r})$ form the basis of irreducible representations of the symmetry group appropriate to the nuclear configuration Q_0.

The criterion of validity of the Born-Oppenheimer approximation can be obtained from the requirement that the first-order perturbation corrections to the wave functions (2.1.14) be sufficiently small. It is obvious that here the off-diagonal matrix elements $U_{nm}(Q)$ have to be taken as perturbations, and since $U_{nm}(Q_0) = \varepsilon_n(Q_0)\delta_{nm}$, the expansion of $U_{nm}(Q)$, $m \neq n$, in a power series with respect to the displacements $Q - Q_0 \sim \kappa$ begins with the linear term. Taking, as above, the oscillator solutions for the $\chi_n(Q)$ functions in (2.1.14), we come to the same criterion (2.1.10) as in the case of the adiabatic approximation.

Finally, for any polyatomic system there are electronic states for which the inequality (2.1.10) is invalid and the adiabatic approximation (as well as the Born-Oppenheimer approximation) is inapplicable. In such systems the electronic term energies are very close or coincide at the point Q_0 of the configurational space of the nuclei. The degeneracy of the terms in the electronic subsystem is usually due to the high symmetry of the nuclear configuration at the point Q_0. In a polyatomic system of n atoms the adiabatic potentials are functions of $s = 3n - 6$ interatomic distances. From the geometrical point of view, each of these adiabatic potentials is a surface in $s + 1$ dimensions, energy and s nuclear coordinates. Generally speaking, these surfaces may be degenerate not only at a single point, but also along a line or a manifold of higher dimensionality ($\mathrm{r} \leq s$).

The question of the dimension of the manifold on which there may be, in principle, degeneracy of the adiabatic potentials was investigated by *Wigner* and *von Neumann* [2.7]. In particular, they showed that two adiabatic potentials of the same symmetry may be degenerate on the manifold in $s - 2$ dimensions, but if the adiabatic potentials have different symmetry, then the degeneracy exists on the manifold in $s - 1$ dimensions.[3] Thus for a two-dimensional subspace of displacements ($s = 2$) the adiabatic potentials may be represented by surfaces in the three-dimensional system of coordinates. The degeneracy of these surfaces occurs along lines in the case of different symmetry ($s - 1 = 1$), and at a single point ($s - 2 = 0$) when the surfaces have the same symmetry.

The method developed by Wigner and von Neumann can be used to go beyond the results of [2.7] and to investigate the shape of the adiabatic potential surface in the immediate neighborhood of the point of degeneracy [2.8]. In order to do this, it is necessary to expand the electronic Hamiltonian $H(\mathrm{r}) + V(\mathrm{r}, Q)$ in

[3] Since a polyatomic system may have internal (dynamic) symmetry in addition to the obvious space symmetry, this rule may have accidental exceptions. They can be removed if the electronic terms are classified by the irreducible representations of the extended group which takes into account the dynamic symmetry as well.

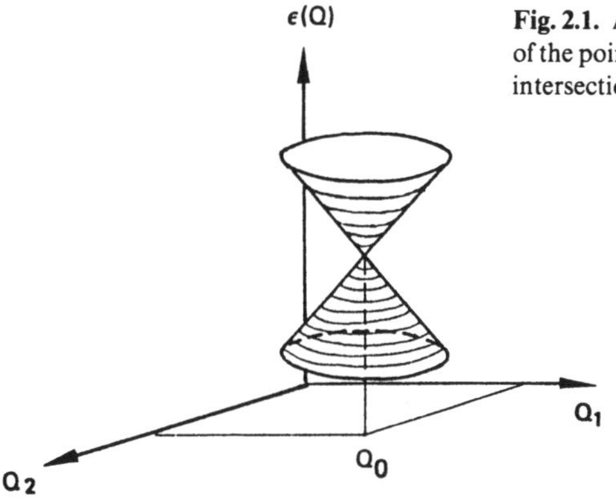

Fig. 2.1. Adiabatic potentials in the immediate vicinity of the point of degeneracy Q_0 have the shape of a conic intersection

a Taylor series with respect to the displacements $Q_i - Q_{0i}$, breaking off the series at the linear term, and to find the first-order corrections to the degenerate values of the adiabatic potential $\varepsilon_n(Q_0) = \varepsilon_m(Q_0)$ considering the displacements $Q_i - Q_{0i}$ as small perturbation parameters. Since all the nonzero (by symmetry) matrix elements of the perturbation are linear in $Q_i - Q_{0i}$, it is clear that the roots of the secular equation are also linearly dependent on $Q_i - Q_{0i}$. In particular, for the example $s = 2$ discussed above, the $\varepsilon_n(Q)$ and $\varepsilon_m(Q)$ functions are represented near the point of degeneracy by an arbitrarily spaced double elliptic *conical surface* [2.8]. As can be seen from Fig. 2.1, which illustrates this example, the adiabatic potentials have no extremum at the point of degeneracy Q_0.

A more detailed investigation of the shape of the adiabatic potential surfaces near the point of degeneracy was performed by Jahn and Teller [2.9]. Using space-time symmetry considerations, they showed that in all cases of non-accidental degeneracies (see footnote 3) except two (see below), the point of degeneracy cannot be an extremum point of the adiabatic potential. On the basis of this result Jahn and Teller reached the conclusion that polyatomic systems in nuclear configurations with electronic degeneracy are unstable.

Postponing detailed discussion of this conclusion to later sections here we just note that this conclusion is valid solely in the area far away from the point Q_0 where the energy distance between the adiabatic potential surfaces is large enough and the contribution of the nonadiabaticity operator Λ_{nm}, $n \neq m$, may be neglected. If the forces of chemical bonding are not large enough to compensate the Jahn-Teller (JT) instability, the polyatomic system dissociates. Such phenomena are the subject of investigation in the theory of nonadiabatic chemical reactions [2.5]. But if the chemical bond is strong enough and the polyatomic system remains stable, then the above instability induced by the electronic degeneracy at the point Q_0 results in a complicated nuclear dynamics. A similar situation occurs in the case of close-in-energy (quasi-degenerate) electronic terms (Sect. 3.4). All these phenomena are often met in literature under the general title of the *Jahn-Teller* or *pseudo-Jahn-Teller effect*.

2.2 The Vibronic Hamiltonian

2.2.1 Reference Nuclear Configuration

For the formulation of the problem in the case of electronic degeneracy or quasi-degeneracy, let us determine more precisely the notions of Jahn-Teller center, reference space configuration, symmetrized displacements and normal coordinates of nuclei, and introduce appropriate simplifications into the Hamiltonian. First it should be noted that for systems with a degenerate electronic term separated from the other electronic states by large enough energy gaps, the nonadiabatic mixing with the excited states may be neglected within the framework of the adiabatic approximation (Sect. 2.1). Since a discrete term describes localized electronic states, the group of atoms involved in the degenerate or quasi-degenerate electronic states is usually called the *Jahn-Teller center*. The latter may be only a part of the polyatomic system which shows the Jahn-Teller effect. Then we assume that the polyatomic system under consideration is chemically stable and therefore the expected nuclear displacements are relatively small.

If the reference configuration is determined correctly, it allows one to expand the Hamiltonian in a power series with respect to nuclear displacements $Q_i - Q_{0i}$ and, within the approximation of quadratic terms, to introduce normal coordinates, thus considerably simplifying the solution of the problem.

It is obvious that the nuclear configuration for which the electron degeneracy occurs has to be chosen as the starting point Q_0. In principle this configuration is known from experimental data. It is also obvious that the electronic degeneracy is due to the maximum symmetry of the system, and therefore it is sometimes possible to choose the Q_0 configuration by means of symmetry considerations. The exact Hamiltonian of a polyatomic system is invariant with respect to all permutations of the same nuclei and to inversion. These symmetry elements form the group G which is usually called the Longuet-Higgins group [2.10]. It contains the operations R which when applied to the Q_0 configuration result in rotations and inversion of the nuclear framework.

Now we take into account that different configurations of the polyatomic system which transform into one another by nonrotational operations of the symmetry group G, are separated, as a rule, by very high potential barriers, whereas those related by operations R are not separated by potential barriers at all. Therefore, in most cases the nonrotational elements of symmetry may be excluded from consideration. (Exceptions are the so-called nonrigid molecules [2.3].) Thus the problem is reduced to the selection of an appropriate molecular point group which contains the maximum number of elements of the group G. Usually this selection is not difficult.

For example, for a molecule X_3 which contains three identical atoms, the group $G = \pi(3)$ is isomorphous to the group of symmetry of an equilateral triangle D_{3h}, and therefore the reference configuration has to be chosen appropriate to an equilateral triangle.

For a complex ML_4 (here M is a metal atom or ion and L is a ligand) the Longuet-Higgins group is a direct product $\pi(4) \times C_i$, where $\pi(4)$ is the group of permutations of four identical nuclei of the ligands and C_i is the group containing two elements: the identity element and the inversion. Since in the system ML_4 there should be a special point (the position of the metal ion) which takes no part in the symmetry operations, the group $\pi(4)$ is isomorphous to the group of the tetrahedron T_d, and it is clear that the configuration Q_0 has to correspond to the tetrahedron with the metal ion at the center and ligands at the vertices [the existence of quadratic complexes can be explained by the (pseudo-)Jahn-Teller instability of the tetrahedral structure in these cases, due to which the symmetry is reduced from T_d to D_{4h}].

The permutation groups $\pi(6)$ and $\pi(8)$ which describe the symmetry of the complexes ML_6 and ML_8 are not isomorphic with any of the point groups. But of all the point groups which are homomorphic to them, O_h would have to be chosen as the one richest in rotational operations of symmetry. The configuration of an octahedron for ML_6 and a cube for ML_8 correspond to this group.

Determining in this way the point symmetry group of the nuclear configuration Q_0 (we denote this group by G_0) we thus determine almost all the coordinates of the point Q_0 in the configurational space of nuclear displacements. Only those coordinates which do not lower the symmetry G_0 of the molecule remain vague. The nuclear motions along these degrees of freedom are not related to the Jahn-Teller effect, as will be seen later, and can easily be separated. Therefore we assume that these coordinates of the point Q_0 are also known, they can be defined in the process of the solution of the problem when necessary (Sect. 2.2.3).

In accordance with the estimate (2.1.11) the nuclear displacements $Q_i - Q_{0i}$ from the point Q_0 are relatively small, and therefore the operator $V(\mathrm{r}, Q)$ in the exact Hamiltonian (2.1.1) can be replaced by the first terms of its power-series expansion,

$$V(\mathrm{r},Q) \simeq V(\mathrm{r},Q_0) + \sum_i \left(\frac{\partial V}{\partial Q_i}\right)_0 (Q_i - Q_{0i})$$
$$+ \frac{1}{2}\sum_{ij} \left(\frac{\partial^2 V}{\partial Q_i \partial Q_j}\right)_0 (Q_i - Q_{0i})(Q_j - Q_{0j}) \; . \qquad (2.2.1)$$

Note that the exact Hamiltonian (2.1.1) and the operator $V(\mathrm{r}, Q)$ on the left side of (2.2.1) are invariants of the group G, and therefore they are scalars of the subgroup G_0 not only at the point Q_0, but also anywhere in its neighborhood. The approximate Hamiltonian obtained from (2.1.1) by replacing the operator $V(\mathrm{r}, Q)$ by the right-hand side of (2.2.1) is an invariant of the group G_0, but owing to the cutoff of the expansion it is devoid of all the other symmetry properties which supplement G_0 to form G.

Since the Jahn-Teller displacements are small compared with the interatomic distances, the limitation of the expansion (2.2.1) to quadratic terms is usually satisfactory (see Chap. 3).

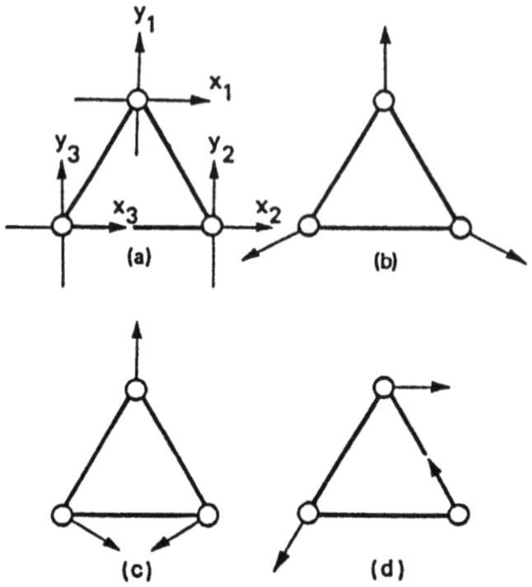

Fig. 2.2a–d. Symmetrized displacements of atoms in a triangular molecule X_3. (**a**) Labeling of Cartesian displacements; (**b**) totally symmetrized displacement of the type A_1; (**c**) symmetrized displacements Q_y of the type E'; (**d**) symmetrized displacement Q_x of the type E'

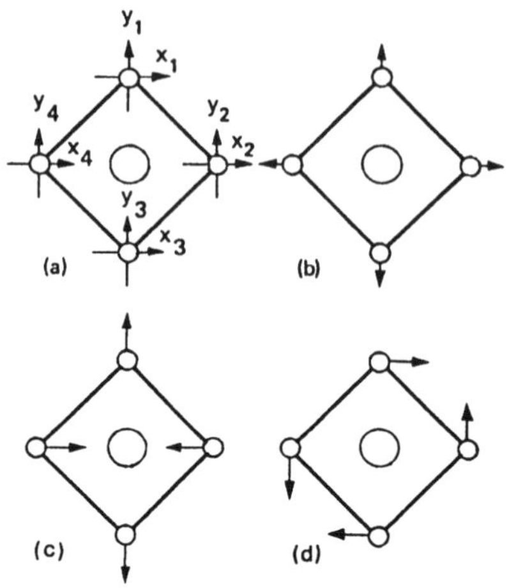

Fig. 2.3a–d. Symmetrized displacements of atoms in a square molecule. (**a**) Labeling of Cartesian displacements; (**b**) totally symmetric displacement of the type A_1; (**c**) B_{1g} displacements; (**d**) B_{2g} displacements

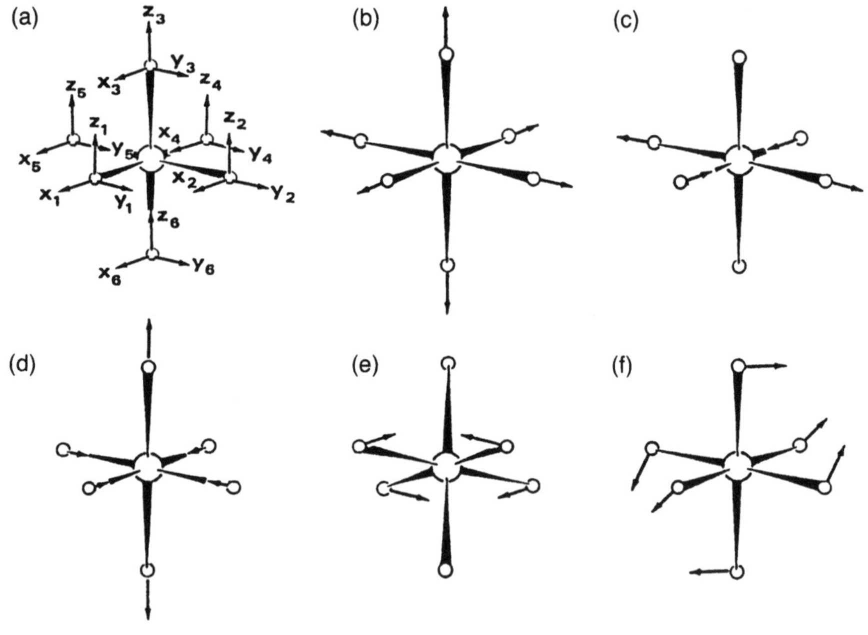

Fig. 2.4a–f. Shape of symmetrized displacements of atoms in a octahedral molecule ML_6. Numbering and orientation of Cartesian displacements (**a**); totally symmetric A_{1g} (**b**); E_g type Q_ε (**c**); E_g type Q_θ (**d**); T_{2g} type Q_ζ (**e**) displacements. In the case of degenerate displacements any linear combination of them can be realized, e.g. $(Q_\xi + Q_\eta + Q_\zeta)/\sqrt{3}$ for T_{2g} displacements (**f**)

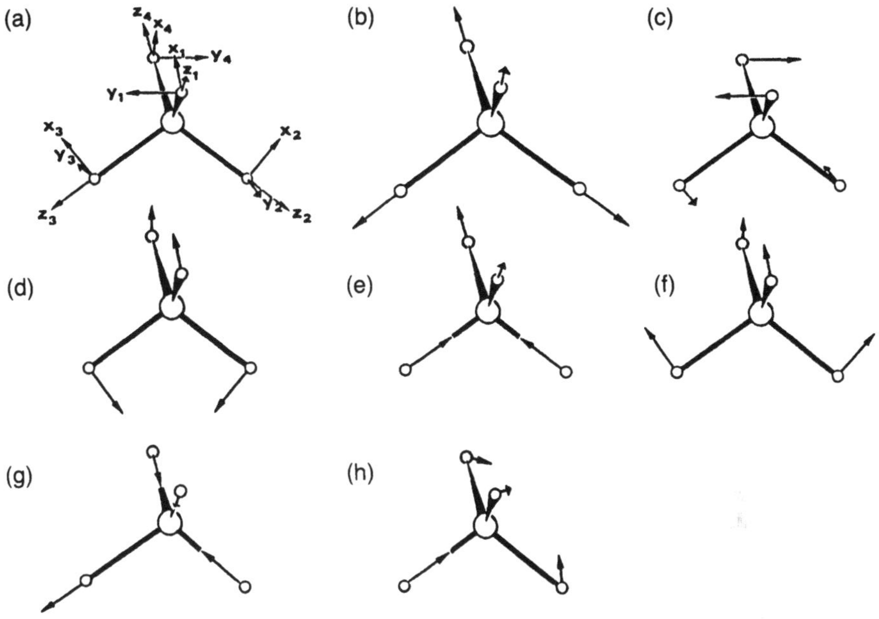

Fig. 2.5a–h. Shape of symmetrized displacements of atoms in a tetrahedral molecule ML_4: numbering and orientation of Cartesian coordinates (**a**), totally symmetric A_1 (**b**), E type Q_ε (**c**), E type Q_θ (**d**), T_{2g} type Q_ζ (**e**), T_{2g}' type $\tilde{Q}_\zeta$ (**f**) displacements. In the case of degeneracy any combination of component displacements can be realized, e.g. $(Q_\xi + Q_\eta + Q_\zeta)/\sqrt{3}$ (**g**), $(\tilde{Q}_\xi + \tilde{Q}_\eta + \tilde{Q}_\zeta)/\sqrt{3}$ (**h**)

The nuclear displacements of a symmetrical polyatomic system can be classified by the irreducible representations Γ of the appropriate symmetry group. In other words, one can transform from the displacements $Q_i - Q_{0i}$ to their linear combinations $Q_{\Gamma\gamma}$ which transform according to the row γ of the irreducible representation Γ of the group G_0 [2.11]. The technique of such a transformation using projection operators is described in any group theory manual, the corresponding matrices of the irreducible representations for all the molecular point groups being given in [2.12]. For more details see [2.11]. The most important *symmetrized displacements of the atoms* of triangular (X_3), quadratic, tetrahedral (ML_4) and octahedral (ML_6) molecules are shown in Figs. 2.2–5. Expressions for symmetrized displacements in terms of Cartesian displacements of the nuclei are given in Table 2.1.

Useful information about the expansion of the full vibrational representation in terms of irreducible ones for different molecular structures is given in [2.9], see Appendix A.

There is no need to use complicated group-theoretical considerations for the classification of symmetrized displacements of the atoms in the case of linear molecules. The total number of vibrational degrees of freedom of an n-atomic

Table 2.1 Symmetrized displacements $Q_{\Gamma\gamma}$ expressed in terms of Cartesian coordinates of the nuclei. x_0, y_0 and z_0 are the Cartesian coordinates of the central atom

Symmetrized displacements	Irreducible representation	Transformation properties	Expression in Cartesian coordinates
Trigonal system X_3. Symmetry group D_{3h}			
Q_1	A'_1	$x^2 + y^2$	$(2y_1 + \sqrt{3}x_2 - y_2 - \sqrt{3}x_3 - y_3)/\sqrt{12}$
Q_2	A'_2	S_z	$(2x_1 - x_2 - \sqrt{3}y_2 - x_3 + \sqrt{3}y_3)/\sqrt{12}$
Q_x	E'	x	$(x_1 + x_2 + x_3)/\sqrt{3}$
Q_y		y	$(y_1 + y_2 + y_3)/\sqrt{3}$
Q_x	E'	$2xy$	$(2x_1 - x_2 + \sqrt{3}y_2 - x_3 - \sqrt{3}y_3)/\sqrt{12}$
Q_y		$x^2 - y^2$	$(2y_1 - \sqrt{3}x_2 - y_2 + \sqrt{3}x_3 - y_3)/\sqrt{12}$
Tetragonal system ML_4. Symmetry group D_{4h}			
Q_a	A_{1g}	$x^2 + y^2$	$1/2(y_1 + x_2 - y_3 - x_4)$
Q_a	A_{2g}	S_z	$1/2(x_1 - y_2 - x_3 + y_4)$
Q_x	E_{1u}	x	$1/2(x_1 + x_2 + x_3 + x_4)$
Q_y		y	$1/2(y_1 + y_2 + y_3 + y_4)$
$\tilde{Q}_x$	E_{1u}	x	x_0
$\tilde{Q}_y$		y	y_0
Q_1	B_{1g}	$x^2 - y^2$	$1/2(y_1 - x_2 - y_3 + x_4)$
Q_2	B_{2g}	$2xy$	$1/2(x_1 + y_2 - x_3 - y_4)$
Tetrahedral system ML_4. Symmetry group T_d			
Q_1	A_1	$x^2 + y^2 + z^2$	$1/2(z_1 + z_2 + z_3 + z_4)$
Q_θ	E	$2z^2 - x^2 - y^2$	$1/2(x_1 - x_2 - x_3 + x_4)$
Q_ε		$\sqrt{3}(x^2 - y^2)$	$1/2(y_1 - y_2 - y_3 + y_4)$
Q_ξ		x	$1/2(z_1 - z_2 + z_3 - z_4)$
Q_η	T_2	y	$1/2(z_1 + z_2 - z_3 - z_4)$
Q_ζ		z	$1/2(z_1 - z_2 - z_3 + z_4)$
$\tilde{Q}_\xi$		x	$1/4(-x_1 + x_2 - x_3 + x_4)$ $+ \sqrt{3}/4(-y_1 + y_2 - y_3 + y_4)$
$\tilde{Q}_\eta$	T_2	y	$1/4(-x_1 - x_2 + x_3 + x_4)$ $+ \sqrt{3}/4(y_1 + y_2 - y_3 - y_4)$
$\tilde{Q}$		z	$1/2(x_1 + x_2 + x_3 + x_4)$
Q_x		x	x_0
Q_y	T_2	y	y_0
Q_z		z	z_0
Octahedral system ML_6. Symmetry group O_h			
Q_1	A_{1g}	$x^2 + y^2 + z^2$	$(y_2 - y_5 + z_3 - z_6 + x_1 - x_4)/\sqrt{6}$
Q_θ	E_g	$2z^2 - x^2 - y^2$	$(2z_3 - 2z_6 - x_1 + x_4 - y_2 + y_5)/\sqrt{12}$
Q_ε		$\sqrt{3}(x^2 - y^2)$	$1/2(x_1 - x_4 - y_2 + y_5)$
Q_ξ		yz	$1/2(z_2 - z_5 + y_3 - y_6)$
Q_η	T_{2g}	xz	$1/2(x_3 - x_6 + z_1 - z_4)$
Q_ζ		xy	$1/2(y_1 - y_4 + x_2 - x_5)$
Q_x		x	$1/2(x_2 + x_3 + x_5 + x_6)$
Q_y	T_{1u}	y	$1/2(y_1 + y_3 + y_4 + y_6)$
Q_z		z	$1/2(z_1 + z_2 + z_4 + z_5)$
$\tilde{Q}_x$		x	$(x_1 + x_4)/\sqrt{2}$
$\tilde{Q}_y$	T_{1u}	y	$(y_2 + y_5)/\sqrt{2}$

Table 2.1 (*cont.*)

Symmetrized displacements	Irreducible representation	Transformation properties	Expression in Cartesian coordinates
$\tilde{Q}_z$		z	$(z_3 + z_6)/\sqrt{2}$
$\tilde{\tilde{Q}}_x$		x	x_0
$\tilde{\tilde{Q}}_y$	T_{1u}	y	y_0
$\tilde{\tilde{Q}}_z$		z	z_0
$\tilde{Q}_\xi$		$x(y^2 - z^2)$	$1/2(x_2 + x_5 - x_3 - x_6)$
$\tilde{Q}_\eta$	T_{2u}	$y(z^2 - x^2)$	$1/2(y_3 + y_6 - y_1 - y_4)$
$\tilde{Q}_\zeta$		$z(x^2 - y^2)$	$1/2(z_1 + z_4 - z_2 - z_5)$

linear molecule is $3n - 5$. Among the n degrees of freedom of the n different types of nuclear motion along the axis of the molecule, one degree corresponds to the translational displacement of the molecule as a whole, the remaining $n - 1$ degrees correspond in general to $n - 1$ different eigenfrequencies. The remaining $(3n - 5) - (n - 1) = 2(n - 2)$ symmetrized displacements violate the linearity of the molecule and transform as $n - 2$ twofold-degenerate representations of the type E_L, where L is the quantum number of the angular momentum of rotations around the axis of the molecule. For all the representations L is equal to 1.

2.2.2 Matrix Vibronic Hamiltonian

If the symmetrized displacements $Q_{\Gamma\gamma}$ are taken as parameters of the expansion (2.2.1) the approximate Hamiltonian obtained from (2.1.1) by means of (2.2.1) may be written in the form

$$H(\mathrm{r}, Q) = T(Q) + U(\mathrm{r}, Q) \;, \tag{2.2.2}$$

where

$$U(\mathrm{r}, Q) = H(\mathrm{r}, Q_0) + \sum_{\Gamma\gamma} V_{\Gamma\gamma}(\mathrm{r}) Q_{\Gamma\gamma} + \tfrac{1}{2} \sum_{\Gamma_1\gamma_1} \sum_{\Gamma_2\gamma_2} W_{\Gamma_1\gamma_1\Gamma_2\gamma_2}(\mathrm{r}) Q_{\Gamma_1\gamma_1} Q_{\Gamma_2\gamma_2} \;. \tag{2.2.3}$$

Here $H(\mathrm{r}, Q_0)$ has the form given in (2.1.12),

$$V_{\Gamma\gamma}(\mathrm{r}) = \frac{\partial V(\mathrm{r}, Q_0)}{\partial Q_{\Gamma\gamma}} \;; \qquad W_{\Gamma_1\gamma_1\Gamma_2\gamma_2}(\mathrm{r}) = \frac{\partial^2 V(\mathrm{r}, Q_0)}{\partial Q_{\Gamma_1\gamma_1} \partial Q_{\Gamma_2\gamma_2}} \;, \tag{2.2.4}$$

and the summation in (2.2.3) is taken over the rows of the irreducible representation contained in the full vibrational representation.

The quantities $V_{\Gamma\gamma}(\mathrm{r})$ possess the transformation properties of the irreducible representation Γ^*, which is the complexconjugate of the representation Γ of the nuclear displacement $Q_{\Gamma\gamma}$. However owing to the symmetry with respect to time reversal, the displacements $Q_{\Gamma\gamma}$ and $Q_{\Gamma^*\gamma}$ which transform according to representations related by complex conjugation should have the same frequency of vibration. The same is true for electronic states $\psi_{\Gamma\gamma}(\mathrm{r})$ and $\psi_{\Gamma^*\gamma}(\mathrm{r})$, which have the same energy provided there is no external magnetic field (or the appropriate magnetic phenomena are not taken into consideration). For the same reason, a representation and its complex conjugate have to be considered together as one physically irreducible real representation of double dimensionality [2.8].

Taking into account these joint representations, the quantities $V_{\Gamma\gamma}(\mathrm{r})$ possess the transformation properties of the irreducible representation Γ, the same as the coordinates $Q_{\Gamma\gamma}$, while the quantities $W_{\Gamma_1\gamma_1\Gamma_2\gamma_2}(\mathrm{r})$ have the properties of the second rank tensor $\Gamma_1 \times \Gamma_2$.

The Hamiltonian (2.2.2), as mentioned above, is a scalar of the point group G_0. From this point of view the last sum in (2.2.3) is a scalar convolution of two tensors of second rank, $W_{\Gamma_1\gamma_1\Gamma_2\gamma_2}(\mathrm{r})$ and $Q_{\Gamma_1\gamma_1}Q_{\Gamma_2\gamma_2}$. The components of these tensors form the basis of a reducible representation $\Gamma_1 \times \Gamma_2$. This means that one can go over to those convolutions of these tensors that transform according to the irreducible representations of the group G_0:

$$\{W(\Gamma_1 \times \Gamma_2)\}_{\Gamma\gamma} = \sum_{\gamma_1\gamma_2} W_{\Gamma_1\gamma_1\Gamma_2\gamma_2}(\mathrm{r})\langle\Gamma_1\gamma_1\Gamma_2\gamma_2|\Gamma\gamma\rangle \ , \tag{2.2.5}$$

$$\{Q_{\Gamma_1} \times Q_{\Gamma_2}\}_{\Gamma\gamma} = \sum_{\gamma_1\gamma_2} Q_{\Gamma_1\gamma_1}Q_{\Gamma_2\gamma_2}\langle\Gamma_1\gamma_1\Gamma_2\gamma_2|\Gamma\gamma\rangle. \tag{2.2.6}$$

Here $\langle\Gamma_1\gamma_1\Gamma_2\gamma_2|\Gamma\gamma\rangle$ are the Clebsch-Gordan coefficients of the point group G_0 [2.13]. The terms of the last sum in (2.2.3) can be rearranged in such a way as to represent a scalar convolution of irreducible tensors:

$$U(\mathrm{r}, Q) = H(\mathrm{r}, Q_0) + \sum_{\Gamma\gamma} V_{\Gamma\gamma}(\mathrm{r})Q_{\Gamma\gamma} + \tfrac{1}{2} \sum_{\Gamma_1\Gamma_2} \sum_{\Gamma\gamma} \{Q_{\Gamma_1} \times Q_{\Gamma_2}\}_{\Gamma\gamma}\{W(\Gamma_1 \times \Gamma_2)\}_{\Gamma\gamma} \ . \tag{2.2.7}$$

The equivalency of (2.2.3) and (2.2.7) can be easily proved by taking into account (2.2.5) and (2.2.6) and the orthogonality relations of the Clebsch-Gordan coefficients[4]:

$$\sum_{\Gamma\gamma} \langle\Gamma_1\gamma_1\Gamma_2\gamma_2|\Gamma\gamma\rangle\langle\Gamma\gamma|\Gamma_1\bar{\gamma}_1\Gamma_2\bar{\gamma}_2\rangle = \delta_{\gamma_2\bar{\gamma}_2}\delta_{\gamma_1\bar{\gamma}_1} \ .$$

Let us now turn to the matrix representation of the Hamiltonian (2.2.2) in the electronic basis of Born-Oppenheimer states $\psi_n(\mathrm{r})$. The Schrödinger equation with such a Hamiltonian, as said above, takes the form of a system of differential equations (2.1.4), where $T_{mn} = T(Q)\delta_{mn}$, and $U_{mn}(Q)$ is represented by the matrix

---

[4] Owing to the reality of the representations mentioned above, the Clebsch-Gordan coefficients can be made real by means of a special choice of the phase, so that

$$\langle\Gamma\gamma|\Gamma_1\gamma_1\Gamma_2\gamma_2\rangle = \langle\Gamma\gamma|\Gamma_1\gamma_1\Gamma_2\gamma_2\rangle^* = \langle\Gamma_1\gamma_1\Gamma_2\gamma_2|\Gamma\gamma\rangle \ .$$

(2.1.15) in which the approximate expression (2.2.7) for the operator $H(\mathrm{r}) + V(\mathrm{r}, Q)$ is used. Since within the basis set under consideration the operator $H(\mathrm{r}, Q_0)$ is diagonal, only the linear and quadratic terms in $Q_{\Gamma\gamma}$ of (2.2.7) contribute to the off-diagonal elements of the Hamiltonian matrix $\|H_{mn}\| = \|T_{mn} + U_{mn}\|$. According to the estimate (2.1.11), these contributions are small and can be included by perturbation theory methods. The first approximation of the perturbation theory for a nondegenerate electronic term is discussed in Sect. 2.1 and results in the Born-Oppenheimer approximation.

Consider now perturbation theory in the case of electronic degeneracy or quasi-degeneracy. Let us number the electronic states $\psi_n(\mathrm{r})$ so as to reserve the first numbers for the group of close-in-energy (degenerate or quasi-degenerate) electronic states under consideration. Assume that the energy gap which separates these states from all the other ones is sufficiently large. Following the rules of perturbation theory, in the case of degeneracy or quasi-degeneracy one has to consider in the first order only that block of the Hamiltonian matrix which corresponds to the degenerate (quasi-degenerate) states. This leads to an $f \times f$ matrix, the elements of which are differential operators in the space of nuclear coordinates. The appropriate eigenvalue equation is a system of f coupled differential equations of the type (2.1.4), where the summation index runs over the values $1, 2, \ldots, f$. The unknowns are f functions $\chi_k(Q)$ which determine the "correct zeroth-order functions"

$$\Psi(\mathrm{r}, Q) = \sum_{k=1}^{f} \psi_k(\mathrm{r})\chi_k(Q) \ . \tag{2.2.8}$$

The coefficients $\chi_k(Q)$, $k = 1, 2, \ldots$, in this sum are comparable in magnitude, and the wave functions (2.2.8), unlike those of (2.1.14), cannot be factorized into electronic and nuclear variables. Small energy intervals in the electronic spectrum mean relatively slow electronic motions comparable in velocity with the nuclear motions. The latter are no longer separable from the electronic motions and cannot be reduced to simple vibrations within the minimum of the potential created by the electron mean field. Therefore the wave functions of the type (2.2.8) are called *vibronic states*, and the appropriate energy spectrum is called *vibronic*, as distinguished from the electron-vibrational states (2.1.14) and corresponding electron-vibrational spectrum considered above. The differential equations determining the f functions $\chi_k(Q)$ are called *vibronic*. Within this terminology, the wave function of the full adiabatic approximation having formally the multiplicative form (2.1.8) is a particular case of the vibronic function (2.2.8). This can be easily seen if one replaces the function $\varphi_n(\mathrm{r}, Q)$ in (2.1.8) by its expansion over the basis electronic functions $\psi_k(\mathrm{r})$: $\varphi_n(\mathrm{r}, Q) = \sum_k a_k^{(n)}(Q)\psi_k(\mathrm{r})$, where $a_k^{(n)}(Q)$ are the expansion coefficients. The representation (2.2.8), even in this case, is much more convenient, since it requires the electronic wave functions $\psi_k(\mathrm{r})$ of a certain (one) nuclear configuration instead of the function $\varphi_n(\mathrm{r}, Q)$, which, as a function of Q, is unknown in practice for most cases.

The off-diagonal elements $H_{nm} = U_{nm}$, $n \neq m$, of the vibronic Hamiltonian represented by the $f \times f$ matrix $\|H_{nm}\|$ are of primary importance; without them the system of vibronic equations of the type (2.1.4) decouples into a set of independent Schrödinger equations with the Hamiltonian (2.1.16) for each electronic state separately. In order to clarify the question of when these matrix element are nonzero the group-theoretical selection rules can be used, provided the transformational properties of the basis electronic functions $\psi_n(\mathrm{r})$ under the symmetry operations of the group G are known, i.e. the classification of the electronic basis functions into the irreducible representations Γ and their rows γ is performed.

In accordance with the known theorem of selection rules, the matrix element $\langle \psi_{\Gamma_i\gamma_i} | V_{\Gamma\gamma}(\mathrm{r}) | \psi_{\Gamma_k\gamma_k} \rangle$ is nonzero if, and only if, the representation Γ is present in the decomposition of the product $\Gamma_i \times \Gamma_k$ into irreducible representations. In cases when the two wave functions of the matrix element under consideration belong to the same irreducible representation, $\Gamma_i = \Gamma_k$, the representation Γ has to be present in the symmetric square $[\Gamma_i^2]$, if Γ_i is a normal representation, and in the antisymmetric square $\{\Gamma_i^2\}$, if Γ_i is a double representation of the group G_0. This also applies to the matrix elements of the operators $\{W(\Gamma_1 \times \Gamma_2)\}_{\Gamma\gamma}$ in (2.2.7).

This rule allows us to determine the vibrations which are responsible for the vibronic mixing of close-in-energy electronic states, provided the transformation properties of these latter are known, and, vice versa, the electronic terms mixed by given normal displacements can be revealed.

The case when all f electronic wave functions belong to the same irreducible representation Γ of the group G_0 is of special interest. Here a known group-theoretical result can be used, owing to which different matrix elements of any operator $V_{\Gamma\gamma}(\mathrm{r})$ calculated from these functions are not independent. According to the Wigner-Eckart theorem [2.13] we have $\langle \psi_{\bar{\Gamma}\gamma_1}(\mathrm{r}) | V_{\Gamma\gamma}(\mathrm{r}) | \psi_{\bar{\Gamma}\gamma_2}(\mathrm{r}) \rangle = \langle \bar{\Gamma} \| V_\Gamma \| \bar{\Gamma} \rangle \langle \Gamma\gamma\bar{\Gamma}\gamma_2 | \Gamma\gamma_1 \rangle$, where $\langle \bar{\Gamma} \| V_\Gamma \| \bar{\Gamma} \rangle$ is the so-called reduced matrix element (the same for all γ_1, γ and γ_2) and the $\langle \Gamma\gamma\bar{\Gamma}\gamma_2 | \Gamma\gamma_1 \rangle$ are Clebsch-Gordan coefficients.

Using this theorem, the potential energy matrix in the basis of degenerate electronic functions $\psi_{\bar{\Gamma}\gamma}(\mathrm{r})$ can be written in the form

$$\hat{U} = \sum_{\Gamma\gamma} V_\Gamma Q_{\Gamma\gamma} \hat{C}_{\Gamma\gamma} + \tfrac{1}{2} \sum_{\Gamma_1\Gamma_2} \sum_{\Gamma\gamma} W_\Gamma(\Gamma_1 \times \Gamma_2) \{Q_{\Gamma_1} \times Q_{\Gamma_2}\}_{\Gamma\gamma} \hat{C}_{\Gamma\gamma} . \tag{2.2.9}$$

Here V_Γ and $W_\Gamma(\Gamma_1 \times \Gamma_2)$ are reduced matrix elements of the operators $V_{\Gamma\gamma}(\mathrm{r})$ and $\{W(\Gamma_1 \times \Gamma_2)\}_{\Gamma\gamma}$ from (2.2.7), and $\hat{C}_{\Gamma\gamma}$ are the matrices composed of appropriate Clebsch-Gordan coefficients active in the space of degenerate electronic states $\psi_{\bar{\Gamma}\gamma}(\mathrm{r})$. The zero of the energy in (2.2.9) is taken from the electronic term under consideration, i.e., $\langle \psi_{\bar{\Gamma}\gamma} | H(\mathrm{r}, Q_0) | \psi_{\bar{\Gamma}\gamma} \rangle = 0$.

2.2.3 Primary Force Constants. Linear and Quadratic Vibronic Coupling

Let us separate the totally symmetric terms with $\Gamma = A_1$ from the second sum in (2.2.9). They all contain the unit matrix $\hat{C}_{A_1}$. Adding to them the kinetic energy operator, which is also proportional to the matrix $\hat{C}_{A_1}$, we obtain the part of the Hamiltonian which is a linear form of coordinates and momenta,

$$\hat{H}_0 = \left(\sum_{\Gamma\gamma} \frac{1}{2M(\Gamma)} P^2_{\Gamma\gamma} + \frac{1}{2} \sum_{\Gamma\gamma} \sum_{\alpha\beta} W_{A_1}(\Gamma_\alpha \times \Gamma_\beta) Q_{\Gamma_\alpha} Q_{\Gamma_\beta} \right) \hat{C}_{A_1} \ . \tag{2.2.10}$$

Here α and β label the repeated irreducible representations with the same transformation properties, and $M(\Gamma)$ are the appropriate reduced masses. In (2.2.10) the relation $\langle \Gamma_1\gamma_1\Gamma_2\gamma_2|A_1\rangle \sim \delta_{\Gamma_1\Gamma_2}\delta_{\gamma_1\gamma_2}$ is also used; due to this, (2.2.6) acquires a simple form $\{Q_{\Gamma_\alpha} \times Q_{\Gamma_\beta}\}_{A_1} = \sum_{\alpha\beta} Q_{\Gamma_\alpha\gamma} Q_{\Gamma_\beta\gamma}$.

We change now to mass-weighted coordinates $q_{\Gamma\gamma} = \sqrt{M(\Gamma)}\, Q_{\Gamma\gamma}$ and perform an orthogonal transformation to normal coordinates which diagonalize the quadratic form

$$\sum_{\alpha,\beta} W_{A_1}(\Gamma_\alpha \times \Gamma_\beta) Q_{\Gamma_\alpha\gamma} Q_{\Gamma_\beta\gamma} = \sum_{\alpha,\beta} \frac{W_A(\Gamma_\alpha \times \Gamma_\beta)}{\sqrt{M(\Gamma_\alpha)M(\Gamma_\beta)}} q_{\Gamma_\alpha\gamma} q_{\Gamma_\beta\gamma} \ .$$

As a result the Hamiltonian (2.2.10) takes a simple oscillator form

$$\hat{H}_0 = \tfrac{1}{2} \sum_{\Gamma\gamma} (P^2_{\Gamma\gamma} + K_\Gamma Q^2_{\Gamma\gamma}) \hat{C}_{A_1} \ . \tag{2.2.11}$$

Here the previous notation $Q_{\Gamma\gamma}$ is used for new normal coordinates, and it is assumed that they are also the parameters of the expansion (2.2.1). Index Γ in (2.2.11) labels both different and repeated identical irreducible representations, while K_Γ is the eigenvalue of the matrix $\| W_{A_1}(\Gamma_\alpha \times \Gamma_\beta)/\sqrt{M(\Gamma_\alpha)M(\Gamma_\beta)} \|$. In principle K_Γ may be either positive or negative. If the latter is the case, the higher-order terms in the expansion (2.2.1) have to be taken into account in order to provide the stability of the system. In what follows, only positive values of K_Γ are considered, that is, the Hamiltonian $\hat{H}_0$ describes harmonic vibrations with frequencies $\omega_\Gamma = \sqrt{K_\Gamma}$ near the point of equilibrium Q_0 (all the $Q_{\Gamma\gamma} = 0$).

Note that in fact the nuclear dynamics is much more complicated than simple harmonic vibrations near a single minimum Q_0 of the potential energy. As will be shown (Chap. 3), inclusion of the other terms of (2.2.9) may lead to a complicated shape of the adiabatic potential energy surface with more than one equivalent minimum. In some cases, in the first rough approximation, the nuclear dynamics can be reduced to small vibrations at the bottom of these minima. Their curvatures (and hence vibration frequencies) are different from the primary ones. Therefore hereafter the above K_Γ values will be called *primary force constants*, to distinguish them from those obtained by considering the vibronic interaction.

By means of the normal coordinates $Q_{\Gamma\gamma}$ defined above, the *vibronic Hamiltonian* of the degenerate electronic term may be written as

$$\hat{H} = \hat{H}_0 + \sum_{\Gamma \neq A_1} \sum_{\gamma} V_\Gamma Q_{\Gamma\gamma} \hat{C}_{\Gamma\gamma} + \tfrac{1}{2} \sum_{\Gamma_1\Gamma_2} \sum_{\Gamma \neq A_1} \sum_{\gamma} W_\Gamma(\Gamma_1 \times \Gamma_2) \{Q_{\Gamma_1} \times Q_{\Gamma_2}\}_{\Gamma\gamma} \hat{C}_{\Gamma\gamma} \ . \tag{2.2.12}$$

The terms linear in Q_{A_1} are excluded from the Hamiltonian (2.2.12) by taking the zero of the values from the equilibrium coordinate $Q_{A_1}^{(0)} = -V_{A_1}/K_{A_1}$. This procedure defines the previously uncertain totally symmetric coordinates of the point Q_0.

When $V_\Gamma = 0$ and $W_\Gamma(\Gamma_1 \times \Gamma_2) = 0$ ($\Gamma \neq A_1$), the vibronic Hamiltonian (2.2.12) is reduced to H_0 from (2.2.11), and the motion of the nuclei is again reduced to harmonic oscillations, while the wave function takes the factorizable form (2.1.14). It follows that all the vibronic effects are connected with the last two terms in (2.2.12), called the *operators of linear* and *quadratic vibronic interactions*; the parameters V_Γ and $W_\Gamma(\Gamma_1 \times \Gamma_2)$ are called the *constants of linear* and *quadratic vibronic coupling*, respectively.

In the particular case of linear vibronic coupling, when for some reason or other all the quadratic vibronic interaction constants are negligible [$W_\Gamma(\Gamma_1 \times \Gamma_2) = 0$], the vibronic Hamiltonian (2.2.12) takes the simple form

$$\hat{H} = \tfrac{1}{2}\sum_{\Gamma\gamma}(P_{\Gamma\gamma}^2 + \omega_\Gamma^2 Q_{\Gamma\gamma}^2)\hat{C}_{A_1} + \sum_{\Gamma\gamma} V_\Gamma Q_{\Gamma\gamma}\hat{C}_{\Gamma\gamma} \ . \tag{2.2.13}$$

This Hamiltonian can be written as a sum of two commutative terms, one of which contains the degrees of freedom $Q_{\Gamma\gamma}$, for which $V_\Gamma \neq 0$, the other containing all the remaining nuclear coordinates (remember that $\hat{C}_{A_1}$ is the unit matrix, which commutes with all the other matrices $\hat{C}_{\Gamma\gamma}$). This means that the variables separate and the problem of nuclear motion in the subspace of the coordinates for which $V_\Gamma = 0$ can be solved separately and can be reduced to trivial harmonic vibrations.

The main problem is to evaluate the nuclear dynamics in the subspace of those coordinates $Q_{\Gamma\gamma}$ for which $V_\Gamma \neq 0$. These coordinates are called Jahn-Teller-active coordinates, or, for short, active coordinates. The corresponding Hamiltonian has the form (2.2.13) where the index Γ now labels only the active coordinates. The methods of solving the eigenvalue problem of the vibronic Hamiltonian will be considered in Chap. 4.

The vibronic Hamiltonian of the degenerate or quasi-degenerate electronic term is considerably simplified compared with the exact Hamiltonian, and therefore it corresponds to a simplified model of a polyatomic system. The main failure of this model is the neglect of nonadiabatic transitions between the group of (relatively close-in-energy) electronic states under consideration and other electronic states separated from the first group by large energy gaps. But the model takes into account the essential nonadiabatic nature of the electronic motions within the electronic states of the separated degenerate or quasi-degenerate term and the possibility of quantum transitions between them. Another limitation of the model is the harmonic approximation; the vibronic Hamiltonian contains only linear and quadratic terms in $Q_{\Gamma\gamma}$. Note that the harmonic approximation in the vibronic Hamiltonian does not mean that the potential energy surfaces are described only by multidimensional paraboloids. As will be shown in Chap. 3, even the linear terms of the vibronic interaction which mix different sheets of an adiabatic potential result in significant anharmonicity. Therefore it might be worthwhile to distinguish the *proper anharmonicity*, which is due to the higher-order terms in the expansion of the potential energy matrix (of the type $\{Q_{\Gamma_1} \times Q_{\Gamma_2} \times Q_{\Gamma_3} \times \cdots\}_{A_1}$), from the *vibronic anharmonicity* resulting from the mixing of the electronic states. The latter occurs even when only the linear terms of the expansion are considered.

2.3 The Jahn-Teller Theorem

The possible methods of approximately solving the Schrödinger equation with the vibronic Hamiltonian (2.2.12) are determined to a large extent by the magnitude of the vibronic constants. When V_Γ and W_Γ are relatively small, perturbation theory with respect to these parameters can be used, taking as the initial approximation the Born-Oppenheimer states [for which the vibrational part of the Hamiltonian (2.2.12) is diagonal].

Another limiting case is provided by the strong vibronic coupling when the inverse magnitudes V_Γ^{-1} are small. Here it is convenient to change to another basis of electronic states in order to write the potential energy (2.2.9) (which contains the most important terms of the vibronic Hamiltonian) as a diagonal matrix and include it in the zeroth-order approximation of the perturbation theory. In other words, in the case of strong vibronic coupling one has to perform a unitary transformation to the adiabatic electronic basis $\varphi_n(\mathrm{r}, Q)$. In this basis (and with an accuracy up to small admixtures from other electronic terms) the eigenvalues of the matrix $\hat{U}$ coincide with the adiabatic potentials $\varepsilon_n(Q)$ while the matrix of kinetic energy $\hat{T}$ acquires a nondiagonal form $\|T_{nm}\| = \|T(Q)\delta_{nm} + \Lambda_{nm}(Q)\|$, the nonadiabaticity Λ_{nm} being given by (2.1.5).

Note that the energy intervals between the eigenvalues of the zeroth-order approximation $|\varepsilon_n(Q) - \varepsilon_m(Q)|$ are of the order of magnitude of $|V_\Gamma Q_{\Gamma\gamma}|$, and in a case of strong enough coupling ($V_\Gamma \to \infty$) may be very large except in the region of small $Q_{\Gamma\gamma}$ values. This fact, together with the criterion (2.1.10), justifies the application of the adiabatic approximation everywhere except in the region of very small values of $Q_{\Gamma\gamma}$, i.e. except in the immediate neighborhood of the point Q_0.

Since they are manifested in the majority of experiments on the systems under consideration, the lowest vibronic state localized near the minima of the potential $\varepsilon_n + \Lambda_{nn}$ seem to be most interesting. The dependence of Λ_{nn} on Q may, as a rule, be neglected compared with the stronger dependence of $\varepsilon_n(Q)$. Just this function $\varepsilon_n(Q)$ determines the existence and position of the extrema of the potential $\varepsilon_n + \Lambda_{nn}$, which allow correct conclusions to be drawn about the nature of the nuclear motion without detailed solution of the Schrödinger equation with the Hamiltonian (2.1.7).

Let us investigate the surface $\varepsilon_n(Q)$ in a small region surrounding the point of degeneracy Q_0 (all the $Q_{\Gamma\gamma} = 0$, $\Gamma \neq A_1$). To do this it is enough to take into account only the terms of the matrix (2.2.9) linear in $Q_{\Gamma\gamma}$, i.e. to assume that

$$\hat{U} \approx \sum_{\Gamma\gamma} V_\Gamma Q_{\Gamma\gamma} \hat{C}_\Gamma \ . \tag{2.3.1}$$

It can easily be seen that all the f eigenvalues of this matrix have conically shaped surfaces (which are not necessarily solids of revolution) with the apex at the point Q_0. Indeed, consider the cross section of the surfaces $\varepsilon_n(Q)$ along some arbitrary direction (from the point Q_0) specified by a unit vector $\boldsymbol{e} = \{e_{\Gamma\gamma}\}$ in the space of nuclear coordinates $Q_{\Gamma\gamma}$. Along this section $Q_{\Gamma\gamma} = qe_{\Gamma\gamma}$ and the matrix (2.3.1) takes the form $q\sum_{\Gamma\gamma} V_\Gamma e_{\Gamma\gamma} C_{\Gamma\gamma}$. It is obvious that all the eigenvalues of this matrix are linear with respect to q, $\varepsilon_n = \lambda_n q$, where λ_n are the eigenvalues of the numerical matrix $\sum_{\Gamma\gamma} V_\Gamma e_{\Gamma\gamma} Q_{\Gamma\gamma}$.

All the matrices $\hat{C}_{\Gamma\gamma}$, $\Gamma \neq A_1$, included in (2.3.1) have one common property: their traces are equal to zero. This follows directly from the properties of the Clebsch-Gordan coefficients. In other words, the signs and magnitudes of all eigenvalues of the matrix (2.3.1) are such that the sum of these eigenvalues equals zero for any value of $Q_{\Gamma\gamma}$, $\Gamma \neq A_1$. It follows that among the conically shaped surfaces $\varepsilon_n(Q)$ there are always some oriented with the apex up and the mouth of the cone down (Fig. 2.6, cf. Fig. 2.2).

This result emerges as a direct consequence of the electronic degeneracy at the point Q_0. Indeed, any low-symmetry nuclear displacement $Q_{\Gamma\gamma}$, $\Gamma \neq A_1$, lifts the electronic degeneracy, provided the latter is due to the space symmetry of the nuclear configuration Q_0. The only exception is Kramers degeneracy, which is

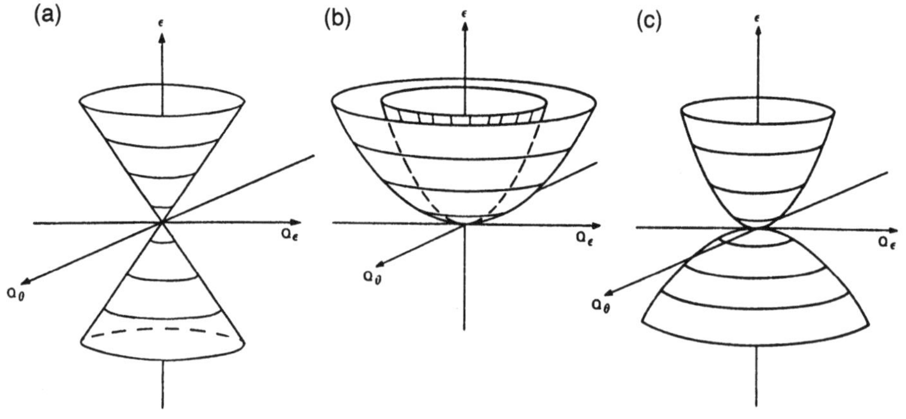

Fig. 2.6a–c. Splitting of the adiabatic potentials in the immediate vicinity of the point of degeneracy (**a**) Jahn-Teller case; (**b**) linear molecules, case of stability; (**c**) linear molecules, case of instability

due to symmetry with respect to inversion of time and which cannot be violated by any nuclear displacement. If the splitting occurs in the first order with respect to the small $Q_{\Gamma\gamma}$ values, the adiabatic potential in the neighborhood of the point of degeneracy consists of conically shaped forms, at least one of them being oriented with the wider end down.

The question of whether the splitting of the adiabatic potentials always takes place in the first order with respect to the $Q_{\Gamma\gamma}$ values seems to be more complicated. In fact it can be reduced to the question of whether there exist cases in which all the linear vibronic constants V_Γ, $\Gamma \neq A_1$, in (2.2.12 and 13) are equal to zero. If this is so, the adiabatic potentials in the neighborhood of the point Q_0 depend on $Q_{\Gamma\gamma}$ quadratically, i.e. they have an extremum at this point.

By means of examining all the point groups and all the possible types of molecular systems belonging to a certain point group, Jahn and Teller [2.9, 14] showed that in all possible cases except linear molecules, and for all degenerate terms except twofold Kramers degeneracy, there are nuclear displacements $Q_{\Gamma\gamma}$ for which the reduced matrix element $V_\Gamma = \langle \bar{\Gamma} \| V_\Gamma \| \bar{\Gamma} \rangle$ calculated with the electronic functions $\psi_{\bar{\Gamma}\gamma}(\mathrm{r})$ is nonzero.

According to a well-known theorem of group theory [2.8], the matrix elements V_Γ are nonzero if in the decomposition of $[\bar{\Gamma}^2]$ (the symmetric square of $\bar{\Gamma} \times \bar{\Gamma}$) into irreducible representations of the group G_0 there are representations Γ (when Γ is a two-valued representation of the double group the antisymmetric part $\{\bar{\Gamma}^2\}$ of the product $\bar{\Gamma} \times \bar{\Gamma}$ has to be considered instead of the symmetry square $[\bar{\Gamma}^2]$).

The expansion of the symmetric and antisymmetric products of degenerate irreducible representations of all the possible molecular symmetry groups can be obtained by means of character tables. The results of such expansions can be found in, for example, [2.12] and are listed in the Table B.2 of Appendix B.

As a result it was shown that in all cases for any polyatomic system there are always low-symmetry nuclear displacements which lift the electronic degeneracy in first-order perturbation theory. The only exception besides the Kramers degeneracy mentioned above is when the point Q_0 corresponds to a linear position of the atoms in the polyatomic system, i.e., for a linear molecule [2.9, 14].

Thus the following *Jahn-Teller theorem* is valid: *For any polyatomic system, among the adiabatic potentials intersecting at the point Q_0, there is at least one which has no extremum at this point. Two kinds of cases are exceptions:*

a) *a linear configuration of the atoms at the point Q_0,* i.e. *linear molecules, and*
b) *twofold spin (Kramers) degeneracy of the electronic term.*

More constructive analytic proofs of this theorem are given in [2.15, 16], see Appendix C. Neither the direct examination of all the different possible cases, given by Jahn and Teller [2.9, 14], nor the analytical proofs obtained in [2.15, 16] are exhaustive. They deal only with molecular point groups.

A proof of the Jahn-Teller theorem for both molecular and space crystal groups based on the Frobenius theorem was suggested in [2.17–19]. A direct verification of the Jahn-Teller theorem for several specific space groups is given in [2.20, 21]. An algebraic proof of the theorem for molecules, also including some situations of accidental degeneracy, is presented in [2.22].

Note that the symmetry considerations used for the proof of the Jahn-Teller theorem are sufficient but not necessary conditions for the lack of extremum of the adiabatic potential at the point of degeneracy Q_0. The cases of accidental degeneracy of electronic states remain outside the above considerations. Other cases are also possible when all the reduced matrix elements V_Γ are equal to zero accidentally (see, however, footnote 3 on page 12).

The lack of extremum of the adiabatic potential at the point of degeneracy is often interpreted as *instability* of the symmetrical polyatomic system in degenerate electronic states resulting in its spontaneous distortion and removal of degeneracy. This interpretation, literally understood, is incorrect. In the neighborhood of the point of degeneracy Q_0 (small $Q_{\Gamma\gamma}$, $\varepsilon_m \approx \varepsilon_n$) the nonadiabaticity corrections of the order of $\Lambda_{nm}/(\varepsilon_n - \varepsilon_m) \sim (V_\Gamma Q_{\Gamma\gamma})^{-1}$ are not sufficiently small, and hence the adiabatic potential loses its physical meaning of the potential energy of the nuclei in the field of the electrons. The real behavior of the nuclei in this case follows the Schrödinger equation with the vibronic matrix Hamiltonian (2.2.12), whereas the adiabatic Hamiltonian (2.1.7) is invalid. Nevertheless, for sufficiently strong vibronic coupling ($V_\Gamma \to \infty$) the area of applicability of the approximation

$$\hat{U} \approx \sum_{\Gamma\gamma} V_\Gamma Q_{\Gamma\gamma} \hat{C}_{\Gamma\gamma}$$

may be larger than the region of small $Q_{\Gamma\gamma}$ values where the adiabatic approximation is invalid, and the conclusion about the instability of the nuclear configuration may have physical grounds. But even in these cases the nuclear configuration, if correctly calculated as quantum-mechanical averaged nuclear coordinates, is not distorted, and the degeneracy of the ground state is not removed.

Indeed, a more careful study of the surfaces $\varepsilon_n(Q)$ shows that in all Jahn-Teller systems there are always several equivalent minima of the adiabatic potential appropriate to several (indistinguishable in symmetry) distorted nuclear configurations. For instance, a molecule of the ML_6 type with a two-fold degenerate electronic E term (in the configuration Q_0 corresponding to a regular octahedron), as a result of the Jahn-Teller effect becomes elongated along one of its four-fold axes of symmetry C_4. There are three equivalent distortions of this kind corresponding to three equivalent axes of symmetry C_4 (for details see Sect. 3.1). The whole wave function is a superposition of the states in the minima, and therefore the quantum—mechanical average value $\bar{Q}_{\Gamma\gamma}$ is equal to zero, the nuclear configuration, determined quantum-mechanically, is not distorted, and the degeneracy of the ground state is not lifted (Chap. 4).

The last conclusion follows from the fact that the operator of vibronic interaction in (2.2.12) is a scalar of the group G_0, and therefore it cannot remove the degeneracy of the ground state in perturbation theory of any order.

For further visual interpretation of the origin of the Jahn-Teller effect we take into account that in the case of electronic degeneracy the symmetry of the electron wave functions is lower than that of the field of the nuclei. These low-symmetry wave functions belong to the same degenerate energy level and transform into each other under the symmetry operations forming the basis of the degenerate irreducible representation. An example is illustrated in Fig. 2.7a where two degenerate wave functions of a square molecule of the type ML_4 (D_{4h} symmetry) localized mainly on the central atom are shown. These functions

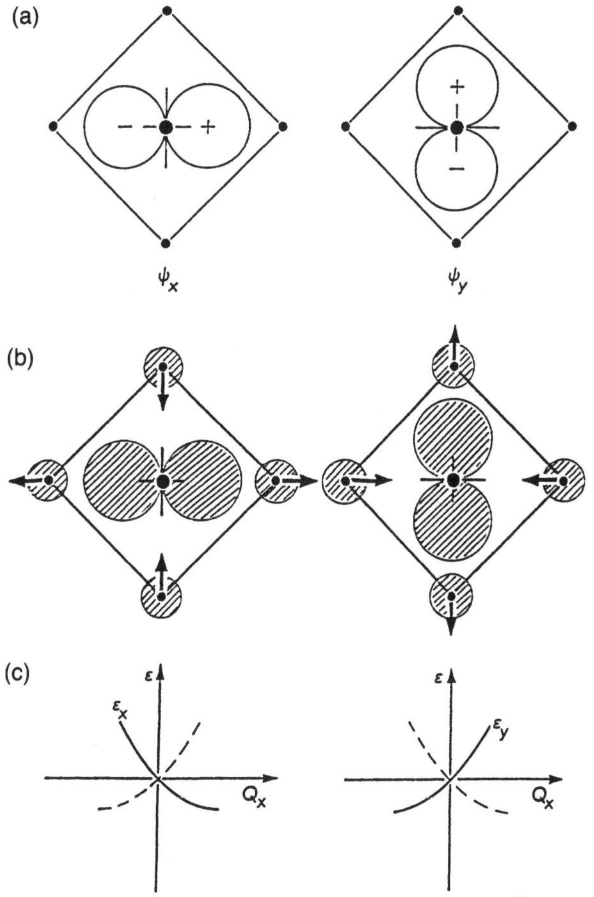

Fig. 2.7a–c. The Jahn-Teller distortion of a square molecule as a result of strengthening of the chemical bonding. (**a**) Degenerate electronic states ψ_x and ψ_y. (**b**) Low symmetry displacement Q_x that strengthens the bonding (the case of repulsion between the electronic distribution on the central atom and negative charges on the ligands is shown). (**c**) The adiabatic potential dependence on Q_x near the point $Q_x = 0$

belong to the E_u representation and transform into each other under rotations around the fourfold axis by an angle $\pi/4$.

The interaction of the charge distribution in any of these states with the electronic shells of the ligands at the apexes of the squares results in a nonzero force distorting the molecule (see also [2.23]) in order to adapt the nuclear configuration to the low symmetry electronic distribution (Fig. 2.7b). Depending on the nature of the chemical bonding, this force can be either positive or negative, distorting the molecule in the corresponding direction. Thus the cause of the Jahn-Teller distortion is the gain of energy by the chemical bonding, which is stronger in the distorted configuration than in the regular one (Fig. 2.7c).

The two exceptions to the Jahn-Teller theorem are quite different. Kramers degeneracy is determined by the symmetry with respect to time reversal. This symmetry cannot be destroyed by any type of nuclear configuration distortion. Kramers degeneracy cannot be removed by nuclear displacements; it remains over the whole range of $Q_{\Gamma\gamma}$ values.

On the other hand, the exception of linear molecules is due to the fact that the orbital angular momentum also determines the parity of the irreducible representation. As a result of the inversion operation the basis functions $\psi_{LM}(\mathrm{r})$ are multiplied by $(-1)^L$. Hence all the irreducible representations with even values of L are even and those with odd values of L are odd. Since doubly degenerate vibrations of a linear molecules are of E_L symmetry with $L = 1$ (see Sect. 2.2), i.e., they are odd, the selection rule on parity results in $V_E = 0$ and there is no splitting of the adiabatic potential to first order with respect to the nuclear displacements $Q_{E\gamma}$. However, there is such a splitting in the second-order approximation. Indeed, the $Q_{E\gamma}Q_{E\gamma'}$ transform according to $[E_1^2] = A_1 + E_2$, and the mentioned prohibition on parity is removed. Usually, this is due to the fact that for distorted bent configurations of the molecule the L value ceases to be a good quantum number. In other words, the adiabatic potential of the linear molecule may be split due to different curvatures of different sheets (Fig. 2.6b, c). It follows also that for linear molecules, as distinguished from the nonlinear case, the adiabatic potential sheets have an extremum at the point of degeneracy. If the vibronic coupling is strong enough, the lowest surface has negative curvature at the point of degeneracy (Fig. 2.6c), which means that the latter is a point of unstable equilibrium. This instability of linear molecules in degenerate electronic states with respect to bending distortions is known as the *Renner effect* [2.24]. Note that in the case of weak coupling there is no instability but adiabatic potential splitting (Fig. 2.6b). This case originates from the same vibronic Hamiltonian and therefore it is often called the *weak Renner effect*.

3. Adiabatic Potentials

In this chapter the adiabatic potential surfaces which are formal solutions of the electronic part of the Schrödinger equation are investigated.

Knowledge of the adiabatic potentials is necessary for the solution of the vibronic problem, i.e. for the determination of the energy spectrum and wave functions of the system to be performed in Chap. 4. However, some qualitative predictions of the expected vibronic effects based on the shape of the adiabatic potential can be made without solving the vibronic equations. Indeed, although in the neighborhood of the point of degeneracy these potentials have no physical meaning, far from this point the energy gaps between different sheets of the adiabatic potential are large enough for each of them to have approximately the meaning of the potential energy of nuclei in the electron mean field. This allows the nuclear motion for each of these sheets to be considered separately and the nature of nuclear dynamics to be evaluated directly.

The approximate consideration of the nuclear motion on one of the sheets of the potential surface corresponds to the adiabatic approximation, but the shape of this sheet (as of the whole multisheet surface) becomes anharmonic with several or an infinite number of minima when the vibronic mixing of different electronic states is considered.

The possibility of obtaining qualitative results at an earlier stage justifies the separation of the comparatively simple problem of determining adiabatic potentials from the rather complicated problem of solving vibronic equations. The investigation of the adiabatic potential surfaces (in most cases only the extrema points and their curvature are determined) has its own characteristic features depending on the number of mixing electronic states, the symmetry of the polyatomic system and the number of active vibrational modes involved in the vibronic effects under consideration. Some of the most important cases are considered in this chapter.

3.1 The Orbital Doublet (E Term)

The motion of the polyatomic system near the nuclear configuration Q_0 with a degenerate electronic state is described by the matrix Hamiltonian (2.2.13). The Schrödinger equation with this Hamiltonian is a system of coupled differential equations of the type (2.2.13). The complexity of its solution is determined first by the number of equations in the system and second by the number of Jahn-Teller active vibrational degrees of freedom. The simplest Jahn-Teller system is realized in the case of two electronic states mixed by one or two vibrations. This is the case of a non-Kramers orbital doublet which has the transformation properties of the E representation of the point group.

As can be shown [3.1], the degeneracy of electronic states owing to the symmetry of the nuclear configuration exists only in polyatomic systems with at least one rotational (C_n) or rotoflection (S_n) symmetry axis of order $n > 2$. When rotated around this axis by an arbitrary angle φ, the electronic states transform as $\exp(\pm ik\varphi)$, where $k = 1, 2, \ldots, n$. Due to the symmetry with respect to time reversal, wave functions related by complex conjugation belong to the same representation. Thus the symmetry group G_0 which contains an n-fold rotational axis may have different doublet irreducible representations E_k [$k = 1, 2, \ldots, (n-2)/2$, if n is even, and $k = 1, 2, \ldots, (n-1)/2$, for odd n values] as well as the singlet totally symmetric A ($k = 0$) and low-symmetry B representations, the latter being possible for even n values only ($k = n/2$); the corresponding basis functions have the transformation properties $\psi_{\pm k} \sim \exp(\pm ik\varphi)$. The basis of the symmetric square $[E_k^2]$ is provided by three functions: $\psi_k^* \psi_k$, $\psi_k^* \psi_{-k}$ and $\psi_{-k}^* \psi_k$. The last two functions possess the transformation properties of $\exp(\pm 2ik\varphi)$ and correspond to the doublet representation E_{2k}, except when n is a multiple of 4 (tetragonal symmetry), $2k = n/2$, and then these two functions correspond to two singlet representations B_1 and B_2. This means that $[E_k^2] = A + E_{2k}$ when $2k \neq n/2$ and $[E_k^2] = A + B_1 + B_2$ when $2k = n/2$.

Thus, for electronic orbital doublets two types of Jahn-Teller effect are possible. The first, more widespread, occurs when the electronic degeneracy of the E_k term ($k \neq n/2$) is removed by nuclear displacements which transform according to the doublet E_{2k} representation. This is the so-called *$E \otimes e$ case.* The simplest polyatomic systems in which there is one pair of symmetrized E displacements interacting with the electronic orbital doublet are triangular molecules X_3, tetrahedral molecules ML_4, and octahedral molecules ML_6.

The second type of Jahn-Teller effect for an orbital doublet is possible only in polyatomic systems with C_n (or S_n) axes of symmetry of order $n = 4k$ ($k = 1, 2, \ldots, 3, \ldots$) for electronic states transforming as $\exp(\pm in\varphi/4)$. In this case the singlet low-symmetry displacements of B_1 or B_2 type are active in the linear vibronic interaction. This is the *$E \otimes (b_1 + b_2)$ case.* The simplest system for which such a Jahn-Teller effect is possible is a square planar or pyramidal molecule ML_4. In more complicated polyatomic systems the number of Jahn-Teller-active nuclear displacements may be larger and this complicates the problem significantly (for details see Sect. 3.5).

3.1.1 *$E \otimes e$* Case: The Mexican Hat and the Tricorn

Consider the first type of Jahn-Teller effect, in which the degenerate electronic states of the E_k term interact with the E_{2k} vibrations (the $E \otimes e$ case). First restrict the problem to the linear vibronic coupling terms only. If one chooses the complex coordinates transforming as $Q_+ \sim \exp(2ik\varphi)$ and $Q_- \sim \exp(-2ik\varphi)$ as the normal coordinates of the E_{2k} type, then the operator of the linear vibronic coupling can be written in the form of a scalar convolution $V_+(\mathrm{r})Q_- + V_-(\mathrm{r})Q_+$. Going over to the matrix representation of the electronic operators $V_\pm(\mathrm{r})$ in the basis of the states $\psi_\pm \sim \exp(\pm ik\varphi)$ and considering the transformation properties $V_\pm(\mathrm{r}) \sim \exp(\pm 2ik\varphi)$, we obtain for this convolution $V_E\sqrt{2}(Q_+\hat{\sigma}_- + Q_-\hat{\sigma}_+)$, where V_E is the reduced matrix element (the factor $\sqrt{2}$ is introduced for convenience) and $\hat{\sigma}_+$ and $\hat{\sigma}_-$ are Pauli operators,

$$\hat{\sigma}_+ = \begin{pmatrix} 0 & 1 \\ 0 & 0 \end{pmatrix} , \qquad \hat{\sigma}_- = \begin{pmatrix} 0 & 0 \\ 1 & 0 \end{pmatrix} , \tag{3.1.1}$$

active in the space of the two states $\psi_\pm(r)$. If we now change to real normal coordinates Q_θ and Q_ε,

$$\begin{aligned} Q_\theta &= \frac{1}{\sqrt{2}}(Q_+ + Q_-) , \qquad & Q_+ &= \frac{1}{\sqrt{2}}(Q_\theta + \mathrm{i}Q_\varepsilon) , \\ Q_\varepsilon &= -\frac{\mathrm{i}}{\sqrt{2}}(Q_+ - Q_-) , \qquad & Q_- &= \frac{1}{\sqrt{2}}(Q_\theta - \mathrm{i}Q_\varepsilon) , \end{aligned} \tag{3.1.2}$$

the operator of linear vibronic interaction takes the form

$$V_E\begin{pmatrix} 0 & Q_\theta - \mathrm{i}Q_\varepsilon \\ Q_\theta + \mathrm{i}Q_\varepsilon & 0 \end{pmatrix} = V_E(Q_\theta\hat{\sigma}_x + Q_\varepsilon\hat{\sigma}_y) , \tag{3.1.3}$$

where the σ_i are Pauli matrices,

$$\begin{aligned} \hat{\sigma}_0 &= \begin{pmatrix} 1 & 0 \\ 0 & 1 \end{pmatrix} , \qquad & \hat{\sigma}_x &= \begin{pmatrix} 0 & 1 \\ 1 & 0 \end{pmatrix} , \\ \hat{\sigma}_y &= \begin{pmatrix} 0 & -\mathrm{i} \\ \mathrm{i} & 0 \end{pmatrix} , \qquad & \hat{\sigma}_z &= \begin{pmatrix} 1 & 0 \\ 0 & -1 \end{pmatrix} . \end{aligned} \tag{3.1.4}$$

Note, that in the real basis of electronic states,

$$\begin{aligned} \psi_\theta &= \frac{1}{\sqrt{2}}(\psi_+ + \psi_-) , \qquad & \psi_+ &= \frac{1}{\sqrt{2}}(\psi_\theta + \mathrm{i}\psi_\varepsilon) , \\ \psi_\varepsilon &= -\frac{\mathrm{i}}{\sqrt{2}}(\psi_+ - \psi_-) , \qquad & \psi_- &= \frac{1}{\sqrt{2}}(\psi_\theta - \mathrm{i}\psi_\varepsilon) , \end{aligned} \tag{3.1.5}$$

the operator (3.1.3) has the form

$$V_E(Q_\theta\hat{\sigma}_z + Q_\varepsilon\hat{\sigma}_x) . \tag{3.1.6}$$

If one introduces the appropriate values of the Clebsch-Gordan coefficients into the Hamiltonian (2.2.13) written for the particular case of an electronic E term interacting with E vibrations of a certain molecule with some particular symmetry, then for the operator of linear vibronic interaction one obtains an expression of the type (3.1.6).

All the foregoing is also valid for the case of an E term of a tetrahedral (octahedral, cubic) molecule with the only distinction that the sign of Q_θ in (3.1.6) has to be changed. This is due to the fact that cubic groups contain threefold axes of symmetry and therefore the above discussion is applicable to these systems. The rotations around fourfold axes are not connected with the E term under consideration since the twofold E representation in the subgroup C_4 decomposes into two one-dimensional representations. The difference in the sign of Q_θ results from the difference in the conventional right- and left-handed systems of coordinates used for trigonal and cubic molecules, respectively.

Thus the linear vibronic interaction of the type (3.1.6) is general for any polyatomic system in which the Jahn-Teller effect of the $E \otimes e$ type is possible. Hereafter when considering the $E \otimes e$ case, if the particular type of molecule is not specified, the triangular X_3 molecule will be implied.

For approximate solution of the $E \otimes e$ problem in the limiting case of strong vibronic coupling and, in particular, for a qualitative understanding of the vibronic effects, the basis of the adiabatic approximation is preferred instead of the basis of electronic states $\psi_\pm(\mathrm{r})$. For the adiabatic basis $\hat{U}\varphi = \varepsilon\varphi$, and the matrix of the potential energy

$$\hat{U} = \tfrac{1}{2}\omega_E^2(Q_\theta^2 + Q_\varepsilon^2)\hat{\sigma}_0 + V_E(Q_\theta\hat{\sigma}_x + Q_\varepsilon\hat{\sigma}_y) \tag{3.1.7}$$

should be diagonal. Since the operator of the elastic energy $\frac{1}{2}\omega_E^2(Q_\theta^2 + Q_\varepsilon^2)\sigma_0$ is a multiple of the unit matrix σ_0 (which does not change its form under any unitary transformation of the basis), the eigenvectors of the matrix (3.1.7) can be found from the condition that the operator of the vibronic interaction is diagonal

$$V_E(Q_\theta\hat{\sigma}_x + Q_\varepsilon\hat{\sigma}_y) = V_E\varrho\begin{pmatrix} 0 & \mathrm{e}^{-\mathrm{i}\varphi} \\ \mathrm{e}^{\mathrm{i}\varphi} & 0 \end{pmatrix}. \tag{3.1.8}$$

(in polar coordinates $Q_\theta = \varrho\cos\varphi$, $Q_\varepsilon = \varrho\sin\varphi$). Up to an arbitrary phase factor, the operator of the unitary transformation $\hat{S}$ which diagonalizes the matrix (3.1.8) [and hence the matrix of the potential energy (3.1.7)] has the form

$$\hat{S} = \frac{1}{\sqrt{2}}\begin{pmatrix} \mathrm{e}^{-\mathrm{i}\varphi/2} & \mathrm{e}^{-\mathrm{i}\varphi/2} \\ \mathrm{e}^{\mathrm{i}\varphi/2} & -\mathrm{e}^{\mathrm{i}\varphi/2} \end{pmatrix}. \tag{3.1.9}$$

Indeed,

$$\hat{S}^+(Q_\theta\hat{\sigma}_x + Q_\varepsilon\sigma_y)\hat{S} = \varrho\hat{\sigma}_z\ ,$$

and the matrix $\hat{U}$ in the new basis is diagonal,

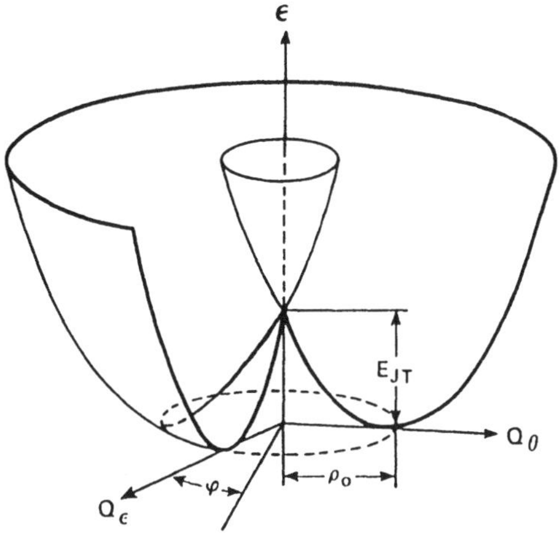

Fig. 3.1. Shape of the adiabatic potential of the electronic E term linearly interacting with E displacements (the "Mexican hat").

$$\hat{\hat{U}} = \hat{S}^{+}\hat{U}\hat{S} = \tfrac{1}{2}\omega_E^2\varrho^2\hat{\sigma}_0 + V_E\varrho\hat{\sigma}_z \; . \tag{3.1.10}$$

The adiabatic potentials of the E term, which are eigenvalues of the matrix $\hat{U}$, as can be seen from (3.1.10), have the form

$$\varepsilon_{\pm} = \tfrac{1}{2}\omega_E^2\varrho^2 \pm |V_E|\varrho \tag{3.1.11}$$

and can be represented by the figure of revolution called the *Mexican hat* (sombrero), Fig. 3.1. Note the conic-like shape of the surfaces $\varepsilon_{\pm}$ in the immediate vicinity of the point where the two sheets stick together (cf. Fig. 2.6). As shown by Fig. 3.1, the lowest sheet

$$\varepsilon_{-} = \tfrac{1}{2}\omega_E^2\varrho^2 - |V_E|\varrho \tag{3.1.12}$$

has a characteristic equipotential continuum of minima (a *trough*). The radius of the trough and its depth (*the Jahn-Teller stabilization energy* E_{JT}) can easily be found from (3.1.12):

$$\varrho_0 = \frac{|V_E|}{\omega_E^2} \, , \qquad \varepsilon_{-}(\varrho_0) = -\frac{1}{2}\frac{V_E^2}{\omega_E^2} \, , \qquad E_{\mathrm{JT}} = \frac{1}{2}\frac{V_E^2}{\omega_E^2} \; . \tag{3.1.13}$$

The energy gap between the sheets of the adiabatic potential at the point of the minima is

$$\varepsilon_{+}(\varrho_0) - \varepsilon_{-}(\varrho_0) = 2V_E^2/\omega_E^2 = 4E_{\mathrm{JT}} \; . \tag{3.1.14}$$

The operator of the unitary transformation (3.1.9) which diagonalizes the matrix of potential energy does not commute with the matrix of the kinetic energy $\hat{T} = \tfrac{1}{2}(P_\theta^2 + P_\varepsilon^2)\hat{\sigma}_0$, which as a result of this transformation becomes nondiagonal. The nondiagonal matrix elements Λ_{+-} and Λ_{-+} of the operator $\hat{T}$ result directly from the nonadiabaticity (Sect. 2.1). The need to take this nonadiabaticity into account is a characteristic feature of Jahn-Teller polyatomic systems. Exceptions are found when the constant of vibronic interaction is large and the nuclear motion is localized near the bottom of the trough. In this region the energy gap (3.1.14) between the sheets of the adiabatic potential can be significantly larger than the characteristic energy of E vibrations $\hbar\omega_E$ [inequality (2.1.10) is fulfilled], the contribution of the operators of nonadiabaticity Λ_{+-} and Λ_{-+} can be neglected and hence the adiabatic approximation is valid. The adiabatic potential in this case acquires the physical meaning of the potential energy of the nuclei moving in the average field of the electrons.

Knowledge of the potential energy surfaces, and especially the position and the curvature of its extrema, allows both a qualitative understanding of the nature of the nuclear motions and the solution of the Schrödinger equation in the simplest cases. For instance, in the lowest vibronic states of the $E \otimes e$ problem with linear vibronic coupling considered above (Fig. 3.1), the motion of the nuclei should be localized near the bottom of the trough of the lowest sheet of the adiabatic potential $\varepsilon_-(Q)$, and owing to the axial symmetry of the potential energy surface, this motion has to be free rotations in the trough about the central axis of symmetry.

It is interesting to follow the motion of individual atoms of the triangular molecule X_3 when the system rotates in the trough of the adiabatic potential.[1] In order to do this, we assume that all the $Q_{\Gamma\gamma}$ coordinates are equal to zero, except $Q_\theta = \varrho\cos\varphi$ and $Q_\varepsilon = \varrho\sin\varphi$, and use the relations of $Q_{\Gamma\gamma}$ with the Cartesian displacements x_i, y_i and z_i of the atoms of the triangular system (Table 2.1). Solving this system for x_i, y_i and z_i, we obtain

$$x_1 = \frac{1}{\sqrt{3}}\varrho\cos\varphi\ , \quad x_2 = \frac{1}{\sqrt{3}}\varrho\cos\left(\varphi - \frac{4\pi}{3}\right), \quad x_3 = \frac{1}{\sqrt{3}}\varrho\cos\left(\varphi - \frac{2\pi}{3}\right),$$

$$y_1 = \frac{1}{\sqrt{3}}\varrho\sin\varphi\ , \quad y_2 = \frac{1}{\sqrt{3}}\varrho\sin\left(\varphi - \frac{4\pi}{3}\right), \quad y_3 = \frac{1}{\sqrt{3}}\varrho\sin\left(\varphi - \frac{2\pi}{3}\right).$$

As one can see, when moving along the bottom of the trough the atoms of the triangular molecule circumscribe circles of radius $\varrho_0/\sqrt{3}$. The movement of different atoms is coherent. The vectors of atomic displacements from the equilibrium positions are shifted in phase by $2\pi/3$, and if reduced to the same origin of coordinates, they form a three-pronged star with apexes producing an equilateral triangle with sides of length ϱ_0. Figure 3.2a shows how the form of a triangular molecule changes during such a motion. The motion along the trough

[1] We discuss here the formal description of the nuclear motion which is adequate for the case when the adiabatic potentials have the physical sense of the potential energy of the nuclei. A more accurate description of the behavior of the nuclear subsystem is given by solutions of the system of vibronic equations (Chap. 4).

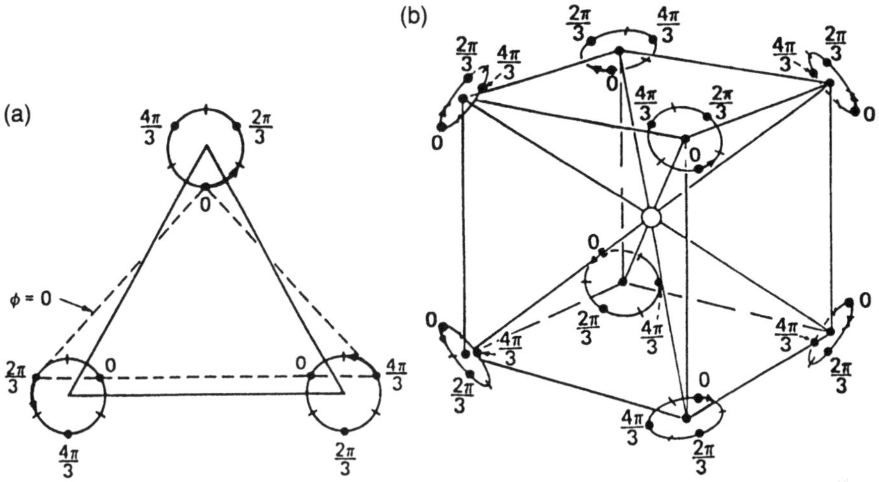

Fig. 3.2. Distortions of triangular (**a**) and cubic (**b**) molecules in the motion round the trough of the lowest sheet of the adiabatic potential in the linear $E \otimes e$ problem. The circles represent the trajectories of the atoms of the molecular framework. Heavy points indicate atomic configurations corresponding to the minima of the adiabatic potential arising when the quadratic terms of vibronic interactions are considered. (After [3.2])

can be thought of as a *wave of distortions* propagating around the geometric center of the equilateral triangle and transforming the equilateral triangle into an isosceles one. There are analogous waves of tetragonal distortions in the case of linear $E \otimes e$ coupling in tetrahedral and cubic systems (Fig. 3.2b), [3.2, 3].

Let us now consider the influence of the quadratic (in $Q_{E\gamma}$) terms of the vibronic interaction. Substituting the appropriate values of Clebsch-Gordan coefficients for a trigonal group D_{3h} (correct to a common factor that is unimportant here) into (2.2.6) we find

$$\{Q_E \otimes Q_E\}_{E\theta} = Q_\theta^2 - Q_\varepsilon^2 \ , \qquad \{Q_E \otimes Q_E\}_{E\varepsilon} = -2Q_\theta Q_\varepsilon \ . \tag{3.1.15}$$

Considering (3.1.15), the matrix of the potential energy (2.2.9) written in the basis of the electronic states $\psi_\pm(\mathrm{r})$ of the orbital doublet has the form

$$\begin{aligned}\hat{U}(Q) = \tfrac{1}{2}\omega_E^2(Q_\theta^2 + Q_\varepsilon^2)\hat{\sigma}_0 + V_E(Q_\theta\hat{\sigma}_x + Q_\varepsilon\hat{\sigma}_y) \\ + W_E[(Q_\theta^2 - Q_\varepsilon^2)\hat{\sigma}_x - 2Q_\theta Q_\varepsilon\hat{\sigma}_y] \ ,\end{aligned} \tag{3.1.16}$$

Where W_E is the constant of quadratic vibronic coupling, or, in polar coordinates,

$$\hat{U} = \frac{1}{2}\omega_E^2\varrho^2\hat{\sigma}_0 + V_E\varrho\left[1 + 2\frac{W_E}{V_E}\varrho\cos 3\varphi + \left(\frac{W_E}{V_E}\varrho\right)^2\right]^{1/2}\begin{pmatrix} 0 & e^{-i\Omega} \\ e^{i\Omega} & 0 \end{pmatrix} , \tag{3.1.17}$$

where the angle Ω is determined by the relation

$$\tan\Omega = \frac{V_E \sin\varphi - W_E\varrho\sin 2\varphi}{V_E\cos\varphi + W_E\varrho\cos 2\varphi} . \tag{3.1.18}$$

The matrix part of the operator (3.1.17) which includes the quadratic vibronic interaction is analogous to the matrix part of the operator of linear vibronic coupling (3.1.8), the only distinction being that the polar angle φ in (3.1.8) is replaced by the angle Ω determined by the relation (3.1.18). Similarly, the matrix of the unitary transformation which diagonalizes the matrix (3.1.17) [and hence the matrix of the potential energy (3.1.16)] has the same form as that in (3.1.9) with the angle Ω instead of φ. Therefore, the adiabatic potentials (the eigenvalues of the matrix $\hat{U}$) are

$$\varepsilon_\pm(\varrho,\varphi) = \tfrac{1}{2}\omega_E^2\varrho^2 \pm \sqrt{V_E^2\varrho^2 + 2V_E W_E\varrho^3\cos(3\varphi) + W_E^2\varrho^4} . \tag{3.1.19}$$

As can be seen from (3.1.19), the equipotential cross sections of the surfaces $\varepsilon_\pm$, as distinguished from the case of linear vibronic coupling, depend not only on ϱ, but also on φ. Thus when the quadratic terms are taken into account the surfaces of the adiabatic potential $\varepsilon_\pm$ become warped. Three wells alternating with three humps occur along the bottom of the trough, thus transforming the Mexican hat into a *cocked hat* (*tricorn*) as shown in Fig. 3.3. The extrema of the surface $\varepsilon_-(\varrho,\varphi)$

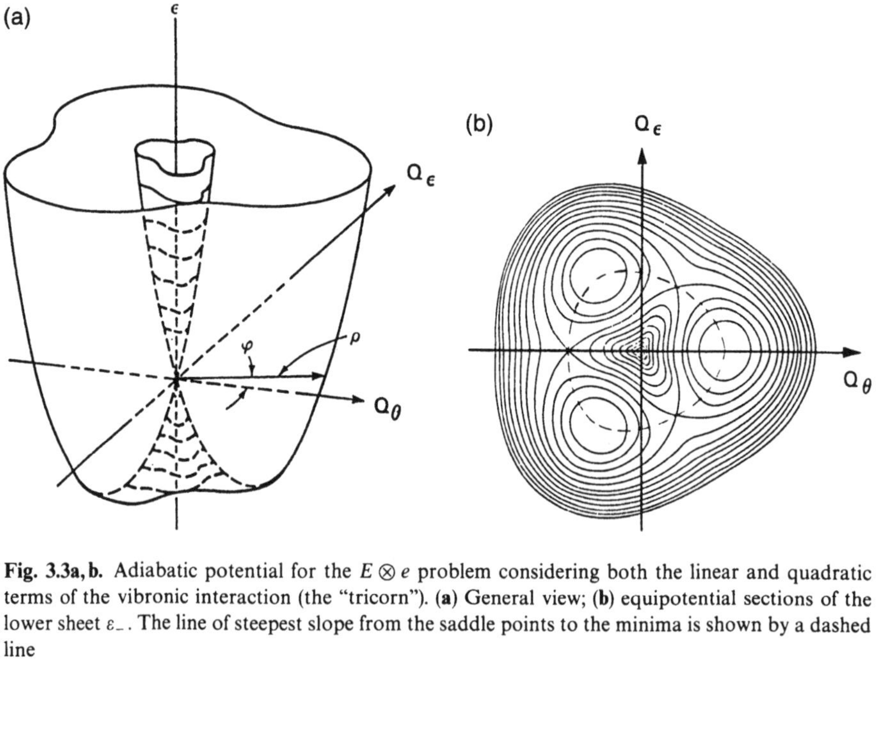

Fig. 3.3a,b. Adiabatic potential for the $E\otimes e$ problem considering both the linear and quadratic terms of the vibronic interaction (the "tricorn"). **(a)** General view; **(b)** equipotential sections of the lower sheet ε_-. The line of steepest slope from the saddle points to the minima is shown by a dashed line

are given by

$$\varrho_n = \frac{V_E^2}{|V_E|\omega_E^2 - (-1)^n 2V_E W_E}\ , \qquad \varphi_n = \frac{\pi n}{3}\ , \qquad n = 0, 1, 2, \ldots \tag{3.1.20}$$

the points $n = 1, 3, 5$ being absolute minima and the points $n = 0, 2, 4$ being saddle points when $W_E V_E < 0$, and vice versa when $W_E V_E > 0$. The depth of the minima (the Jahn-Teller stabilization energy) and the barrier height Δ between these minima are given by

$$E_{\mathrm{JT}} = \frac{1}{2}\frac{V_E^2}{\omega_E^2 - 2|W_E|}\ , \qquad \Delta = \frac{4|W_E|E_{\mathrm{JT}}}{\omega_E^2 + 2|W_E|}\ . \tag{3.1.21}$$

The *curve of steepest slope* connecting all the extrema of the surface ε_- is shown in Fig. 2.3b by a dashed line. It can be represented by the following function given parametrically:

$$\begin{aligned} Q_\theta &= \frac{V_E}{\omega_E^2 - 4W_E^2}(\omega_E^2 \cos\alpha + 2W_E \cos 2\alpha)\ , \\ Q_\varepsilon &= \frac{V_E}{\omega_E^2 - 4W_E^2}(\omega_E^2 \sin\alpha - 2W_E \sin 2\alpha)\ . \end{aligned} \tag{3.1.22}$$

It can be easily shown that along this line and in its neighborhood the energy gap between the adiabatic potentials is $\varepsilon_+ - \varepsilon_- \sim V_E^2/\omega_E^2$. This means that, similarly to the case of linear vibronic coupling, for large enough values of V_E the contribution of the operators of nonadiabaticity Λ_{+-} and Λ_{-+} to the wave functions of the ground and the lowest excited states is negligible. The adiabatic potential $\varepsilon_-(\varrho, \varphi)$ in this case acquires the physical meaning of the potential energy of nuclei in the field of the electrons and this allows us to draw some qualitative conclusions about the nature of the nuclear motion in the lowest vibronic states. Thus, for example, it is obvious that for small values of Δ in (3.1.21) the rotation of the system along the trough of the Mexican hat described above becomes hindered and the motion of the nuclei is localized in the minima with the coordinates (3.1.20). The nuclear configurations corresponding to these three minima are tetragonally distorted cubes, octahedrons or tetrahedrons, elongated or compressed along one of their three axes of fourfold symmetry. The symmetry of the molecule in these minima is lowered to D_{4h} in the case of an octahedron or cube, and to D_{2d} in the case of a tetrahedron. The equilateral triangle (for X_3, AX_3 molecules, etc.) in the configurations of the minima becomes isosceles. All these configurations are shown in Fig. 3.2 by means of heavy points.

The curvature of the adiabatic potential ε_- at the minima is given by

$$\tilde{\omega}_A^2 = \omega_E^2 - 2|W_E|\ , \qquad \tilde{\omega}_B^2 = 9|W_E|\frac{\omega_E^2 - 2|W_E|}{\omega_E^2 - |W_E|}\ , \tag{3.1.23}$$

from which it follows that if $|W_E| \geq \omega_E^2/2$ the system is unstable and decomposes into different parts, provided the chemical bond does not remain stable on account of the higher-order even terms of the expansion of the operator of the vibronic interaction in a series with respect to $Q_{E\gamma}$. This explains, for instance, the instability of the H_3 molecule, the lowest electronic term of which in the nuclear configuration of an equilateral triangle is an orbital doublet [3.2, 4].

The existence of two different values $\tilde{\omega}_A^2$ and $\tilde{\omega}_B^2$ of the curvature of the surface ε_- in the tetragonal minima configurations is due to the splitting of the E_g vibrations when the symmetry is lowered from O_h to D_{4h}: $E_g = A_{1g} + B_{1g}$. Nevertheless, generally speaking, the motion of the polyatomic system in this case is not reduced to small vibrations near the minima with the corresponding frequencies $\tilde{\omega}_A$ and $\tilde{\omega}_B$. Postponing more detailed discussion of the physical consequences of this special form of the potential surface ε_-, we note here that for a limited barrier height between the minima the system can tunnel between them, which significantly alters the vibronic properties of the system (Sect. 4.3).

In linear molecules, owing to the selection rules for matrix elements, linear vibronic coupling is absent, i.e. $V_E = 0$. The adiabatic potentials (3.1.19) in this case again acquire an axially symmetric form (Fig. 2.6b, c)

$$\varepsilon_\pm = \tfrac{1}{2}\omega_E^2\varrho^2 \pm |W_E|\varrho^2 = \tfrac{1}{2}(\omega_E^2 \pm 2|W_E|)\varrho^2 \ , \tag{3.1.24}$$

In spite of the fact that both surfaces have an extremum at the point of degeneracy (Sect. 2.3, the Renner case), if $W_E > \omega_E^2/2$ the surface ε_- has negative curvature at this point. This means that (considering the exceptions mentioned at the end of Sect. 2.3) $Q_\theta = Q_\varepsilon = 0$ is a point of unstable equilibrium. In this case the stability can be maintained by the higher-order even terms of the expansion of the vibronic interaction. Note, that in Renner-type molecules the axial symmetry of the adiabatic potentials is not violated, even when one takes into account these high-order terms (Sect. 3.2). Hence the adiabatic potential ε_- in cases when $|W_E| > \omega_E^2/2$ has a circle of minimum points (a trough) similar to the case of the linear $E \otimes e$ problem.

3.1.2 $E \otimes (b_1 + b_2)$ Case. The Method of Öpik and Pryce

Consider now the E term Jahn-Teller effect of the second type, the $E \otimes (b_1 + b_2)$ case. The operator of the linear vibronic coupling in this case has the form $V_1(\mathrm{r})Q_1 + V_2(\mathrm{r})Q_2$, where Q_1 and Q_2 are normal coordinates which transform according to the representations B_1 and B_2, respectively. The matrix of the potential energy defined in the space of the electronic functions $\psi_\pm(\mathrm{r}) \sim \exp(\pm in\varphi/2)$ of the orbital doublet has the form [cf. (3.1.7)]

$$\hat{U} = \tfrac{1}{2}(\omega_1^2 Q_1^2 + \omega_2^2 Q_2^2)\hat{\sigma}_0 + V_1 Q_1 \hat{\sigma}_x + V_2 Q_2 \hat{\sigma}_y \ , \tag{3.1.25}$$

where V_1 and V_2 are the reduced matrix elements, the constants of the linear vibronic coupling. The adiabatic potentials, the eigenvalues of the matrix $\hat{U}$, can

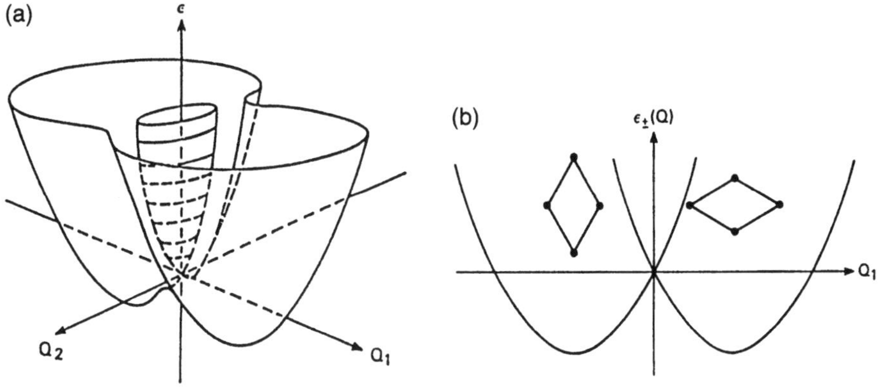

Fig. 3.4a, b. Adiabatic potential of a tetragonal molecular system with an electronic *E* term linearly coupled to B_1 and B_2 vibrations. (**a**) General view; (**b**) section along the Q_1 coordinate (in the plane $Q_2 = 0$) and corresponding distortions at the minima

be found directly:

$$\varepsilon_{\pm} = \tfrac{1}{2}(\omega_1^2 Q_1^2 + \omega_2^2 Q_2^2) \pm \sqrt{V_1 Q_1^2 + V_2 Q_2^2} \ . \tag{3.1.26}$$

The potential energy surfaces representing these adiabatic potentials (3.1.26) are shown in Fig. 3.4. Near the point of degeneracy they have the form of a double cone (cf. Figs 2.1 and 2.6a).

The extrema of the surface ε_- can be found directly by equating the derivatives of the function (3.1.26) to zero and solving the resulting equations for Q_1 and Q_2. This approach implies that we know the analytic expression for the adiabatic potential. However, there is another way to find the coordinates of the extrema of the adiabatic potential which does not require a preliminary diagonalization of the matrix of the potential energy. This method, suggested by *Öpik* and *Pryce* [3.5], is widely used, especially when the derivation of analytic expression for the adiabatic potentials is difficult. The method is based on the idea that when one diagonalizes the matrix $\hat{U}$ and solves the transcendental equations for the extrema, one solves in fact the following system of coupled equations:

$$\hat{U}(Q)\hat{a}(Q) = \varepsilon(Q)\hat{a}(Q) \ , \qquad \hat{a}^{\dagger}(Q)\hat{a}(Q) = 1 \ , \qquad \frac{\partial\varepsilon(Q)}{\partial Q_{\Gamma\gamma}} = 0 \ . \tag{3.1.27}$$

Here $\hat{a}(Q)$ is the eigenvector of the matrix $\hat{U}(Q)$ and represents the electronic adiabatic wave function $\psi(\mathrm{r}, Q)$ in the basis of the states $\psi_{\Gamma\gamma}(\mathrm{r})$ of the degenerate electronic term. The last equation of (3.1.27) can be transformed using the Hellmann-Feynman theorem [3.6, 7].

$$\frac{\partial\varepsilon}{\partial Q_{\Gamma\gamma}} = \frac{\partial}{\partial Q_{\Gamma\gamma}}[\hat{a}^{\dagger}(Q)\hat{U}(Q)\hat{a}(Q)] = \hat{a}^{\dagger}(Q)\frac{\partial\hat{U}}{\partial Q_{\Gamma\gamma}}\hat{a}(Q) \ .$$

Then the system (3.1.27) acquires the form

$$\hat{U}\hat{a}^\dagger = \varepsilon\hat{a} \ , \qquad \hat{a}^\dagger\hat{a} = 1 \ , \qquad \hat{a}^\dagger \frac{\partial \hat{U}}{\partial Q_{\Gamma\gamma}} a = 0 \ . \tag{3.1.28}$$

Let us illustrate the *method of Öpik and Pryce* with the $E \otimes (b_1 + b_2)$ example. It is convenient to start not with the complex basis $\psi_\pm(\mathrm{r})$ (and complex components of the column vector $\hat{a}$), but with the real basis $\psi_\theta(\mathrm{r})$ and $\psi_\varepsilon(\mathrm{r})$, see (3.1.5). In this basis the matrix (3.1.25) has the form

$$\hat{U} = \tfrac{1}{2}(\omega_1^2 Q_1^2 + \omega_2^2 Q_2^2)\hat{\sigma}_0 + V_1 Q_1 \hat{\sigma}_z + V_2 Q_2 \hat{\sigma}_x \ . \tag{3.1.29}$$

Remembering that $\hat{a}^\dagger\hat{a} = a_\theta^2 + a_\varepsilon^2 = 1$, $\hat{a}^\dagger\hat{\sigma}_x\hat{a} = 2a_\theta a_\varepsilon$, $\hat{a}^\dagger\hat{\sigma}_z\hat{a} = a_\theta^2 - a_\varepsilon^2$, where a_θ and a_ε are the components of the column vector $\hat{a}$, the equations $\hat{a}^\dagger\partial\hat{U}/\partial Q_{\Gamma\gamma}\hat{a} = 0$ can be written in the form

$$\omega_1^2 Q_1 + V_1(a_\theta^2 - a_\varepsilon^2) = 0 \ , \qquad \omega_2^2 Q_2 + V_2 2a_\theta a_\varepsilon = 0 \ ; \tag{3.1.30}$$

from which we find

$$Q_1 = -\frac{V_1}{\omega_1^2}(a_\theta^2 - a_\varepsilon^2) \ , \qquad Q_2 = -\frac{V_2}{\omega_2^2} 2a_\theta a_\varepsilon \ . \tag{3.1.31}$$

Substituting these values into (3.1.29) and writing out the matrix equation $\hat{U}\hat{a} = \varepsilon\hat{a}$ explicitly, together with the condition of normalization $a_\theta^2 + a_\varepsilon^2 = 1$ we obtain a system of three equations,

$$\begin{aligned} &-\frac{V_1^2}{\omega_1^2}(a_\theta^2 - a_\varepsilon^2)a_\theta - \frac{V_2^2}{\omega_2^2} 2a_\theta a_\varepsilon^2 = \lambda a_\theta \ , \\ &-\frac{V_1^2}{\omega_1^2}(a_\theta^2 - a_\varepsilon^2)a_\varepsilon - \frac{V_2^2}{\omega_2^2} 2a_\theta^2 a_\varepsilon = \lambda a_\varepsilon \ , \\ &a_\theta^2 + a_\varepsilon^2 = 1 \end{aligned} \tag{3.1.32}$$

for three unknowns a_θ, a_ε, and $\lambda = \varepsilon - \frac{1}{2}(\omega_1^2 Q_1^2 + \omega_2^2 Q_2^2)$.

Four quite obvious solutions of the system (3.1.32) are given in Table 3.1. Four other solutions which differ from those given in Table 3.1 by a simultaneous change of the signs of a_θ and a_ε do not result in new extrema of the adiabatic potential since the multiplication of the column vector $\hat{a}$ by $\exp(i\pi)$ does not change the wave function of the system. Substituting a_θ and a_ε into (3.1.31) we find the following extrema for the above roots:

$$Q_1^{(0)} = \pm\frac{V_1}{\omega_1^2} \ , \qquad Q_2^{(0)} = 0 \ , \qquad \varepsilon_-(Q^{(0)}) = -\frac{1}{2}\frac{V_1^2}{\omega_1^2} \ , \qquad E_{\mathrm{JT}}^{(1)} = \frac{1}{2}\frac{V_1^2}{\omega_1^2} \ ;$$

Table 3.1 Solutions of the Öpik-Pryce equations for the $E \otimes (b_1 + b_2)$ case

a_θ	$\frac{1}{\sqrt{2}}$	$\frac{1}{\sqrt{2}}$	1	0
a_ε	$\frac{1}{\sqrt{2}}$	$-\frac{1}{\sqrt{2}}$	0	1
λ	$-\frac{V_2^2}{\omega_2^2}$	$-\frac{V_2^2}{\omega_2^2}$	$-\frac{V_1^2}{\omega_1^2}$	$-\frac{V_1^2}{\omega_1^2}$

$$Q_1^{(0)} = 0 , \quad Q_2^{(0)} = \pm\frac{V_2}{\omega_2^2} , \quad \varepsilon_-(Q^{(0)}) = -\frac{1}{2}\frac{V_2^2}{\omega_2^2} , \quad E_{\text{JT}}^{(2)} = \frac{1}{2}\frac{V_2^2}{\omega_2^2} . \tag{3.1.33}$$

The question of whether these extrema are minima or saddle points can be solved, following *Öpik* and *Pryce* [3.5], by means of perturbation theory. In order to do this, the matrix $\hat{U}$ is written as

$$\hat{U}(Q) = U(Q^{(0)}) + \hat{U}_1(Q) + \hat{U}_2(Q) , \tag{3.1.34}$$

$$\hat{U}_1(Q) = \sum_{\Gamma\gamma} \frac{\partial \hat{U}(Q^{(0)})}{\partial Q_{\Gamma\gamma}} (Q_{\Gamma\gamma} - Q_{\Gamma\gamma}^{(0)}) , \tag{3.1.35}$$

$$\hat{U}_2(Q) = \frac{1}{2} \sum_{\Gamma_1\gamma_1} \sum_{\Gamma_2\gamma_1} \frac{\partial^2 \hat{U}(Q^{(0)})}{\partial Q_{\Gamma_1\gamma_1} \partial Q_{\Gamma_2\gamma_2}} (Q_{\Gamma_1\gamma_1} - Q_{\Gamma_1\gamma_1}^{(0)})(Q_{\Gamma_2\gamma_2} - Q_{\Gamma_2\gamma_2}^{(0)}) . \tag{3.1.36}$$

Considering the displacements $q_{\Gamma\gamma} = Q_{\Gamma\gamma} - Q_{\Gamma\gamma}^{(0)}$ as small parameters of the theory, one can evaluate the second-order correction to the eigenvalue $\varepsilon_-(Q^{(0)})$ as

$$\varepsilon_-(Q) \approx \varepsilon_-(Q^{(0)}) + \hat{a}^\dagger \hat{U}_2(Q)\hat{a} + \sum_i{}' \frac{|\hat{a}_i^\dagger \hat{U}_1(Q)\hat{a}|^2}{\varepsilon_-(Q^{(0)}) - \varepsilon_i(Q^{(0)})} . \tag{3.1.37}$$

Here a_i and $\varepsilon_i(Q^{(0)})$ are the eigenvectors and eigenvalues of the matrix $\hat{U}(Q^{(0)})$, and the prime on the summation sign means, as usual, that the singular term in the sum is omitted. Expression (3.1.37) is the bilinear form of the displacements $q_{\Gamma\gamma}$. The linear term in $q_{\Gamma\gamma}$ in (3.1.37) is absent since at the point $Q^{(0)}$ the function $\varepsilon_-(Q)$ has an extremum.

For the $E \otimes (b_1 + b_2)$ case under consideration, the quadratic form (3.1.37) reads

$$\varepsilon_-(q) \approx \varepsilon_-(Q^{(0)}) + \frac{1}{2}(\omega_1^2 q_1^2 + \omega_2^2 q_2^2) + \frac{|\hat{a}_-^\dagger (V_1 q_1 \hat{\sigma}_z + V_2 q_2 \hat{\sigma}_x)\hat{a}_+|^2}{\varepsilon_-(Q^{(0)}) - \varepsilon_+(Q^{(0)})} , \tag{3.1.38}$$

and in the particular case of the extremum with the coordinates $Q_1^{(0)} = V_1/\omega_1^2$ and

$Q_2^{(0)} = 0$ it reduces to

$$\varepsilon_- \approx -E_{JT}^{(1)} + \frac{1}{2}(\omega_1^2 q_1^2 + \omega_2^2 q_2^2) - \frac{1}{2}\left(\frac{V_2}{V_1}\right)^2 \omega_1^2 q_2^2 \ . \tag{3.1.39}$$

From this we can find the curvature of the lowest sheet of the adiabatic potential at the extremum, $\tilde{\omega}_1^2 = \omega_1^2$, $\tilde{\omega}_2^2 = \omega_2^2(1 - E_{JT}^{(2)}/E_{JT}^{(1)})$. Similarly, in a small neighborhood around the extremum with the coordinates $Q_1^{(0)} = 0$, $Q_2^{(0)} = V_2/\omega_2^2$, we have

$$\varepsilon_- \approx -E_{JT}^{(2)} + \frac{1}{2}(\omega_1^2 q_1^2 + \omega_2^2 q_2^2) - \frac{1}{2}\left(\frac{V_1}{V_2}\right)^2 \omega_2^2 q_1^2 \tag{3.1.40}$$

and hence $\tilde{\omega}_1^2 = \omega_1^2(1 - E_{JT}^{(1)}/E_{JT}^{(2)})$, $\tilde{\omega}_2^2 = \omega_2^2$. This means that if $E_{JT}^{(1)} > E_{JT}^{(2)}$, then the extrema displaced in the Q_1 direction are absolute minima and those shifted along Q_2 are saddle points, and vice versa when $E_{JT}^{(1)} < E_{JT}^{(2)}$. In the two minima of the first type the square molecule is distorted into a rhombus, whereas in the two other minima it is distorted into a rectangle (Fig. 2.3c,d).

When the quadratic terms of the vibronic interaction are taken into account in the $E \otimes (b_1 + b_2)$ problem, coexistence of the two types of minima becomes possible for some values of the vibronic coupling constants [3.8].

The case of the $E \otimes (b_1 + b_2)$ problem with negligibly small vibronic coupling to one of the B vibrations seems to be of particular interest. Then the whole Hamiltonian of the system consists of two commuting terms $\hat{H}(Q_1) + \hat{H}(Q_2)$ and the variables can be separated. The motion along the coordinate which does not interact with the orbital doublet (for definiteness let it be Q_2) is simple harmonic vibrations near the totally symmetric configuration of the nuclei. The Hamiltonian for the other coordinate in the basis of the electronic states ψ_θ and ψ_ε has the form

$$\hat{H}(Q_1) = \tfrac{1}{2}(P_1^2 + \omega_1^2 Q_1^2)\hat{\sigma}_0 + V_1 Q_1 \hat{\sigma}_z \ . \tag{3.1.41}$$

The matrix of the potential energy in this basis is diagonal, and therefore the adiabatic potential of such a $E \otimes b_1$ *case* is

$$\varepsilon_\pm = \tfrac{1}{2}\omega_1^2 Q_1^2 \pm V_1 Q_1 \ . \tag{3.1.42}$$

It can easily be seen from (3.1.5) that the unitary transformation of the electronic basis ψ_+, ψ_- into the basis of states ψ_θ, ψ_ε is independent of Q and hence commutes with the kinetic energy operator. Therefore the Hamiltonian (3.1.41) written in the basis which diagonalizes the potential energy matrix does not contain the operators of nonadiabaticity Λ_{+-} and Λ_{-+}. Thus the $E \otimes b_1$ ($E \otimes b_2$) problem is one of the few in the theory of vibronic interactions for which the adiabatic potential has the physical meaning of the potential energy of the nuclei for arbitrary values of the vibronic constant and nuclear coordinate.

3.2 Symmetry of Jahn-Teller Systems

This section presents some results based on general considerations and obtained by group-theoretical methods [3.9–11]. These results allow the determination of the integrals of motion without solving the system of vibronic equations and, even more importantly, they clarify the origin of the symmetry properties of the adiabatic potentials (the number of extrema, the symmetry of the system in each of them, the group of symmetry of equivalent extrema, etc.).

3.2.1 Lee Symmetry of the Jahn-Teller Hamiltonian

The use of different approximations for analytical solution of the Schrödinger equation and the development of appropriate models aim at complete separation of the variables and reduction of the complicated many-dimensional problem to a set of simpler one-dimensional ones. Complete or partial separation of the variables means that there is an additional symmetry in the Hamiltonian which permits this separation. From this point of view the approximations used for the solution of the Schrödinger equation are usually accompanied by an increase in the symmetry of the Hamiltonian.

Consider the changes in symmetry due to the approximations used above, i.e. due to limiting the electronic spectrum to several states of the degenerate electronic term and to replacing the exact operator $V(\mathrm{r}, Q)$ by its quadratic expansion (2.2.1). The method of investigating the symmetry of the Hamiltonian will be illustrated by the $E \otimes e$ example. Its generalization to other cases of the Jahn-Teller effect, as will be shown later, is trivial.

The Hamiltonian of the linear $E \otimes e$ problem (i.e., without considering the quadratic terms of the vibronic interaction) written in the basis of electronic states $\psi_\pm(\mathrm{r})$ has the form [cf. (3.1.7)]

$$\hat{H} = E_0\hat{\sigma}_0 + \tfrac{1}{2}[P_\theta^2 + P_\varepsilon^2 + \omega_E^2(Q_\theta^2 + Q_\varepsilon^2)]\hat{\sigma}_0 + V_E(Q_\theta\hat{\sigma}_x + Q_\varepsilon\hat{\sigma}_y) . \tag{3.2.1}$$

Here E_0 is the energy of the orbital doublet, which is usually excluded from consideration by means of an appropriate choice of the zero of the energy. In the absence of vibronic interactions, i.e. when $V_E = 0$, the Hamiltonian (3.2.1) consists of two terms that commute, one of which ($\hat{H}_e = E_0\hat{\sigma}_0$) corresponds to the energy of the electrons in the degenerate state and the other to the energy of the harmonic E-type vibrations,

$$\hat{H}_\mathrm{v} = \tfrac{1}{2}[P_\theta^2 + P_\varepsilon^2 + \omega_E^2(Q_\theta^2 + Q_\varepsilon^2)]\hat{\sigma}_0 . \tag{3.2.2}$$

The eigenfunctions of the Hamiltonian $\hat{H}_e + \hat{H}_\mathrm{v}$ have a multiplicative form $\Psi = \psi_\pm(\mathrm{r})\chi(Q)$, where $\psi_\pm(\mathrm{r})$ and $\chi(Q)$ are the eigenfunctions of the Hamiltonians $\hat{H}_e$ and $\hat{H}_\mathrm{v}$, respectively. As is well known, in such cases the total symmetry group of the Hamiltonian is the direct product $G_e \times G_\mathrm{v}$, where G_e and G_v are the groups of symmetry of the Hamiltonians $\hat{H}_e$ and $\hat{H}_\mathrm{v}$, respectively.

The electronic Hamiltonian H_e is a multiple of the second-order unit matrix $\hat{\sigma}_0$ which does not change its form under any unitary transformation in the two-dimensional space of the basis functions $\psi_{\pm}(\mathrm{r}) = |\pm\rangle$. Therefore, the group G_e is identical to the second-order unitary group $U(2)$. The properties of the group $U(n)$ are well studied [3.12]. The number of parameters that determine an arbitrary element of this group is n^2. The operators $\hat{I}_{ij} = |i\rangle\langle j|$, where $i, j = 1, 2, \ldots, n$, form a manifold of n^2 elements closed with respect to the operation of commutation,

$$[\hat{I}_{ij}, \hat{I}_{kl}] = \hat{I}_{il}\delta_{kj} - \hat{I}_{kj}\delta_{il} , \tag{3.2.3}$$

i.e., they form an algebra. It can easily be seen that they all commute with the unit matrix representing the electronic Hamiltonian of the n-fold degenerate electronic term, and therefore they can be chosen as the generators of the group $U(n)$. In the case of an orbital doublet they can be represented by

$$\begin{aligned} \hat{I}_{++} &= \begin{pmatrix} 1 & 0 \\ 0 & 0 \end{pmatrix} , \quad & \hat{I}_{+-} &= \begin{pmatrix} 0 & 1 \\ 0 & 0 \end{pmatrix} , \\ \hat{I}_{-+} &= \begin{pmatrix} 0 & 0 \\ 1 & 0 \end{pmatrix} , \quad & \hat{I}_{--} &= \begin{pmatrix} 0 & 0 \\ 0 & 1 \end{pmatrix} . \end{aligned} \tag{3.2.4}$$

The eigenfunctions of the Hamiltonian $\hat{H}_e = E_0\hat{\sigma}_0$ may be given by two-component column matrices, i.e., by spinors,

$$\hat{\psi}_+ = \begin{pmatrix} 1 \\ 0 \end{pmatrix} , \quad \hat{\psi}_- = \begin{pmatrix} 0 \\ 1 \end{pmatrix} . \tag{3.2.5}$$

Formally they can be regarded as two different states of the same particle which differ in the value of the projection of the vector $\frac{1}{2}\hat{\boldsymbol{\sigma}}$, the latter being analogous (in its properties) to the spin 1/2 vector. This new quantity, usually called the *energy spin* or *pseudo-spin* (see for example [3.13]), is a vector in some auxiliary "energy" space (which of course has nothing in common with real space). The projection of the energy spin on the z axis has only two values: $\pm 1/2$. The Hamiltonian $\hat{H}_e$ expressed in terms of the components of the energy spin has the form

$$\hat{H}_e = \tfrac{1}{3}E_0(\hat{\sigma}_x^2 + \hat{\sigma}_y^2 + \hat{\sigma}_z^2) . \tag{3.2.6}$$

It is seen that the unitary group $U(2)$ should have the rotation group $O(3)$ in three dimensions of spin space as a subgroup. With respect to these rotations the operators $\hat{I}_{ij} = \hat{\psi}_i\hat{\psi}_j^\dagger$ are components of a second-rank tensor which can be arranged in linear combinations transforming according to the irreducible representations $D^{(1/2)} \times D^{(1/2)} = D^{(0)} + D^{(1)}$ of the group $O(3)$. The components of the corresponding irreducible tensors are given by

$$\{\hat{\psi}^{(1/2)} \times \hat{\psi}^{(1/2)}\}^l_m = \sum_{p,q} \hat{I}_{pq}(\tfrac{1}{2}p;\tfrac{1}{2}q|lm) , \tag{3.2.7}$$

where q, $p = \pm 1/2$ and $(\frac{1}{2}p;\frac{1}{2}q|lm)$ are the Clebsch-Gordan coefficients of the group $O(3)$. Substituting their numerical values into (3.2.7) we find

$$\{\hat{\psi}^{(1/2)} \times \hat{\psi}^{(1/2)}\}^{(0)}_0 = \sqrt{2}\hat{\sigma}_0 , \qquad \{\psi^{(1/2)} \times \psi^{(1/2)}\}^{(1)} = \sqrt{2}\hat{\boldsymbol{\sigma}} . \tag{3.2.8}$$

Thus, having in mind the subsequent reduction of the group $U(2)$, it is conveinent to use the linear combinations (3.2.8) irreducible in the group $O(3)$ instead of four generators from (3.2.4). The operators (3.2.8) are Hermitian and also have the advantage that the corresponding algebra decomposes into subalgebras with respect to the operation of commutation. For instance, three operators $\hat{\sigma}_x$, $\hat{\sigma}_y$ and $\hat{\sigma}_z$ form a manifold closed in the above sense. An arbitrary operator of the group $U(2)$ dependent on four ($n^2 = 4$) parameters φ_0, φ_x, φ_y, φ_z can be written in the form

$$\hat{G}(\varphi_0, \varphi_x, \varphi_y, \varphi_z) = \exp\{\mathrm{i}(\varphi_0\hat{\sigma}_0 + \varphi_x\hat{\sigma}_x + \varphi_y\hat{\sigma}_y + \varphi_z\hat{\sigma}_z)\} . \tag{3.2.9}$$

It is not difficult to see that the parameter φ_0 describes a trivial multiplication of the basis functions by a phase factor. Indeed, $\hat{G}(\varphi_0) = \exp(\mathrm{i}\varphi_0\hat{\sigma}_0) = \exp(\mathrm{i}\varphi_0)\hat{\sigma}_0$, and therefore

$$\hat{G}(\varphi_0)\hat{\psi}_+ = \mathrm{e}^{\mathrm{i}\varphi_0}\begin{pmatrix}1 & 0\\ 0 & 1\end{pmatrix}\begin{pmatrix}1\\ 0\end{pmatrix} = \mathrm{e}^{\mathrm{i}\varphi_0}\hat{\psi}_+ ,$$

$$\hat{G}(\varphi_0)\hat{\psi}_- = \mathrm{e}^{\mathrm{i}\varphi_0}\begin{pmatrix}1 & 0\\ 0 & 1\end{pmatrix}\begin{pmatrix}0\\ 1\end{pmatrix} = \mathrm{e}^{\mathrm{i}\varphi_0}\hat{\psi}_- .$$

Excluding from consideration these trivial operations, we can confine ourselves to the three-parameter subgroup $SU(2)$ of the group $U(2)$, which consists of the unitary matrices

$$\hat{G}(\varphi_x, \varphi_y, \varphi_z) = \exp\{\mathrm{i}(\varphi_x\hat{\sigma}_x + \varphi_y\hat{\sigma}_y + \varphi_z\hat{\sigma}_z)\} = \exp(\mathrm{i}\boldsymbol{\varphi}\hat{\boldsymbol{\sigma}}) , \tag{3.2.10}$$

with a determinant equal to 1.

It is more convenient to investigate the symmetry of the vibrational Hamiltonian in the formulation of second quantization. Let us introduce creation and annihilation operators

$$b^+_{\Gamma\gamma} = -\frac{\mathrm{i}}{\sqrt{2\hbar\omega_\Gamma}}(P_{\Gamma\gamma} + \mathrm{i}\omega_\Gamma Q_{\Gamma\gamma}) , \qquad b_{\Gamma\gamma} = \frac{\mathrm{i}}{\sqrt{2\hbar\omega_\Gamma}}(P_{\Gamma\gamma} - \mathrm{i}\omega Q_{\Gamma\gamma}) , \tag{3.2.11}$$

which have the common commutation relations

$$[b_{\Gamma\gamma}, b^+_{\bar{\Gamma}\bar{\gamma}}] = \delta_{\Gamma\bar{\Gamma}}\delta_{\gamma\bar{\gamma}} . \tag{3.2.12}$$

The Hamiltonian (3.2.2) can be written as

$$\hat{H}_v = \hbar\omega_E(b_\theta^+ b_\theta + b_\varepsilon^+ b_\varepsilon + 1)\hat{\sigma}_0 \ . \tag{3.2.13}$$

This Hamiltonian is invariant under an arbitrary unitary transformation (in the space of two operators b_θ and b_ε), thus not changing the commutation relations (3.2.12):

$$\tilde{b}_\theta = U_{\theta\theta} b_\theta + U_{\theta\varepsilon} b_\varepsilon \ , \qquad \tilde{b}_\varepsilon = U_{\varepsilon\theta} b_\theta + U_{\varepsilon\varepsilon} b_\varepsilon \ , \qquad \sum_{\bar{\gamma}} U_{\gamma\bar{\gamma}} U^*_{\gamma'\bar{\gamma}} = \delta_{\gamma\gamma'} \ . \tag{3.2.14}$$

In other words, the Hamiltonian (3.2.13) of the two-dimensional harmonic oscillator is invariant under the group $U(2)$. Similarly, it can be shown that for an n-dimensional oscillator the appropriate symmetry group is $U(n)$. The operators $J_{ij} = b_i^+ b_j$, where $i, j = 1, 2, \ldots, n$, form a manifold of n^2 elements closed with respect to the commutation operations [cf. (3.2.3)]:

$$[J_{ij}, J_{kl}] = J_{il}\delta_{kj} - J_{kj}\delta_{il} \ . \tag{3.2.15}$$

One can verify directly that all these operators commute with the Hamiltonian of the n-dimensional isotropic oscillator and therefore that they can be chosen as generators of the group $U(n)$. As in the case considered above, under the operations of the group $O(3)$, which is a subgroup of group $U(n)$, the operators b_i transform as components of an n-dimensional vector, i.e., as the components of an irreducible tensor $b^{(l)}$, where $2l + 1 = n$. From this point of view the operators $b_i^+ b_j$ transform as second-rank tensors, from which linear combinations irreducible in the group O(3) can be formed [3.9] as follows:

$$\{b^{(l)} \times b^{(l)}\}^{(L)}_M = \sum_{m_1, m_2} b^+_{m_1} b_{m_2} (lm_1 lm_2/LM) \ , \tag{3.2.16}$$

where $L = 0, 1, \ldots, 2l$; $m_1, m_2 = -l, (-l+1), \ldots, l$. In particular, for a two-dimensional oscillator ($n = 2, l = \frac{1}{2}$) we have

$$\begin{aligned} \{b^{(1/2)} \times b^{(1/2)}\}^{(0)} &= \sqrt{2}(J_{\theta\theta} + J_{\varepsilon\varepsilon}) \ , \\ \{b^{(1/2)} \times b^{(1/2)}\}^{(1)}_0 &= -\mathrm{i}\sqrt{2}(J_{\theta\varepsilon} - J_{\varepsilon\theta}) \ , \\ \{b^{(1/2)} \times b^{(1/2)}\}^{(1)}_{\pm 1} &= = \sqrt{2}[(J_{\theta\theta} - J_{\varepsilon\varepsilon}) \pm \mathrm{i}(J_{\theta\varepsilon} + J_{\varepsilon\theta})] \ . \end{aligned} \tag{3.2.17}$$

Thus, having in mind the subsequent reduction of the group $U(2)$, it is convenient to take the following operators as generators of the group:

$$\begin{aligned} L_0 &= b_\theta{}^\dagger b_\theta + b_\varepsilon{}^\dagger b_\varepsilon \ , \qquad & L_x &= b_\theta{}^\dagger b_\theta - b_\varepsilon{}^\dagger b_\varepsilon \ , \\ L_y &= b_\theta^\dagger b_\varepsilon + b_\varepsilon^\dagger b_\theta \ , \qquad & L_z &= -\mathrm{i}(b_\theta{}^\dagger b_\varepsilon - b_\varepsilon{}^\dagger b_\theta) \ . \end{aligned} \tag{3.2.18}$$

Note that the components of the irreducible tensor operators form submanifolds closed with respect to the commutation operations, and this is especially convenient for the reduction of the group $U(2)$ into subgroups. Thus, for instance, the operators L_x, L_y and L_z have the same commutation relations as the Pauli matrices $\hat{\sigma}_x$, $\hat{\sigma}_y$ and $\hat{\sigma}_z$, i.e., they form an algebra.

An arbitrary element of the four-parameter group $U(2)$ which describes the symmetry of the Hamiltonian $\hat{H}_v$ can be written as

$$G(\alpha_0, \alpha_x, \alpha_y, \alpha_z) = \exp[\mathrm{i}(\alpha_0 L_0 + \alpha_x L_x + \alpha_y L_y + \alpha_z L_z)] \ , \qquad (3.2.19)$$

where α_i are real parameters. As in the case considered above, it is easy to verify that the parameter α_0 corresponds to the trivial operation of multiplication of the wave functions by a phase factor, and one can restrict oneself to the three-parameter subgroup $SU(2)$ of the group $U(2)$ by assuming that $\alpha_0 = 0$ in (3.2.19).

Thus, in the absence of the vibronic interaction the symmetry of the $(E \otimes e)$-type Jahn-Teller system is described at least by the group $SU(2) \times SU(2)$. An arbitrary element of this group,

$$\hat{G} = \exp(\mathrm{i}\boldsymbol{\varphi}\hat{\boldsymbol{\sigma}} + \mathrm{i}\boldsymbol{\alpha}\boldsymbol{L}) \ , \qquad (3.2.20)$$

depends on six parameters, φ_x, φ_y, φ_z and α_x, α_y, α_z. Taking the vibronic interaction into account lowers the initial symmetry of the problem so that the six parameters cease to be independent. The relations between them can be found from the condition that the operator (3.2.20), where φ_i and α_i are independent, should commute with the full Hamiltonian and, in particular, with the operator of the vibronic interaction. For small $\boldsymbol{\varphi}$ and $\boldsymbol{\alpha}$ values, from (3.2.20) we have

$$\hat{G} \approx \hat{I} + \mathrm{i}\boldsymbol{\varphi}\hat{\boldsymbol{\sigma}} + \mathrm{i}\boldsymbol{\alpha}\boldsymbol{L}$$

and the condition that it commutates with the Hamiltonian (3.2.1) takes the form

$$[(\boldsymbol{\varphi}\hat{\boldsymbol{\sigma}} + \boldsymbol{\alpha}\boldsymbol{L}), (Q_\theta \sigma_x + Q_\varepsilon \sigma_y)] = 0 \ . \qquad (3.2.21)$$

Substituting

$$Q_\theta = \sqrt{\frac{\hbar}{2\omega_E}}(b_\theta^\dagger + b_\theta) \ , \qquad Q_\varepsilon = \sqrt{\frac{\hbar}{2\omega_E}}(b_\varepsilon^\dagger + b_\varepsilon) \ , \qquad (3.2.22)$$

calculating all the necessary commutators and setting the coefficients of the independent combinations of operators equal to zero (remembering the independency of the Pauli matrices and of the operators $b_{\Gamma\gamma}$ and $b_{\Gamma\gamma}^\dagger$) we obtain

$$\alpha_x = \alpha_y = 0 \ , \qquad \varphi_x = \varphi_y = 0 \ , \qquad \alpha_z = 2\varphi_z \ .$$

Substituting these values into (3.2.20) we have

$$\hat{G}(\alpha) = \exp[\mathrm{i}\alpha(L_z + \tfrac{1}{2}\hat{\sigma}_z)] \ . \qquad (3.2.23)$$

Thus, by taking into account the linear vibronic interaction the symmetry of the system is lowered from the six-parameter group $SU(2) \times SU(2)$ to the one-parameter axial group $O(2)$ of two-dimensional rotations. An arbitrary element of this group has the form of (3.2.23) and the corresponding infinitesimal operator [3.14],

$$\hat{J}_z = L_z + \tfrac{1}{2}\hat{\sigma}_z \ , \tag{3.2.24}$$

as one can easily verify, commutes with the Hamiltonian (3.2.1), i.e., it is an integral of motion.

This result has a clear-cut physical meaning. In the absence of vibronic interactions the motion of the electrons is independent of that of the nuclei, the energy spin $\frac{1}{2}\hat{\boldsymbol{\sigma}}$ and vibrational momentum $\boldsymbol{L}$ being conserved separately. The states $\psi_\pm(\mathrm{r})$ with a certain value of the projection of the energy spin on the z-axis correspond to the wave of electronic density propagating along the perimeter of the molecule (in the case of the X_3 molecule) clockwise or counterclockwise, respectively. Similarly, the states with a certain value of L_z correspond to the wave of distortions of an equilateral triangle running clockwise or counterclockwise. The vibronic interaction results in a special coupling of the vibrational momentum $\boldsymbol{L}$ and energy spin $\hat{\boldsymbol{\sigma}}/2$ due to which they are not conserved (each of them) separately, and the energy of the system depends not only on the magnitudes of the vectors $\boldsymbol{L}$ and $\hat{\boldsymbol{\sigma}}/2$ but also on their relative positions. The projection of the total momentum $\hat{\boldsymbol{J}} = \boldsymbol{L} + \frac{1}{2}\hat{\boldsymbol{\sigma}}$ on the z axis is conserved. The wave of electronic density and the wave of distortions are no longer independent, but are interconnected (coherently). In a sense this state of the molecule is similar to the polaron state well known in solid-state physics.

It is convenient to write the operator L_z in the coordinate representation. By means of (3.2.11) and (18) we have

$$L_z = \frac{1}{\hbar}(Q_\theta P_\varepsilon - Q_\varepsilon P_\theta) \ , \tag{3.2.25}$$

or, in polar coordinates, $L_z = -\mathrm{i}\partial/\partial\varphi$. The unitary transformation (3.1.9) to the adiabatic basis changes the explicit form of the integral of motion:

$$\hat{\tilde{L}}_z = \hat{S}^+ L_z \hat{S} = L_z - \tfrac{1}{2}\hat{\sigma}_x \ , \qquad \hat{\tilde{\sigma}}_z = \hat{S}^+ \hat{\sigma}_z \hat{S} = \hat{\sigma}_x \ , \tag{3.2.26}$$

and therefore

$$\hat{\tilde{J}}_z = \hat{S}^+ \hat{J}_z \hat{S} = \hat{\tilde{L}}_z + \tfrac{1}{2}\hat{\tilde{\sigma}}_z = L_z \ . \tag{3.2.27}$$

Thus, in the adiabatic basis the operator $-\mathrm{i}\partial/\partial\varphi$ serves as an integral of motion. This explains the axial symmetry of the adiabatic potentials in the $E \otimes e$ example.

Besides the arbitrary rotations around the z-axis the Hamiltonian of the $E \otimes e$ case is an invariant of two more symmetry operations [3.15]:

$$\hat{P} = R_\varepsilon \hat{\sigma}_x \ , \qquad \hat{\Theta} = K\hat{\sigma}_x \ , \tag{3.2.28}$$

where $R_\varepsilon = \exp(i\pi b_\varepsilon^+ b_\varepsilon)$ is an operator that changes the sign of Q_ε, and K is an operator of complex conjugation. One can easily verify that $\hat{P}$ and $\hat{\Theta}$ commute with the Hamiltonian (3.2.1) and satisfy the relations

$$\begin{aligned} &\hat{P}^2 = \hat{I} , \quad \hat{\Theta}^2 = \hat{I} , \quad \hat{J}_z\hat{P} + \hat{P}\hat{J}_z = 0 , \\ &\hat{J}_z\hat{\Theta} + \hat{\Theta}\hat{J}_z = 0 , \quad K\hat{P} - \hat{P}K = 0 . \end{aligned} \tag{3.2.29}$$

The fact that the integrals of motion $\hat{P}$ and $\hat{\Theta}$ anti-commute with $\hat{J}_z$ means that the energy does not change on the substitution of $-\hat{J}_z$ for $\hat{J}_z$, and hence, it depends only on the $|j|$ values, where j is the eigenvalue of the operator $\hat{J}_z$. If one takes into account that according to (3.2.24) the quantum number j may have only half-integer values, $j = \pm\frac{1}{2}, \pm\frac{3}{2}, \ldots$, then it is obvious that all the vibronic levels of the linear $E \otimes e$ example have to be doubly degenerate. This result is analogous to Kramers theorem. Indeed, the Hamiltonian (3.2.1) contains spin variables of one "particle" with a half-integer energy spin. It means also that the above-mentioned axial symmetry group of the linear $E \otimes e$ system includes the improper rotations, thus being the $O(2)$ group, and not $SO(2)$.

The axial symmetry of the $E \otimes e$ example results from the fact that the components of the E representation of the trigonal groups have the same transformation properties as the components of the E representation of the axial group. This determines the special features of the linear vibronic coupling. The explicit form of the higher-order terms of the operator of vibronic interactions, i.e. the explicit form of the appropriate convolution of the irreducible tensor operators, is determined by the Clebsch-Gordan coefficients for the molecular point group (not the axial one), and therefore the axial symmetry is reduced when the higher-order terms in $Q_{E\gamma}$ are considered. As can be seen from Sect. 3.1, even

Table 3.2 Lie symmetries of the Jahn-Teller Hamiltonians with linear vibronic coupling

System	Symmetry group
Trigonal, pentagonal, hexagonal and cubic	
$E \otimes e$	O(2)
Cubic	
$\Gamma_8 \otimes t_2$	O(3)
$(s + p) \otimes t_{1u}$	O(3)
$T \otimes (e + t_2)$[a]	SO(3)
$\Gamma_8 \otimes (e + t_2)$[a]	O(5)
$(s + p) \otimes (a_{1g} + t_{1u} + e_g + t_{2g})$[a]	O(4)
Icosahedral	
$T \otimes h$	SO(3)
$H \otimes (g + 2h)$[a]	SO(5)
$H \otimes h$	SO(3)
$G \otimes (g + h)$[a]	O(4)

[a] For special coupling and frequency conditions

the quadratic terms of the vibronic interaction in the $E \otimes e$ case result in such a lowering of the symmetry.

A similar group-theoretical analysis may be performed for other Jahn-Teller systems. Accidental symmetry higher than the reference molecular one has been found for several cases of linear vibronic interaction [3.16–18]. A list of these cases and the corresponding Lie groups for them are given in Table 3.2. Full discussion of *Lie symmetries* in Jahn-Teller systems and the influence of these symmetries on vibronic energy spectra and wave functions, as well as the further development of the group-theoretical approach in vibronic problems, including the use of the noncompact group $O(1,2)$, are given in the review article [3.10].

3.2.2 Symmetry Properties of Adiabatic Potential Energy Surfaces

The lowest possible symmetry of the adiabatic potential for a given electronic term in the space of normal coordinates $Q_{\Gamma\gamma}$ was considered in [3.11]. In the space of all normal coordinates of the polyatomic system, the adiabatic potential as a whole, of course, has to be an invariant of the group G_0 of the initial nuclear configuration. However, not all the elements of the group G_0 correspond to nontrivial transformations in the subspace of the coordinates $Q_{\Gamma\gamma}$. Some of the elements of the group G_0 in the Γ representation are unit matrices. It is obvious that these elements leave the coordinates $Q_{\Gamma\gamma}$ unchanged. The manifold of such elements of symmetry is called the kernel G_Γ of the representation Γ. The other elements of the group G_0 together with the identity element form the factor group G_0/G_Γ which is just the *symmetry group of the adiabatic potential* in the subspace of the coordinates $Q_{\Gamma\gamma}$. For instance, for the octahedral molecule (G_0 equals O_h) the kernel of the representation E_g is D_{2h}. Therefore the symmetry group of the adiabatic potential in the two-dimensional subspace of the normal coordinates Q_θ and Q_ε is the factor group O_h/D_{2h}, which is isomorphous to C_{3v}. This is just the symmetry of the adiabatic potential of the ($E \otimes e$)-type JT effect in octahedral systems obtained by including the quadratic terms of vibronic interaction (Fig. 3.3).

This result holds for all adiabatic potentials of ($E \otimes e$)-type JT effects for the majority of polyatomic systems, provided their symmetry G_0 allows E representations. Analysis of the matrices of the irreducible representations of different molecular symmetry groups shows that in all cases only rotations around the threefold axis and reflections in the planes σ_v occur as nontrivial transformations in the subspace of E displacements (provided these transformations are elements of the group G_0). Therefore the factor-group C_{3v} (or C_3) is a general symmetry group of the adiabatic potential in the subspace of E displacements for any molecular system, independent of its initial symmetry.

Just this fact forms the basis of the mathematical equivalency of the ($E \otimes e$)-type JT effect for systems with different initial symmetry G_0.

A similar result can be obtained for different systems for the $E \otimes (b_1 + b_2)$ case. The symmetry of the adiabatic potential in the subspace of $(B_1 + B_2)$ vibrations in all these cases is C_{2v} (Fig. 3.4).

In cubic systems with Jahn-Teller-active T_2 vibrations (Sect. 3.3) the adiabatic potential in the T_2 subspace has O symmetry (isomorphous to T_d), if $G_0 = O, O_h, T_d$, and T symmetry when $G_0 = T, T_h$. The same symmetry is inherent to the adiabatic potential in the combined space of E and T_2 vibrations.

The group-theoretical considerations also allow the solution of the problem of how many terms should be kept in the expansion of the vibronic interaction operator as a power series with respect to $Q_{\Gamma\gamma}$ in order to obtain a qualitatively correct result. It is obvious that one has to take into account all the terms which give the lowest possible symmetry of the adiabatic potential. The inclusion of terms of higher order in $Q_{\Gamma\gamma}$ cannot change this result qualitatively, although quantitatively these changes may be considerable, especially in the case of strong vibronic coupling when the displacements $Q_{\Gamma\gamma}$ are relatively large. However, when the expansion (2.2.1) is cut off too early, the apparent symmetry of the potential surface may be too high, leading in particular to "accidental" degeneracy of the vibronic terms.

The number of types of extrema points that are in principle possible is determined by the number of different subgroups of the factor group G_0/G_Γ. The adiabatic potential points belong to a certain type determined by the point position with respect to the symmetry elements of the factor-group. Under a rotation or reflection belonging to this factor group the initial point under consideration is mapped onto another equivalent point. The set of equivalent points of the same type forms what is known as the transitive set. For example, in the case of the $(E \otimes e)$-type JT effect for which the adiabatic potential has C_{3v} symmetry there are three transitive sets, i.e., three types of equivalent points, *a*, *b* and *c* (Fig. 3.5).

Six points of the type *a* are obtained by the action of the six elements of the group C_{3v} upon a general initial point, i.e. one that does not belong to an axis or plane of symmetry. From the local point of view the adiabatic potential at this

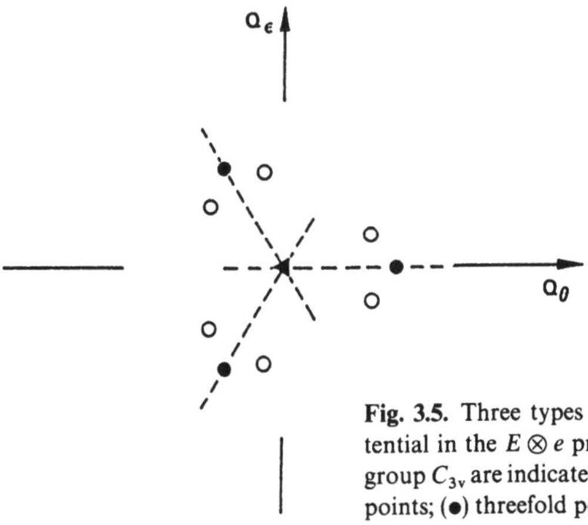

Fig. 3.5. Three types of equivalent points of the adiabatic potential in the $E \otimes e$ problem. The three symmetry planes in the group C_{3v} are indicated by broken lines. (o) low symmetry sixfold points; (•) threefold points; (▲) high symmetry point

point has no symmetry elements, i.e., its symmetry group is C_1. The symmetry of the nuclear configuration corresponding to this point is thus $C_1 \times D_{2h} = D_{2h}$.

Three points of type b lie on the reflection planes σ_v contained in the C_{3v} group. The local symmetry of the adiabatic potential in these points is thus C_s. The symmetry of the molecule at this point is $C_s \times D_{2h} = D_{4h}$.

The point of type c (only one point) lies at the intersection of all the symmetry elements of the group C_{3v}, and its local symmetry is C_{3v}. The symmetry of the adiabatic potential is $C_{3v} \times D_{2h} = O_h$.

Thus the adiabatic potential of an octahedral molecule in the subspace of E vibrations may have extrema points of three types: (a) six equivalent extrema in which the symmetry of the molecule is lowered to D_{2h}; (b) three extrema of D_{4h} symmetry and (c) one extremum of O_h symmetry. The latter case is realized in the $E \otimes e$ problem in the absence of vibronic coupling.

The above considerations about the types of adiabatic potential extrema in the subspace of low-symmetry normal coordinates are not related to the multiplicity of the electronic degeneracy, and thus they are valid for all Jahn-Teller cases where normal vibrations of given symmetry are active. For example, the same three types of extrema as in the $E \otimes e$ example, considered above, may also occur, in principle, in the $T \otimes e$ case, i.e. in the case of the Jahn-Teller effect for a T term coupled to E vibrations (Sect. 3.3) and in the case of a singlet electronic term in the pseudo-Jahn-Teller effect with E-active vibrations (Sect. 3.4).

These group-theoretical considerations of the symmetry properties of the adiabatic potential are equally valid in case of the so-called multimode vibronic systems (multimode systems are discussed in detail in Sect. 3.5 and 4.5). For instance, for the *multimode* $E \otimes (e + e + \cdots)$ *problem* with linear vibronic coupling described by the Hamiltonian

$$\hat{H} = \tfrac{1}{2} \sum_n [P_{n\theta}^2 + P_{n\varepsilon}^2 + \omega_n^2(Q_{n\theta}^2 + Q_{n\varepsilon}^2)]\hat{\sigma}_0 + \sum_n V_n(Q_{n\theta}\hat{\sigma}_x + Q_{n\varepsilon}\hat{\sigma}_y), \tag{3.2.30}$$

as in the one-mode case there is an integral of motion, $\hat{J}_z = L_z + \tfrac{1}{2}\hat{\sigma}_z$, where [cf. (3.2.25)]

$$L_z = \frac{1}{\hbar} \sum_n (Q_{n\theta}P_{n\varepsilon} - Q_{n\varepsilon}P_{n\theta}) = -\mathrm{i} \sum_n \frac{\partial}{\partial \varphi_n} \,. \tag{3.2.31}$$

The unitary transformation leading to the adiabatic electronic basis changes the explicit form of the integral of motion, similarly to the one-mode case (3.2.26, 27), so that $\hat{\tilde{J}}_z = L_z$. Thus, in the adiabatic basis the Hamiltonian (3.2.30) is invariant under the symmetry operations of the one-parametric Lie group with the infinitesimal operator determined by (3.2.31). It is obvious that the adiabatic potential in the linear $E \otimes (e + e + \cdots)$ case has one-dimensional variety of equipotential points, in particular, a circle of minima (a trough).

On considering the quadratic terms of the vibronic interaction the trough becomes warped, resulting in three equivalent minima divided by three potential barriers quite similar to the two-dimensional case. The reasons for such a coincidence of the extremum properties of the adiabatic potentials for multimode and one-mode vibronic systems are discussed in Sect. 3.5.

3.3 Triplet and Quadruplet Terms

It follows from the character tables of molecular symmetry groups that threefold irreducible representations are inherent to the cubic groups T, T_d, O, T_d, O_h and to the icosahedral groups I and I_h. In cubic groups without inversion centers, $[T_1^2] = [T_2^2] = A_1 + E + T_2$. In the presence of inversion, $[T_{1g}^2] = [T_{1u}^2] = [T_{2g}^2] = [T_{2u}^2] = A_{1g} + E_g + T_{2g}$. This means that for triplet electronic states the doubly degenerate E-type vibrations and the three-fold degenerate T-type vibrations are Jahn-Teller active. This is known as the $T \otimes (e + t_2)$ *problem*. The adiabatic potentials of all the cubic systems in the subspace of $(E + T_2)$ vibrations possess the same symmetry, O or T (Sect. 3.2). The simplest cubic molecule, the full vibrational representation of which contains only one E and one T_2 irreducible representation, is the octahedral molecule of type ML_6 (symmetry group O_h). Therefore, hereafter when considering the properties of an electronic T term for cubic systems, the T_{1u} term of an octahedral molecule of the type ML_6 is implied, provided no other particular systems are specified.

It is convenient to discuss the transformation properties of the irreducible representations of the molecular groups starting with the spherical group $O(3)$ (the group of rotations in three-dimensional space) and then reducing it to the appropriate molecular group. Thus the quintet representation $D^{(2)}$ of the group $O(3)$, to which five hydrogen d functions belong, splits into a doublet and triplet, $D^{(2)} = E_g + T_{2g}$, when the symmetry is reduced to O_h. The representation $D^{(1)}$ of the group $O(3)$, to which the hydrogen p functions belong, is not split by the reduction $O(3) \rightarrow O_h$ but is transformed into the representation T_{1u}. It follows that the $P \otimes d$ case of a triplet electronic P term coupled to d vibrations is a high-symmetry particular case of the $T \otimes (e + t_2)$ problem in a cubic polyatomic system when the cubic splitting of the vibrational d mode is negligible. The Jahn-Teller effect for $2p$ states of the F^+ center in the crystal CaO can be taken as an example. In this case, the electronic orbitals, owing to their diffuseness, do not feel the cubic symmetry of the local surroundings in the lattice [3.19] (see however [3.20]). Other examples of small splittings of the d mode are given in [3.21].

3.3.1 $T \otimes d$ problem (d Mode Approximation)

Following the discussion in the preceding section let us first consider a particular case of the $T \otimes (e + t_2)$ problem, namely, the $T \otimes d$ *problem* [3.22–24]. The Hamiltonian (2.2.13) with only the linear terms of the vibronic interaction (the linear JT effect) is

$$\hat{H} = \tfrac{1}{2}(P_\theta^2 + P_\varepsilon^2 + P_\xi^2 + P_\eta^2 + P_\zeta^2)\hat{C}_A + \hat{U} \ ,$$

$$\hat{U} = \tfrac{1}{2}\omega^2(Q_\theta^2 + Q_\varepsilon^2 + Q_\xi^2 + Q_\eta^2 + Q_\zeta^2)\hat{C}_A + V\left(Q_\theta\hat{C}_\theta + Q_\varepsilon\hat{C}_\varepsilon + \frac{\sqrt{3}}{2}Q_\xi\hat{C}_\xi + \frac{\sqrt{3}}{2}Q_\eta\hat{C}_\eta + \frac{\sqrt{3}}{2}Q_\zeta\hat{C}_\zeta\right) . \tag{3.3.1}$$

Here Q_γ are the real normal coordinates transforming as the $D^{(2)}$ representation, i.e., possessing the transformation properties $Q_\theta \propto (3z^2 - r^2)/\sqrt{3}$, $Q_\varepsilon \propto x^2 - y^2$, $Q_\xi \propto 2yz$, $Q_\eta \propto 2xz$, $Q_\zeta \propto 2xy$. The P_γ are the conjugated momenta and $\hat{C}_\gamma$ are the matrices of Clebsch-Gordan coefficients [they are the same as in the case of the $O(3)$ group] determined by the real electronic basis of the T_{1u} term, $\psi_x \propto x$, $\psi_y \propto y$, $\psi_z \propto z$:

$$\hat{C}_A = \begin{pmatrix} 1 & 0 & 0 \\ 0 & 1 & 0 \\ 0 & 0 & 1 \end{pmatrix}, \quad \hat{C}_\theta = \begin{pmatrix} \frac{1}{2} & 0 & 0 \\ 0 & \frac{1}{2} & 0 \\ 0 & 0 & -1 \end{pmatrix}, \quad \hat{C}_\varepsilon = \begin{pmatrix} -\sqrt{3/2} & 0 & 0 \\ 0 & \sqrt{3/2} & 0 \\ 0 & 0 & 0 \end{pmatrix},$$

$$\hat{C}_\xi = \begin{pmatrix} 0 & 0 & 0 \\ 0 & 0 & -1 \\ 0 & -1 & 0 \end{pmatrix}, \quad \hat{C}_\eta = \begin{pmatrix} 0 & 0 & -1 \\ 0 & 0 & 0 \\ -1 & 0 & 0 \end{pmatrix}, \quad \hat{C}_\zeta = \begin{pmatrix} 0 & -1 & 0 \\ -1 & 0 & 0 \\ 0 & 0 & 0 \end{pmatrix}. \tag{3.3.2}$$

In the absence of vibronic interaction, i.e., when $V = 0$, the symmetry of the Hamiltonian (3.3.1) is described by a 32-parameter Lie group $SU(3) \times SU(5)$. Inclusion of the vibronic interaction is accompanied by a reduction of the symmetry to $SO(3)$, and this can be found directly using the method described in Sect. 3.2. It is convenient to perform the symmetry reduction stepwise and to include the vibronic interaction only at the last step. The eight integrals of motion of the group $SU(3)$ can be divided into two sets containing three components of the tensor convolution $\{\psi^{(1)} \times \psi^{(1)}\}^{(1)}$, see (3.2.7), and five components of the tensor $\{\psi^{(1)} \times \psi^{(1)}\}^{(2)}$, respectively. When the symmetry is reduced from $SU(3)$ to $SO(3)$ the components of the tensor $\{\psi^{(1)} \times \psi^{(1)}\}^{(2)}$ cease to be integrals of motion, and hence only three components $\{\psi^{(1)} \times \psi^{(1)}\}^{(1)}$ forming a vector of the energy spin $S = 1$ remain:

$$\hat{S}_x = \hat{C}_{T,x} = \begin{pmatrix} 0 & 0 & 0 \\ 0 & 0 & -\mathrm{i} \\ 0 & \mathrm{i} & 0 \end{pmatrix}, \quad \hat{S}_y = \hat{C}_{T,y} = \begin{pmatrix} 0 & 0 & \mathrm{i} \\ 0 & 0 & 0 \\ -\mathrm{i} & 0 & 0 \end{pmatrix},$$

$$\hat{S}_z = C_{T,z} = \begin{pmatrix} 0 & -\mathrm{i} & 0 \\ \mathrm{i} & 0 & 0 \\ 0 & 0 & 0 \end{pmatrix}. \tag{3.3.3}$$

The matrices (3.3.2) are related to the spin matrices (3.3.3) by the simple relations $\hat{C}_\theta = (3\hat{S}_z^2 - \hat{S}^2)/2$, $\hat{C}_\varepsilon = (S_x^2 - S_y^2)\sqrt{3/2}$, $\hat{C}_\xi = \hat{S}_y\hat{S}_z + \hat{S}_z\hat{S}_y$, $\hat{C}_\eta = \hat{S}_x\hat{S}_z + \hat{S}_z\hat{S}_x$, $\hat{C}_\zeta = \hat{S}_x\hat{S}_y + \hat{S}_y\hat{S}_x$; they follow the transformation properties of the matrices $\hat{C}_\gamma$, transforming under rotations as the hydrogen d functions. The integrals of motion of the group $SU(5)$ (their number is $5^2 - 1 = 24$) can also be divided into four sets by means of (3.2.16): $\{b^{(2)} \times b^{(2)}\}^{(k)}$, $k = 1, 2, 3, 4$. As a result of the reduction $SU(5) \to O(5)$ only ten integrals of motion, the components of the tensors $\{b^{(2)} \times b^{(2)}\}^{(1)}$ and $\{b^{(2)} \times b^{(2)}\}^{(3)}$, remain. The subsequent reduction $O(5) \to O(3)$ leaves only three components of the tensor $\{b^{(2)} \times b^{(2)}\}^{(1)}$:

$$\begin{aligned} L_x &= \sqrt{3}(Q_\theta P_\xi - Q_\xi P_\theta) + (Q_\xi P_\varepsilon - Q_\varepsilon P_\xi) + (Q_\eta P_\zeta - Q_\zeta P_\eta) \;, \\ L_y &= \sqrt{3}(Q_\eta P_\theta - Q_\theta P_\eta) + (Q_\eta P_\varepsilon - Q_\varepsilon P_\eta) + (Q_\zeta P_\xi - Q_\xi P_\zeta) \;, \\ L_z &= 2(Q_\varepsilon P_\zeta - Q_\zeta P_\varepsilon) + (Q_\xi P_\eta - Q_\eta P_\xi) \;. \end{aligned} \tag{3.3.4}$$

They are the components of the angular momentum vector $\boldsymbol{L}$ that describes the rotation of the distortion wave. By direct verification one can ascertain that of the 32 integrals of motion of the Hamiltonian (3.3.1) for $V = 0$ only three integrals of motion remain after the introduction of the vibronic interaction, and they are three components of the vector of the total angular momentum $\hat{\boldsymbol{J}} = \boldsymbol{L} + \hat{\boldsymbol{S}}$. This just means that the symmetry of the problem is reduced to $O(3)$. In this sense, as for the $E \otimes e$ case, the vibronic interaction is analogous to the spin-orbit interaction which reduces the group $O(3) \times O(3)$ of independent rotations in the spin and orbital spaces to the group $O(3)$ of the free rotations of the spin-orbital system as a whole (for which only the total momentum $\hat{\boldsymbol{J}} = \boldsymbol{L} + \hat{\boldsymbol{S}}$ is conserved; $\boldsymbol{L}$ and $\boldsymbol{S}$ separately are not conserved). The total momentum is integer, because it is a sum of integers ($\boldsymbol{S} = 1$ and $\boldsymbol{L} = 0, 2, \ldots$), and therefore we do not need to consider improper rotations. The symmetry group of the vibronically coupled $T \otimes d$ system is thus $SO(3)$ and not $O(3)$ (Table 3.2).

The extrema of the adiabatic potential can be investigated by the method of Öpik and Pryce (Sect. 3.1). Substituting the matrix $\hat{U}$ from (3.3.1) into the system (3.1.28) we obtain

$$\begin{aligned} &Q_\theta^{(0)} = -\frac{V}{2\omega^2}(a_x^2 + a_y^2 - 2a_z^2) \;, \qquad Q_\varepsilon^{(0)} = -\frac{V\sqrt{3}}{2\omega^2}(a_x^2 - a_y^2) \;, \\ &Q_\xi^{(0)} = \frac{V\sqrt{3}}{\omega^2}a_y a_z \;, \qquad Q_\eta^{(0)} = \frac{V\sqrt{3}}{\omega^2}a_x a_z \;, \qquad Q_\zeta^{(0)} = \frac{V\sqrt{3}}{\omega^2}a_x a_y \;, \end{aligned} \tag{3.3.5}$$

where $a_x^2 + a_y^2 + a_z^2 = 1$, and the other equations of the system (3.1.28) transform into identities. All the possible values of a_x, a_y and a_z determining the coordinates $Q_{\Gamma\gamma}^{(0)}$ from (3.3.5) should be solutions of the system of equations (3.1.28), which includes the equation $a^\dagger a = 1$, which in our case is $a_x^2 + a_y^2 + a_z^2 = 1$. This is the equation of the unit sphere in the space of a_x, a_y, a_z. Therefore the solutions of (3.1.28) should in this case map on the unit sphere. Using spherical coordinates

$a_x = \sin\theta\cos\varphi$, $a_y = \sin\theta\sin\varphi$ and $a_z = \cos\theta$, we find from (3.3.5)

$$Q_\theta^{(0)} = \frac{V}{2\omega^2}(3\cos^2\theta - 1)\ , \qquad Q_\varepsilon^{(0)} = \frac{V\sqrt{3}}{2\omega^2}\sin^2\theta\cos 2\varphi\ ,$$

$$Q_\xi^{(0)} = \frac{V\sqrt{3}}{2\omega^2}\sin 2\theta\sin\varphi\ , \qquad Q_\eta^{(0)} = \frac{V\sqrt{3}}{2\omega^2}\sin 2\theta\cos\varphi\ , \tag{3.3.6}$$

$$O_\zeta^{(0)} = \frac{V\sqrt{3}}{2\omega^2}\sin^2\theta\sin 2\varphi$$

(for arbitrary values of θ and φ), and the Jahn-Teller stabilization energy is

$$E_{\mathrm{JT}} = V^2/2\omega^2\ . \tag{3.3.7}$$

This result means that the minima of the lowest sheet of the adiabatic potential corresponding to the linear $T \otimes d$ problem form a two-dimensional continuum (a trough) of equipotential points with the coordinates (3.3.6) [3.22].

The fact that the vibronic Hamiltonian (3.3.1) has three invariants $\hat{J}_x$, $\hat{J}_y$ and $\hat{J}_z$ whereas the equipotential manifolds are two-dimensional should not be a surprise. Indeed, only two commuting combinations $\hat{J}^2$ and $\hat{J}_z$ can be formed from three integrals of motion $\hat{J}_x$, $\hat{J}_y$ and $\hat{J}_z$, quite similarly to the well-known results in atomic theory.

The energy gap between the lowest sheet of the adiabatic potential and the first excited one at the minima points is $\Delta\varepsilon = 3V^2/2\omega^2 = 3E_{\mathrm{JT}}$. From this it follows that in the case of strong vibronic coupling the criterion (2.1.10) for the adiabatic approximation holds. This means that the lowest sheet of the adiabatic potential acquires the physical meaning of the potential energy of the nuclei. The motion of the nuclei along the two-dimensional trough of the lowest sheet can be reduced in this case to free two-dimensional rotations.

3.3.2 $T \otimes (e + t_2)$ Problem

We now consider the linear $T \otimes (e + t_2)$ problem, i.e. the Jahn-Teller effect for an electronic T term of an octahedral molecule taking into account the linear vibronic interaction with E and T_2 vibrations. The reduction of the $O(3)$ symmetry to O_{h} reduces the representation $D^{(2)}$ of the five-dimensional vibration to E_g and T_{2g}, $D^{(2)} = E_g + T_{2g}$, and this means that the force constant of the normal vibration is split into ω_E^2 and ω_T^2, and the vibronic coupling with these vibrations is different. The Hamiltonian of such a system has the form [cf. (3.3.1)]

$$\hat{H} = \tfrac{1}{2}[P_\theta^2 + P_\varepsilon^2 + P_\xi^2 + P_\eta^2 + P_\zeta^2 + \omega_E^2(Q_\theta^2 + Q_\varepsilon^2) + \omega_T^2(Q_\xi^2 + Q_\eta^2 + Q_\zeta^2)]\hat{C}_A$$
$$+ V_E(Q_\theta\hat{C}_\theta + Q_\varepsilon\hat{C}_\varepsilon) + V_T(Q_\xi\hat{C}_\xi + Q_\eta\hat{C}_\eta + Q_\zeta\hat{C}_\zeta)\ . \tag{3.3.8}$$

The extrema of the adiabatic potentials, i.e., of the eigenvalues of the matrix of the potential energy of the nuclei, can be found by the method of Öpik and Pryce (Sect. 3.1). Considering the condition of normalization, $a_x^2 + a_y^2 + a_z^2 = 1$, one can find the equilibrium coordinates $Q_{\Gamma\gamma}^{(0)}$ from the equations $\hat{a}^\dagger(\partial\hat{U}/\partial Q_{\Gamma\gamma})\,\hat{a}$ of the system (3.1.28). They are

$$Q_\theta^{(0)} = -\frac{V_E}{\omega_E^2}\left(\frac{1}{2}a_x^2 + \frac{1}{2}a_y^2 - a_z\right), \qquad Q_\varepsilon^{(0)} = \frac{V_E}{\omega_E^2}\cdot\frac{\sqrt{3}}{2}(a_x^2 - a_y^2) \,,$$
$$Q_\xi^{(0)} = \frac{2V_T}{\omega_T^2}a_y a_z \,, \qquad Q_\eta^{(0)} = \frac{2V_T}{\omega_T^2}a_x a_z \,, \qquad Q_\zeta^{(0)} = \frac{2V_T}{\omega_T^2}a_x a_y \,. \tag{3.3.9}$$

Substituting these expressions into (3.3.8) for $Q_{\Gamma\gamma}$ and writing out the matrix equation $\hat{U}\hat{a} = \varepsilon\hat{a}$ explicitly, one can obtain a system of four equations for four unknowns a_x, a_y, a_z and ε. This system can be solved in a trivial way. Its roots are given in Table 3.3. Other roots which differ from these by a simultaneous

Table 3.3 Solutions of the Öpik-Pryce equations for the linear $T\otimes(e+t_2)$ case [3.5]

	Tetragonal extrema			Trigonal extrema			
a_x	1	0	0	$\frac{1}{\sqrt{3}}$	$-\frac{1}{\sqrt{3}}$	$\frac{1}{\sqrt{3}}$	$\frac{1}{\sqrt{3}}$
a_y	0	1	0	$\frac{1}{\sqrt{3}}$	$\frac{1}{\sqrt{3}}$	$-\frac{1}{\sqrt{3}}$	$\frac{1}{\sqrt{3}}$
a_z	0	0	1	$\frac{1}{\sqrt{3}}$	$\frac{1}{\sqrt{3}}$	$\frac{1}{\sqrt{3}}$	$-\frac{1}{\sqrt{3}}$
ε	$-\frac{1}{2}\frac{V_E^2}{\omega_E^2}$			$-\frac{2}{3}\frac{V_T^2}{\omega_T^2}$			
	Orthorhombic extrema						
a_x	$\frac{1}{\sqrt{2}}$	$\frac{1}{\sqrt{2}}$	0	$-\frac{1}{\sqrt{2}}$	$-\frac{1}{\sqrt{2}}$	0	
a_y	$\frac{1}{\sqrt{2}}$	0	$\frac{1}{\sqrt{2}}$	$\frac{1}{\sqrt{2}}$	0	$\frac{1}{\sqrt{2}}$	
a_z	0	$\frac{1}{\sqrt{2}}$	$\frac{1}{\sqrt{2}}$	0	$\frac{1}{\sqrt{2}}$	$-\frac{1}{\sqrt{2}}$	
ε	$-\frac{1}{8}\frac{V_E^2}{\omega_E^2}-\frac{1}{2}\frac{V_T^2}{\omega_T^2}$						

change of the signs of a_x, a_y and a_z do not result in new extrema points since the multiplication of the column vector $\hat{a}$ by a phase factor $\exp(i\pi)$ has no influence on the eigenvalues of the matrix $\hat{U}$. Substituting these values of a_x, a_y and a_z into (3.3.9) we obtain the coordinates of the extrema. Thus the values of the first three columns of Table 3.3 result in three equivalent extrema points which transform into each other by rotations around the threefold axes of symmetry, the Jahn-Teller stabilization energy being

$$E_{\mathrm{JT}}(E) = \frac{1}{2}\frac{V_E^2}{\omega_E^2} \ . \tag{3.3.10}$$

Since at these points the symmetrized displacements just of E type are nonzero and the symmetry of the octahedral molecule is reduced to the tetragonal one $D_{4\mathrm{h}}$ (Sect. 3.2), these three extrema are called *tetragonal* ones. Four other equivalent extrema points can be obtained by inserting the values

$$|a_x| = |a_y| = |a_z| = \frac{1}{\sqrt{3}} \ .$$

They transform into one another by rotations around the fourfold axes of symmetry. Their Jahn-Teller stabilization energy is

$$E_{\mathrm{JT}}(T) = \frac{2}{3}\frac{V_T^2}{\omega_E^2} \ . \tag{3.3.11}$$

At these points just the T_2 type coordinates are displaced and the octahedral molecule is elongated or compressed along one of its threefold axes of symmetry. The symmetry of the molecule is reduced to D_{3d} and the appropriate extrema are called *trigonal*. Finally, the last six columns of Table 3.3 result in six equivalent extrema for which the nontrivial symmetry operations form the D_3 group. The Jahn-Teller stabilization energy is

$$E_{\mathrm{JT}}(OR) = \tfrac{1}{4}E_{\mathrm{JT}}(E) + \tfrac{3}{4}E_{\mathrm{JT}}(T) \ . \tag{3.3.12}$$

The coordinates of one of these extrema are

$$Q_\theta^{(0)} = -\frac{1}{2}\frac{V_E}{\omega_E^2} \ , \qquad Q_\varepsilon^{(0)} = Q_\xi^{(0)} = Q_\eta^{(0)} = 0 \ , \qquad Q_\zeta^{(0)} = \frac{V_T}{\omega_T^2} \ . \tag{3.3.13}$$

At these points the octahedral molecule is distorted along one of its six twofold axes of symmetry. Simultaneously it is distorted along the perpendicular fourfold axis, resulting in orthorhombic symmetry $D_{2\mathrm{h}}$. Accordingly, the last six extrema are called *orthorhombic*.

The curvature of the adiabatic potential at the above extrema can also be investigated by the method of Öpik and Pryce (Sect. 3.1). In each of the three equivalent tetragonal extrema the symmetry of the system is reduced to D_{4h}. In this group the representations E_g and T_{2g} to which the Jahn-Teller-active vibrations belong are reduced: $E_g = A_{1g} + B_{1g}$, $T_{2g} = B_{2g} + E_g$. At the tetragonal points it appears that the A_{1g} and B_{1g} vibrations are degenerate whereas the B_{2g} and E_g vibrations have different force constants:

$$\tilde{\omega}_{A_1}^2 = \tilde{\omega}_{B_1}^2 = \omega_E^2 , \qquad \tilde{\omega}_{B_2}^2 = \omega_T^2 ,$$
$$\tilde{\omega}_E^2 = \omega_T^2(1 - \eta^{-1}) , \qquad \eta = \frac{E_{\mathrm{JT}}(E)}{E_{\mathrm{JT}}(T)} . \tag{3.3.14}$$

Similarly, at the trigonal extrema the symmetrized displacements which transform as the irreducible representations E_g and $A_{1g} + E_g$ of the D_{3d} group have new force constants $\tilde{\omega}_\Gamma^2$, where $\tilde{\omega}_A = \omega_T$, and in order to evaluate the force constants of the E_g vibrations the matrix

$$\begin{pmatrix} \omega_E^2\left(1 - \frac{2}{3}\eta\right) & -\frac{\sqrt{2}}{3}\omega_E\omega_T\sqrt{\eta} \\ -\frac{\sqrt{2}}{3}\omega_E\omega_T\sqrt{\eta} & \frac{2}{3}\omega_T^2 \end{pmatrix} . \tag{3.3.15}$$

must be diagonalized.

At the orthorhombic extrema (the symmetry group is D_{2h}) we have $E_g = A_{1g} + B_{1g}$ and $T_{2g} = A_{1g} + B_{2g} + B_{3g}$, and for the force constants $\tilde{\omega}_\Gamma^2$ we obtain

$$\tilde{\omega}_A^2 = \omega_E^2 , \qquad \tilde{\omega}_{A'}^2 = \omega_T^2$$
$$\tilde{\omega}_{B_1}^2 = \omega_E^2(1 - \eta) , \qquad \tilde{\omega}_{B_2}^2 = \omega_T^2\frac{\eta - 1}{\eta + 1} , \qquad \tilde{\omega}_{B_3}^2 = \omega_T^2 \tag{3.3.16}$$

Comparing the expressions (3.3.14), (3.3.15) and (3.3.16), one can conclude that in the linear $T \otimes (e + t_2)$ problem, depending on the value of $\eta = E_{\mathrm{JT}}(E)/E_{\mathrm{JT}}(T)$, either tetragonal or trigonal extrema can be absolute minima, whereas the orthorhombic extrema points, intermediate in energy, can never be absolute minima. The analysis of the curvature of the extrema shows that all the points which are not absolute minima are saddle points. Indeed, it follows from (3.3.14), that if $\eta < 1$ then the trigonal minima of the adiabatic potential are the deepest, $\tilde{\omega}_E^2 < 0$, and the tetragonal extrema are saddle points.

A certain conclusion about the nature of the trigonal extrema can be drawn from the criterion that the quadratic form (3.1.37) be positive definite. This is equivalent to the requirement that the determinant of the matrix (3.3.15) be positive:

$$\tfrac{2}{3}\omega_E^2\omega_T^2(1 - \eta) > 0 . \tag{3.3.17}$$

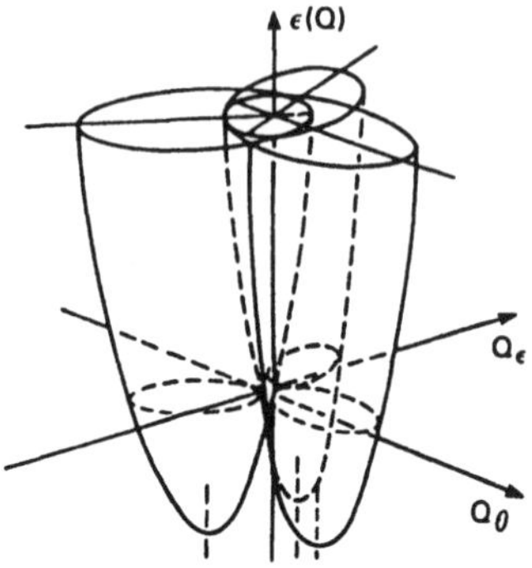

Fig. 3.6. Adiabatic potentials of the $T \otimes e$ problem: three displaced paraboloids

It follows that if $\eta > 1$ then the tetragonal minima are absolute ones, the inequality (3.3.17) is invalid, and the trigonal extrema are saddle points.

Finally, it follows from (3.3.16) that for any values of $\eta \neq 1$ the orthorhombic extrema are saddle points.

In the particular cases when the vibronic interaction with one of the two types of vibrations, E or T_2, is negligible, the Hamiltonian of the degrees of freedom involved in the vibronic interaction depends on a smaller number of variables and the corresponding adiabatic potentials are surfaces in the space of lower dimensionality (Sect. 2.2). For instance, when $V_T = 0$ and $V_E \neq 0$ (the *linear* $T \otimes e$ *problem*), the matrix $\hat{U}(Q)$ is diagonal and its eigenvalues may be represented by rotational paraboloids centered at the tetragonal extrema points (Fig. 3.6). Note that, as in the case of the $E \otimes b_1$ problem (Sect. 3.1), the operators of nonadiabaticity in the linear $T \otimes e$ problem are equal to zero and the adiabatic potentials presented in Fig. 3.6 represent the potential energy surfaces for any values of V_E and at any point of the configurational space of nuclear displacements.

For $V_E = 0$ and $V_T \neq 0$ (*the* $T \otimes t_2$ *problem*), the adiabatic potentials can be represented by surfaces in the four-dimensional space $(\varepsilon, Q_\xi, Q_\eta, Q_\zeta)$. The three tetragonal extrema are joined together at the coordinate origin, and the lowest sheet of the adiabatic potential has four equivalent trigonal minima divided by six orthorhomic saddle points. The curvature of the trigonal minima can be found from the matrix (3.3.15), which is diagonal for $\eta = 0$. At each of the trigonal minima the T_2 vibrations are split into A_{1g} and E_g vibrations, $\tilde{\omega}_A = \omega_T$ and $\tilde{\omega}_E = \omega_T\sqrt{2/3}$.

Note that if $\eta = 1$, i.e., if $E_{JT}(E) = E_{JT}(T)$, then the depth of all three types of extrema is the same, and the curvature with respect to two of five degrees of freedom at each extremum point is equal to zero. This means that there is a two-dimensional trough [3.22]. It follows that the above two conditions ($\omega_E = \omega_T$, $V_T = V_E\sqrt{3/2}$) are not necessary for the Hamiltonian (3.3.12) to take the form of

(3.3.1). Only one condition, $E_{JT}(E) = E_{JT}(T)$, is sufficient for the occurrence of a two-dimensional trough in the linear $T \otimes (e + t_2)$ problem. In this sense the $T \otimes d$ problem and the particular case of the $T \otimes (e + t_2)$ problem with $\eta = 1$ are not completely equivalent.

It is interesting to follow the displacements of the atoms of an octahedral molecule during the motion of the system along the two-dimensional trough of the adiabatic potential. This can be done in the same way as for the triatomic molecule in Sect. 3.1 [3.25]. First, we take into account that all the $Q_{\Gamma\gamma}$ are zero except the $Q_{E\gamma}$ and $Q_{T_2\gamma}$ [whose values at the points of the trough are given by (3.3.9)]. Then we consider the relations between the $Q_{\Gamma\gamma}$ and the Cartesian displacements x_i, y_i and z_i of the atoms of the octahedral molecule (Table 2.1). Solving this system of equations with respect to x_i, y_i and z_i we obtain the trajectories of the ligands lying on spheres with their centers at the vertices of a regular octahedron. The dimension of the latter differs from the initial one by $V\sqrt{3}/(12\omega^2)$. The motion of the ligands on the spherical surfaces is coherent. The radius vectors of the ligands from the centers of these spheres, when reduced to a common origin, compose a star whose vertices form a regular octahedron rotating around its geometric center. The Euler angles describing this rotation are $(-\varphi, 2\theta, \varphi)$.

Thus in the case of strong vibronic coupling in the $T \otimes (e + t_2)$ problem with $E_{JT}(E) = E_{JT}(T)$, the motion of the system can be reduced to free rotations similar to the corresponding case in the linear $E \otimes e$ problem.

Consider now the quadratic terms of the vibronic interaction. Following the discussion in Sect. 3.2, if just qualitative information is required one can cut off the series expansion in $Q_{\Gamma\gamma}$ of the vibronic interaction operator after those terms which provide for the formation of all possible types of extrema of the adiabatic potential surface.

The group O describing the symmetry of the adiabatic potential in the subspace of $(E \oplus T_2)$ vibrations contains 24 elements. Hence there have to be 24 equivalent extrema of the most general position on the adiabatic potential surface of the $T \otimes (e + t_2)$ problem. However, in the linear approximation only 3-fold, 4-fold and 6-fold (i.e. 3, 4 and 6 equivalent) extrema occur, which means that this approximation may be insufficient even for a qualitative analysis of all the features of the adiabatic potential. As will be shown below, the inclusion of quadratic terms of vibronic interaction results in 12-fold and 24-fold extrema of the adiabatic potential. The reduced matrix elements $W_\Gamma(\Gamma_1 \times \Gamma_2)$ in the Hamiltonian (2.2.12), the constants of quadratic coupling, are nonzero if $\Gamma \in \Gamma_1 \times \Gamma_2$. Therefore in the space of the Jahn-Teller-active E- and T_2-type vibrations the quadratic forms of the types $[E^2] = A_1 + E$, $E \times T_2 = T_1 + T_2$ and $[T_2^2] = A_1 + E + T_2$ are possible. Taking into account that the totally symmetric combinations $\{Q_\Gamma \times Q_\Gamma\}_{A_1}$ are not included in (2.2.12) and $W_{T_1}(E \times T_2) = 0$ [owing to the real and Hermitian character of the corresponding operator $\{W(E \times T_2)\}_{T_1\gamma}$, see (2.2.5,6)], one finds from (2.2.12) that

$$\hat{U} = \frac{1}{2}\sum_{\Gamma\gamma} \omega_\Gamma^2 Q_{\Gamma\gamma}^2 \hat{C}_A + \sum_{\Gamma\gamma} V_\Gamma Q_{\Gamma\gamma} \hat{C}_{\Gamma\gamma} + W_E(E \times E)[(Q_\varepsilon^2 - Q_\theta^2)\hat{C}_\theta$$
$$+ 2Q_\theta Q_\varepsilon \hat{C}_\varepsilon] + W_{T_2}(E \times T_2)\Bigg[\left(-\frac{1}{2}Q_\theta + \frac{\sqrt{3}}{2}Q_\varepsilon\right)Q_\xi \hat{C}_\xi$$
$$+ \left(-\frac{1}{2}Q_\theta - \frac{\sqrt{3}}{2}Q_\varepsilon\right)Q_\eta \hat{C}_\eta + Q_\theta Q_\zeta \hat{C}_\zeta\Bigg]$$
$$+ W_E(T_2 \times T_2)\Bigg[\left(-\frac{1}{2}Q_\xi^2 - \frac{1}{2}Q_\eta^2 + Q_\zeta^2\right)\hat{C}_\theta + \left(\frac{\sqrt{3}}{2}Q_\xi^2 - \frac{\sqrt{3}}{2}Q_\eta^2\right)\hat{C}_\varepsilon\Bigg]$$
$$+ W_{T_2}(T_2 \times T_2)(Q_\eta Q_\zeta \hat{C}_\xi + Q_\xi Q_\zeta \hat{C}_\eta + Q_\xi Q_\eta \hat{C}_\zeta) \ . \qquad (3.3.18)$$

The adiabatic potentials of the $T \otimes (e + t_2)$ problem including the quadratic vibronic terms (3.3.18) can be investigated by the method of Öpik and Pryce (Sect. 3.1). Since the qualitative features induced by the quadratic terms are displayed even when one considers the quadratic terms just of the $E \times T_2$ type, we confine ourselves to a discussion of the role of the latter [3.26, 27], i.e., we assume

$$W_E(E \times E) = W_E(T_2 \times T_2) = W_{T_2}(T_2 \times T_2) = 0 \ ,$$
$$W_{T_2}(E \times T_2) = W \ . \qquad (3.3.19)$$

With this simplification, using the method of Öpik and Pryce one can obtain from (3.3.18) a system of four equations for four unknowns a_x, a_y, a_z and ε. In spite of its complicated form one can easily verify that the values of a_x, a_y and a_z listed in Table 3.3 as solutions of the appropriate system of equations in the linear $T \otimes (e + t_2)$ problem remain the roots also in the case when the quadratic terms of the vibronic interaction are included. Moreover, the coordinates and the Jahn-Teller stabilization energies for the tetragonal and trigonal extrema points remain the same as in the linear approximation, provided the quadratic terms just of the $E \otimes T_2$ type are taken into account. This result is quite obvious. The quadratic terms of $E \times T_2$ type contain only "cross" terms of the type $Q_{E\gamma}Q_{T_2\gamma'}$. Therefore, the extrema points displaced in only one of the subspaces (and not displaced in the other) are not affected by the quadratic terms of the $Q_{E\gamma}Q_{T_2\gamma'}$ type (which are zero at these extrema points). For the same reason it is obvious that the orthorhombic extrema points which are displaced in both subspaces, see (3.3.13), distinguished from the tetragonal and trigonal ones, should change their parameters significantly when $W \neq 0$. For instance, the extremum point with the coordinates (3.3.13) is displaced to the point with the coordinates

$$Q_\theta^{(0)} = -\frac{V_E}{2\omega_E^2}\frac{B - 2A^2}{B(1 - A^2)} \ , \quad Q_\varepsilon^{(0)} = Q_\xi^{(0)} = Q_\eta^{(0)} = 0 \ , \quad Q_\zeta^{(0)} = \frac{V_T}{\omega_T^2}\frac{2 - B}{2(1 - A^2)} \ . \qquad (3.3.20)$$

where the dimensionless parameters

$$A = W/\omega_E \omega_T \quad \text{and} \quad B = WV_E/\omega_E^2 V_T \tag{3.3.21}$$

have been introduced which are zero in the absence of quadratic vibronic coupling, i.e., when $W = 0$. The Jahn-Teller stabilization energy at the orthorhombic extrema is also different from that of (3.3.12):

$$E_{\mathrm{JT}}(OR) = \frac{1}{8}\frac{V_E^2}{\omega_E^2}\frac{B^2 + 4A^2 - 4A^2 B}{B^2(1 - A^2)} \quad . \tag{3.3.22}$$

It follows from (3.3.20 and 22) that if $A \to \pm 1$ the expressions for $Q_{\Gamma\gamma}^{(0)}$ and $E_{\mathrm{JT}}(OR)$ become divergent, i.e., the polyatomic system becomes unstable. The condition of stability is $|A| < 1$, that is.

$$|W| < \omega_E \omega_T \ . \tag{3.3.23}$$

Note that $4A^2/3B^2 = E_{\mathrm{JT}}(T)/E_{\mathrm{JT}}(E)$. This means that if $4A^2 < 3B^2$, the tetragonal extrema are lower in energy than the trigonal ones, and vice versa in the case of $4A^2 > 3B^2$. In Fig. 3.7 the area of stability of the system $|A| < 1$ is divided by the lines $A = \pm B\sqrt{3/2}$ into four parts. In two of them the trigonal extrema

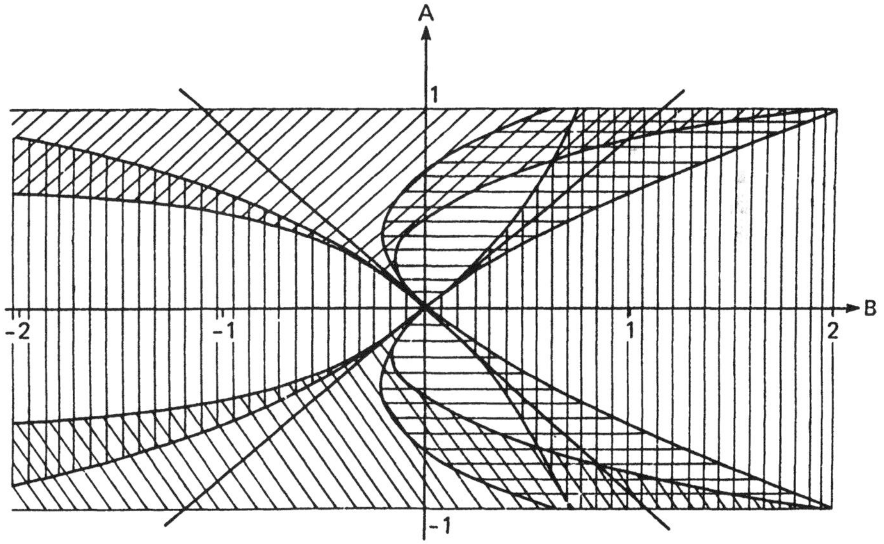

Fig. 3.7. Domains of existence (occurrence zones) of different types of minima points in the quadratic $T \otimes (e + t_2)$ problem with $E \times T_2$ type quadratic terms of vibronic interactions, $W_{T_2}(E \times T_2) = W \neq 0$, in dimensionless parameters $A = W/\omega_E\omega_T$, $B = WV_E/\omega_E^2 V_T$. The area of stability is limited by the inequality $|A| < 1$. Along the lines $A = \pm B\sqrt{3/2}$ the depths of tetragonal and trigonal minima are the same. The vertical, horizontal and inclined hatching indicate the domains of tetragonal, trigonal and orthorhombic minima points, respectively

are lower than the tetragonal ones, whereas in the other two the tetragonal extrema are lower. Along the lines $A = \pm B\sqrt{3/2}$ the depths of tetragonal and trigonal extrema coincide, and at the point of intersection of these two lines, i.e., when $W = 0$, the two-dimensional trough of minima discussed above is realized. If $W \neq 0$, the parameters A and B are also nonzero and along the lines $A = \pm B\sqrt{3/2}$ the depth of the orthorhombic extrema, as can be seen from (3.3.22), does not coincide with $E_{\mathrm{JT}}(E) = E_{\mathrm{JT}}(T)$. In other words, when the quadratic terms of vibronic interaction are included the two-dimensional trough of minima becomes warped and alternating wells and humps occur along its bottom, similarly to the case of the $E \otimes e$ problem (Sect. 3.1).

The most important consequence of the inclusion of quadratic terms in the $T \otimes (e + t_2)$ problem is the emergence of a range of the parameters A and B where the orthorhombic extrema become absolute minima of the adiabatic potential [3.26, 27]. In this area the following two inequalities should be valid simultaneously:

$$E_{\mathrm{JT}}(OR) > E_{\mathrm{JT}}(E)\ , \qquad E_{\mathrm{JT}}(OR) > E_{\mathrm{JT}}(T)\ . \tag{3.3.24}$$

Inserting here the expressions for $E_{\mathrm{JT}}(E)$, $E_{\mathrm{JT}}(T)$ and $E_{\mathrm{JT}}(OR)$ from (3.3.10, 11 and 22) and using the relation $4A^2/3B^2 = E_{\mathrm{JT}}(T)/E_{\mathrm{JT}}(E)$, we find the boundaries

$$\begin{aligned} &B = \frac{[-2A^2 \pm 2\sqrt{3(1-A^2)}]}{(3-4A^2)}\ , \qquad \begin{cases} B < 0\ , \\ B > 1\ , \end{cases} \\ &B = 2A^2 \pm \frac{2}{3}A\sqrt{3(1-A^2)}\ , \qquad \frac{\sqrt{3}}{3} - 1 < B < 1\ . \end{aligned} \tag{3.3.25}$$

Besides the tetragonal, trigonal and orthorhombic extrema considered above, the procedure of Öpik and Pryce also indicates the possible existence of two additional types of extrema, 12-fold and 24-fold, and for each of these types of equivalent extrema points there may be several transitive sets [3.26, 27].

The quadratic terms of the vibronic interaction significantly change the curvature at the tetragonal, trigonal and orthorhombic points as compared with the linear case. Using the method of Öpik and Pryce, we obtain the curvature of the adiabatic potential at the extrema points. Considering also the contribution of the quadratic terms of the $E \times T_2$ type, we find for the tetragonal extrema [cf. (3.3.14)]

$$\tilde{\omega}_{A_1}^2 = \tilde{\omega}_{B_1}^2 = \omega_E^2,\ \tilde{\omega}_{B_2}^2 = \omega_T^2,\ \tilde{\omega}_E^2 = \omega_T^2[1 - \eta^{-1}(1 - \tfrac{1}{2}B)^2]. \tag{3.3.26}$$

Similarly, for the trigonal extrema, instead of (3.3.15) we obtain

$$\begin{pmatrix} \omega_E^2\left[1 - \frac{2}{3}\eta\left(1 - \frac{2A^2}{3B}\right)^2\right] & \omega_E\omega_T\frac{8A^2 - 3B}{3\sqrt{6}A} \\ \omega_E\omega_T\frac{8A^2 - 3B}{3A\sqrt{6}} & \frac{2}{3}\omega_T^2 \end{pmatrix}. \tag{3.3.27}$$

The force constants for two A_1 vibrations in the orthorhombic minimum can be found by diagonalizing the matrix

$$\begin{pmatrix} \omega_E^2 & -W \\ -W & \omega_T^2 \end{pmatrix}, \tag{3.3.28}$$

while for B_1, B_2 and B_3 vibrations we have [cf. (3.3.16)]

$$\tilde{\omega}_{B_1}^2 = \omega_E^2\left(1 - \frac{3B^2(1-A^2)^2}{A^2(2-B)^2}\right), \qquad \tilde{\omega}_{B_3}^2 = \omega_T^2 ,$$
$$\tilde{\omega}_{B_2}^2 = \omega_T^2\left(1 - \frac{A^2(4+B-6A^2)^2}{2(3B^2+4A^2-10A^2B-2A^2B^2+6A^4B)}\right). \tag{3.3.29}$$

It follows from these results that there may be parameter values for which the lowest sheet of the adiabatic potential possesses *coexisting minima* of different types, tetragonal with trigonal, tetragonal with orthorhombic and trigonal with orthorhombic. There is also a range of A and B values for which all three types of minima coexist (Fig. 3.7). In this case the points at the top of the barriers which divide these minima are 12 equivalent saddle points.

The quadratic terms of the type $[T^2]$ in the $T \otimes t_2$ problem and the related possibility of the orthorhombic extrema becoming absolute minima of the adiabatic potential were first considered in [3.28]. The $[E^2]$- and $[T^2]$-type quadratic terms in the $T \otimes (e + t_2)$ problem and the possible coexistence of minima of different types were investigated in [3.29, 30]. In all these studies similar results were obtained with minor quantitative differences. In [3.30] cubic terms of the vibronic interaction of the $E \times E \times E$ type and $T_2 \times T_2 \times T_2$ type were also included, besides the quadratic ones (3.3.18), and the A_1 coordinate was also taken into account in the calculations. Besides the changes of the coordinates and curvatures at tetragonal, trigonal and orthorhombic extrema in the subspace of E and T vibrations discussed above, the equilibrium value of the totally symmetric coordinate $Q_{A_1}^{(0)}$ also becomes different at the extrema points of different types.

3.3.3 Cubic Quadruplet Terms: $\Gamma_8 \otimes (e + t_2)$ Problem

Quadruplet terms are inherent to cubic systems with spin-orbital coupling and to icosahedral systems.

The simplest and best studied cases of electronic quadruplets are found in cubic systems with an odd number of electrons. The electronic states of such a quadruplet transform as the two-valued irreducible representation Γ_8 of the double group O'[2]. Since $\{\Gamma_8^2\} = A_1 + E + T_2$, the Jahn-Teller-active coordi-

[2] In the notation of the book by *Landau* and *Lifshitz* [3.1] this irreducible representation is G'. Sometimes it is denoted by $G_{3/2}$ in order to emphasize its relation to the representation $D^{(3/2)}$ of the group $O(3)$

nates for this case are the E and T_2 ones [3.31]. For the $\Gamma_8 \otimes (e + t_2)$ *problem* the Hamiltonian of the system with just the linear terms of the vibronic interaction has the same form as (3.3.8) for the $T \otimes (e + t_2)$ problem, but with the difference that the matrices of the Clebsch-Gordan coefficients $\hat{C}_{\Gamma\gamma}$ are now given by the basis of electronic states of the Γ_8 term, viz. $(\psi_{1/2} + \mathrm{i}\psi_{-3/2})/\sqrt{2}$, $(-\psi_{-1/2} - \mathrm{i}\psi_{3/2})/\sqrt{2}$, $(-\psi_{1/2} + \mathrm{i}\psi_{-3/2})/\sqrt{2}$, $(\psi_{-1/2} - \mathrm{i}\psi_{3/2})/\sqrt{2}$:

$$\hat{C}_\theta = \begin{pmatrix} 0 & 0 & -1 & 0 \\ 0 & 0 & 0 & -1 \\ -1 & 0 & 0 & 0 \\ 0 & -1 & 0 & 0 \end{pmatrix}, \quad \hat{C}_\varepsilon = \begin{pmatrix} 0 & 0 & -\mathrm{i} & 0 \\ 0 & 0 & 0 & -\mathrm{i} \\ \mathrm{i} & 0 & 0 & 0 \\ 0 & \mathrm{i} & 0 & 0 \end{pmatrix},$$

$$\hat{C}_\xi = \begin{pmatrix} 0 & 1 & 0 & 0 \\ 1 & 0 & 0 & 0 \\ 0 & 0 & 0 & -1 \\ 0 & 0 & -1 & 0 \end{pmatrix}, \quad \hat{C}_\eta = \begin{pmatrix} 0 & -\mathrm{i} & 0 & 0 \\ \mathrm{i} & 0 & 0 & 0 \\ 0 & 0 & 0 & \mathrm{i} \\ 0 & 0 & -\mathrm{i} & 0 \end{pmatrix}, \tag{3.3.30}$$

$$\hat{C}_\zeta = \begin{pmatrix} 1 & 0 & 0 & 0 \\ 0 & -1 & 0 & 0 \\ 0 & 0 & -1 & 0 \\ 0 & 0 & 0 & 1 \end{pmatrix}.$$

The matrices (3.3.30) are identical to some of the Dirac ones.

The eigenvalues of the potential energy matrix of (3.3.8) in which $\hat{C}_{\Gamma\gamma}$ are determined by the matrices (3.3.30) can be found directly [3.2]. The adiabatic potentials can be represented by two Kramers doublets with energy

$$\varepsilon_\pm = \tfrac{1}{2}(\omega_E^2 \varrho^2 + \omega_T^2 Q^2) \pm \sqrt{V_E^2 \varrho^2 + V_T^2 Q^2}\ , \tag{3.3.31}$$

where

$$\varrho^2 = Q_\theta^2 + Q_\varepsilon^2\ , \qquad Q^2 = Q_\xi^2 + Q_\eta^2 + Q_\zeta^2\ .$$

One can easily see that $\varepsilon_\pm$ is independent of the angle φ of the polar coordinates $Q_\theta = \varrho\cos\varphi$, $Q_\varepsilon = \varrho\sin\varphi$, and of the angles α and β of the spherical coordinates $Q_\xi = Q\cos\alpha\cos\beta$, $Q_\eta = Q\sin\alpha\cos\beta$, $Q_\zeta = Q\sin\beta$. This means that the equipotential cross sections of the surfaces $\varepsilon_\pm$ are three-dimensional manifolds. On the other hand, (3.3.31) has the same form as the expressions (3.1.26) for the adiabatic potentials of the linear $E \otimes (b_1 + b_2)$ problem (Sect. 3.1). This means that in the (ϱ, Q) subspaces the adiabatic potentials of the linear $\Gamma_8 \otimes (e + t_2)$

problem have the shape presented in Fig. 3.2 (the regions of negative values of $Q_1 = \varrho$ and $Q_2 = Q$ in this case have no physical meaning). Hence, depending on the value of $\eta = E_{JT}(E)/E_{JT}(T)$, where

$$E_{JT}(E) = \frac{1}{2}\frac{V_E^2}{\omega_E^2}\ , \qquad E_{JT}(T) = \frac{1}{2}\frac{V_T^2}{\omega_T^2} \tag{3.3.32}$$

[cf. (3.1.33)], the absolute minima of the surface $\varepsilon_-(\varrho, Q)$ may be either tetragonal or trigonal extrema. If $\eta > 1$, the absolute minima are displaced only in the tetragonal subspace,

$$\varrho = \frac{V_E}{\omega_E^2}\ , \qquad Q = 0\ , \tag{3.3.33}$$

whereas the trigonal extrema with the coordinates

$$\varrho = 0\ , \qquad Q = \frac{V_T}{\omega_T^2} \tag{3.3.34}$$

are saddle points. Conversely, if $\eta < 1$, the trigonal extrema with the coordinates (3.3.34) are absolute minima. As in the case of the $E \otimes e$ problem, the origin of the trough is not accidental. An analysis of the appropriate group-theoretical reasons and the deduction of the integrals of motion by a method similar to that presented is Sect. 3.2 is given in [3.10, 17, 18].

Note that if $E_{JT}(E) = E_{JT}(T)$, a one-dimensional trough occurs at the surface ε_- in the (ϱ, Q) subspace, and the three-dimensional trough mentioned above transforms into a four-dimensional one in the full five-dimensional space of E and T_2 vibrations.

Another example of the Jahn-Teller effect for an electronic quadruplet can be found in polyatomic systems with the symmetry of an icosahedron. The extrema of the adiabatic potential of this quadruplet term interacting with quadruplet and quintet vibrations, the *$U \otimes (u + v)$ problem*, can be investigated by the method of Öpik and Pryce [3.5, 24].

Besides the quadruplet terms, triplet and quintet terms coupled to U and V vibrations are also inherent to the icosahedron. The linear *$T \otimes v$ and $V \otimes (u + v)$ problems* were considered in [3.24]. The group-theoretical analysis of the symmetry properties of the adiabatic potential for these cases is given in [3.17].

3.4 Pseudodegenerate Electronic Terms (Pseudo-Jahn-Teller Effect)

3.4.1 The Two-Level Case

We consider first a simple case of two close-in-energy nondegenerate electronic states ψ_g and ψ_u (for a system with inversion symmetry) divided by an energy gap

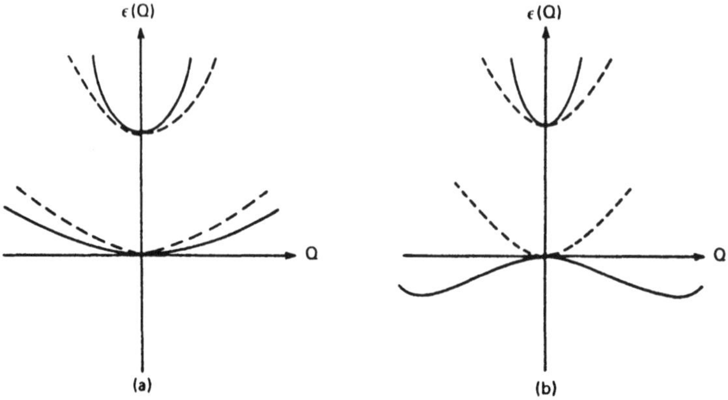

Fig. 3.8. Adiabatic potentials for two singlet electronic terms mixed by one singlet vibration in the case of weak (**a**) and strong (**b**) pseudo-Jahn-Teller effects. The same adiabatic potentials without vibronic mixing are shown by dashed curves

2Δ and mixed by one nondegenerate odd vibration [3.5]. The Hamiltonian of the system has the form

$$\hat{H} = \tfrac{1}{2}(P^2 + \omega^2 Q^2)\hat{\sigma}_0 + \Delta\hat{\sigma}_z + VQ\hat{\sigma}_x \ . \tag{3.4.1}$$

The eigenvalues of the matrix of the potential energy can be found directly:

$$\varepsilon_\pm(Q) = \tfrac{1}{2}\omega^2 Q^2 \pm \sqrt{\Delta^2 + V^2 Q^2} \ . \tag{3.4.2}$$

The curves (3.4.2) are presented in Fig. 3.8. Both sheets of the adiabatic potential have an extremum at $Q = 0$, its curvature being $\omega^2 \pm V^2/\Delta$. This means that the vibronic coupling lowers the curvature of the $\varepsilon_-(Q)$ sheet and increases the curvature of the upper sheet, and if

$$\Delta < \frac{V^2}{\omega^2} \tag{3.4.3}$$

the curvature of the lower sheet becomes negative (vibronic instability, Fig. 3.8b). In this case two minima at the points $Q_\pm^{(0)} = \pm[(V/\omega^2)^2 - (\Delta/V)^2]^{1/2}$ with the curvature $\omega^2 - \omega^6\Delta^2/V^4$ occur in the lower sheet of the adiabatic potential.

Thus the vibronic mixing of different electronic terms may result in the instability of the reference nuclear configuration. In some sense it is similar to the Jahn-Teller instability [although the latter is related to the linear dependence $\varepsilon(Q)$ in the neighborhood of the point $Q = 0$] and therefore it is usually called the *pesudo-Jahn-Teller effect*. Because it is determined by the second-order term in the expansion of ε in Q (while the first-order term, contrary to the usual Jahn-Teller effect, is zero), it is sometimes also called the second-order Jahn-Teller effect (see, for example [3.32]). The term pseudo-Jahn-Teller effect seams to be more appropriate since it is determined by the first order terms of the vibronic interaction.

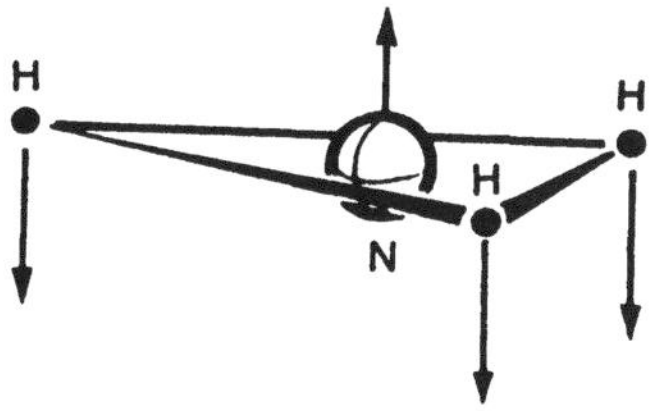

Fig. 3.9. Symmetrized A_2'' atomic displacement in the NH_3 molecule equivalent to out-of-plane displacement of the nitrogen atom

If the condition (3.4.3) is fulfilled, the effect is considered to be *strong*, and if the opposite condition is valid, it is called the *weak pseudo-Jahn-Teller effect.*

Note that the condition (3.4.3) contains three parameters and can be obeyed also in the cases when the energy gap 2Δ between the sheets of the adiabatic potential is large. For large 2Δ values the criterion of the adiabatic approximation (2.1.10) is fulfilled, and in these cases the vibronic instability at the point $Q = 0$ has a real physical meaning.

Consider, for example, the NH_3 molecule, for which an ab initio electronic structure calculation was performed in order to reveal the vibronic nature of the instability of its planar equilateral triangle D_{3h} nuclear configuration with respect to A_2'' out-of-plane nuclear displacement (Fig. 3.9) [3.33]. In this case the lowest two excited electronic states of A_2'' symmetry are admixed with the ground A_1' state by the A_2'' displacements, the corresponding energy gaps being 4.6 eV and 14 eV. The calculated ω^2 value is ≈ 43 Nm^{-1} (0.43 m dyn/Å), whereas the curvature of the adiabatic potential at the point $Q = 0$ obtained by direct point-by-point Hartree-Fock calculations is known to be equal to -26 Nm^{-1} (-0.26 m dyn/Å). The lower-energy excited A_2'' state is formed by electron excitation from the lone pair $|1a_2''\rangle \approx |2p_z(\mathrm{N})\rangle$ molecular orbital (MO) to the Rydberg $|3s(\mathrm{N})\rangle$ atomic orbital (AO) of the nitrogen atom. Its admixture to the ground state gives a small negative contribution to the curvature equal to -6 Nm^{-1} (-0.06 m dyn/Å). The most significant negative contribution to the curvature, equal to -62 Nm^{-1} (-0.62 m dyn/Å), results from the non-Rydberg excited state corresponding to electronic excitation from the $|1a_2''\rangle$ MO to the antibonding $|a_1'\rangle = C_1|2s(\mathrm{N})\rangle - C_2(|1s(\mathrm{H}_1)\rangle + |1s(\mathrm{H}_2)\rangle + |1s(\mathrm{H}_3)\rangle)$ MO. This contribution is an order of magnitude greater than that of the lower-energy A_2'' state. The resulting negative curvature of the adiabatic potential equals -25 Nm^{-1} (-0.25 m dyn/Å), which is close to the value obtained by direct Hartree-Fock calculations. Other examples of the same nature are given in [3.33].

Thus ab initio calculations of the molecular systems show explicitly that, contrary to a widely held belief, the pseudo-Jahn-Teller effect may be significant (in the sense that it may lead to the instability of molecular systems with all the consequences for the observable properties of such systems) even for electronic states with large energy level gaps of about 10–15 eV and more. The vibronic effects are most significant when the mixing states (which are not necessarily nearest in energy) involve strongly bonding or antibonding orbitals.

The origin of instability of the high symmetry nuclear configuration in systems with a strong pseudo-Jahn-Teller effect admits a visual interpretation. For instance, in the example of the molecule NH_3 in planar configuration, mentioned above, the unoccupied MO

$$|a_1'\rangle = C_1|2s(\mathrm{N})\rangle - C_2(|1s(\mathrm{H}_1)\rangle + |1s(\mathrm{H}_2)\rangle + |1s(\mathrm{H}_3)\rangle)$$

does not change its sign on reflection with respect to the plane of the triangle, whereas the highest occupied nonbonding MO $|1a_2''\rangle = |2p_z(\mathrm{N})\rangle$ does. This means that the overlap integral between these two orbitals is zero.

When the nitrogen atom undergoes an out-of-plane displacement the symmetry with respect to this reflection disappears, the symmetry group lowers, $D_{3h} \to C_{3v}$, and the overlap integral becomes nonzero, resulting in a new bonding MO. The nonorthogonality of the atomic orbitals, as follows from perturbation theory, is proportional to VQ/Δ, where $V = \langle 1a_2''|V_{A_2''}(\mathrm{r})|a_1'\rangle$ is the vibronic constant, $\Delta \approx 14$ eV is the energy gap between the atomic states in question, and Q is the normal coordinate for the out-of-plane displacement of the nitrogen atom.

Thus, as a result of the distortion of the high symmetry (planar) configuration of the NH_3 molecule, a new covalent bond between its highest occupied (mainly hydrogen) and low-lying unoccupied (nitrogen) orbitals forms. The meaning of the criterion (3.4.3) is that if for the displacement Q the energy of the new bond is larger than the loss of energy due to elastic distortion of other bonds, then the system becomes unstable with respect to this displacement.

In the case of the two-level pseudo-Jahn-Teller effect under consideration, the properties of the system are determined, generally speaking, by the vibronic Hamiltonian (3.4.1) or, if condition (2.1.10) is fulfilled, by the simpler adiabatic Hamiltonian (2.1.7). If the two mixing electronic states are nondegenerate, the Hamiltonian (2.1.7) transforms into [3.34, 35]

$$H = \frac{1}{2}(P^2 + \omega^2 Q^2) \pm \sqrt{\Delta^2 + V^2 Q^2} + \frac{\hbar^2 V^2 \Delta^2}{8(\Delta^2 + V^2 Q^2)^2} \,. \tag{3.4.4}$$

The last term in (3.4.4) is the diagonal matrix element of the nonadiabaticity operator (2.1.5, 6) obtained by the electronic eigenfunctions of the potential energy matrix in (3.4.1). This term decreases rapidly with Q from its maximum value $\hbar^2 V^2/8\Delta^2$ at $Q = 0$ to zero at $Q = \pm\infty$, and therefore it is negligibly small in the neighborhood of the minima, provided the vibronic instability is strong and the minima coordinates $Q_\pm^{(0)}$ are large enough. This is the reason why an even more simplified nuclear motion Hamiltonian for the ground electronic state,

$$H = \tfrac{1}{2}(P^2 + \omega^2 Q^2) - \sqrt{\Delta^2 + V^2 Q^2} \,, \tag{3.4.5}$$

can be employed in such cases. However, at $Q = 0$ the diagonal nonadiabaticity can be considerable and modify the shape of the adiabatic potential energy surfaces. In particular, it contributes to the destabilization of the ground state $Q = 0$ nuclear configuration, the corrected value of the curvature of the potential energy surfaces at this point being $\omega^2 \pm (V^2/\Delta) - \hbar^2 V^4/\Delta^4$.

Note, however, that in most cases this contribution is negligible due to the smallness of the Planck constant $\hbar$. It can be considerable only in cases of rather small values of Δ, when $2\Delta \lesssim \hbar\omega$, i.e., in cases of near-degeneracy of the electronic states concerned.

It is also necessary to include the off-diagonal nonadiabaticity, which is of the same order of magnitude as the diagonal one. Therefore, from the quantitative point of view, the above consideration of nonadiabaticity is, strictly speaking, insufficient. Its only aim is to give a better qualitative understanding of possible diabatic effects in the pseudo-Jahn-Teller systems with close-in-energy electronic states.

Vibronic coupling in the case of the pseudo-Jahn-Teller effect may also significantly influence the nuclear dynamics in the subspace of totally symmetric coordinates. Consider, for instance, the system possessing an inversion center discussed at the beginning of this section. Besides the vibronic coupling with the odd vibration Q_u described by the Hamiltonian (3.4.1), the mixing electronic states ψ_g and ψ_u are also coupled to the totally symmetric even displacements Q_g, the coupling constant in general being different for the ψ_g and ψ_u states. The operator of linear vibronic coupling in this case acquires the form [3.36–40]

$$\begin{pmatrix} V_1 Q_g & V_u Q_u \\ V_u Q_u & V_2 Q_g \end{pmatrix} . \tag{3.4.6}$$

Taking the value $Q_g^{(0)} = -(V_1 + V_2)/2\omega_g^2$ as the zero for the coordinates Q_g, we arrive at the Hamiltonian [3.41]

$$H = \tfrac{1}{2}(P_u^2 + P_g^2 + \omega_u^2 Q_u^2 + \omega_g^2 Q_g^2)\hat{\sigma}_0 + (\delta + V_g Q_g)\hat{\sigma}_z + V_u Q_u \hat{\sigma}_x . \tag{3.4.7}$$

Here ω_u and ω_g are the frequencies of the corresponding vibrations, P_u and P_g being their conjugated momenta, $V_g = (V_1 - V_2)/2$ and $\delta = \Delta + (V_2^2 - V_1^2)/4\omega_g^2$.

The displacements Q_g and Q_u are essentially different in a physical sense: Q_g simply adjusts the nuclear framework, not changing its symmetry and not mixing the electronic states ψ_g and ψ_u, whereas Q_u distorts the system, removing its inversion center and mixing the electronic states. Also in Jahn-Teller systems of the type $E \otimes e$ or $E \otimes (b_1 + b_2)$ there are mixing and nomixing modes, but in these cases their separation is conventional, since a simple transformation of the electronic basis transforms the mixing mode into a nonmixing one, and vice versa [see, for example, the Q_θ mode in (3.1.3) and in (3.1.6)]. In the pseudo-Jahn-Teller case being considered in this section such a transformation of the basis is impossible, since the two states ψ_g and ψ_u are no longer degenerate. As a result, the coupling mode Q_u is significantly different from the tuning mode Q_g.

The adiabatic potentials (the eigenvalues of the potential energy matrix) can be obtained straightforwardly [3.41]:

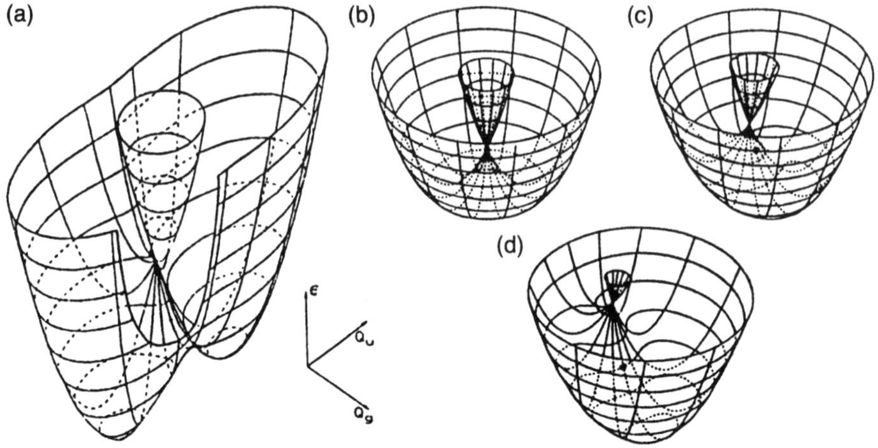

Fig. 3.10. Shape of adiabatic potential in the case of the strong pseudo-Jahn-Teller effect of the type $E \otimes (a_{1g} + b_{1u})$ at $\omega_g \neq \omega_u$, $V_g \neq V_u$ (**a**) and $\omega_g = \omega_u$, $V_g = V_u$ (**b–d**) with different energy gap values: $\delta = 0$ (**b**), E_{JT} (**c**), $2E_{JT}$ (**d**). (After [3.41])

$$\varepsilon_{\pm} = \tfrac{1}{2}(\omega_g^2 Q_g^2 + \omega_u^2 Q_u^2) \pm \sqrt{(\delta + V_g Q_g)^2 + V_u^2 Q_u^2} \ . \tag{3.48}$$

They are illustrated in Fig. 3.10. Note the conical intersection of the surfaces at the point with coordinates $(Q_g, Q_u) = (-V_g/\delta, 0)$. The lowest sheet has two extrema points at $(Q_g, Q_u) = (\pm V_g/\omega_g, 0)$, and under the condition

$$|(V_u^2/\omega_u^2) - (V_g^2/\omega_g^2)| > \delta \tag{3.4.9}$$

the lower sheet has two more equivalent minima at the points

$$\begin{aligned} Q_g^{(0)} &= \frac{V_g}{\omega_g^2}\,\delta\left(\frac{V_u^2}{\omega_u^2} - \frac{V_g^2}{\omega_g^2}\right)^{-1} , \\ Q_u^{(0)} &= \pm\frac{V_u}{\omega_u^2}\,\frac{[(V_u^2\omega_g^2 - V_g^2\omega_u^2)^2 - \delta^2\omega_g^2\omega_u^2]^{1/2}}{V_u^2\omega_g^2 - V_g^2\omega_u^2} \end{aligned} \tag{3.4.10}$$

with stabilization energy

$$E_{JT} = \frac{1}{2}\frac{V_u^2}{\omega_u^2} + \frac{\delta^2\omega_g^2\omega_u^2}{2(V_u^2\omega_g^2 - V_g^2\omega_u^2)} \ . \tag{3.4.11}$$

The case when the inequality (3.4.9) is not satisfied corresponds to the weak pseudo-Jahn-Teller effect (Fig. 3.10b–d) whereas when this inequality is fulfilled we have the case of the strong pseudo-Jahn-Teller effect (Fig. 3.10a).

If $V_u^2/\omega_u^2 \approx V_g^2/\omega_g^2 - \delta$, then the anharmonicity of the lower sheet is considerable, and the motion of the nuclei is spread over all extrema points encircling the point of conical intersection. But if the stabilization energy is comparable with kT, then the vibronic states, for which the nonadiabatic mixing of different sheets of the potential surface is important, are also populated. In this sense, the problem under consideration has features similar to that of the Jahn-Teller $E \otimes e$ problem, but it is much more complicated because of lower symmetry.

There are quite a number of examples of polyatomic systems with potential surfaces of the kind illustrated in Fig. 3.10. The most studied are the butatriene cation $C_4H_4^+$ [3.42], the ethylene cation $C_2H_4^+$ [3.43], NO_2 [3.44], NCH^+ [3.45] and others [3.41]. In all these cases good agreement with the experimental data has been reached with the model of the pseudo-Jahn-Teller effect including the totally symmetric nuclear displacements discussed above [3.41].

3.4.2 Dipole Instability in Systems with an Inversion Center

One of the special features of pseudodegeneracy, as distinguished from degeneracy, is that the mixing electronic states $\psi_{\Gamma\gamma}(\mathrm{r})$ and $\psi_{\bar{\Gamma}\bar{\gamma}}(\mathrm{r})$ may belong to different irreducible representations of the symmetry group of the system, whereas in the case of degeneracy $\Gamma = \bar{\Gamma}$. This may significantly change the space of normal nuclear displacements in which the instability and the complicated nuclear dynamics occur. In particular, for a system with a center of inversion Γ and $\bar{\Gamma}$ may possess opposite parity, as in the case considered in the previous section. Then the constant of vibronic mixing is nonzero only for odd nuclear displacements $Q_{\Gamma\gamma}$ which reduce the inversion symmetry and lead to the formation of a dipole moment in each of the minima of the adiabatic potential (*dipole instability* [3.46–48], see Chap. 6). It is quite obvious that for systems with an inversion center the interaction of degenerate electronic states cannot result in dipole instability, since in this case $\Gamma = \bar{\Gamma}$ and the Jahn-Teller-active displacements can only be of the even type.

Consider now an example of a more complicated case that is important for applications, namely, the case of the pseudo-Jahn-Teller effect in a cluster of tetravalent titanium surrounded by an octahedron of oxygens, TiO_6^{8-} (electronic configuration d^0), which is the main group of titanate crystals, for example in the ferroelectric crystal $BaTiO_3$.

In order to correctly formulate the pseudo-Jahn-Teller problem one has first to find the group of electronic states which mix strongly with the ground state and are well separated by a considerable energy gap from all other electronic terms. This can be done by evaluating the electronic structure of the octahedral complex TiO_6^{8-}.

The qualitative one-electron MO LCAO energy level scheme for this case is illustrated in Fig. 3.11 [3.46]. It is seen that for a complete treatment of the problem the vibronic mixing of at least nine molecular orbitals, t_{2g}, t_{1u} and t_{2g}^*, occupied by twelve electrons has to be taken into account [3.46, 49].

The ground multielectron state of the system is the orbital singlet ${}^1A_{1g}$. The nearest excited terms of the same multiplicity formed by the excitation of one electron from the t_{1u} and t_{2g} orbitals to the antibonding t_{2g}^* orbitals are ${}^1A_{2u}$, ${}^1E_u$,

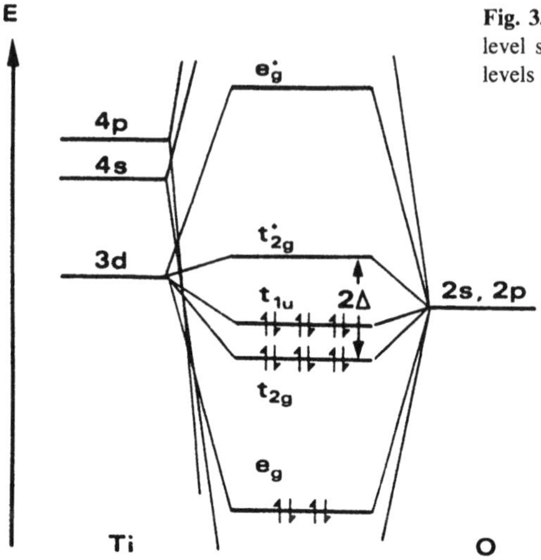

Fig. 3.11. One-electron molecular orbital energy level scheme and the electron population of the levels in the TiO_6^{8-} cluster

${}^1T_{1u}$, ${}^1T_{2u}$, ${}^1A_{1g}$, ${}^1E_g$, ${}^1T_{1g}$, and ${}^1T_{2g}$; in total 18 states. All these states are intermixed and mixed with the ground state by the odd nuclear displacements of T_{1u} type. In order to get a qualitative picture of the behavior of the lowest sheet of the adiabatic potential, and taking into account that the energies and wave functions of the above excited terms are known only in the one-electron approximation (without a covalency contribution), one can simplify the calculations by constructing the matrix of vibronic interaction in the basis of nine initial atomic wave functions, three $3d_\pi$ functions of the titanium atom and six $2p_\pi$ functions of the oxygen atoms, which form the molecular orbitals t_{1u}, t_{2g} and t_{2g}^*.

In the linear approximation, according to the Wigner-Eckart theorem only one constant of one-electron vibronic coupling has to be introduced:

$$V = \langle 2p_y | V_{T_1x}(\mathrm{r}) | 3d_{xy} \rangle = \langle 2p \| V_{T_1} \| 3d \rangle \ . \tag{3.4.12}$$

All the matrix elements of the operator of linear vibronic coupling of the self-consistent electron with T_{1u} vibrations can be approximated by this constant (3.4.12). Another constant of the Hamiltonian is the energy gap 2Δ between the atomic states $3d_{xy}$ (Ti) and $2p_y$ (O). Omitting here the secular equation we present directly its solutions, the one-electron adiabatic potentials [3.46, 50]

$$\begin{aligned} \varepsilon_i &= \tfrac{1}{2}\omega^2(Q_x^2 + Q_y^2 + Q_z^2) + \lambda_i(Q_x, Q_y, Q_z) \ , \\ \lambda_{1,2} &= \pm\sqrt{\Delta^2 + V^2(Q_x^2 + Q_y^2)} \ , \\ \lambda_{3,4} &= \pm\sqrt{\Delta^2 + V^2(Q_y^2 + Q_z^2)} \ , \\ \lambda_{5,6} &= \pm\sqrt{\Delta^2 + V^2(Q_x^2 + Q_z^2)} \ , \\ \lambda_{7,8,9} &= -\Delta \ . \end{aligned} \tag{3.4.13}$$

If we assume the six lowest adiabatic potential sheets to be populated by twelve electrons we obtain the following approximate expression for the adiabatic potential of the ground state of the multielectron system as a whole:

$$\varepsilon = \tfrac{1}{2}\omega^2(Q_x^2 + Q_y^2 + Q_z^2) - 6\Delta - 2[\sqrt{\Delta^2 + V^2(Q_x^2 + Q_y^2)} + \sqrt{\Delta^2 + V^2(Q_y^2 + Q_z^2)} + \sqrt{\Delta^2 + V^2(Q_x^2 + Q_z^2)}\,] \ . \tag{3.4.14}$$

The shape of this surface depends on the relations between the constants V, Δ and ω. If $\Delta > 4V^2/\omega^2$ the surface has one minimum at the point $Q_x = Q_y = Q_z = 0$, at which the system has the undistorted reference configuration. This case corresponds to the weak pseudo-Jahn-Teller effect (Fig. 3.8a). But if

$$\Delta < 4V^2/\omega^2 \tag{3.4.15}$$

then the surface (3.4.14) acquires a complicated shape with four types of extrema points:

1) One maximum at the point $Q_x = Q_y = Q_z = 0$ (the dynanic instability).
2) Eight minima at the points $|Q_x| = |Q_y| = |Q_z| = Q_0^{(1)}$, where

$$Q_0^{(1)} = \sqrt{8\left(\frac{V}{\omega^2}\right)^2 - \frac{1}{2}\left(\frac{\Delta}{V}\right)^2} \ , \tag{3.4.16}$$

with the stabilization energy

$$E_{\mathrm{JT}}^{(1)} = 3\left[4\frac{V^2}{\omega^2} + \frac{\omega^2\Delta^2}{4V^2} - 2\Delta\right] . \tag{3.4.17}$$

At these minima the Ti atom is displaced along the trigonal axis simultaneously approaching three oxygen atoms and moving away from the other three.
3) Twelve saddlepoints at $Q_p = Q_q \neq 0$, $Q_r = 0$, $p, q, r = x, y, z$ with a maximum in the section along Q_r and minima along Q_p and Q_q. At these points the Ti atom is displaced towards two oxygen atoms lying on the axes p and q.
4) Six saddlepoints at $Q_p = Q_q = 0$, $Q_r = Q_0^{(2)} \neq 0$,

$$Q_0^{(2)} = \sqrt{\left(\frac{4V}{\omega^2}\right)^2 - \left(\frac{\Delta}{V}\right)^2} \ , \tag{3.4.18}$$

with the stabilization energy

$$E_{\mathrm{JT}}^{(2)} = 2\left[4\frac{V^2}{\omega^2} + \frac{\omega^2\Delta^2}{4V^2} - 2\Delta\right] . \tag{3.4.19}$$

The instability of the octahedral system $[TiO_6]^{8-}$ with respect to the out-of-center displacement of the Ti^{4+} ion because of the strong pseudo-Jahn-Teller effect, as in the previous cases, admits a visual interpretation. The highest occupied

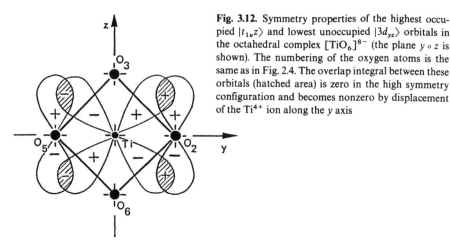

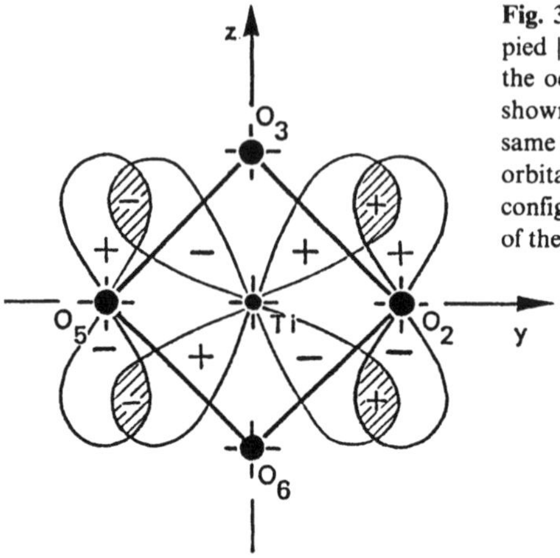

Fig. 3.12. Symmetry properties of the highest occupied $|t_{1u}z\rangle$ and lowest unoccupied $|3d_{yz}\rangle$ orbitals in the octahedral complex $[TiO_6]^{8-}$ (the plane $y \circ z$ is shown). The numbering of the oxygen atoms is the same as in Fig. 2.4. The overlap integral between these orbitals (hatched area) is zero in the high symmetry configuration and becomes nonzero by displacement of the Ti^{4+} ion along the y axis

nonbonding three-fold degenerate t_{2u} orbitals of the octahedral complex are symmetrized combinations of oxygen $2p_\pi$ functions. One of them is

$$|t_{1u}z\rangle = \tfrac{1}{2}[|2p_z(O_1)\rangle + |2p_z(O_2)\rangle + |2p_z(O_4)\rangle + |2p_z(O_5)\rangle] \ ,$$

and the other two can be constructed by appropriate symmetry operations (compare with the expressions for the vibrations in Table 2.1). The function $|t_{1u}z\rangle$ is invariant with respect to the transformation $y \to -y$ (reflection in the plane $x \circ z$, Fig. 3.12). The lowest unoccupied t_{2g} orbital is a weak-antibonding one composed mainly of $3d$ functions of the Ti^{4+} ion. One of them, $|3d_{yz}\rangle$, lies in the plane $y \circ z$ and changes its sign under reflection in the plane $x \circ z$ (Fig. 3.12).

In the high symmetry configuration, where the Ti^{4+} ion occupies the center of inversion, the overlap integral between the orbitals $|t_{1u}z\rangle$ and $|3d_{yz}\rangle$ is a zero because of symmetry restrictions (see the dashed area in Fig. 3.12), and hence in this case no chemical bonding between these two orbitals is possible. But if the Ti^{4+} ion undergoes a low symmetry displacement, for instance along the metal–ligand bonding line (along the y axis in Fig. 3.12), then in the group C_{4v} the symmetry restriction is removed and the orbital overlap is nonzero, resulting in a bonding MO LCAO.

Thus, under this displacement an additional π-bonding between the Ti and O atoms occurs, lowering the energy. If the gain of energy by this new bonding is larger than the loss of energy due to distortion of other bonds (in the case in question these are the σ bonds), then the system becomes unstable with respect to the displacement in question.

An analogous treatment is possible for tetrahedral complexes [3.51]. In this case the strong pseudo-Jahn-Teller effect under certain conditions results in four equivalent minima in each of which there is one nonequivalent bond between the central atom and the ligand which is longer or shorter than the other three equal-length bonds.

3.4.3 Vibronic Origin of Molecular Dynamic Instability

As stated above, the vibronic mixing of the ground state with the excited states may be the reason for dynamic instability of the polyatomic system [3.52–54]. The elucidation of the nature of this instability is of primary importance in chemistry, determining, in particular, the origin of the unstable transition states of chemical reactions.

Let us define, as usual, the instability of a given configuration of a molecular system as the absence of a minimum of the adiabatic potential of the electronic state. There are two types of instabilities in real systems: (i) the adiabatic potential has no extremum in the region of the configurations under consideration (this is the case of *static instability*), and (ii) the adiabatic potential surface has in this area a maximum, or a saddlepoint, but no a minimum; this is the case of *dynamic instability*. The static instability is obviously due to the nonequilibrium between the attractive and repulsive forces acting between the electrons and nuclei. In the case of a dynamic instability these attractive and repulsive forces compensate one another at the extremum of the adiabatic potential, but become decompensated by any nuclear displacement from the extremum point, the repulsion becoming predominant. The dynamic instability thus means that the adiabatic potential energy surface has a negative curvature at the extremum.

In order to investigate the curvature of the adiabatic potential at the extremum point Q_0 one can use perturbation theory with respect to small displacements $Q_{\Gamma\gamma}$ from this point, similarly to the procedure used in the method of Öpik and Pryce (Sect. 3.1). For the most general case the matrix of the potential energy of the pseudo-Jahn-Teller system can be obtained from (2.2.7) by means of the Wigner-Eckart theorem, as was done when (2.2.9) was obtained. Unlike the Jahn-Teller case, where there is only one value E_Γ, here the diagonal matrix elements of the Hamiltonian $H(\mathrm{r}, Q_0)$, $E_\Gamma = \langle \Gamma\gamma | H(\mathrm{r}, Q_0) | \Gamma\gamma \rangle$, cannot be reduced to zero simultaneously by the choice of an appropriate energy zero. For the sake of simplicity consider the case when the ground electronic term in the configuration Q_0 is a singlet (this is the most common case). We denote its energy by $E_{\Gamma_0} = E_0$ and its electronic wave function by $\psi_{\Gamma_0} = |0\rangle$. If the matrix of the potential energy is taken in the form of (3.1.34–36) with $Q_{\Gamma\gamma}^{(0)} = 0$, then the second-order correction with respect to $Q_{\Gamma\gamma}$ takes the form of (3.1.37). Note that in this case the quadratic form (3.1.37) is diagonal due to the high symmetry of the nuclear configuration Q_0, and hence $Q_{\Gamma\gamma}$ are the normal coordinates of the system. The coefficient at $\frac{1}{2}Q_{\Gamma\gamma}^2$,

$$\tilde{K}_\Gamma = K_\Gamma^{(0)} - 2 \sum_{\Gamma\gamma}' \frac{|\langle 0 | V_\Gamma(\mathrm{r}) | \bar{\Gamma}\bar{\gamma} \rangle|^2}{E_{\bar{\Gamma}} - E_0} \quad . \tag{3.4.20}$$

determines the curvature of the adiabatic potential in the direction $Q_{\Gamma\gamma}$. An important feature of (3.4.20) is that it contains two terms of different types. The first one,

$$K_\Gamma^{(0)} = \left\langle 0 \left| \frac{\partial^2 V(\mathrm{r}, Q_0)}{\partial Q_{\Gamma\gamma}^2} \right| 0 \right\rangle \quad , \tag{3.4.21}$$

is the mean value of the operator of curvature in the ground state of the system. It is equivalent to the force constant arising due to nuclear displacement in the direction $Q_{\Gamma\gamma}$ while the electronic cloud remains fixed. The second term is always negative. It describes the decrease of the force constant due to the fact that the electronic cloud moves some way towards the nuclei, i.e., due to the electronic "relaxation". This term describes the effect of the vibronic mixing of the ground state with all the excited states.

It can be shown that in chemically bonded systems the primary force constants $K_{\Gamma}^{(0)}$ are always positive [3.33, 52–54]. From this result it follows that when the vibronic interactions are not taken into account the polyatomic system is always stable and that in the formulation of the problem under consideration *the only source of instability of the system or of the negative curvature of the adiabatic potential is the vibronic mixing of the electronic ground state with the excited ones.* This result is of general importance in the physics and chemistry of molecules and crystals. Indeed, it explains the origin of a dynamic instability of any kind, including chemically unstable systems, stable systems in unstable configurations (for examples see [3.32]), transition states of chemical reactions, the so-called off-center impurities in crystals (see, for example, [3.55]), alternating bonds in conjugated hydrocarbons, spontaneous distortions and structural phase transitions in crystals (Sect. 6.2), etc. [3.56–60].

Within the two-level approximation, the instability of the ground state means that the excited state is stable. Due to the interaction with the ground state the curvature of the excited state increases, and this increase is equal in value to the decrease of that of the ground state. Taking into account the mentioned positiveness of the primary force constant $K_{\Gamma}^{(0)}$ we conclude that the excited state in the configuration with an unstable ground state is stable, and hence that it may be possible to observe it experimentally.

The fact that there are stable electronic excited states of a molecular system in a nuclear configuration for which the ground state is unstable suggests the idea of the possible existence of stable bonding electronic states for any group of atoms and any nuclear configuration. This idea is discussed in [3.52, 53] where the validity of the following theorem is demonstrated:

For any nuclear configuration of an arbitrary neutral polyatomic system there are bonding stationary electronic density distributions in the ground or excited state.

Here the term "bonding" implies a state for which the nuclear configuration corresponds to a point of the adiabatic potential well, but not necessarily to the point of the minimum itself. In other words, the above theorem states that for any nuclear configuration there are electronic states for which this nuclear configuration falls within the amplitude of stationary (or quasi-stationary) vibrations.

It follows from this that the transition states of chemical reactions which are unstable in the ground state may have stable excited states. This allows us to hope that they may be checked and investigated experimentally. The discussion of experimental methods which can be used for this purpose goes beyond the scope of this book, but it can be assumed that, in principle, any spectroscopic method which can detect quasi-stationary excited states may be used.

Of the most important information which can be obtained from such an experiment we mention first the nuclear configuration of the activated state determining the mechanism of the chemical reaction. Information about the magnitude and the shape of the potential barrier of the reaction can also be obtained. Electronic absorption in such molecular systems which are unstable in the ground state but stable in an excited state is used in excimer lasers.

The pseudo-Jahn-Teller origin of the transition state of chemical reactions proved above allows an additional theoretical basis to be given to the so-called *orbital symmetry rules* in the mechanisms of chemical reactions [3.61, 62]. Consider for example two molecular systems A and B approaching each other in a given way. Each system is assumed to be stable in itself. On approaching each other these two systems form common states of a joint system AB, the transformation properties of which are determined by the symmetry group of the activated complex AB. As an intermediate state of the chemical reaction, the system AB is unstable, i.e., the corresponding point Q_0 of the adiabatic potential in the configurational space of nuclear coordinates is a maximum or a saddlepoint. The problem is to find the direction of the line of steepest descent from this point. It is obvious that this direction determines the mechanism of the reaction. Taking into account that the dynamic instability of the system AB is of vibronic nature, this problem can be solved by means of the selection rules for the matrix elements of the vibronic interaction operator in the pseudo-Jahn-Teller effect [3.32, 63].

In cases where there are several allowed reaction paths in the sense discussed above, the question of the preferred one can be solved quantitatively by means of inequalities of the type (3.4.3). The reaction proceeds along the coordinate for which this inequality is strongest.

At present, the orbital symmetry rules are widely used in organic, inorganic and coordination chemistry, as well as for elucidation of mechanisms of catalytic reactions. In the last case participation of the catalyst allows forbidden reaction mechanisms [3.64–66].

A further development of this trend led to the realization that electronic rearrangement of the reagents by coordination to the catalyst, owing to the vibronic interaction, results in a certain distortive displacement of the nuclei, as well as in a decrease of their elastic interactions, and this in turn lowers the activation energy of the corresponding reactions. The ideas of vibronic activation in chemical reactions and catalysis were developed in [3.61, 67–70].

3.5 Adiabatic Potentials for Multimode Systems

3.5.1 Ideal and Multimode Vibronic Systems

All the cases considered so far are constrained by the assumption that the polyatomic system has a minimum number of Jahn-Teller-active vibrational degrees of freedom. For instance, in the $E \otimes e$ problem these are two normal coordinates which transform according to the E representation; for the $T \otimes (e + t_2)$ there are five such degrees of freedom, two E vibrations and three T_2 vibrations, and so on. Simple polyatomic systems in which this assumption is realized (molecules of the type X_3, ML_6, etc.) are called *ideal vibronic systems.*

Investigation of the irreducible representations of the full vibrational representation for polyatomic molecules shows that ideal vibronic systems make up a very small fraction of their number. More complicated polyatomic systems usually have more than one set of Jahn-Teller normal vibrations of a given symmetry. For instance, even for the simplest tetrahedral molecule of the type ML_4 the full vibrational representation contains the irreducible representations $A_1 + E + 2T_2$, and hence the Jahn-Teller effect for the electronic T term involves eight vibrational degrees of freedom, two of the E type and six of the T_2 type.

The polyatomic systems which have more than one set of Jahn-Teller normal coordinates of given symmetry are called *multimode Jahn-Teller systems.*

An important group of multimode Jahn-Teller systems is provided by crystals with point defects, the energy spectrum of which contains a discrete (pseudo)-degenerate electronic term well separated from the others. A discrete energy level, as is widely known, corresponds to a limited motion of the electrons. This means that these electrons are to a greater or lesser extent localized near the point defect and interact with the vibrations of the atoms of the nearest coordination spheres only. Owing to the elastic interaction between the atomic displacements in the crystal these vibrations are not normal modes but represent a wave packet of all the normal vibrations of the crystal. Thus, through the elastic interaction in the crystal the localized electrons of the defect interact with all the normal vibrations of the crystal lattice.

The extremal properties of the adiabatic potentials of multimode Jahn-Teller systems are, as shown below, the same as for the ideal vibronic systems, that is, all the qualitative conclusions about the adiabatic potentials obtained in the previous sections of this chapter are valid for multimode Jahn-Teller sytems.

3.5.2 The Two-Mode $E \otimes (b_1 + b_1)$ Problem

For the sake of simplicity consider first the two-mode analogue of the $E \otimes b_1$ problem, the $E \otimes (b_1 + b_1)$ problem. The Hamiltonian of the system is [cf. (3.1.41)]

$$\hat{H} = \tfrac{1}{2}(P_1^2 + P_2^2 + \omega_1^2 Q_1^2 + \omega_2^2 Q_2^2)\hat{\sigma}_0 + (V_1 Q_1 + V_2 Q_2)\hat{\sigma}_z \ . \tag{3.5.1}$$

The matrix of the potential energy is diagonal and its adiabatic potentials can be found directly (Fig. 3.13):

$$\varepsilon_{\pm} = \tfrac{1}{2}(\omega_1^2 Q_1^2 + \omega_2^2 Q_2^2) \pm (V_1 Q_1 + V_2 Q_2) \ . \tag{3.5.2}$$

The coordinates of the two minima and the corresponding Jahn-Teller stabilization energy can be deduced by differentiation:

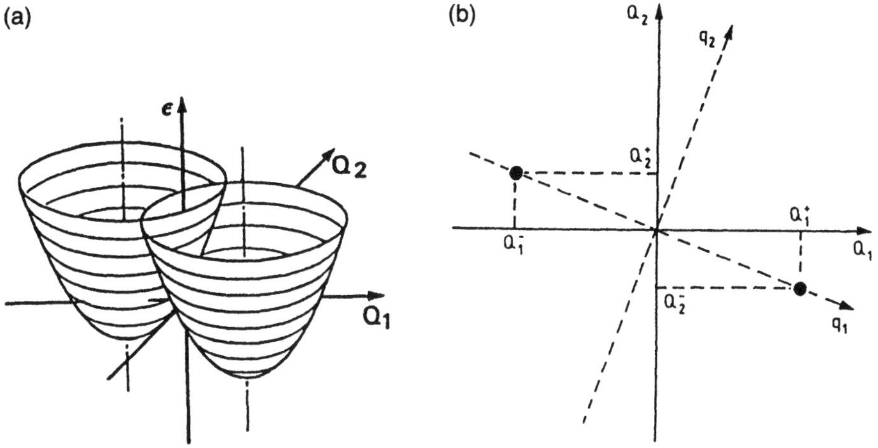

Fig. 3.13a, b. Adiabatic potential for the $E \otimes (b_1 + b_1)$ problem. **a)** General view; **b)** minima coordinates. The dashed line indicates the position of the rotated coordinate system

$$Q_1^{(\pm)} = \pm\frac{V_1}{\omega_1^2}\,, \qquad Q_2^{(\pm)} = \pm\frac{V_2}{\omega_2^2}\,; \qquad E_{\mathrm{JT}} = \frac{1}{2}\left(\frac{V_1^2}{\omega_1^2} + \frac{V_2^2}{\omega_2^2}\right)\,. \tag{3.5.3}$$

It is obvious that by means of a rotation in the space of the normal coordinates Q_1 and Q_2 one can pass to new coordinates q_1 and q_2 such that the minima will be displaced along only one of the coordinates, say q_1 (Fig. 3.13b). Since the coordinates of the minima (3.5.3) are known, the appropriate orthogonal transformation can easily be found. In the particular case when $\omega_1 = \omega_2$ it has the form

$$\begin{aligned} q_1 &= \frac{1}{V}(V_1 Q_1 + V_2 Q_2)\,, \\ q_2 &= \frac{1}{V}(-V_2 Q_1 + V_1 Q_2)\,, \\ V &= \sqrt{V_1^2 + V_2^2}\,. \end{aligned} \tag{3.5.4}$$

If $\omega_1 = \omega_2 = \omega$ the operator of the potential energy of elastic deformations of the molecule is invariant with respect to orthogonal transformations in the (Q_1, Q_2) space, and hence in the new coordinates q_1, q_2 it preserves its initial form,

$$\tfrac{1}{2}\omega^2(Q_1^2 + Q_2^2) = \tfrac{1}{2}\omega^2(q_1^2 + q_2^2)\,.$$

As a result of the orthogonal transformation (3.5.4), the potential energy matrix from (3.5.1) takes the form

$$\hat{U} = \tfrac{1}{2}\omega^2(q_1^2 + q_2^2)\hat{\sigma}_0 + Vq_1\sigma_z = \hat{U}_{\mathrm{JT}}(q_1) + \tfrac{1}{2}\omega^2 q_2^2\hat{\sigma}_0\,, \tag{3.5.5}$$

where

$$\hat{U}_{\mathrm{JT}}(q_1) = \tfrac{1}{2}\omega^2 q_1^2 + V q_1 \hat{\sigma}_z \ . \tag{3.5.6}$$

Thus the matrix $\hat{U}$ decomposes into a sum of two commuting terms, one of which, $\hat{U}_{\mathrm{JT}}$, depends on one *interacting mode* q_1 and coincides with that of an ideal vibronic system [cf. (3.1.41)], while the other depends on q_2 and is a multiple of the unit matrix σ_0.

This separation of one interacting mode is possible if $\omega_1 = \omega_2 = \omega$ and hence the potential energy of the elastic deformations is isotropic. This constraint, however, can be avoided, if, before the rotation (3.5.4), one performs a scale transformation of the coordinates [3.71]

$$q_i' = \omega_i Q_i \ . \tag{3.5.7}$$

The matrix $\hat{U}$ in the new variables has the form

$$\hat{U} = \frac{1}{2}(q_1'^2 + q_2'^2)\hat{\sigma}_0 + \left(\frac{V_1}{\omega_1} q_1' + \frac{V_2}{\omega_2} q_2'\right)\hat{\sigma}_z \ . \tag{3.5.8}$$

From this it follows that the potential energy of the elastic deformations acquires an isotropic form with an effective force constant $\omega^2 = 1$ and effective constants of vibronic coupling

$$v_i = \frac{V_i}{\omega_i} \ . \tag{3.5.9}$$

Now performing the rotation (3.5.4) we arrive at the potential energy matrix of an ideal vibronic system (3.5.6), where

$$\omega = 1 \ , \qquad V = \sqrt{v_1^2 + v_2^2} = \sqrt{\left(\frac{V_1}{\omega_1}\right)^2 + \left(\frac{V_2}{\omega_2}\right)^2} \ . \tag{3.5.10}$$

For the coordinates of the two minima, and the Jahn-Teller stabilization energy, we have

$$\begin{aligned} q_1^{(\pm)} &= \pm V = \pm\sqrt{\frac{V_1^2}{\omega_1^2} + \frac{V_2^2}{\omega_2^2}} \ ; \qquad q_2^{(\pm)} = 0 \ ; \\ E_{\mathrm{JT}} &= \frac{1}{2}V^2 = \frac{1}{2}\left(\frac{V_1^2}{\omega_1^2} + \frac{V_2^2}{\omega_2^2}\right) \ . \end{aligned} \tag{3.5.11}$$

As can be seen, the number of extrema (two), their nature (minima) and the Jahn-Teller stabilization energy are not changed as a result of the above transformations. If an inverse transformation is performed for the coordinates of the minima (3.5.11), we arrive at (3.5.3).

The above method of separating interacting Jahn-Teller modes is valid only for the investigation of adiabatic potentials. The real separation of variables in this case is not achieved, since the kinetic energy operator $\frac{1}{2}(P_1^2 + P_2^2)\hat{\sigma}_0$ loses its isotropic form as a result of the scale transformation (3.5.7). After the rotation (3.5.4) a nondiagonal term bilinear in the momenta $\propto p_1 p_2$ occurs in the expression for the kinetic energy and this term does not allow the variables to be separated.

3.5.3 General Case

The method of separation of one interacting mode presented here is not limited by the number of degrees of freedom and the nature of the Jahn-Teller effect. In the general case of the linear Jahn-Teller effect, for a multimode system the matrix of potential energy is

$$\hat{U} = \frac{1}{2}\sum_{\Gamma\gamma}\sum_{n=1}^{N_\Gamma} \omega_{n\Gamma}^2 Q_{n\Gamma\gamma}^2 \hat{C}_A + \sum_{\Gamma\gamma}\sum_{n=1}^{N_\Gamma} V_{n\Gamma} Q_{n\Gamma\gamma} \hat{C}_{\Gamma\gamma} \ . \tag{3.5.12}$$

Here, as before, the matrices $\hat{C}_{\Gamma\gamma}$ are composed of the Clebsch-Gordan coefficients and determined by the basis electronic states of the degenerate electronic term; $\hat{C}_A$ is the unit matrix. Performing a scale transformation [analogous to (3.5.7)] of the coordinates

$$q'_{n\Gamma\gamma} = \omega_{n\Gamma} Q_{n\Gamma\gamma} \tag{3.5.13}$$

we transform the operator of the potential energy of the elastic deformation to an isotropic form. In so doing the matrix (3.5.12) becomes

$$\hat{U} = \frac{1}{2}\sum_{\Gamma\gamma}\sum_{n=1}^{N_\Gamma} q'^2_{n\Gamma\gamma} \hat{C}_A + \sum_{\Gamma\gamma}\sum_{n=1}^{N_\Gamma} \frac{V_{n\Gamma}}{\omega_{n\Gamma}} q'_{n\Gamma\gamma} \hat{C}_{\Gamma\gamma} \ . \tag{3.5.14}$$

Performing a rotation in the subspace of each of the groups of N_Γ coordinates transforming by the same row γ of the same irreducible representation Γ in such a way as to make one of the unit vectors of the system of coordinates be directed along the coordinate

$$q_{1\Gamma\gamma} = \frac{1}{\mathscr{V}_\Gamma}\sum_{n=1}^{N_\Gamma} \frac{V_{n\Gamma}}{\omega_{n\Gamma}} q'_{n\Gamma\gamma} \ , \qquad \text{where} \tag{3.5.15}$$

$$\mathscr{V}_\Gamma = \left(\sum_{n=1}^{N_\Gamma} \frac{V_{n\Gamma}^2}{\omega_{n\Gamma}^2}\right)^{1/2} \ , \tag{3.5.16}$$

we get

$$\hat{U} = \hat{U}_{\rm JT}(q_1) + \frac{1}{2}\sum_{\Gamma\gamma}\sum_{n=2}^{N_\Gamma} q_{n\Gamma\gamma}^2 \hat{C}_A \qquad \text{with} \tag{3.5.17}$$

$$\hat{U}_{\mathrm{JT}}(q_1) = \tfrac{1}{2}\sum_{\Gamma\gamma} q_{1\Gamma\gamma}^2 \hat{C}_A + \sum_{\Gamma\gamma} \mathscr{V}_\Gamma q_{1\Gamma\gamma} \hat{C}_{\Gamma\gamma} \ . \tag{3.5.18}$$

The normalization factor $\mathscr{V}_\Gamma^{-1}$ in (3.5.15) fulfils the condition of transformation orthogonality

$$q_{1\Gamma\gamma} = \sum_n C_{1n} q'_{n\Gamma\gamma} \ , \qquad \sum_n C_{1n}^2 = 1 \ .$$

Thus, as a result of the above transformation the potential energy matrix decomposes into a sum of two commuting terms, one of which, $\hat{U}_{\mathrm{JT}}$, corresponds to an ideal vibronic system, while the other, which is proportional to the unit matrix $\hat{C}_A$, depends on all the remaining non-Jahn-Teller coordinates. The diagonalization of the matrix $\hat{U}$ is hence reduced to the diagonalization of the matrix $\hat{U}_{\mathrm{JT}}$, and the eigenvalues can be presented as the sum

$$\varepsilon_j(\cdots q \cdots) = \varepsilon_j^{(\mathrm{JT})}(q_1) + \tfrac{1}{2}\sum_{\Gamma\gamma}\sum_{n=2}^{N_\Gamma} q_{n\Gamma\gamma}^2 \ , \tag{3.5.19}$$

where the $\varepsilon_j^{(\mathrm{JT})}(q_1)$ are the eigenvalues of the matrix $\hat{U}_{\mathrm{JT}}$. It can easily be seen that the nature of the extrema, their number and the Jahn-Teller stabilization energy are determined just by the term $\varepsilon_j^{(\mathrm{JT})}$ corresponding to the ideal vibronic system with the Jahn-Teller interacting modes $q_{1\Gamma\gamma}$. The coordinates of the extrema points are

$$q_{1\Gamma\gamma}^{(0)} \neq 0 \ , \qquad q_{n\Gamma\gamma}^{(0)} = 0 \ , \qquad n \neq 1 \ . \tag{3.5.20}$$

The inverse transformation from the $q_{n\Gamma\gamma}$ to the $q'_{n\Gamma\gamma}$ coordinates requires the knowledge of all the elements of the transformation matrix, while (3.5.15) indicates only the first row of the matrix. This means that in the inverse matrix we know only the first column:

$$q'_{n\Gamma\gamma} = \frac{1}{\mathscr{V}_\Gamma}\frac{V_{n\Gamma}}{\omega_{n\Gamma}} q_{1\Gamma\gamma} + \sum_{m=2}^{N_\Gamma} C_{nm} q_{m\Gamma\gamma} \ . \tag{3.5.21}$$

However, taking into account that all the $q_{m\Gamma\gamma}^{(0)} = 0$, $m \neq 1$, we obtain for the coordinates of the extrema

$$q_{n\Gamma\gamma}^{\prime(0)} = \frac{1}{\mathscr{V}_\Gamma}\frac{V_{n\Gamma}}{\omega_{n\Gamma}} q_{1\Gamma\gamma}^{(0)} \ .$$

Going back to the initial coordinates $Q_{n\Gamma\gamma} = q'_{n\Gamma\gamma}/\omega_{n\Gamma}$ we have

$$Q_{n\Gamma\gamma}^{(0)} = \frac{1}{\mathscr{V}_\Gamma}\frac{V_{n\Gamma}}{\omega_{n\Gamma}} q_{1\Gamma\gamma}^{(0)} \ . \tag{3.5.22}$$

Thus the coordinates of the extrema points $Q_{n\Gamma\gamma}^{(0)}$ of the adiabatic potential of a multimode Jahn-Teller system can be expressed as the coordinates $q_{1\Gamma\gamma}^{(0)}$ of the extrema points of the adiabatic potential of an ideal vibronic system, and the Jahn-Teller stabilization energy is the same as in the ideal vibronic system with effective vibronic coupling constants determined by (3.5.16). For instance, in the multimode $E \otimes (e + e + \cdots)$ problem the coordinates of the trough on the surface of the lowest sheet of the adiabatic potential depend on one arbitrary parameter, the angle φ of the polar coordinates $q_{1\theta} = \varrho \cos\varphi, q_{1\varepsilon} = \varrho \sin\varphi$. From (3.1.13, 3.5.16 and 22) we obtain

$$Q_{n\theta}^{(0)} = \frac{V_n}{\omega_n^2} \cos\varphi \ , \qquad Q_{n\varepsilon}^{(0)} = \frac{V_n}{\omega_n^2} \sin\varphi \ , \qquad E_{\mathrm{JT}} = \frac{1}{2} \sum_n \frac{V_n^2}{\omega_n^2} \ . \tag{3.5.23}$$

As in the case of the ideal vibronic system with an E-term linear Jahn-Teller effect, in the multimode $E \otimes (e_1 + e_2 + \cdots)$ problem the trough is a one-dimensional manifold of minima.

Note that in multimode Jahn-Teller systems with equal frequencies of normal vibrations $\omega_{n\Gamma} = \omega_\Gamma$ the scale transformation (3.5.13) is not necessary and the rotation (3.5.15) results in a real separation of the variables. One variable is a Jahn-Teller one with the effective coupling constant $V_\Gamma = (\sum_n V_{n\Gamma}^2)^{1/2}$, and all the others describe harmonic vibrations near the point with the coordinates $q_{n\Gamma\gamma}^{(0)} = 0$, $n \neq 1$. If the frequencies are different, then as a result of the transformations (3.5.13 and 15) the kinetic energy assumes a nondiagonal bilinear form with respect to the momenta, and the variables cannot be separated.

In the case of point defects in crystals, the $Q_{n\Gamma\gamma}$ represent the normal mode coordinates of the lattice formed by the symmetrized displacements of the atoms of different coordination spheres. These are the so-called crystalline harmonics which diverge from the defect analogously to diverging spherical waves in the isotropic scattering problem.

Using an impurity center of small radius (with electrons localized within the first coordination sphere) as an example, we shall show that the Hamiltonian of such a system can be reduced to the one considered above with the potential energy matrix in the form (3.5.12). The electronic-vibrational Hamiltonian of a crystal with a defect in a degenerate electronic state, when only the linear terms of vibronic interaction are considered, has the form

$$\hat{H} = \tfrac{1}{2} \sum_\kappa (p_\kappa^2 + \omega_\kappa^2 q_\kappa^2) + \sum_{\Gamma\gamma} V_\Gamma Q_{\Gamma\gamma} \hat{C}_{\Gamma\gamma} \ , \tag{3.5.24}$$

where $Q_{\Gamma\gamma}$ are the symmetrized displacements of the atoms of the first coordination sphere (which are the only ones with which the electrons of the defect interact), and the index κ labels all the normal coordinates of the crystal q_κ. If one neglects the alteration of the mass and force constants introduced by the impurity, then κ labels the wave vector values in the first Brillouin zone (and the vibrational branches), while ω_κ represents the phonon dispersion law of the ideal crystal. The symmetrized displacements $Q_{\Gamma\gamma}$ of the atoms of the first coordination sphere can be expanded in terms of plane waves (or normal coordinates of the pure lattice)

$$Q_{\Gamma\gamma} = \sum_{\kappa} a_{\kappa}(\Gamma\gamma) q_{\kappa} \; , \tag{3.5.25}$$

where $a_{\kappa}(\Gamma\gamma)$ are the known Van Vleck coefficients [3.72] (their properties and explicit form for some simple cases are given in [3.73–76]).

The summation over κ can be carried out in two steps. First, all the values of the wave vector at an isofrequency surface with $\omega_{\kappa} = \omega_{n}$, then, different isofrequency surfaces with different ω_{n} have to be considered. One can perform the orthogonal transformation (3.5.15) of the normal coordinates belonging to the same isofrequency surface resulting in a separation of the variables, and separate from them one set of Jahn-Teller interacting modes for each Jahn-Teller-active irreducible representation. The Hamiltonian (3.5.24) decomposes into a sum of two commuting terms [3.77],

$$\hat{H} = \hat{H}'_{\mathrm{ph}} + \hat{H}_{\mathrm{JT}} \; , \tag{3.5.26}$$

the first of which describes the harmonic vibrations (free phonons) with the initial dispersion law, and the second has the form

$$\hat{H}_{\mathrm{JT}} = \tfrac{1}{2} \sum_{n} \sum_{\Gamma\gamma} (P_{n\Gamma\gamma}^{2} + \omega_{n\Gamma}^{2} Q_{n\Gamma\gamma}^{2}) + \sum_{n} \sum_{\Gamma\gamma} V_{n\Gamma} Q_{n\Gamma\gamma} \hat{C}_{\Gamma\gamma} \; , \tag{3.5.27}$$

where

$$V_{n\Gamma} = V_{\Gamma} \left[\sum_{\kappa} a_{\kappa}^{2}(\Gamma\gamma) \delta(\omega_{\kappa} - \omega_{n}) \right]^{1/2} . \tag{3.5.28}$$

The prime an H'_{ph} in (3.5.26) means that this term does not contain the Jahn-Teller dynamic variables $Q_{n\Gamma\gamma}$ and $P_{n\Gamma\gamma}$ included in $\hat{H}_{\mathrm{JT}}$. It can easily be seen that the potential energy matrix in (3.5.27) is identical to (3.5.12).

That the extremal properties of the adiabatic potential of multimode and ideal vibronic systems are identical can also be proved by the method of Öpik and Pryce (Sect. 3.1). Indeed, for impurity centers of small radius described by the Hamiltonian (3.5.24) the potential energy operator is

$$\hat{U} = \tfrac{1}{2} \sum_{\kappa} \omega_{\kappa}^{2} q_{\kappa}^{2} + \sum_{\Gamma\gamma} \sum_{\kappa} V_{\Gamma} a_{\kappa}(\Gamma\gamma) q_{\kappa} \hat{C}_{\Gamma\gamma} \; . \tag{3.5.29}$$

Taking into account (3.1.28) we have

$$q_{\kappa}^{(0)} = -\omega_{\kappa}^{-2} \sum_{\Gamma\gamma} V_{\Gamma} a_{\kappa}(\Gamma\gamma) (\hat{a}^{\dagger} \hat{C}_{\Gamma\gamma} \hat{a}) \; . \tag{3.5.30}$$

Substituting (3.5.30) into (3.5.29) and using the orthogonality of the Van Vleck coefficients [3.72],

$$\sum_{\kappa} a_{\kappa}(\Gamma\gamma) a_{\kappa}(\bar{\Gamma}\bar{\gamma}) f(\omega_{\kappa}) = \delta_{\Gamma\bar{\Gamma}} \delta_{\gamma\bar{\gamma}} \sum_{\kappa} a_{\kappa}^{2} f(\omega_{\kappa}) \; , \tag{3.5.31}$$

where $f(\omega_{\kappa})$ is an arbitrary function of ω_{κ}, we have

$$\hat{U} = \frac{1}{2}\sum_{\Gamma\gamma}\left(\frac{V_\Gamma}{\omega_\Gamma}\right)^2 (\hat{a}^\dagger \hat{C}_{\Gamma\gamma}\hat{a})^2 - \sum_{\Gamma\gamma}\left(\frac{V_\Gamma}{\omega_\Gamma}\right)^2 (\hat{a}^\dagger \hat{C}_{\Gamma\gamma}\hat{a})\hat{C}_{\Gamma\gamma}\ , \qquad (3.5.32)$$

where we have introduced the notation

$$\omega_\Gamma^{-2} = \sum_\kappa \frac{a_\kappa^2(\Gamma\gamma)}{\omega_\kappa^2}\ . \qquad (3.5.33)$$

The matrix of the potential energy in the equilibrium configuration (i.e., at the extremum) has the same form (3.5.32) for multimode and ideal vibronic systems. Accordingly, the system of Öpik and Pryce equations (3.1.28) for the components of the eigenvector $\hat{a}$ at the extrema has the same form in both cases and leads to the same results. For instance, for the multimode $T \otimes (e + e + \cdots + t_2 + t_2 + \cdots)$ problem the system of solutions is the same as for the ideal $T \otimes (e + t_2)$ problem, i.e. it is the same as the results given in Table 3.3, provided the ω_Γ values are taken as those determined by (3.5.33) and $Q_{\Gamma\gamma}^{(0)}$ are taken as the symmetrized displacements of the atoms of the first coordination sphere.

The equilibrium coordinates $q_\kappa^{(0)}$ have a simple physical meaning. The adiabatic wave function in the minimum of the adiabatic potential under consideration has a lower symmetry than the reference symmetry group G_0 of the impurity center. Due to the vibronic coupling, the atoms of the crystal lattice are subjected to this lowering of the symmetry of the electron density and tend to follow it by shifting their equilibrium positions. In particular, for Jahn-Teller impurity centers in ionic crystals, where the vibronic coupling has to a large degree an electrostatic nature, (3.5.30) describes the polarization of the crystal lattice by the low-symmetry distribution of the electron density at the impurity center. In other words, the low-symmetry adiabatic state is dressed in a "fur coat" of initial phonons, resulting in a *multiphonon structure of the polaron type* localized at the impurity center.

All the above is equally valid for multimode systems with the pseudo-Jahn-Teller effect, provided the difference in the initial force constants $K_\Gamma^{(0)} = \omega_\Gamma^2$ in different electronic states mixed by the vibronic coupling is neglected. By means of the transformations (3.5.13–16) one can separate one interacting mode corresponding to some ideal vibronic system. The extrema of the adiabatic potential of the latter are discussed in Sect. 3.4. Neglecting the kinetic energy, one can separate the remaining degrees of freedom, the corresponding contribution to the potential energy being represented by a paraboloid of revolution with the minimum point at $q_{\Gamma\gamma}^{(0)} = 0,\ n \neq 1$ [cf. (3.5.20)].

Crystals with Jahn-Teller impurities in a low-symmetry environment form a very large class of multimode vibronic systems with a pseudo-Jahn-Teller effect. The low-symmetry ligand field partly removes the electronic degeneracy. If the corresponding splitting of the electronic term is not very large [see the criterion (3.4.3)], the adiabatic potential energy surface possesses a complicated many-well shape with all the consequences for the nuclear dynamics. Similar properties are inherent in systems with a considerable spin-orbital interaction $H_{\mathrm{SO}} \gtrsim \hbar\omega$. The latter results in spin-orbital energy levels of the fine structure being mixed by the vibronic coupling. If some of the energy levels of a given multiplet term are degenerate with non-Kramers degeneracy, a complicated mixture of combined Jahn-Teller and pseudo-Jahn-Teller effects takes place (see also Sect. 4.4).

As mentioned in Sect. 3.4.2, one of the important features of systems with a pseudo-Jahn-Teller effect is the possibility of vibronic mixing of electronic states of opposite parity. In this mixing the odd nuclear displacements are active. For impurity centers with inversion symmetry the odd nuclear displacements remove it, shifting the impurity from the lattice site. If the effect is strong enough, the system at the high symmetry point $Q_{\Gamma\gamma} = 0$ of the adiabatic potential surface becomes unstable and low-symmetry minima at $Q_{\Gamma\gamma}^{(0)} \neq 0$ occur, resulting in the off-site distortion of the impurity, that is, in the formation of the so-called off-center impurity. The possible pseudo-Jahn-Teller origin of off-center impurities has been discussed earlier [3.55, 78], and the vibronic nature of any dynamic instability proved in Sect. 3.4.3 allows us to state that the pseudo-Jahn-Teller effect is the only reason for the noncentral position of impurity ions.

The interpretation of some features of the off-center Cu^{2+} : SrO in terms of the pseudo-Jahn-Teller effect $(E_g + T_{1u}) \otimes (e_g + t_{1u})$ is given in [3.79]. The multimode pseudo-Jahn-Teller problem $(A_{1g} + T_{1u}) \otimes (e_g + t_{2g} + t_{1u})$ is also taken as a basis for the consideration of the off-center impurity Li^+ in alkali halide matrices [3.80] (see also [3.81]). Another interesting example of the multimode pseudo-Jahn-Teller effect is provided by the central proton dynamics in the porphin ring considered as a vibronic $(A_{1g} + E_g) \otimes e_g$ problem [3.82].

3.6 The Jahn-Teller Effect in Polynuclear Clusters

So far we have considered molecular systems with one Jahn-Teller center, usually an atom or ion in a degenerate or pseudodegenerate electronic state with its near-neighbor environment. However, in some cases, within the same polyatomic system there are two or more such Jahn-Teller centers with a strong enough interaction between them significantly to influence the energy spectrum and wave functions of the system as a whole. If there are direct chemical interactions (no bridging atoms) between the Jahn-Teller centers, the direct exchange interaction between them is very strong, resulting in corresponding energy gaps (in the energy spectrum) of the order of several electronvolts (in the simplest cases this is the well-known singlet-triplet splitting). Here the vibronic coupling can be considered for each of the exchange terms separately, thus resulting in one of the vibronic problems discussed above, the only distinction being that the electronic state involves not one but all the Jahn-Teller centers participating in the direct exchange. Usually these states are coupled to a relatively large number of vibrational degrees of freedom and hence the vibronic problem is a multimode one (Sect. 3.5).

The situation is rather different when there are bridging atoms between the Jahn-Teller ions. In these cases we have the so-called indirect exchange with corresponding relatively small splittings of the order of 10–100 cm^{-1} in the energy spectrum. If the vibronic coupling at each center is weak enough, the exchange interaction has to be considered first, and then the possibility of vibronic mixing of exchange terms with the same spin multiplicity must be included. Note that even when there is no orbital degeneracy at each center, such a degeneracy may occur as a result of the exchange interaction of pure spin multiplets. For instance, the exchange interaction within a system of three electronic spins positioned at three corners of an equilateral triangle results in the multiplets $^4A_2$ and ^2E [3.83]. The latter corresponds to a wave of spin density propagating around the perimeter of the triangle clockwise or counterclockwise. In this case the ($E \otimes e$)-type JT effect is a possibility for the exchange $^2E_{\text{term}}$, but the vibronic coupling is expected to be rather small. The vibronic mixing of such spin states in three- and four-center systems is considered in [3.84–86].

Later in this section the opposite limiting case of strong vibronic coupling on each center is considered. In this case the vibronic interaction has to be taken into account first and then the indirect exchange coupling considered as a small perturbation.

Various many-center Jahn-Teller systems differ in both the number of interacting Jahn-Teller centers and the mode of their coordination. For instance, two octahedral complexes of the type ML_6 can form a two-center system (bioctahedron) in three different ways with the two central atoms lying on the common axes of symmetry of the fourth, third and second order, respectively (Fig. 3.14)

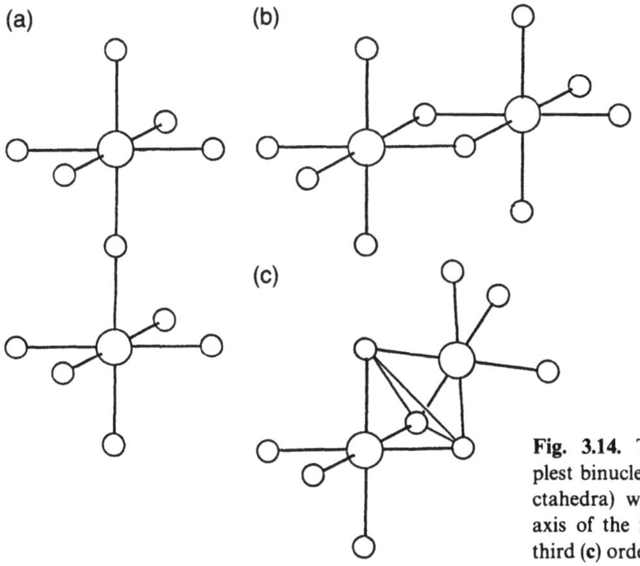

Fig. 3.14. Three types of the simplest binuclear octahedral cluster (bioctahedra) with a common symmetry axis of the fourth (**a**), second (**b**) and third (**c**) orders

(there may be more than one bridging atom). A considerable number of three- and four-center Jahn-Teller systems are well known, some of them containing several identical transition metal atoms (exchange clusters) [3.87, 88]. Other examples are provided by crystals where two or more Jahn-Teller impurities occupy near-neighbor lattice cells.

There are two main reasons for the interest in vibronic interactions in polynuclear systems. Because they have a relatively small number of degrees of freedom and a discrete energy spectrum, the polynuclear clusters serve as a model system for crystals with a cooperative Jahn-Teller (or pseudo-Jahn-Teller) effect and structural phase transitions (Chap. 6). In addition, these systems are of special interest in coordination chemistry and biology.

Among the polynuclear clusters, the binuclear ones, the *bioctahedron* with a common vertex (Fig. 3.14a), are the most studied from the point of view of vibronic coupling. The adiabatic potentials of a bioctahedron with a $(E \otimes e)$-type JT effect at each of the two centers were considered first by *Lohr* [3.89]. In [3.90], which is devoted to the cooperative Jahn-Teller effect in spinels, this problem occurs as an auxiliary one (see also [3.91]). A different approach, based on the analysis of vibrational dynamics in these systems, is suggested in [3.92]. In [3.93, 94] the energy of the centers with respect to nuclear displacements is minimized, within the framework of a simple model with Born-Mayer potentials and Coulomb interaction between nearest-neighbor ions, the problem being reduced to electronic interaction of the orbitally degenerate states of the centers. The molecular field approximation is employed in [3.95] in order to explain the temperature dependence of the exchange interaction.

The results obtained in these studies differ in some respects. As shown below, these differences are due to different ways of accounting for the translational degrees of freedom of the octahedrons, in particular, for the differences in the masses of the central atom and ligands.

In Sect. 3.6.1 we consider also the case of a $(T \otimes e)$-type JT effect at each center of the bioctahedron [3.96–99]. The more complicated systems, e.g., tetrahedral tetraclusters with twofold degeneracy at each center [3.100, 101], are discussed only in general terms. Section 3.6.2 treats the vibronic interaction in homonuclear mixed-valence clusters. It is shown that the localization-delocalization problem in these compounds is closely related to the vibronic mixing of the electronic states.

3.6.1 Vibronic Interactions in a Bioctahedron

Consider the example of the 13-atom system of D_{4h} symmetry shown in Fig. 3.14a. We assume first that each of the two Jahn-Teller ions in an octahedral environment has an orbital doublet ^2E_g state formed by a single electron (or a hole) in the highest occupied orbital e_g. Denoting the two electronic states by θ and ε, the two centers by 1 and 2 and the two spin states by α and β, one can form 28 two-electron determinants describing the possible symmetrized electronic states of the two centers. The group-theoretical classification according to the representations of the D_{4h} group and total spin $S_1 + S_2$ results in the following possible terms: $4\,{}^1A_{1g} + 2\,{}^1B_{1g} + 2\,{}^1B_{2u} + 2\,{}^1A_{2u} + 2\,{}^3B_{1g} + 2\,{}^3B_{2u} + 2\,{}^3A_{2u}$. We thus arrive at a very difficult problem of the vibronic mixing of 10 singlet and 18 triplet states by a considerable number of nuclear displacements, which it is hardly possible to solve without simplifications.

The first approximation that can be introduced is based on the fact that the interaction energy of the electrons is much larger when they are localized on one center than when they are delocalized over two centers. This allows us to use the Heitler-London approach and to exclude the high-energy states, i.e. those where both electrons are localized on one center (the vibronic mixing of these states will be considered in Sect. 3.6.2). Then the number of mixing electronic states is reduced to sixteen, $2\,{}^1A_{1g} + {}^1B_{1g} + {}^1B_{2u} + 2\,{}^3A_{2u} + {}^3B_{1g} + {}^3B_{2u}$, and the modes of nuclear displacements mixing these states are of the types A_{1g}, A_{2u}, B_{1g} and B_{2u}. Note that the states with the same spin and spin projection quantum numbers only are mixed by the vibronic interaction.

If the exchange interaction (and hence the overlap of the wave functions of different centers) is neglected, the symmetrized determinant representation may be omitted, and any linear combination of the above 16 degenerate wave functions can be chosen as a basis set. In particular, the multiplicative states on the centers $\psi^{(1)}_{E\gamma_1}(\mathrm{r}_1)\sigma_1\psi^{(2)}_{E\gamma_2}(\mathrm{r}_2)\sigma_2$, where r_1 and r_2 are the electron (hole) coordinates, and σ_1 and σ_2 are the spin states of the two centers, 1 and 2, respectively, can serve as such a set. Since different spin states are not mixed by the vibronic interaction, these 16 multiplicative states can be divided into four groups with different orbital parts only. The matrix of vibronic interaction is the same for all these groups. Thus we have reduced the problem to the investigation of the vibronic mixing of four states $\psi^{(1)}_\theta(\mathrm{r}_1)\psi^{(2)}_\theta(\mathrm{r}_2)$, $\psi^{(1)}_\theta(\mathrm{r}_1)\psi^{(2)}_\varepsilon(\mathrm{r}_2)$, $\psi^{(1)}_\varepsilon(\mathrm{r}_1)\psi^{(2)}_\theta(\mathrm{r}_2)$ and $\psi^{(1)}_\varepsilon(\mathrm{r}_1)\psi^{(2)}_\varepsilon(\mathrm{r}_2)$ by the displacements of A_{1g}, A_{2u}, B_{1g}, B_{2u} symmetry.

The second simplification can be introduced in the vibrational subsystem. The full vibrational representation of the 13-atom molecule shown in Fig. 3.14a contains the following irreducible representations of the D_{4h} group: $4\,A_{1g} + A_{1u} + 4\,A_{2u} + 2\,B_{1g} + B_{1u} + B_{2g} + 2B_{2u} + 4\,E_g + 5\,E_u$. The number of active modes, even if the totally symmetric one A_{1g} is excluded, is too large ($4\,A_{2u} + 2\,B_{1g} + 2\,B_{2u} = 8$). It could be reduced by the method described in Sect. 3.5 passing to the so-called interaction modes, however, here an easier way is possible. Remembering that E_g terms interact only with local E_g vibrations (Sect. 3.5.1), the elastic energy of isolated octahedrons can be presented as a sum with the main contributions from the E_g vibrations,

$$U_0 = \tfrac{1}{2} \sum_{n=1,2} K_E[Q_\theta^2(n) + Q_\varepsilon^2(n)] \,, \tag{3.6.1}$$

where n labels the two octahedrons, and K_E is the appropriate force constant for the E_g vibrations. Using the group projection operator one can form the following linear combinations of the $Q_{E\gamma}(n)$ coordinates transforming as the irreducible

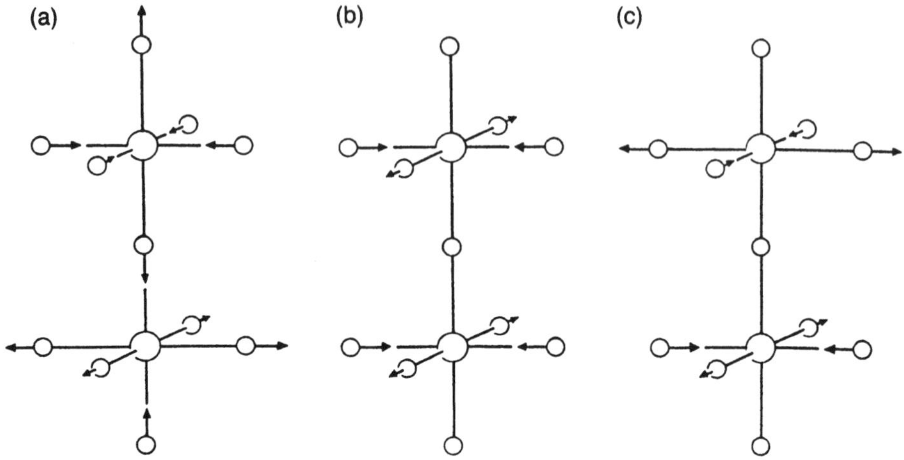

Fig. 3.15. Symmetrized atomic displacements in a bioctahedron (symmetry group D_{4h}) transforming as the irreducible representations A_{2u} (**a**), B_{1g} (**b**) and B_{2u} (**c**)

representations of the D_{4h} group:

$$Q(A_{1g}) = \frac{1}{\sqrt{2}}[Q_\theta(1) + Q_\theta(2)] \; ; \qquad Q(B_{1g}) = \frac{1}{\sqrt{2}}[Q_\varepsilon(1) + Q_\varepsilon(2)] \; ;$$
$$Q(A_{2u}) = \frac{1}{\sqrt{2}}[Q_\theta(1) - Q_\theta(2)] \; ; \qquad Q(B_{2u}) = \frac{1}{\sqrt{2}}[Q_\varepsilon(1) - Q_\varepsilon(2)] \; . \tag{3.6.2}$$

The resulting symmetrized displacements $Q(A_{2u})$, $Q(B_{1g})$ and $Q(B_{2u})$ of the bioctahedron as a whole are illustrated in Fig. 3.15. In these coordinates the elastic energy (3.6.1) is

$$U_0 = \tfrac{1}{2}K_E[Q^2(A_{1g}) + Q^2(A_{2u}) + Q^2(B_{1g}) + Q^2(B_{2u})] \; . \tag{3.6.3}$$

Now one has to take into account the shared (bridging) atom due to which the two octahedrons are not isolated. Its displacements introduce additional coupling conditions which can be evaluated in the following way. The Cartesian displacements of this atom can be expressed by the symmetrized displacements of either the first octahedron $Q_{\Gamma\gamma}(1)$, or the second $Q_{\Gamma\gamma}(2)$. Equating these two expressions for each component x, y and z, we get three conditions of coupling. If only E_g and translational degrees of freedom are considered (the latter have no elasticity), then all the $Q_{\Gamma\gamma}(n)$ coordinates are zero except $Q_\theta(n)$, $Q_\varepsilon(n)$, $Q_x(n)$, $Q_y(n)$, and $Q_z(n)$. Separating the motion of the bioctahedron center of mass we find that two of these coupling relations become identities, while the third one can be reduced to

$$Q_\theta(1) + Q_\theta(2) = \left(\frac{3m}{M + 6m}\right)^{1/2} [Q_z(1) - Q_z(2)] \ , \tag{3.6.4}$$

where m is the mass of the ligand and M is the mass of the central atom. The transition to mass-weighted coordinates is performed, as in Sect. 2.2. If the mass of the central atom is much larger than the mass of the ligands, i.e., $M \gg 6m$, then the right-hand side of (3.6.4) is close to zero, and to a good approximation one can assume that $Q_\theta(1) + Q_\theta(2) = 0$. This obvious result means that if $M \gg 6m$, then the translational motion of the octahedra can be neglected due to the high inertia of the central atoms, and the axial elongation of one of the octahedra has to be accompanied by equal compression of the other. However, in general, this is not the case, since the octahedra can adjust to each other at the expense of translational degrees of freedom.

For crystals with an impurity pair the intermediate case when the translational motion of the octahedra is not completely excluded, but is not free, becomes important. Appropriate results in this case can be obtained by adding to (3.6.3) the term

$$\tfrac{1}{2} K_{T_{1u}} [Q_z^2(1) + Q_z^2(2)] \ ,$$

which describes the elastic energy of the shift of the octahedra.

To take into account the condition (3.6.4), let us change to new coordinates

$$\begin{cases} q(A_{1g}) = \dfrac{\lambda^2}{1+\lambda^2} Q(A_{1g}) + \dfrac{\lambda}{1+\lambda^2} \tilde{Q}(A_{1g}) \ , & q(A_{2u}) = Q(A_{2u}) \ , \\ \tilde{q}(A_{1g}) = -\dfrac{1}{\lambda} Q(A_{1g}) + \tilde{Q}(A_{1g}) \ , & q(B_{1g}) = Q(B_{1g}) \ , \\ & q(B_{2g}) = Q(B_{2g}) \ , \end{cases} \tag{3.6.5}$$

where $\lambda = [3m/(M + 6m)]^{1/2}$ and $\tilde{Q}(A_{1g}) = [Q_z(2) - Q_z(1)]/\sqrt{2}$ is the coordinate describing the relative motion of the octahedra, the other $Q(\Gamma\gamma)$ being determined by (3.6.2). Since $\tilde{q}(A_{1g}) = 0$ (3.6.2 and 4) the elastic energy acquires the form

$$U_0 = \tfrac{1}{2} K_E [q^2(A_{1g}) + q^2(A_{2u}) + q^2(B_{1g}) + q^2(B_{2u})] \ , \tag{3.6.6}$$

in the new coordinates. After the transformation (3.6.5) the coordinate $q(A_{1g})$ acquires a reduced mass equal to $(1 + \lambda^2)/\lambda^2$. Expression (3.6.6) in the old coordinates is more complicated:

$$U_0 = \frac{1}{2} K_E \left[\frac{1}{2}\left(\frac{\lambda^4}{(1+\lambda^2)^2} + 1\right) Q_\theta^2(1) + Q_\varepsilon^2(1) + \frac{\lambda^2}{2(1+\lambda^2)^2} Q_z^2(1) \right.$$
$$+ \frac{\lambda^3}{(1+\lambda^2)^2} Q_\theta(1) Q_z(1) + \frac{1}{2}\left(\frac{\lambda^4}{(1+\lambda^2)^2} + 1\right) Q_\theta^2(2) + Q_\varepsilon^2(2)$$

$$+ \frac{\lambda^2}{2(1+\lambda^2)^2} Q_z^2(2) - \frac{\lambda^3}{(1+\lambda^2)^2} Q_\theta(2)Q_z(2) + \frac{\lambda^3}{(1+\lambda^2)^2} [Q_z(1)Q_\theta(2)$$

$$- Q_\theta(1)Q_z(2)] - \frac{\lambda^2}{(1+\lambda^2)^2} Q_z(1)Q_z(2)$$

$$\left. - \frac{1+2\lambda^2}{(1+\lambda^2)^2} Q_\theta(1)Q_\theta(2) \right] . \tag{3.6.7}$$

Thus, as a result of the coupling condition (3.6.4) the degeneracy of the E vibrations of the octahedrons is removed; the translational motion of the latter in the axial direction is no longer free but elastic with an elastic constant $\lambda^2 K_E/2(1+\lambda^2)^2$; and the nuclear displacements $Q_\theta(n)$ and $Q_z(n)$, belonging to the same irreducible representation A_1 of the local symmetry group of the center, are no longer independent but mixed. The last three terms in (3.6.7) describe the elastic interactions of the octahedra.

Expression (3.6.7) can be rewritten in the matrix form

$$U_0 = \tfrac{1}{2} \sum_{n,m} \hat{Q}^+(n)\hat{K}(n,m)\hat{Q}(m) , \tag{3.6.8}$$

where $\hat{Q}(n)$ is a column vector with components $Q_\theta(n)$, $Q_\varepsilon(n)$ and $Q_z(n)$, and $\hat{K}(n,m)$ is a 3×3 matrix,

$$\hat{K}(1,1) = \frac{K_E}{2(1+\lambda^2)^2} \begin{pmatrix} 1+2\lambda^2+2\lambda^4 & 0 & -\lambda^3 \\ 0 & 2(1+\lambda^2)^2 & 0 \\ -\lambda^3 & 0 & \lambda^2 \end{pmatrix} , \tag{3.6.9}$$

$$\hat{K}(1,2) = \frac{K_E}{2(1+\lambda^2)^2} \begin{pmatrix} -1-2\lambda^2 & 0 & \lambda^3 \\ 0 & 0 & 0 \\ -\lambda^3 & 0 & \lambda^2 \end{pmatrix} , \tag{3.6.10}$$

and the matrices $\hat{K}(2,2)$ and $\hat{K}(2,1)$ can be obtained from the expressions for $\hat{K}(1,1)$ and $\hat{K}(1,2)$, respectively, by the substitution $\lambda \to -\lambda$.

Expression (3.6.8), together with the coupling conditions of the type (3.6.4), presents a general description of the elastic energy of many-center systems. For all the 21 degrees of freedom of a seven-atom octahedron the vectors $\hat{Q}(n)$ have 21 components $Q_{\Gamma\gamma}(n)$, while $\hat{K}(n,m)$ is a 21×21 matrix. The 42 degrees of freedom of two isolated octahedrons are reduced to 39, including three translational and three rotational degrees of freedom, by the above three coupling conditions. The above approximation is then equivalent to a reduction of the 21×21 matrix $\hat{K}(n,m)$ to a 3×3 block (3.6.9, 10) corresponding just to the active nuclear displacements of the octahedra; all the other matrix elements of $\hat{K}(n,m)$, $n \neq m$ are assumed to be negligibly small. As a result, the symmetrized displacements $q(A_{1g})$, $q(A_{2u})$, $q(B_{1g})$ and $q(B_{2u})$ become normal coordinates and the corresponding A_{1g}, A_{2u}, B_{1g} and B_{2u} vibration frequencies become equal.

The operator of the linear vibronic coupling in the two-center system can be presented in the form of a sum [cf. (3.1.6)]

$$V = \sum_{n=1,2} V_E[Q_\theta(n)\hat{\sigma}_z(n) + Q_\varepsilon(n)\hat{\sigma}_x(n)] \ , \tag{3.6.11}$$

where the Pauli matrices $\hat{\sigma}_x(n)$ and $\hat{\sigma}_z(n)$ are given in the basis of degenerate E states of the nth Jahn-Teller center. The matrix of this operator in the basis of the above four multiplicative states can easily be converted to its diagonal form, provided the wave functions which diagonalize the linear vibronic coupling in each isolated octahedron (the adiabatic states of each center) are used as a basis. The operator (3.6.11) in this basis has the form [cf. (3.1.10)]

$$V = |V_E| \sum_{n=1,2} \varrho(n)\hat{\sigma}_z(n) \ ,$$

where $\varrho(n) = [Q_\theta^2(n) + Q_\varepsilon^2(n)]^{1/2}$. The resulting four adiabatic potential surfaces are

$$\varepsilon(q) = \frac{K_E}{2}[q^2(A_{1g}) + q^2(A_{2u}) + q^2(B_{1g}) + q^2(B_{2u})] \pm |V_E|[\varrho(1) \mp \varrho(2)] \ , \tag{3.6.12}$$

where, taking into account (3.6.2, 4 and 5),

$$\begin{aligned} \varrho(1) &= \{\tfrac{1}{2}[q(A_{1g}) + q(A_{2u})]^2 + \tfrac{1}{2}[q(B_{1g}) + q(B_{2u})]^2\}^{1/2} \ , \\ \varrho(2) &= \{\tfrac{1}{2}[q(A_{1g}) - q(A_{2u})]^2 + \tfrac{1}{2}[q(B_{1g}) - g(B_{2u})]^2\}^{1/2} \ . \end{aligned} \tag{3.6.13}$$

The extrema of the lowest sheet [corresponding to the lower signs in (3.6.12)] form a two-dimensional equipotential manifold (a trough):

$$\begin{aligned} q^{(0)}(A_{1g}) &= \frac{1}{\sqrt{2}}(\cos\varphi_1 + \cos\varphi_2) \ , \qquad & q^{(0)}(B_{1g}) &= \frac{1}{\sqrt{2}}(\sin\varphi_1 + \sin\varphi_2) \ , \\ q^{(0)}(A_{2u}) &= \frac{1}{\sqrt{2}}(\cos\varphi_1 - \cos\varphi_2) \ , \qquad & q^{(0)}(B_{2u}) &= \frac{1}{\sqrt{2}}(\sin\varphi_1 - \sin\varphi_2) \ , \end{aligned} \tag{3.6.14}$$

where φ_1 and φ_2 are arbitrary parameters varying between 0 and 2π, and $\varrho_0 = |V_E|/K_E$ [cf. (3.1.13)]. The Jahn-Teller stabilization energy equals $E_{JT} = V_E^2/K_E$.

The physical meaning of this result can easily be understood if one notes that E_{JT} is equal to the sum of the Jahn-Teller stabilization energies of the isolated octahedra [cf. (3.1.13)]. Taking account of (3.6.2–5) the extrema point coordinates (3.6.14) can be written as

$$\begin{aligned} Q_\theta^{(0)}(1) &= \varrho_0\cos\varphi_1 \ , \qquad & Q_\theta^{(0)}(2) &= \varrho_0\cos\varphi_2 \ , \\ Q_\varepsilon^{(0)}(1) &= \varrho_0\sin\varphi_1 \ , \qquad & Q_\varepsilon^{(0)}(2) &= \varrho_0\sin\varphi_2 \ . \end{aligned} \tag{3.6.15}$$

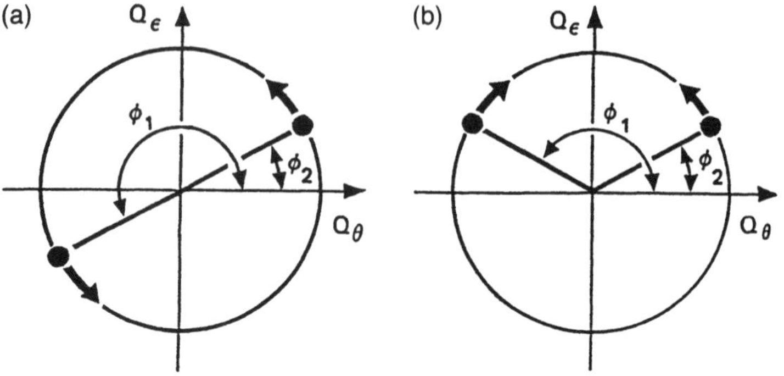

Fig. 3.16a, b. Correlated motions of the waves of $E \otimes e$ deformation of each of the two octahedra in the hioctahedron. (**a**) In-phase motion corresponding to ferrodistortive ordering; (**b**) antiphased motion in the case of antiferrodistortive packing

These relations mean that in a bioctahedron with one common (bridging) atom the rotations of the waves of deformation around each Jahn-Teller center are not correlated, i.e., they proceed independently in each octahedron. This is due to the inclusion of the translational degrees of freedom, which removes any constraints on the packing of the two octahedra with distortions [3.89]. However, if the mass of the central atoms is much larger than the mass of the ligands ($M \gg 6m$), then $\lambda \approx 0$ and the coupling condition (3.6.4) imposes restrictions on the free angular parameters φ_1 and φ_2, namely, $\varphi_1 \pm \varphi_2 = \pi$, i.e., one of them is no longer independent. It follows from this, in particular, that $\dot{\varphi}_1 \pm \dot{\varphi}_2 = 0$. This means that in the bioctahedron with one bridging atom and heavy central atoms the two waves of deformations propagate around each center not independently, but with a certain phase shift, the motion along the bottom of the trough occurring either in the same direction ($\dot{\varphi}_1 = \dot{\varphi}_2$), or in opposite directions ($\dot{\varphi}_1 = -\dot{\varphi}_2$). These two possibilities are shown schematically in Fig. 3.16.

The presence of a trough in the potential energy surface means that the symmetry of the adiabatic potential is higher than that of the bioctahedron in its initial high-symmetry configuration (D_{4h}). This is due to the approximations used above: neglect of exchange interactions and tetragonal crystal fields, special choice of the dynamic matrix $\hat{K}(n, m)$ in (3.6.8), the limitation to one-center vibronic coupling, and the linear vibronic interaction. The removal of any one of these approximations, except the last, restores the D_{4h} symmetry, resulting in the corresponding warping of the trough.

Taking account of the quadratic terms of vibronic interactions leads to a more general result [3.90, 91]. Indeed, replacing the expressions (3.6.13) by the square roots from (3.1.19) we obtain four types of extrema points with coordinates given in Table 3.4. All the coordinates are proportional to ϱ_0 [cf. (3.1.20)],

$$\varrho_0 = \frac{|V_E|}{\omega_E^2 - 2|W_E|}, \tag{3.6.16}$$

Table 3.4 The coordinates of the adiabatic potential extrema points of a bioctahedron with a Jahn-Teller ($E \otimes e$)- or ($T \otimes e$)-type effect on each of the two centers (in units of ρ_0)

No.	$q(A_{1g})$	$q(A_{2u})$	$q(B_{1g})$	$q(B_{2u})$	$Q_\theta(1)$	$Q_\varepsilon(1)$	$Q_\theta(2)$	$Q_\varepsilon(2)$	Symmetry group
1	$\sqrt{2}$	0	0	0	1	0	1	0	D_{4h}
2	$-\frac{\sqrt{2}}{2}$	0	$\frac{\sqrt{6}}{2}$	0	$-\frac{1}{2}$	$\frac{\sqrt{3}}{2}$	$-\frac{1}{2}$	$\frac{\sqrt{3}}{2}$	D_{2h}
3	$-\frac{\sqrt{2}}{2}$	0	$-\frac{\sqrt{6}}{2}$	0	$-\frac{1}{2}$	$-\frac{\sqrt{3}}{2}$	$-\frac{1}{2}$	$-\frac{\sqrt{3}}{2}$	D_{2h}
4	$-\frac{\sqrt{2}}{2}$	0	0	$\frac{\sqrt{6}}{2}$	$-\frac{1}{2}$	$\frac{\sqrt{3}}{2}$	$-\frac{1}{2}$	$-\frac{\sqrt{3}}{2}$	D_{2d}
5	$-\frac{\sqrt{3}}{2}$	0	0	$-\frac{\sqrt{6}}{2}$	$-\frac{1}{2}$	$-\frac{\sqrt{3}}{2}$	$-\frac{1}{2}$	$\frac{\sqrt{3}}{2}$	D_{2d}
6	$\frac{\sqrt{2}}{4}$	$\frac{3\sqrt{2}}{4}$	$\frac{\sqrt{6}}{4}$	$-\frac{\sqrt{6}}{4}$	1	0	$-\frac{1}{2}$	$\frac{\sqrt{3}}{2}$	C_{2v}
7	$\frac{\sqrt{2}}{4}$	$\frac{3\sqrt{2}}{4}$	$-\frac{\sqrt{6}}{4}$	$\frac{\sqrt{6}}{4}$	1	0	$-\frac{1}{2}$	$-\frac{\sqrt{3}}{2}$	C_{2v}
8	$\frac{\sqrt{2}}{4}$	$-\frac{3\sqrt{2}}{4}$	$\frac{\sqrt{6}}{4}$	$\frac{\sqrt{6}}{4}$	$-\frac{1}{2}$	$\frac{\sqrt{3}}{2}$	1	0	C_{2v}
9	$\frac{\sqrt{2}}{4}$	$-\frac{3\sqrt{2}}{4}$	$-\frac{\sqrt{6}}{4}$	$-\frac{\sqrt{6}}{4}$	$-\frac{1}{2}$	$-\frac{\sqrt{3}}{2}$	1	0	C_{2v}

and therefore in Table 3.4 only the corresponding coefficients are given. The Jahn-Teller stabilization energies for all nine extrema points coincide and are equal to

$$E_{JT} = \frac{V_E^2}{K_E - 2|W_E|}, \tag{3.6.17}$$

which is exactly twice the E_{JT} value for an isolated octahedron. This allows us to give the following physical interpretation of the results [3.89]. As in the case of linear vibronic coupling at each center, the octahedron distortions are not correlated, and the resulting nuclear configuration is determined only by the packing of differently distorted octahedra; as in the linear case, this becomes possible due to inclusion of the translational degrees of freedom of the octahedra.

However, if $M \gg 6m(\lambda \approx 0)$ the translations become rather inertial and the condition (3.6.4) is reduced to the requirement $Q_\theta(1) = -Q_\theta(2)$, i.e., $q(A_{1g}) = 0$. With this condition none of the nine extrema obtained above remain. Instead, others appear, but with the same types of symmetry, D_{2h}, D_{2d} and C_{2v} [3.92]. At the extrema of the type D_{2h}, $Q_\varepsilon(1) = Q_\varepsilon(2)$, while for the D_{2d} type, $Q_\varepsilon(1) = -Q_\varepsilon(2)$. These two types of extrema correspond to two special types of packing of the distorted octahedra. In the case of $Q_\varepsilon(1) = Q_\varepsilon(2)$, when $q^{(0)}(B_{1g}) \neq 0$ and $q^{(0)}(B_{2u}) = 0$, the packing is "parallel" (Fig. 3.15b), or of a "ferro" type, and it is called ferrodistortive, whereas for $Q_\varepsilon^{(0)}(1) = -Q_\varepsilon^{(0)}(2)$, $q^{(0)}(B_{1g}) = 0$ and $q^{(0)}(B_{2u}) \neq 0$ the packing is antiferrodistortive (Fig. 3.15c). Such an ordering of local Jahn-Teller distortions in crystals with regular Jahn-Teller centers results in the so-called cooperative Jahn-Teller effect (Chap. 6).

Consider now the case when the one-center JT effect of the two-center bioctahedral system is a $T \otimes e$ one, i.e. the electronic state of the isolated octahedron is triplet and the vibronic coupling with the T_{2g} vibrations can be neglected. In the same approximation as above, the potential energy operator can be written

$$\hat{U} = \frac{K_E}{2}[q^2(A_{1g}) + q^2(A_{2u}) + q^2(B_{1g}) + q^2(B_{2u})] + V_E \sum_{n=1,2} [Q_\theta(n)\hat{C}_\theta(n) + Q_\varepsilon(n)\hat{C}_\varepsilon(n)] \; , \tag{3.6.18}$$

where $\hat{C}_{E\gamma}(n)$ is the 3×3 matrix (3.3.2) given in the basis of the three degenerate electronic states of the nth center.

In the basis of nine multiplicative states $\psi_{T\gamma_1}^{(1)}(\mathrm{r}_1)\psi_{T\gamma_2}^{(2)}(\mathrm{r}_2)$ the $\hat{C}_{E\gamma}(n)$ operators are 9×9 diagonal matrices, and therefore the adiabatic potentials can be evaluated directly. The electronic contribution to the potential energy is linear with respect to the normal coordinates, and the adiabatic potential consists of nine intersecting paraboloids [3.99]. The minima coordinates are given in Table 3.4. Here, instead of (3.6.16) as in the case of the $E \otimes e$ problem, the ϱ_0 value, is

$$\varrho_0 = |V_E|/K_E \; . \tag{3.6.19}$$

In a similar way, the Jahn-Teller effect in a bioctahedron with $(T \otimes t_2)$-type arrangements at each center can be considered. The minima configurations in this case can be presented as different possible packings of trigonal distortions of the two octahedra.

More complicated four-center Jahn-Teller systems are considered in [3.100, 101]. In these systems with initial nuclear configurations with T_d symmetry, four identical transition metal ions occupy the vertices of a tetrahedron, and the configuration of the atoms nearest to the vertices has C_{3v} symmetry. The local trigonal symmetry of the Jahn-Teller ion allows doublet orbital degeneracy resulting in a $(E \otimes e)$-type situation on each center. As in the previous cases, the Jahn-Teller effect results in a wave of deformation of the local environment of each center, and the waves of deformation of different centers, owing to inter-center interactions, are correlated in phase and magnitude. In the linear approximation of the vibronic interaction the lowest sheet of the adiabatic potential has a continuum of equivalent minima points in the space of E displacements (a trough). The quadratic terms of vibronic interactions result in a warping of the trough with alternating tetragonal minima and saddle points on its bottom. The nuclear configuration at the minima is, in principle, similar to that of the one-center Jahn-Teller tetrahedral $E \otimes e$ problem, also agreeing with the general symmetry requirements (Sect. 3.2).

3.6.2 Electron Localization and Delocalization in Mixed-Valence Cluster Compounds

There are cases when two or more atoms of the same transition metal in different valence (oxidation) states (different number of valence electrons) occupy equivalent sites in exchange-coupled clusters, oxide crystals, etc. [3.102, 103]. The origin of such a localization of one or several "additional" (excess) electrons (or holes) at one of the equivalent centers was the subject of many theoretical investigations [3.104–107], with the conclusion that the exchange interaction accompanied by delocalization of the excess electron results in ferromagnetic ordering of the spins. However, from the experimental data it follows that many *mixed-valence clusters* are antiferromagnetic [3.103]. Besides, different experiments on the same compounds give inconsistent results: from some of them it follows that the excess electron is localized at one center, whereas others show that, under the same conditions, it is delocalized. The origin of all these contradictory results can be clarified if one takes into account the vibronic interaction theory [3.108–111]. The relatively simple case of a dimer mixed-valence cluster formed by two octahedrally coordinated ions Me^{n+} (d^2) and $Me^{(n+1)+}$ (d^1) (with electronic d^2 and d^1 configurations, respectively) having a common 4th order axis of symmetry is considered in this section as an example.

In the strong octahedral crystal field of six ligands the d levels are split into t_{2g} and e_g, the former being lower in energy [3.112]. Therefore, the ground state of the d^1 ion is $^2T_{2g}(t_{2g}^1)$. For the d^2 ion the electronic configuration t_{2g}^2 is the lowest in energy, resulting in $^3T_{1g}$, ^1E_g, $^1T_{2g}$ and $^1A_{1g}$ terms with energy gaps determined by one-center interelectronic interactions (Racah parameters), the two terms $^1T_{2g}$ and ^1E_g remaining accidentally degenerate [3.111].

From general considerations it follows that the energy of intercenter interactions is much smaller than that of the intracenter interelectronic one, and therefore the energy spectrum of the dimer under consideration can be represented by three groups of multiplets [3.111]. The first one contains the terms for which the d^1 center is in the $^2T_{2g}$ state and the d^2 center is in the $^3T_{1g}$ state. Neglecting the intercenter interaction one can write the wave function of the system in a multiplicative form: $|^2T_{2g}\rangle_1|^3T_{1g}\rangle_2$. The second and third groups of multiplets correspond to the states $|^2T_{2g}\rangle_1|^1E_g, {}^1T_{2g}\rangle_2$ and $|^2T_{2g}\rangle_1|^1A_{1g}\rangle_2$, respectively.

The dimer with the excess electron localized at one center has C_{4v} symmetry, and hence the above terms are split: $^2T_{2g} = {}^2B_2 + {}^2E$, $^1E_g = {}^1A_1 + {}^1B_1$, $^1T_{2g} = {}^1B_2 + {}^1E$, $^1A_{1g} = {}^1A_1$. Accordingly, the wave functions become $|^2B_2\rangle_1|^3A_2\rangle_2$, $|^2E\rangle_1|^3A_2\rangle_2$, $|^2B_2\rangle_1|^3E\rangle_2$ and so on. All these terms are additionally two-fold degenerate, since the same energy is inherent to the states $|^3A_2\rangle_1|^2B_2\rangle_2$,

$|^3A_2\rangle_1|^2E\rangle_2$, $|^3E\rangle_1|^2B_2\rangle_2$, etc., provided the tunneling of the excess electron is neglected. Considering the tunneling, the correct zeroth-order functions of D_{4h} symmetry, linear combinations of the above two-center functions, can be constructed by means of projection operations or site-symmetry treatment [3.113]. Another method is used in [3.111]. Note that in the exchange systems, unlike the usual ones, the Pauli principle does not limit the possible multiplets, and all the terms are allowed.

The spin states can be obtained by the usual method of addition of momenta. In the case under consideration, for the $|^2T_{2g}\rangle_1|^3T_{1g}\rangle_2$ multiplet we have $S_a = 1/2$, $S_b = 1$, $S = S_a + S_b = 3/2, 1/2$, and hence the even terms $^4A_{1g}$, $^4A_{2g}$, $^4B_{2g}$, $^4B_{1g}$, ^4E_g, $^4E'_g$, $^2A_{1g}$, $^2A_{2g}$, $^2B_{2g}$, $^2B_{1g}$, $^2B'_{1g}$, ^2E_g, $^2E'_g$ and similar odd terms are possible [3.111]. However, it is convenient to use this symmetrized basis set for the final diagonalization of the Hamiltonian matrix, but not in the intermediate consideration of the vibronic coupling terms.

The operator of vibronic interaction, in general, includes the interaction with all the normal coordinates of the two octahedra

$$V(\mathrm{r}, Q) = \sum_n \sum_{\Gamma\gamma} V_{\Gamma\gamma}(n, \mathrm{r}) Q_{\Gamma\gamma}(n) \ , \tag{3.6.20}$$

where r denotes the electron coordinates, and n, as above, labels the metal atoms and their nearest neighbors [cf. (3.6.11)]. At least one (d^1) center of the two-center $d^1 - d^2$ system under consideration is orbitally degenerate, and therefore the Jahn-Teller effect on the centers (i.e., the vibronic coupling to the Jahn-Teller-active displacements of their coordination spheres) has to be considered. Note that, unlike the one-center case, the active totally symmetric local displacements of the A_{1g} type cannot be separated, since they do not remain totally symmetric in the D_{4h} symmetry of the bioctahedron. Indeed, for the latter the appropriate symmetrized displacements are

$$\begin{aligned} Q(A_{1g}) &= \frac{1}{\sqrt{2}}[Q_{A_{1g}}(1) + Q_{A_{1g}}(2)] \\ Q(A_{2u}) &= \frac{1}{\sqrt{2}}[Q_{A_{1g}}(1) - Q_{A_{1g}}(2)] \end{aligned} \tag{3.6.21}$$

and the inverse transformation is

$$\begin{aligned} Q_{A_{1g}}(1) &= \frac{1}{\sqrt{2}}[Q(A_{1g}) + Q(A_{2u})] \ , \\ Q_{A_{1g}}(2) &= \frac{1}{\sqrt{2}}[Q(A_{1g}) - Q(A_{2u})] \ . \end{aligned} \tag{3.6.22}$$

The coupling condition due to the common atom obtained by the procedure identical to that given in the previous section is ($m \ll M$)

$$Q_{A_{1g}}(1) = -Q_{A_{1g}}(2) \ . \tag{3.6.23}$$

It follows that $Q(A_{1g}) = 0$ and

$$Q_{A_{1g}}(1) = \frac{1}{\sqrt{2}} Q(A_{2u}) \ , \qquad Q_{A_{1g}}(2) = -\frac{1}{\sqrt{2}} Q(A_{2u}) \ . \tag{3.6.24}$$

Thus the coupling to the totally symmetric displacements on each center is equivalent to coupling to the A_{2u} displacements of the bioctahedron. The contribution of this coupling is much more important than that of the coupling to low-symmetry Jahn-Teller-active displacements on the centers. Indeed, because of the charge transfer accompanying the intercenter transition of the excess electron, the equilibrium interatomic distances and hence the $Q_{A_{1g}}(n)$ coordinate of each center change significantly, resulting in an energy gain which is much larger than the Jahn-Teller stabilization energy. Therefore, in the zeroth approximation, the coupling to the low-symmetry displacements (at each octahedron) in the vibronic interaction can be omitted:

$$V(\mathrm{r}, Q) = \sum_{n=1,2} V_{A_{1g}}(\mathrm{r}, n) Q_{A_{1g}}(n) \ , \tag{3.6.25}$$

the $Q_{A_{1g}}(n)$ coordinates being interrelated by (3.6.24).

One can take into account the fact that the coupling to totally symmetric modes is independent of degeneracy. According to the Wigner-Eckart theorem the operator $V_{A_{1g}}(\mathrm{r}, n)$ is a unit matrix in the space of the degenerate states $\psi_{\Gamma\gamma}(\mathrm{r}, n)$. The latter hence remain unmixed and each of them can be considered as a singlet state. This result permits a simpler model of the mixed-valence dimer with all the one-electron states as singlet S-type one [3.114]. For the same reasons all the qualitative features of the vibronic effects can be illustrated by the orbitally singlet states $|^2B_2^{(d^1)}\rangle_1 |^3A_2^{(d^2)}\rangle_2$ and $|^3A_2^{(d^2)}\rangle_1 |^2B_2^{(d^1)}\rangle_2$.

With these states taken as a basis and considering the intracenter interactions predominant, as compared with the intercenter ones, the potential energy matrix can be written in the *approximation of Anderson and Hasegawa* as [3.115]

$$\hat{U}(Q) = \tfrac{1}{2} K Q^2(A_{2u}) + \begin{pmatrix} J s S_1 + \tilde{J}(s+S_1)S_2 + \sqrt{2} V Q_{A_{1g}}(1) & T \\ T & J s S_2 + \tilde{J}(s+S_2)S_1 + \sqrt{2} V Q_{A_{1g}}(2) \end{pmatrix} . \tag{3.6.26}$$

Here $s (s = 1/2)$ is the spin of the excess electron, $\boldsymbol{S}_1$ and $\boldsymbol{S}_2$ are the spins of the atomic cores for centers 1 and 2, respectively (in the case under consideration $S_1 = S_2 = 1/2$), $J = \langle \varphi_1 \psi_1 || \psi_2 \varphi_2 \rangle = \langle \varphi_2 \psi_2 || \psi_2 \varphi_2 \rangle$ and $\tilde{J} = \langle \varphi_1 \varphi_2 || \varphi_2 \varphi_1 \rangle \approx \langle \psi_1 \varphi_2 || \varphi_2 \psi_1 \rangle$ are the intracenter and intercenter exchange parameters, respectively, T is the parameter describing the intercenter transition of the excess electron

$$T = \langle \psi_1 || \psi_2 \rangle + \langle \psi_1 \psi_2 || \varphi_1 \varphi_1 \rangle + \langle \psi_1 \psi_2 || \varphi_2 \varphi_2 \rangle \ ,$$

$V = v/\sqrt{2}$, v being the constant of linear vibronic coupling to the totally symmetric mode at each center, $Q_{A_{1g}}(n)$, with the equilibrium value for the d^1 configuration taken as zero, K is the appropriate force constant which is assumed to be independent of the excess electron, and $\varphi_{1(2)}$ and $\psi_{1(2)}$ are the one-electron functions for the center 1 (2).

As mentioned above, the addition of momenta results in $S = 1/2, 3/2$. For $S = 1/2$ we obtain from (3.6.24 and 26) the following matrix, which is the same for different spin projections M_S [hereafter the notation $Q(A_{2u}) = Q$ is used]:

$$\hat{U}^{(1/2)} = \tfrac{1}{2}KQ^2 + \begin{pmatrix} -J+\tilde{J}+VQ & 0 & T/2 & T\sqrt{3/2} \\ 0 & J-\tilde{J}+VQ & T\sqrt{3/2} & -T/2 \\ T/2 & T\sqrt{3/2} & -J+\tilde{J}-VQ & 0 \\ T\sqrt{3/2} & -T/2 & 0 & J-\tilde{J}-VQ \end{pmatrix}, \tag{3.6.27}$$

while for $S = 3/2$ we have

$$\hat{U}^{(3/2)} = \tfrac{1}{2}KQ^2 + \begin{pmatrix} -J-2\tilde{J}+VQ & T \\ T & -J-2\tilde{J}-VQ \end{pmatrix}. \tag{3.6.28}$$

The intracenter exchange parameter J is a quantity of the order of 1–10 eV, i.e., it is predominant in the matrices (3.6.27) and (28), and therefore all the other parameters can be considered as small perturbations. For instance, in the case of $S = 1/2$ the secular matrix for the ground state with the energy $-J$ is [3.115]

$$\hat{U}^{(1/2)} = \tfrac{1}{2}KQ^2 + \begin{pmatrix} -J+\tilde{J}+VQ & T/2 \\ T/2 & -J+\tilde{J}-VQ \end{pmatrix}. \tag{3.6.29}$$

The tunneling states mentioned above can be obtained by diagonalization of the matrix (3.6.29) at $Q = 0$. In the basis of these states the matrix (3.6.29) is

$$\hat{U}^{(1/2)} = \tfrac{1}{2}KQ^2 + \begin{pmatrix} \Delta & VQ \\ VQ & -\Delta \end{pmatrix} + (\tilde{J}-J)\begin{pmatrix} 1 & 0 \\ 0 & 1 \end{pmatrix}, \tag{3.6.30}$$

where $\Delta = T/2$. The last term is an additive constant which can be excluded from further consideration of the vibronic coupling. The remaining matrix in (3.6.30) describes the pseudo-Jahn-Teller effect for two singlet states of opposite parity mixed by the odd vibration $Q(A_{2u}) = Q$ [cf. (3.4.1)]. The corresponding adiabatic potentials are shown in Fig. 3.8, the extrema coordinates and curvatures being given in Sect. 3.4. A similar result can easily be obtained for the case $S = 3/2$ described by the matrix (3.6.28), but here the splitting due to electron transport is twice that for $S = 1/2$. As follows from Sect. 3.4, there are two types of the pseudo-Jahn-Teller effect, depending on the relation between 2Δ and v^2/K. If

$2\Delta > v^2/K$, the weak pseudo-Jahn-Teller effect is realized. In this case the vibronic effects are weak and the electron, as a rule, is delocalized. Localization is possible here only under low-symmetry and very strong influence of an external field on the mixed valence cluster under consideration.

The opposite inequality $2\Delta < v^2/K$ is more probable and results in a strong pseudo-Jahn-Teller effect with equilibrium positions (minima) at $Q(A_{2u}) = \pm\sqrt{(v/2K)^2 - (2\Delta/v)^2}$. The electronic states of the minima configurations correspond to the localization of the excess electron at one of the centers. The *localization energy* is due to the rearrangement of the nuclear subsystem by the Colomb field of the excess electron. In essence this is nothing but the polaron effect which is well-known in solid-state physics. Because of the larger splitting at $Q = 0$, the pseudo-Jahn-Teller effect, if all other factors are equal, is much smaller for $S = 3/2$ than for $S = \frac{1}{2}$.

However, the real situation is somewhat more complicated than that of pure localization. Indeed, the ability of the excess electron to tunnel between the centers is not removed completely but just reduced by the polaron effect, the effective mass of the excess electron being considerably increased. This results in a significant reduction of the splitting δ (which is here quite similar to the inversion splitting in NH_3, see Sect. 4.3.3) compared with the energy gap T in the electron spectrum in the absence of vibronic coupling [3.110, 116]. Under these conditions the external field which locks the system in one of the minima configurations can be much smaller; it has to be larger than δ, but not larger than T, as in the case of the weak pseudo-Jahn-Teller effect. Since δ is small, and usually $\delta \ll T$, one can assume that in some cases such a "locking" interaction can be just the interaction with the measuring devices (for example, with the magentic field in experiments on magnetic resonance). It follows that the results concerned with the *localization* or *delocalization of the excess electron* in mixed-valence compounds depend on the method of measurement (cf. the relativity rule concerning the means of observation [3.117]). This dependence on the means of measurement explains well the origin of contradictory experimental results obtained for some mixed-valence compounds, because in some experiments (when the "locking" field is large enough) localization of the excess electron is observed, whereas in other experiments (with a small "locking" field) a delocalized state of the electron is seen in the same sample. Another parameter of the system that is important here is the relexation hopping of the localized electron between the centers because of its interaction with the medium. Owing to this relaxation the lifetime of the localized electron is limited, and if the characteristic time of measurement, which depends on the spectral instrument used, is shorter than this lifetime, the localized state is measured in the experiments. In the opposite case the instrument has no time to see the system with the localized electron, so an averaged picture is produced, which can be interpreted as due to the delocalization of the excess electron (see also Sect. 5.5).

As already mentioned above, electron transport promotes the ferromagnetic ordering of spins, whereas the Heisenberg interaction in the many-center systems under consideration leads to antiferromagnetic ground states. The polaron effect discussed above hinders the process of electron transport, hence quenching the ferromagnetic ordering. Thus the vibronic effects result in the localization of the electron and "restoration" of the antiferromagnetic ground state in Heisenberg exchange-coupled clusters [3.108].

The vibronic interaction in mixed-valence clusters significantly influences most of their properties. For instance, the complicated temperature dependence of the magnetic succeptibility cannot be understood without vibronic effects, and attempts to determine the magnitude of the exchange parameter in the framework of the Heisenberg–Dirac–Van Vleck model lead to incorrect values [3.108]; the vibronic effects result also in complicated band shapes for intervalent optical transitions [3.108, 118].

The localization-delocalization problem in trimeric mixed-valence clusters is also discussed in [3.109]. The main distinguishing feature of trimeric system, as compared with dimers, is the possibility of the coexistence of localized and delocalized states, thus breaking the pair of alternatives "localization–delocalization" [3.109]. In particular, the localization of the excess electron on two (out of three) centers is possible, and this effect can be seen directly from Mössbauer spectra.

4. Solution of Vibronic Equations. Tunneling Splitting

The evaluation of the energy spectrum and wave functions is the central problem of the theory of vibronic interactions. In the limiting cases of weak and strong vibronic coupling, this problem can approximately be solved analytically (Sects. 4.1–4.3), while in the general case, numerical solutions can be obtained (Sect. 4.4). Of special interest is the determination of the lowest vibronic levels, which, in the case of strong vibronic coupling, result from the tunneling of the system between the equivalent minima of the adiabatic potential (tunneling splitting). In some cases physical magnitudes determined by the electronic subsystem in only its ground state can be calculated by means of the so-called vibronic reduction factors without solving the vibronic problem (Sect. 4.7).

4.1 Weak Vibronic Coupling. Perturbation Theory

Consider the case of weak vibronic coupling when $E_{JT} \ll \hbar\omega$. In order to solve the eigenvalue problem with the Hamiltonian (2.2.12), or the equivalent system of coupled differential equations of the type (2.1.4), one can use the perturbation theory with the dimensionless vibronic coupling constant $k_\Gamma = |V_\Gamma|(\hbar\omega_\Gamma^3)^{-1/2}$ as a small parameter. Since for a weak Jahn-Teller effect the mean nuclear displacements $\Delta Q_{\Gamma\gamma} \sim \langle Q_{\Gamma\gamma}^2\rangle^{1/2}$ are sufficiently small, the contribution of the quadratic terms of the vibronic interaction can be neglected in the first approximation as being small compared with the contribution of the linear terms. This means that one can start from the simpler Hamiltonian (2.2.13). The latter can be presented in the form $\hat{H}_0 + \hat{V}$, where $\hat{H}_0$ is the oscillator Hamiltonian (2.2.11), and $\hat{V}$ is the linear vibronic interaction operator

$$\hat{V} = \sum_{\Gamma \neq A_1} \sum_{\gamma \in \Gamma} V_\Gamma Q_{\Gamma\gamma} \hat{C}_{\Gamma\gamma} \ . \tag{4.1.1}$$

The eigenfunctions of the Hamiltonian $\hat{H}_0$ have a multiplicative form

$$\Psi^{(0)} = \psi_{\Gamma\gamma}(\mathrm{r})|\ldots n_{\bar{\Gamma}\bar{\gamma}}\ldots\rangle \ , \tag{4.1.2}$$

where $\psi_{\Gamma\gamma}(\mathrm{r})$ are the initial electronic states of the Jahn-Teller term, and $|\ldots n_{\bar{\Gamma}\bar{\gamma}}\ldots\rangle$ are the wave functions of the harmonic vibrations of nuclei, $n_{\bar{\Gamma}\bar{\gamma}}$ being the appropriate quantum numbers for the $\bar{\Gamma}\bar{\gamma}$ mode. A characteristic feature of the perturbation (4.1.1) is that its diagonal matrix elements, calculated by the functions (4.1.2), are zero, and hence the first nonvanishing contribution to the energy occurs in the second order. Instead of evaluating the second-order corrections by the well-known formulae of perturbation theory it seems more convenient here to use the operator version of this theory. Its advantages will be seen immediately.

4.1.1 Operator Version of Perturbation Theory

Let us perform a unitary transformation $\exp(\mathrm{i}\hat{S})$ on the initial Hamiltonian $\hat{H}_0 + \hat{V}$ (see, e.g., [4.1]),

$$\hat{\hat{H}} = \mathrm{e}^{\mathrm{i}\hat{S}}\hat{H}^{-\mathrm{i}\hat{S}} , \tag{4.1.3}$$

where the operator $\hat{S} \sim \hat{V}$ is determined by the condition that the linear terms (in $\hat{V}$) vanish in the Hamiltonian $\hat{\hat{H}}$. In order to do this we expand the exponents in (4.1.3) in a power series with respect to $\hat{S}$ and obtain the following terms up to and including the second order:

$$\hat{H}_2 = \hat{H}_0 + \hat{V} + \mathrm{i}[\hat{S},\hat{H}_0] + \mathrm{i}[\hat{S},\hat{V}] - \tfrac{1}{2}\hat{H}_0\hat{S}^2 - \tfrac{1}{2}\hat{S}^2\hat{H}_0 + \hat{S}\hat{H}_0\hat{S} . \tag{4.1.4}$$

It can be seen easily from this equation that the linear terms vanish if

$$[\hat{S},\hat{H}_0] = \mathrm{i}\hat{V} , \tag{4.1.5}$$

and then the second-order Hamiltonian becomes very simple:

$$\hat{H}_2 = \hat{H}_0 + \frac{\mathrm{i}}{2}[\hat{S},\hat{V}] . \tag{4.1.6}$$

In the case under consideration, when $\hat{H}_0$ is an oscillator Hamiltonian (2.2.11) and the perturbation (4.1.1) is linear in $Q_{\Gamma\gamma}$, (4.1.6) can be solved directly [4.2]:

$$\hat{S} = -\sum_{\Gamma\gamma} \frac{V_\Gamma}{\hbar\omega_\Gamma^2}\hat{C}_{\Gamma\gamma}P_{\Gamma\gamma} . \tag{4.1.7}$$

Here $P_{\Gamma\gamma}$ is the momentum conjugated with $Q_{\Gamma\gamma}$. It can be shown that the operator (4.1.7) transforms (4.1.5) into an identity.

Substituting (4.1.7) and (4.1.1) into (4.1.6) we obtain the Jahn-Teller Hamiltonian of second-order perturbation theory as follows:

$$\hat{H}_2 = \hat{H}_0 - \frac{1}{2}\sum_{\Gamma}\left(\frac{V_\Gamma}{\omega_\Gamma}\right)^2\sum_{\gamma}\hat{C}_{\Gamma\gamma}^2 + \hat{V}_2 , \tag{4.1.8}$$

$$\hat{V}_2 = -\frac{\mathrm{i}}{2\hbar}\sum_{\Gamma\gamma}\sum_{\bar{\Gamma}\bar{\gamma}}\frac{V_\Gamma V_{\bar{\Gamma}}}{\omega_\Gamma^2}Q_{\bar{\Gamma}\bar{\gamma}}P_{\Gamma\gamma}[\hat{C}_{\Gamma\gamma},\hat{C}_{\bar{\Gamma}\bar{\gamma}}] .$$

It is obvious that $\sum_\gamma \hat{C}^2_{\bar{\Gamma}\gamma}$ is a scalar of the reference symmetry point group of the polyatomic system and hence, following the Wigner-Eckart theorem, we have

$$\sum_\gamma \hat{C}^2_{\bar{\Gamma}\gamma} = \text{const} \cdot \hat{C}_A \ , \tag{4.1.9}$$

where $\hat{C}_A$ is a unit matrix. Therefore the second term in (4.1.8) shifts all the vibronic levels by the same magnitude, and hence it can be eliminated by means of a redefinition of the zero of energy.

In order to determine the energy levels in second-order perturbation theory, the perturbation $\hat{V}_2$ has to be taken into account in the first order. This can be done by diagonalizing the matrix of the operator $\hat{V}_2$ built up from the degenerate functions of the Hamiltonian $\hat{H}_0$ for the term under consideration. Only the terms of (4.1.8) with $\Gamma = \bar{\Gamma}$ contribute to the perturbation matrix. Indeed, in the representation of second quantization, using (3.2.11) we have

$$Q_{\bar{\Gamma}\bar{\gamma}} P_{\Gamma\gamma} = \frac{i\hbar}{2}\sqrt{\frac{\omega_\Gamma}{\omega_{\bar{\Gamma}}}}(b^\dagger_{\bar{\Gamma}\bar{\gamma}} b^\dagger_{\Gamma\gamma} - b^\dagger_{\bar{\Gamma}\bar{\gamma}} b_{\Gamma\gamma} + b_{\bar{\Gamma}\bar{\gamma}} b^\dagger_{\Gamma\gamma} - b_{\bar{\Gamma}\bar{\gamma}} b_{\Gamma\gamma}) \ . \tag{4.1.10}$$

The operators $b^\dagger_{\bar{\Gamma}\bar{\gamma}} b^\dagger_{\Gamma\gamma}$ and $b_{\bar{\Gamma}\bar{\gamma}} b_{\Gamma\gamma}$ do not contribute to the perturbation at any Γ and $\bar{\Gamma}$. The operator $b^\dagger_{\bar{\Gamma}\bar{\gamma}} b_{\Gamma\gamma}$ lowers the energy by $\hbar\omega_{\bar{\Gamma}}$ and increases it by $\hbar\omega_\Gamma$, and hence, if $\bar{\Gamma} \neq \Gamma$, it produces a new state out of the group of the vibrational term's degenerate zero-order states under consideration with the energy $\sum_{\Gamma\gamma} \hbar\omega_\Gamma (n_{\Gamma\gamma} + \frac{1}{2})$. Thus the matrix element of this operator is nonzero only for states belonging to different vibrational levels and therefore it does not contribute to the perturbation matrix.[1] Hence only the terms with $\bar{\Gamma} = \Gamma$ have to be kept in (4.1.8):

$$\hat{V}_2 = -\frac{i}{4\hbar}\sum_\Gamma \sum_{\gamma,\bar{\gamma}} \left(\frac{V_\Gamma}{\omega_\Gamma}\right)^2 (Q_{\Gamma\bar{\gamma}} P_{\Gamma\gamma} - Q_{\Gamma\gamma} P_{\Gamma\bar{\gamma}})[\hat{C}_{\Gamma\gamma}, \hat{C}_{\Gamma\bar{\gamma}}] \ , \tag{4.1.11}$$

the second-order contribution to the energy being additive with respect to the index Γ. This allows us to consider separately the vibronic coupling to normal vibrations belonging to different irreducible representations. For instance, the $E \otimes (b_1 + b_2)$ problem can be considered as two separate ones, $E \otimes b_1$ and $E \otimes b_2$, the $T \otimes (e + t_2)$ problem as $T \otimes e$ and $T \otimes t_2$, the $\Gamma_8 \otimes (e + t_2)$ problem as $\Gamma_8 \otimes e$ and $\Gamma_8 \otimes t_2$, and so on. Thus in each case when there is simultaneous vibronic coupling to different types of vibrations the result is the second-order sum of the V_Γ contributions of each of these types of vibrations Γ. Therefore for a given vibrational mode of type Γ the following second-order Hamiltonian can be written:

$$\hat{H}_2(\Gamma) = \frac{1}{2}\sum_\gamma (P^2_{\Gamma\gamma} + \omega^2_\Gamma Q^2_{\Gamma\gamma})\hat{C}_A - \frac{1}{2}\left(\frac{V_\Gamma}{\omega_\Gamma}\right)^2 \sum_\gamma C^2_{\Gamma\gamma} + \hat{V}_2(\Gamma) \ , \tag{4.1.12}$$

[1] These arguments are valid for the ideal vibronic systems only. The perturbation treatment of the multimode Jahn-Teller systems is given below in Sect. 4.5.

where

$$\hat{V}_2(\Gamma) = -\frac{i}{4\hbar}\left(\frac{V_\Gamma}{\omega_\Gamma}\right)^2 \sum_{\gamma,\bar{\gamma}} (Q_{\Gamma\bar{\gamma}}P_{\Gamma\gamma} - Q_{\Gamma\gamma}P_{\Gamma\bar{\gamma}})[\hat{C}_{\Gamma\gamma}, \hat{C}_{\Gamma\bar{\gamma}}] \ . \qquad (4.1.13)$$

The operators $Q_{\Gamma\bar{\gamma}}P_{\Gamma\gamma} - Q_{\Gamma\gamma}P_{\Gamma\bar{\gamma}}$ transform as the lines of the antisymmetric square of the Γ representation, $\{\Gamma^2\}$, and describe rotations in the space of the normal coordinates $Q_{\Gamma\gamma}$ with given Γ. Similarly, the operators $i[\hat{C}_{\Gamma\gamma}, \hat{C}_{\Gamma\bar{\gamma}}]$ transform after the same lines of the representation $\{\Gamma^2\}$ and describe appropriate rotations in the subspace of the electronic states of the initial Jahn-Teller electronic term. From this point of view the scalar convolution (4.1.13) is similar to that of the operator of spin-orbital interaction, in which the energy spin is introduced instead of the usual spin (Sect. 3.2), and the orbital motion of the electrons is represented by rotation of the distortions of the molecular framework.

These symmetry properties of the perturbation operator are very important and allow the construction of symmetry-adapted zero-order wave functions. This may otherwise be a difficult task, especially for highly degenerate excited vibrational energy levels. Indeed, the low-symmetry Jahn-Teller active vibrations, as a rule, are degenerate, and the degeneracy of the vibrational terms increases rapidly with the number of the energy level. For instance, for twofold E vibrations, the degeneracy of the nth level is $n + 1$, while for T_2 vibrations it is $(n + 1)(n + 2)/2$, and so on. These values have also to be multiplied by f_Γ, the electronic degeneracy of the Jahn-Teller term. Thus, the degeneracy of the vibronic energy levels of the Hamiltonian $\hat{H}_0$ and hence the order of the appropriate secular equation increase rapidly when moving up to higher energy levels, making the diagonalization of the perturbation matrix more difficult without employing symmetry considerations.

As stated in Sect. 3.2, the high degeneracy of the states of the Hamiltonian $\hat{H}_0$ is directly related to the dimensionality of the irreducible representations of the group $SU(f_{\bar{\Gamma}})$ describing the symmetry of an isotropic $f_{\bar{\Gamma}}$-dimensional oscillator, and of the $SU(f_\Gamma)$ symmetry group of the orbital f_Γ-fold degenerate electronic term. The group $SU(f_{\bar{\Gamma}})$ contains (as a subgroup) the group $SO(f_{\bar{\Gamma}})$ of $f_{\bar{\Gamma}}$-dimensional rotations of the wave of deformation of the molecular framework around the center of symmetry of the molecule. Similarly, the subgroup $SO(f_{\bar{\Gamma}})$ describes the free rotations of the electron density. The vibronic interaction, in general, reduces this high symmetry to the finite point group of the molecular framework. However, as emerges from (4.1.11), in the second-order perturbation theory the symmetry of the Hamiltonian is still high enough to be described by the continuous group of rotations, namely the group of the total momentum. However, the rotations of the electronic density and the wave of deformation of the molecular framework are no longer independent: only the total momentum (the "orbital" momentum plus the "spin" momentum) is preserved.

The symmetry-adapted wave functions of the zero-order approximation can be obtained in the same way as in the LS coupling scheme for atoms, i.e., by using appropriate Clebsch-Gordan coefficients. To illustrate the foregoing we shall consider some concrete examples.

4.1.2 Weak Coupling Case in the $E \otimes e$ Problem

In the case of the $E \otimes e$ problem the operator of linear vibronic coupling is given by (3.1.3) and in accordance with (4.1.7) the operator $\hat{S}$ is

$$\hat{S} = -\frac{V_E}{\hbar\omega_E}(P_\theta \hat{\sigma}_x + P_\varepsilon \hat{\sigma}_y) \; , \tag{4.1.14}$$

the second-order perturbation Hamiltonian (4.1.12) and (4.1.13) being [4.3]

$$\hat{H}_2 = \left[\frac{1}{2}(P_\theta^2 + P_\varepsilon^2) + \frac{1}{2}\omega_E^2(Q_\theta^2 + Q_\varepsilon^2) - \left(\frac{V_E}{\omega_E}\right)^2\right]\hat{\sigma}_0 - \left(\frac{V_E}{\omega_E}\right)^2 L_z \hat{\sigma}_z \; , \tag{4.1.15}$$

where L_z is the z component of the vibrational momentum (3.2.25).

Since the symmetry group of the two-dimensional oscillator is $SU(2)$, and that of the twofold degenerate electronic term is also $SU(2)$, the full symmetry group of the Hamiltonian $\hat{H}_0$ is $SU(2) \times SU(2)$ (Sect. 3.2). The degeneracy of the energy levels of the Hamiltonian $\hat{H}_0$ is $2(n+1)$ since there are two possible values of the projection of the energy spin 1/2, and $n+1$ values of the component of the vibrational momentum L_z; the appropriate quantum number l runs over the values $-n, -n+2, -n+4, \ldots, n$. The "spin-orbital" interaction $L_z\hat{\sigma}_z$ partly removes this degeneracy, since $\boldsymbol{L}$ and $\boldsymbol{S}$ are no longer preserved separately, only the projection of the total momentum $\hat{J}_z = L_z + \frac{1}{2}\hat{\sigma}_z$ with the quantum number $j = l \pm \frac{1}{2}$ being preserved. Nevertheless (as always in the perturbation treatment) l and ± 1 still remain "good" quantum numbers for L_z and $\hat{\sigma}_z$, and hence these operators in (4.1.15) can be replaced by l and ± 1 [4.3]:

$$\begin{aligned} E_{n,l,\pm 1,j} &= \hbar\omega_E(n+1) - \frac{V_E^2}{\omega_E^2}(1 \pm l) \\ &= \hbar\omega_E(n+1) - \frac{V_E^2}{\omega_E^2}\left(j^2 - l^2 + \frac{3}{4}\right) \; . \end{aligned} \tag{4.1.16}$$

Since for each positive value l in the range $-n, -n+2, \ldots, n$ there is a negative value $-l$, all the energy levels according to (4.1.16) are twofold degenerate. Thus each of the $2(n+1)$-fold degenerate levels of the Hamiltonian is lowered by the vibronic interaction in the second order by the same value $V_E^2/\omega_E^2 = 2E_{\mathrm{JT}}$, where E_{JT} is given by (3.1.13), and split into $n+1$ equally spaced doublets with a spacing of $2V_E^2/\omega_E^2 = 4E_{\mathrm{JT}}$ (Fig. 4.1). The energy interval between the extreme doublets of the split multiplet with a given n is $2nV_E^2/\omega_E^2 = 4nE_{\mathrm{JT}}$. From this the following criterion of applicability of the perturbation theory emerges:

$$4nE_{\mathrm{JT}} \ll \hbar\omega_E \tag{4.1.17}$$

or, equivalently, $4E_{\mathrm{JT}}/\hbar\omega \ll n^{-1}$.

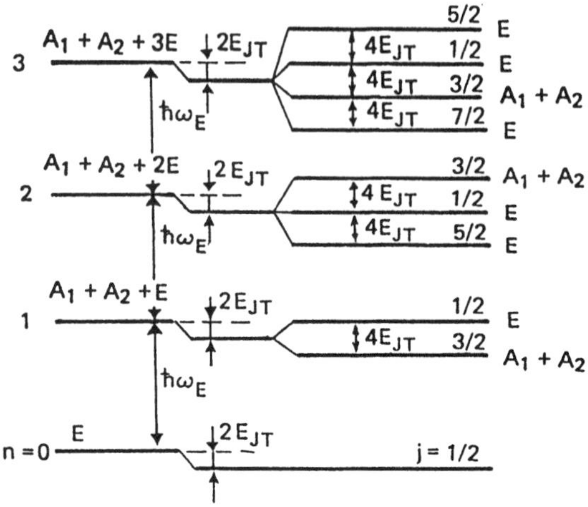

Fig. 4.1. Vibronic splitting of several of the lowest vibrational energy levels of a two-dimensional oscillator in the case of weak ($E \otimes e$)-type Jahn-Teller effect

The twofold degeneracy of the energy levels (4.1.16) has a Kramers origin (Sect. 3.2). In particular, the twofold degenerate ground state of the Hamiltonian $\hat{H}_0$ is not split but shifted by the vibronic interaction (Fig. 4.1). Hence the initial twofold electronic degeneracy of the ground state is not removed by the vibronic interaction but is transformed into vibronic degeneracy.

The results presented by (4.1.15) and (4.1.16) can be given a somewhat different interpretation. The second-order Hamiltonian (4.1.15) does not contain nondiagonal Pauli matrices, and the matrix Schrödinger equation with this Hamiltonian, when written out, can be decoupled into two independent equations with the following Hamiltonians:

$$H_{\pm} = \frac{1}{2}(P_\theta^2 + P_\varepsilon^2) + \frac{1}{2}\omega_E^2(Q_\theta^2 + Q_\varepsilon^2) - 2E_{\text{JT}} \pm \frac{2E_{\text{JT}}}{\hbar}(Q_\theta P_\varepsilon - Q_\varepsilon P_\theta) \ , \quad (4.1.18)$$

or, in dimensionless coordinates $q_\gamma = Q_\gamma\sqrt{\omega_E/\hbar}$,

$$H_{\pm} = \tfrac{1}{2}\hbar\omega_E(p_\theta^2 + p_\varepsilon^2 + q_\theta^2 + q_\varepsilon^2) - 2E_{\text{JT}} \pm 2E_{\text{JT}}(q_\theta p_\varepsilon - q_\varepsilon p_\theta) \ . \quad (4.1.19)$$

This latter is bilinear in coordinates and momenta. The canonical transformation

$$\begin{aligned} q_1 &= \frac{1}{\sqrt{2}}(p_\theta + q_\varepsilon) \ , \quad & p_1 &= \frac{1}{\sqrt{2}}(p_\varepsilon - q_\theta) \ , \\ q_2 &= \frac{1}{\sqrt{2}}(p_\varepsilon + q_\theta) \ , \quad & p_2 &= \frac{1}{\sqrt{2}}(p_\theta - q_\varepsilon) \end{aligned} \quad (4.1.20)$$

diagonalizes this bilinear form [4.2]:

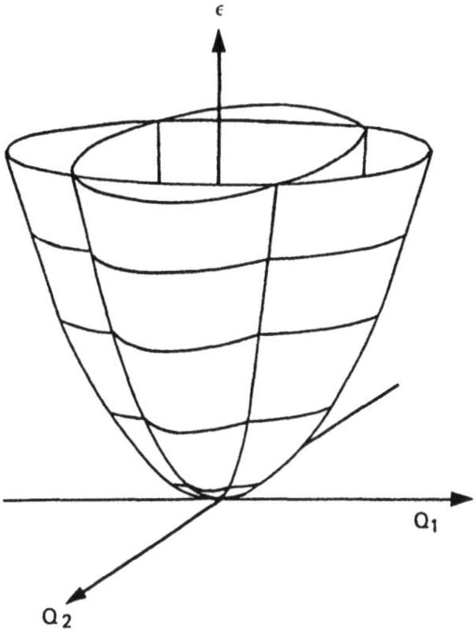

Fig. 4.2. Potential energy surfaces in the linear ($E \otimes e$)-type Jahn-Teller effect with the Hamiltonian (4.1.21) in second-order perturbation theory

$$H_{\pm} = \frac{1}{2}\hbar\omega_E\left(1 \pm \frac{2E_{\mathrm{JT}}}{\hbar\omega_E}\right)(p_1^2 + q_1^2) + \frac{1}{2}\hbar\omega_E\left(1 \mp \frac{2E_{\mathrm{JT}}}{\hbar\omega_E}\right)(p_2^2 + q_2^2) - 2E_{\mathrm{JT}} \; . \tag{4.1.21}$$

In the absence of vibronic interaction the potential energy surface of the Hamiltonian $\hat{H}_0$ is represented by two coinciding (degenerate) paraboloids of revolution. As follows from (4.1.21), the vibronic interaction in second-order perturbation theory compresses these paraboloids along one of the two coordinates, either q_1, or q_2, and elongates them along the other one (Fig. 4.2). The energy spectrum of the Hamiltonian (4.1.21) is a superposition of two equally spaced harmonic oscillator spectra:

$$E_{n_1,n_2} = (\hbar\omega_E \pm 2E_{\mathrm{JT}})n_1 + (\hbar\omega_E \mp 2E_{\mathrm{JT}})n_2 + \hbar\omega_E - 2E_{\mathrm{JT}} \; . \tag{4.1.22}$$

One can easily verify that (4.1.22) agrees with (4.1.16). The twofold degeneracy of each of the levels is due to the coincidence of the two spectra of the distorted paraboloids.

The eigenfunctions of the Hamiltonian (4.1.21) are multiplicative with respect to electronic and nuclear variables:

$$\Psi^{(\pm)}_{n_1,n_2} = (n_1!n_2!)^{-1/2}\psi_{\pm}(\mathrm{r})(b_1^{\dagger})^{n_1}(b_2^{\dagger})^{n_2}|00\rangle = \psi_{\pm}(\mathrm{r})|n_1 n_2\rangle \; , \tag{4.1.23}$$

where $b_1^{\dagger}$ and $b_2^{\dagger}$ are the operators of creation of exact vibrational states of the Hamiltonian (4.1.21), and $|00\rangle$ is the vibrational ground state function. Performing a unitary transformation $\exp(-\mathrm{i}\hat{S}) \approx 1 - \mathrm{i}\hat{S}$ over the function (4.1.23), where in accordance with (4.1.14) and (4.1.20)

$$\hat{S} = -\frac{V_E}{2\sqrt{\hbar^3\omega_E^3}}[(b_1^\dagger + b_1 + ib_2^\dagger - ib_2)\hat{\sigma}_x + (b_2^\dagger + b_2 + ib_1^\dagger - ib_1)\hat{\sigma}_y] \ ,$$

we obtain the first-order perturbation eigenfunctions of the initial Hamiltonian (3.2.1). In particular, for the ground state

$$\Psi_+ = \psi_+(\mathrm{r})|00\rangle - \frac{\mathrm{i}V_E}{\sqrt{\hbar\omega_E^3}}\psi_-(\mathrm{r})|01\rangle \ ,$$
$$\Psi_- = \psi_-(\mathrm{r})|00\rangle + \frac{\mathrm{i}V_E}{\sqrt{\hbar\omega_E^3}}\psi_+(\mathrm{r})|10\rangle \ . \qquad (4.1.24)$$

Thus, when considering the vibronic interaction in the first-order perturbation theory the wave functions are no longer multiplicative with respect to electronic and nuclear functions, and the quantity $V_E^2/\hbar\omega_E^3 = 2E_{\mathrm{JT}}/\hbar\omega_E$ has the physical meaning of the probability of finding (at $T = 0$ K) the Jahn-Teller system in the one-phonon excited state $\psi_+(\mathrm{r})|10\rangle$ or $\psi_-(\mathrm{r})|01\rangle$. The same result also follows from the calculation of the average number of phonons, in the ground state (4.1.24) $\langle\Psi_\pm|b_1^\dagger b_1 + b_2^\dagger b_2|\Psi_\pm\rangle$.

4.1.3 Perturbation Treatment of Other Jahn-Teller Problems ($E\otimes b_1$, $E\otimes b_2$, $T\otimes e$, $T\otimes t_2$, $\Gamma_8\otimes e$, $\Gamma_8\otimes t_2$)

In the case of the $E\otimes b_1$ or $E\otimes b_2$ problem there is only one $\hat{C}_{\Gamma\gamma}$ matrix, $\hat{\sigma}_x$ or $\hat{\sigma}_y$ (Sect. 3.1). This type of vibronic coupling does not change the energy spectrum,

$$\hat{H}_2 = \tfrac{1}{2}(P^2 + \omega^2 Q^2)\hat{\sigma}_0 - E_{\mathrm{JT}}\hat{\sigma}_0 \ , \qquad (4.1.25)$$

but shifts its levels by an amount E_{JT}. This result of the second-order perturbation theory with a small coupling constant remains valid for any values of the vibronic constant, provided the interaction with the vibrations of other symmetries (except A_1) is not taken into account. Actually, the coupling to only the B_1 (or B_2) vibration can be evaluated exactly. Indeed, as mentioned in Sect. 3.1 the matrix of the potential energy in this case can be diagonalized, and since the non-adiabaticity operator does not occur here, the appropriate Schrödinger equation can be decoupled into two independent equations with the following Hamiltonians ($i = +, -$):

$$H_i = \tfrac{1}{2}(P^2 + \omega^2 Q^2) + VQ\langle\psi_i|\sigma_z|\psi_i\rangle \ . \qquad (4.1.26)$$

This can be reduced to an oscillator form by means of a simple substitution:

$$Q = q_i - \frac{V}{\omega^2}\langle\psi_i|\sigma_z|\psi_i\rangle \ . \qquad (4.1.27)$$

This shift can be performed in a matrix form using the unitary transformation (4.1.3) with $\hat{S}$ given by (4.1.7), provided the $\hat{S}$ matrix in this case is diagonal. Indeed, for the Hamiltonian of the linear $E \otimes b_1$ (or $E \otimes b_2$) problem the identity

$$e^{i\hat{S}}\hat{H}e^{-i\hat{S}} = \tfrac{1}{2}(P^2 + \omega^2 Q^2)\hat{\sigma}_0 - E_{JT}\hat{\sigma}_0 \qquad (4.1.28)$$

is valid exactly in any order of V [cf. (4.1.25)].

The Hamiltonian of the linear $T \otimes t_2$ problem has the form (3.3.8) with $V_E = 0$, and hence, in accordance with (4.1.7),

$$\hat{S} = -\frac{V_T}{\hbar\omega_T}(P_\xi \hat{C}_\xi + P_\eta \hat{C}_\eta + P_\zeta \hat{C}_\zeta) \qquad (4.1.29)$$

with the $\hat{C}_{\Gamma\gamma}$ matrices given by (3.3.2). Therefore the second-order perturbation theory Hamiltonian from (4.1.12) and (4.1.13) has the form [4.3]

$$\hat{H}_2(T) = \hat{H}_0 + \tfrac{3}{4}E_{JT}(T)\boldsymbol{L}\hat{\boldsymbol{S}} - \tfrac{3}{2}E_{JT}(T) \ , \qquad (4.1.30)$$

where $E_{JT}(T)$ is given by (3.3.11), the matrices of the energy spin for $S = 1$ are presented in (3.3.3), and the components of the "orbital" momentum describing the rotations in the subspace of T_2 vibrations only can be obtained from (3.3.4) by excluding the motions in the subspace of E vibrations, i.e. assuming $P_\theta = P_\varepsilon = 0$. In other words, the vector L has the usual form of a vector product $\boldsymbol{Q}_{T_2} \times \boldsymbol{P}_{T_2}$:

$$L_x = Q_\eta P_\zeta - Q_\zeta P_\eta \ ; \quad L_y = Q_\zeta P_\xi - Q_\xi P_\zeta \ ; \quad L_z = Q_\xi P_\eta - Q_\eta P_\xi \ . \qquad (4.1.31)$$

As can be seen from (4.1.30), to second order the vibronic coupling problem is fully analogous to the LS-coupling of the orbital and spin momenta in the theory of atoms, and this allows us to obtain directly an expression for the energy levels split by the vibronic interaction [4.3]

$$E_{nlj} = \hbar\omega_T(n + \tfrac{3}{2}) + \tfrac{3}{8}E_{JT}(T)[j(j+1) - l(l+1) - 6] \ . \qquad (4.1.32)$$

Here l is the quantum number of the vibrational momentum L^2 and varies from 0 or 1 to n with a spacing $\Delta l = 2$, j is the quantum number of the total momentum $\hat{\boldsymbol{J}} = \boldsymbol{L} + \hat{\boldsymbol{S}}$, $j = l+1, l, l-1$, if $l > j$, and $j = 1$, if $l = 0$. The degeneracy of each state is $2j + 1$. The energy distance between the extreme components of the split multiplet is $\Delta E = \tfrac{3}{4}E_{JT}(T)(2n+1)$ and thus the criterion for validity of the perturbation theory result (4.1.32) is

$$E_{JT}(T)/\hbar\omega_T \ll \tfrac{4}{3}(2n+1)^{-1} \ . \qquad (4.1.33)$$

As in the case of the $E \otimes e$ problem, in the $T \otimes t_2$ problem the ground state ($n = 0, l = 0, j = 1, 2j + 1 = 3$) is not split by the vibronic interaction: the initial threefold electronic degeneracy is not removed but transformed into the threefold vibronic degeneracy.

The Hamiltonian of the linear $T \otimes e$ problem is given by (3.3.8) with $V_T = 0$. The solution of this problem is quite similar to that given above for the linear $E \otimes b_1$ (or $E \otimes b_2$) problem. It also can be evaluated exactly, provided the operator of nonadiabaticity is zero (Sect. 3.3). The shift transformation (4.1.3) with

$$\hat{S} = -\frac{V_E}{\hbar\omega_E^2}(P_\theta \hat{C}_\theta + P_\varepsilon \hat{C}_\varepsilon) , \tag{4.1.34}$$

where the $\hat{C}_{E\gamma}$ matrices are given by (3.3.2), may be performed exactly:

$$e^{i\hat{S}}\hat{H}e^{-i\hat{S}} = \tfrac{1}{2}[P_\theta^2 + P_\varepsilon^2 + \omega_E^2(Q_\theta^2 + Q_\varepsilon^2)]\hat{C}_A - E_{\mathrm{JT}}(E)\hat{C}_A . \tag{4.1.35}$$

Here $\hat{C}_A$ is 3×3 unit matrix, and $E_{\mathrm{JT}}(E)$ is given by (3.3.10). It is seen that the interaction with E vibrations does not change the energy spectrum but shifts all the energy levels by $E_{\mathrm{JT}}(E)$. The eigenfunctions of the Hamiltonian (4.1.35) can be obtained directly, and have an oscillator form.

This result can be used for calculating the influence of weak coupling to T_2 vibrations in the $T \otimes (e + t_2)$ problem with arbitrary coupling to E vibrations. The solution of this problem which results in simple expressions for the second-order corrections to the energy levels and wave functions of several lowest vibronic states (as well as some other results), is given in [4.4]. As in the above cases, the simultaneous vibronic coupling to E and T_2 vibrations does not split the ground state, and its initial electronic threefold degeneracy transforms into the vibronic one.

Note that all the above perturbation theory solutions of the $T \otimes e$, $T \otimes t_2$ and $T \otimes (e + t_2)$ problems are valid without taking into account the actual spin-orbital interaction, which would lead to a complicated combination of Jahn-Teller and pseudo–Jahn-Teller effects.

Analogous results can be obtained for the Jahn-Teller effect for a cubic electronic quadruplet Γ_8. As discussed in Sect. 3.3, the Hamiltonians of the appropriate $\Gamma_8 \otimes e$ and $\Gamma_8 \otimes t_2$ problems can be obtained from the general Hamiltonian (3.3.8) of the linear $\Gamma_8 \otimes (e + t_2)$ problem with $\hat{C}_{\Gamma\gamma}$ matrices given by (3.3.30).

The $\Gamma_8 \otimes e$ problem is reduced to the $E \otimes e$ one [4.5]. Indeed, the Γ_8 term can be considered as a ^2E term with nonzero spin-orbital interaction. Since the spin states are not related directly to the $E \otimes e$ Jahn-Teller effect, the electronic 4×4 matrix of the $H_2(E)$ Hamiltonian decomposes into two independent blocks of the type (4.1.15), each of which is related to one of the two spin states. Accordingly, all the results presented in (4.1.16) are also valid for the $\Gamma_8 \otimes e$ problem, with the only difference being that all the twofold degenerate vibronic levels are additionally twofold degenerate by spin, and are thus quadruplets Γ_8, instead of E terms, and coincidentally degenerate doublets $\Gamma_6 + \Gamma_7$, instead of $A_1 + A_2$ (Fig. 4.1).

The $\Gamma_8 \otimes t_2$ problem is formally analogous to the $T \otimes t_2$ one, and, in the second-order perturbation theory approximation, can be described by the Hamiltonian (4.1.30) with electronic 4 × 4 matrices $\hat{S}_x$, $\hat{S}_y$ and $\hat{S}_z$ corresponding to the pseudospin 3/2. Again, the problem is reduced to that of $\boldsymbol{L}\hat{\boldsymbol{S}}$-coupling of momenta, and with an accuracy up to an additive constant results in the scheme of energy levels described by the formula (4.1.32) with the quantum number $j = l \pm \frac{3}{2}, l \pm \frac{1}{2}$, if $l > 1$, $j = \frac{3}{2} \pm 1$, if $l = 1$, and $j = \frac{3}{2}$, if $l = 0$ [4.6].

4.1.4 Weak Pseudo–Jahn-Teller Effect

Let us pass now to the pseudo–Jahn-Teller effect. Consider the simplest case of two singlet electronic states mixed by one nondegenerate vibration. The Hamiltonian of the system (3.4.1) can be presented in the form $\hat{H}_0 + V$, where

$$\hat{H}_0 = \tfrac{1}{2}(P^2 + Q^2\omega^2)\hat{\sigma}_0 + \Delta\hat{\sigma}_z \tag{4.1.36}$$

and $\hat{V} = VQ\hat{\sigma}_x$. Solving (4.1.5) again, we find [4.7]:

$$\hat{S} = \frac{V(2\Delta Q\hat{\sigma}_y + \hbar P\hat{\sigma}_x)}{4\Delta^2 - \hbar^2\omega^2}\;, \tag{4.1.3}$$

and substituting this expression in (4.1.6) we obtain the approximate Hamiltonian of the second-order perturbation theory with respect to V:

$$\hat{H}_2 = \frac{1}{2}(P^2 + \omega^2 Q^2)\hat{\sigma}_0 + \Delta\hat{\sigma}_z + \frac{V^2(4\Delta Q^2\hat{\sigma}_z + \hbar^2)}{2(4\Delta^2 - \hbar^2\omega^2)}\;. \tag{4.1.38}$$

This Hamiltonian contains only diagonal matrices and therefore, if one writes out the matrix Schrödinger equation, the system of vibronic equations (2.1.4) decouples into two equations which correspond to harmonic oscillations with redetermined frequencies and a common shift of the energy zero point:

$$H_\pm = \frac{1}{2}(P^2 + \omega_\pm^2 Q^2) + \frac{\hbar^2 V^2}{2(4\Delta^2 - \hbar^2\omega^2)} \pm \Delta\;. \tag{4.1.39}$$

Here we write

$$\omega_\pm^2 = \omega^2 \pm \frac{4V^2\Delta}{4\Delta^2 - \hbar^2\omega^2}\;. \tag{4.1.40}$$

If $2\Delta \gg \hbar\omega$ one can neglect the quantity $\hbar^2\omega^2$ in the denominator. Then this expression coincides with the curvature of the adiabatic potential at the point $Q = 0$ (Sect. 3.4.1). This coincidence is not accidental. It can be explained by the fact that the condition $2\Delta \gg \hbar\omega$ is equivalent to the criterion (2.1.10) for the applicability of the adiabatic approximation in the case under consideration. If this criterion is fulfilled, the nonadiabatic mixing of the electronic states can be neglected and the nuclear dynamics is determined by the shape of the adiabatic potential sheets (Fig. 3.8a) discussed in Sect. 3.4.1.

The Schrödinger equation with the Hamiltonian (4.1.39) results in well-known oscillator-type solutions. The energy spectrum of the system is described by the expression [4.7, 4.8]

$$E_n^{(\pm)} = \hbar\omega_{\pm}(n + \tfrac{1}{2}) \pm \Delta + \frac{\hbar^2 V^2}{2(4\Delta^2 - \hbar^2\omega^2)} \; . \tag{4.1.41}$$

It follows from this equation under the conditions of resonance $2\Delta \approx \hbar\omega$ that the magnitude of $\hat{S}$ can be very large and the cutoff of the expansion (4.1.4) after quadratic terms thus becomes invalid. This conclusion follows also from (4.1.40) and (4.1.41) from which the criterion for the applicability of the perturbation theory developed above can be obtained easily:

$$\frac{E_{JT}}{\hbar\omega}\,\frac{\hbar^2\omega^2}{|4\Delta^2 - \hbar^2\omega^2|} \ll 1 \; , \tag{4.1.42}$$

where $E_{JT} = V^2/2\omega^2$. It is clear that if $2\Delta \approx \hbar\omega$ the case of (quasi) degeneracy occurs (Fig. 4.3), and the perturbation theory has to be applied in another form.

The perturbation matrix in the basis of quasidegenerate states $\psi_-(\mathrm{r})|n\rangle$ and $\psi_+(\mathrm{r})|n-1\rangle$ has the form

$$\begin{pmatrix} \delta/2 & V\sqrt{\dfrac{\hbar n}{2\omega}} \\ V\sqrt{\dfrac{\hbar n}{2\omega}} & -\delta/2 \end{pmatrix} ,$$

where $\delta = \hbar\omega - 2\Delta$ is the distance from resonance (Fig. 4.3). From this we find $(n \neq 0)$

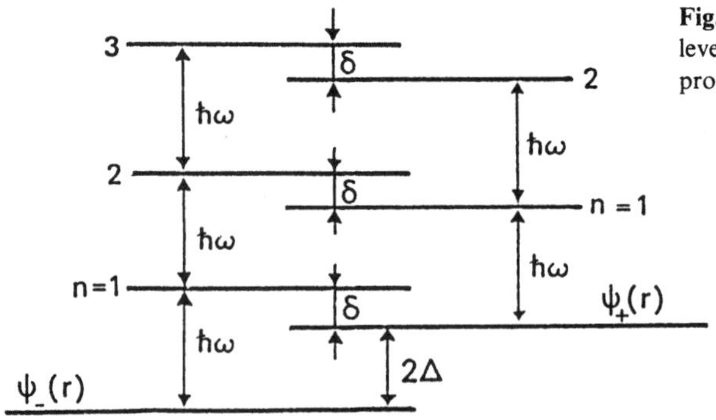

Fig. 4.3. Lowest vibrational energy level scheme for the $(A_1 + B_1) \otimes b_1$ problem without vibronic coupling

$$E_n^{(\pm)} = n\hbar\omega \pm \frac{1}{2}\sqrt{\delta^2 + 2V^2\frac{\hbar n}{\omega}} ,$$

and in the case of exact resonance ($\delta = 0$),

$$E_n^{(\pm)} = n\hbar\omega \pm |V|\sqrt{n\hbar/2\omega},$$

i.e., the correction to the energy is linearly dependent on V.

The case of resonance considered above is of special importance in multimode vibronic systems where among the set of frequencies ω_n there is, as a rule, one frequency coinciding with the appropriate Bohr frequency $2\Delta/\hbar$ (for details see Sect. 4.5).

The vibronic states of a Renner system within the framework of perturbation theory are considered in [4.9, 10].

4.2 Strong Vibronic Coupling. Free Rotation of Distortions

A relatively simple analytic solution and visual physical interpretation is also possible in the other limiting case, that of strong vibronic coupling, which occurs when $E_{JT} \gg \hbar\omega$. In this case the inequality (2.1.10) is valid for a large region near the minima of the adiabatic potential, which thus takes on the meaning of nuclear potential energy surfaces. In other words, in the case of strong vibronic coupling the adiabatic approximation becomes valid again, at least for the lowest vibronic states, and the electronic and nuclear motions can approximately be separated.

The evaluation of the adiabatic electronic states $\varphi(\mathrm{r}, Q)$ is, in general, not difficult (at least in principle). The major difficulties occur in the solution of the Schrödinger equation (2.1.7) determining the nuclear dynamics. The following two cases are possible: (a) the lowest sheet of the adiabatic potential has a continuum of equipotential minima, and (b) there are a finite number of minima divided by potential barriers.

The coordinates of the minima points are proportional to the constant of linear vibronic coupling (Chap. 3), and if the latter is large enough, as assumed in this section, the quadratic terms of vibronic interaction cannot in general be neglected. In some special cases the constant of quadratic vibronic coupling turns out to be very small and then, under certain conditions, the second-order vibronic interaction term may be neglected. For instance, this can be done in the case of the $(E \otimes e)$–type Jahn-Teller effect when due to the negligible quadratic vibronic constant the barrier height between the minima along the bottom of the trough of the adiabatic potential (Sect. 3.1) is much smaller than the energy spacing in the linear problem.

4.2.1 The Adiabatic Separation of Nuclear Motion. Elastically and Centrifugally Stabilized States

The simplest example of the case a) is provided by the $E \otimes e$ problem, its Hamiltonian being given by (3.2.1) with $E_0 = 0$. In polar coordinates $Q_\theta = \varrho \cos\varphi$ and $Q_\varepsilon = \varrho \sin\varphi$ it acquires the following form:

$$\hat{H} = \left\{ -\frac{1}{2}\hbar^2 \left[\frac{1}{\varrho}\frac{\partial}{\partial \varrho}\left(\varrho \frac{\partial}{\partial \varrho}\right) + \frac{1}{\varrho^2}\frac{\partial^2}{\partial \varphi^2} \right] + \frac{1}{2}\omega_E^2 \varrho^2 \right\} \hat{\sigma}_0 + V_E \varrho \begin{pmatrix} 0 & e^{-i\varphi} \\ e^{i\varphi} & 0 \end{pmatrix} . \tag{4.2.1}$$

The eigenfunction (spinor) of this Hamiltonian can be written as $\Psi = \Phi/\sqrt{\varrho}$. Then for $\Phi(\varrho, \varphi)$ the Hamiltonian becomes

$$\hat{H} = \left(-\frac{\hbar^2}{2}\frac{\partial^2}{\partial \varrho^2} - \frac{\hbar^2}{8\varrho^2} + \frac{\hbar^2 L_z^2}{2\varrho^2} + \frac{1}{2}\omega_E^2 \varrho^2 \right) \hat{\sigma}_0 + V_E \varrho \begin{pmatrix} 0 & e^{-i\varphi} \\ e^{i\varphi} & 0 \end{pmatrix} , \tag{4.2.2}$$

where $L_z = -i\partial/\partial\varphi$. Under the unitary transformation (3.1.9) which diagonalizes the matrix of the potential energy, the operator L_z transforms to $L_z - \frac{1}{2}\hat{\sigma}_x$ (see (3.2.26)). Therefore for the Hamiltonian (4.2.2) in the adiabatic electronic basis taking (3.1.10) into consideration we obtain:

$$\hat{S}^+ \hat{H} \hat{S} = \left(-\frac{\hbar^2}{2}\frac{\partial^2}{\partial \varrho^2} + \frac{1}{2}\omega_E^2 \varrho^2 - \frac{\hbar^2}{8\varrho^2} \right) \hat{\sigma}_0 + \frac{\hbar^2}{2\varrho^2}\left(L_z - \frac{1}{2}\hat{\sigma}_x \right)^2 + V_E \varrho \hat{\sigma}_z . \tag{4.2.3}$$

As shown in Sect. 3.2.1, the Hamiltonian of the linear $E \otimes e$ problem has an integral of motion $\hat{J}_z = L_z + \frac{1}{2}\hat{\sigma}_z$. In the adiabatic basis $\hat{S}^+ \hat{J}_z \hat{S} = L_z$. Indeed, one can easily check that the Hamiltonian (4.2.3) commutes with the operator L_z. This means that the eigenfunctions $e^{ij\varphi}/\sqrt{2\pi}$ of the operator L_z are also eigenfunctions of the Hamiltonian (4.2.3). In other words, the angular and radial motions can be separated and the wave function $\Phi(\varrho, \varphi)$ can be taken in the form

$$\Phi(\varrho, \varphi) = \frac{1}{\sqrt{2\pi}} e^{ij\varphi} \chi(\varrho) , \tag{4.2.4}$$

while the operator $L_z = \hat{S}^+ \hat{J}_z \hat{S}$ in (4.2.3) can be substituted by its quantum number j [4.5, 4.11]

$$\hat{H} = \left(-\frac{\hbar^2}{2}\frac{\partial^2}{\partial \varrho^2} + \frac{\hbar^2 j^2}{2\varrho^2} + \frac{1}{2}\omega_E^2 \varrho^2 \right) \hat{\sigma}_0 - \frac{\hbar^2 j}{2\varrho^2}\hat{\sigma}_x + V_E \varrho \hat{\sigma}_z . \tag{4.2.5}$$

Note that when passing from (4.2.1) to (4.2.5) no additional simplifications are introduced, and in this sense (4.2.5) is the exact Hamiltonian. Here the adiabatic approximation is equivalent to neglecting the term $-\hbar^2 j\hat{\sigma}_x/2\varrho^2$ which mixes the two sheets of the adiabatic potential (the criterion is discussed below). Then

$$H_{\pm} = -\frac{\hbar^2}{2}\frac{\partial^2}{\partial \varrho^2} + E_{\pm}(\varrho) \; ; \qquad E_{\pm}(\varrho) = \frac{\hbar^2 j^2}{2\varrho^2} + \frac{1}{2}\omega_E^2 \varrho^2 \pm |V_E| \varrho \; . \tag{4.2.6}$$

The potential energy $E_{\pm}(\varrho)$ differs here from that in (3.1.11) by the *centrifugal energy* term $\hbar^2 j^2/2\varrho^2$. This term is always nonzero due to semi-integer values of the quantum number j, $j = \pm 1/2, \pm 3/2, \pm 5/2, \ldots$. The latter statement is of great importance and will be given additional proof.

The quantum number j occurs as the eigenvalue of the operator $\hat{J}_z = L_z + \frac{1}{2}\hat{\sigma}_z$, and therefore j is always semi-integer. The unitary transformation to the adiabatic basis changing $\hat{J}_z$ to L_z does not change the eigenvalues of $\hat{J}_z$ and hence j remains semi-integer. The fact that j has only semi-integer values can also be proved as follows. The adiabatic electronic functions

$$\psi_{\pm}(\mathrm{r}, Q) = \begin{cases} \dfrac{1}{\sqrt{2}}[\mathrm{e}^{-\mathrm{i}\varphi/2}\psi_+(\mathrm{r}) + \mathrm{e}^{\mathrm{i}\varphi/2}\psi_-(\mathrm{r})] \; , \\ \dfrac{1}{\sqrt{2}}[\mathrm{e}^{-\mathrm{i}\varphi/2}\psi_+(\mathrm{r}) - \mathrm{e}^{\mathrm{i}\varphi/2}\psi_-(\mathrm{r})] \; , \end{cases} \tag{4.2.7}$$

change their sign under the rotation transformation $\varphi \to \varphi + 2\pi$. Since the total wave function

$$|\pm, n, j\rangle = \psi_{\pm}(\mathrm{r}, Q)\mathrm{e}^{\mathrm{i}j\varphi}\chi_n(\varrho)/\sqrt{2\pi\varrho} \tag{4.2.8}$$

has to remain unchanged by this transformation, j must be semi-integer.

The potential curves (4.2.6) are shown in Fig. 4.4. The centrifugal energy is especially important for small values of ϱ, in particular, for the lowest vibronic

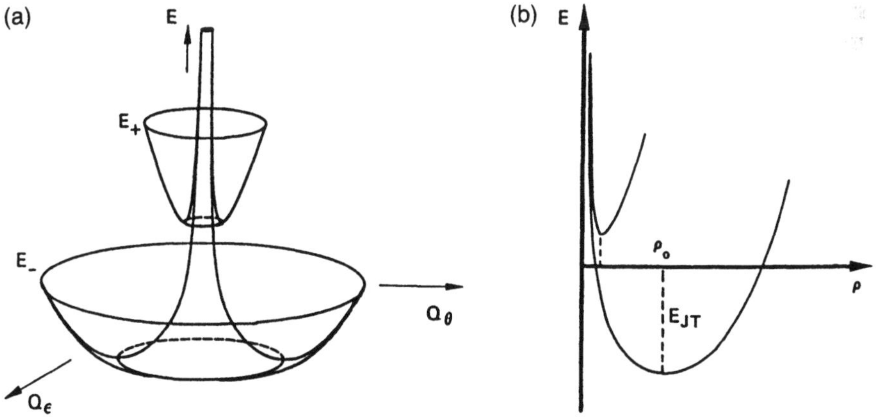

Fig. 4.4a, b. Double-sheet adiabatic potential in the linear ($E \otimes e$)-type Jahn-Teller effect including centrifugal energy. (**a**) General view of E_+ (upper sheet) and E_- (Mexican-hat-type lower sheet). (**b**) Two-valued adiabatic potential of radial motion, i.e., radial cross section of the potential energy sufaces shown in (**a**). (After [4.18])

states of the upper sheet of the adiabatic potential. On the other hand, in the lowest states of the lower sheet the motion of the system is to a great extent localized near the bottom of the adiabatic potential, in the neighbourhood of the minimum point

$$\varrho_0 \cong \frac{|V_E|}{\omega_E^2}\left(1 + \frac{\hbar^2 j^2 \omega_E^6}{V_E^4}\right) , \tag{4.2.9}$$

and for $|V_E|$ values large enough the contribution of the centrifugal energy can be neglected. Note that if $|V_E| \to \infty$, the expression (4.2.9) agrees with the ϱ_0 value from (3.1.13).

Analogously, the lowest vibronic states of the upper sheet of the adiabatic potential are localized near the minimum at the point $\varrho_0^{(+)}$,

$$\varrho_0^{(+)} \approx \left(\frac{\hbar^2 j^2}{|V_E|}\right)^{1/3}\left[1 - \frac{\omega_E^2}{3|V_E|}\left(\frac{\hbar^2 j^2}{|V_E|}\right)^{1/3}\right] . \tag{4.2.10}$$

In the limit of strong vibronic coupling $\varrho_0^{(+)} \approx 0$. This means that in the neighbourhood of the minimum, i.e., at $\varrho \approx \varrho_0^{(+)}$, the elastic energy $\frac{1}{2}\omega_E^2\varrho^2$ in the Hamiltonian H_+ of the upper sheet (see (4.2.6)) can be neglected. The minimum itself results from the compensation of the centrifugal force $\hbar^2 j^2/\varrho^3$ by the Jahn-Teller force $-|V_E|$. From this it follows that the lowest states of the upper sheet are in principle different from those of the lower sheet, where, in contrast to the upper one, the minima occur due to the compensation of the Jahn-Teller distortive force $|V_E|$ by the elastic one $-\omega_E^2\varrho$. Accordingly, the vibronic states localized near the minima of the upper sheet are called *centrifugally stabilized states*, as distinguished from *elastically stabilized states* of the lower sheet [4.12].

For the lowest localized states the potential energy (4.2.6) can be replaced by the first terms of its power series expansion with respect to the displacements $\mathrm{r} = \varrho - \varrho_0$. In the harmonic approximation (i.e., keeping the terms up to second order) with an accuracy of the order of $\sim V_E^{-2}$, one can obtain from (4.2.6) the following Schrödinger equation for the lowest sheet:

$$\left(-\frac{\hbar^2}{2}\frac{\partial^2}{\partial \mathrm{r}^2} + \frac{1}{2}\omega_E^2\mathrm{r}^2 - E_{\mathrm{JT}} + \frac{\hbar^2 j^2}{2\varrho_0^2}\right)\chi(\mathrm{r}) = E\chi(\mathrm{r}) , \tag{4.2.11}$$

where E_{JT} is given by (3.1.13). Since the latter has an oscillator form (the terms E_{JT} and $\hbar^2 j^2/2\varrho_0^2$ are constants), its solutions are known. The energies of the vibronic states are [4.13]

$$E_{nj} = \hbar\omega_E(n + \tfrac{1}{2}) + \frac{\hbar^2 j^2}{2\varrho_0^2} - E_{\mathrm{JT}} , \tag{4.2.12}$$

while the wave functions $\chi(\mathrm{r})$ can be expressed as Hermite polynomials describing harmonic vibrations near the point ϱ_0 with the frequency ω_E.

Analogously, the Schrödinger equation for the upper sheet states in the harmonic approximation is

$$\left[-\frac{\hbar^2}{2}\frac{\partial^2}{\partial r^2}+\frac{3}{2}\left(\frac{V_E^2}{\hbar|j|}\right)^{2/3} r^2+\frac{3}{2}(\hbar^2 j^2 V_E^2)^{1/3}\right]\chi^{(+)}=E\chi^{(+)}\ , \tag{4.2.13}$$

their energy levels being

$$E_{nj}^{(+)}=\hbar\sqrt{3}\left(\frac{V_E^2}{\hbar|j|}\right)^{1/3}(n+\tfrac{1}{2})+\tfrac{3}{2}(\hbar^2 j^2 V_E^2)^{1/3}\ . \tag{4.2.14}$$

As can be seen, e.g., from Fig. 4.4, for the centrifugally stabilized states of the upper sheet the anharmonicity may be significant, especially for high energy levels in the case of strong vibronic coupling. Therefore the harmonic approximation of (4.2.13) may serve as a crude estimation of the ground and first excited states of the upper sheet. The nonadiabatic mixing with the lower sheet of the adiabatic potential introduces additional limitations (see below). The centrifugally stabilized states of the upper sheet are of special interest in the case of multimode Jahn-Teller systems, where they are manifest in peculiar pseudolocal resonances, sometimes called Slonczewski resonances (Sect. 4.5).

The wave functions (4.2.8) and the energies (4.2.12) can be given a clear-cut physical interpretation. As mentioned earlier, the Jahn-Teller distortions of the molecular framework are subject to certain dynamics. In case of the linear $E\otimes e$ problem under consideration, the magnitude of distortion oscillates harmonically with the frequency ω_E appropriate to the radial motion in the trough, and *the wave of distortion moves freely along the trough* (see Fig. 3.2 and the main text of Sect. 3.1). The oscillations are described by the oscillator function (the factor $\chi_n(\varrho-\varrho_0)$ of (4.2.8)), while the appropriate contribution to the energy is given by the term $E_{\text{vib}}=\hbar\omega_E(n+\frac{1}{2})$. The rotation of the wave of distortion around the geometric center of the molecule is described by the factor $\exp(\mathrm{i}j\varphi)/\sqrt{2\pi}$ in (4.2.8), and its contribution to the energy in (4.2.12) is given by the usual centrifugal term $\hbar^2 j^2/2\varrho_0^2$. The electronic wave function $\psi_-(\mathrm{r},Q)$ in (4.2.8) (the corresponding energy being $E_{\text{el}}=-E_{\text{JT}}$) follows adiabatically the molecular framework distortion running around the center of the molecule.

The energy gap ΔE_{el} between the two sheets of the adiabatic potential at the minima is $\sim V_E^2$ (see (3.1.14)), the spacing of the vibrational levels is $\Delta E_{\text{vib}}=\hbar\omega_E$, and the energy intervals between the rotational levels are $\Delta E_{\text{rot}}\sim V_E^{-2}$. Therefore in the case of strong vibronic coupling

$$\Delta E_{\text{el}}\gg\Delta E_{\text{vib}}\gg\Delta E_{\text{rot}}\ .$$

These inequalities reflect the particular partitioning of the energy levels in the case of a Jahn-Teller molecule with strong vibronic coupling and a continuum of minima along a trough of the adiabatic potential. Similar to the case in a normal molecule, the lowest electronic state is accompanied by an equally spaced spectrum of *radial harmonic vibrations*, and each of the vibrational levels is accompanied by a *fine rotational structure*. But unlike the case in normal molecules, the vibronic rotational states under consideration do not describe real rotations of the molecule, when all the atoms move around a common axis, but the rotation of the wave of distortions when each of the atoms moves around its own axis (Fig. 3.2).

Note that the fine rotational structure of the vibronic levels can be obtained also in a more simple (although less rigorous) way. As mentioned in Sect. 3.1, the motion of the wave of distortion of the molecular framework can be represented as rotations of the star formed by the vectors of atomic displacements, or (equivalently) as rotations of a rigid equilateral triangle formed by the apices of this star. The moment of inertia of such a "molecule" is $I = \varrho_0^2$, the masses of the atoms (in the chosen units) being a unit. The appropriate Hamiltonian of these rotations is $H_{\rm rot} = \hbar^2 L_z^2/(2I)$. The rotational energy can be obtained by a simple averaging of this Hamiltonian over a state with a certain value of j.

The fine rotational structure of the vibronic levels in more complicated cases which also have a trough on the lower sheet of the adiabatic potential can be obtained in a similar way. For instance, in the case of the $T \otimes (e + t_2)$ problem in cubic systems with $E_{\rm JT}(E) = E_{\rm JT}(T)$, the motion of the wave of distortion of the molecular framework along the minima of the two-dimensional trough can be represented as rotations of a rigid octahedron. The moment of inertia of such a "spherical top" is $I = 3V^2/4\omega^4$, and the Hamiltonian of these two-dimensional rotations is

$$H_{\rm rot} = \frac{\hbar^2}{2I} \boldsymbol{L}^2 \ , \qquad (4.2.15)$$

where the momentum vector $\boldsymbol{L}$ is determined by (3.3.4) with the relations between $Q_{\Gamma\gamma}$ and the rotation angles θ and φ given by (3.3.6). Substituting (3.3.6) into (3.3.4) we easily find that $\boldsymbol{L} = \frac{1}{2}\boldsymbol{l}$, where $\boldsymbol{l}$ is the usual operator expression for the orbital momentum in spherical coordinates. By averaging (4.2.15) over a state with a certain value of l^2, we obtain [4.14, 4.15, 4.16]

$$E_{\rm rot} = \frac{\hbar^2\omega^4}{6V^2} l(l+1) \ . \qquad (4.2.16)$$

The adiabatic electronic wave function of the lower sheet of the adiabatic potential in this case can be represented as

$$\psi(\mathrm{r}, Q) = \cos\varphi \sin\theta \psi_x(\mathrm{r}) + \sin\varphi \sin\theta \psi_y(\mathrm{r}) + \cos\theta \psi_z(\mathrm{r}) \ ,$$

where $\psi_x(\mathrm{r})$, $\psi_y(\mathrm{r})$ and $\psi_z(\mathrm{r})$ are the initial electronic functions of the T term. Under inversion, $\theta \to \pi - \theta$ and $\varphi \to \varphi + \pi$, the function $\psi(\mathrm{r}, Q)$ changes its sign, while the spherical nuclear functions $Y_{lm}(\theta, \varphi)$ corresponding to the energy levels (4.2.16) are multiplied by $(-1)^l$. Hence in order to keep the total wave function

$\psi(\mathrm{r}, Q) Y_{lm}(\theta, \varphi)$ single-valued, only the terms with odd values of l have to be retained. Each of the energy levels (4.2.16) is $(2l + 1)$-fold degenerate, appropriate to $2l + 1$ values of $l_z = m$ $(2l + 1)$ momentum directions. The same result can be obtained, of course, by directly solving the dynamic problem, i.e., by substituting the spherical coordinates into the Hamiltonian (3.3.1) and then separating the variables, as has been done above for the linear $E \otimes e$ problem [4.14, 15].

It follows from (4.2.12) that all the vibronic states of the linear $E \otimes e$ problem, including the ground state,

$$|-, 0, \pm\tfrac{1}{2}\rangle = \psi_-(\mathrm{r}, Q) \mathrm{e}^{\pm \mathrm{i}\varphi/2} \chi_0(\varrho - \varrho_0)/\sqrt{2\pi\varrho} \ , \tag{4.2.17}$$

are doubly degenerate with respect to the sign of the quantum number j—the sign of the projection of the total electron-vibrational momentum on the axis of symmetry of the molecule. This degeneracy is similar to the Kramers one and is due to the above restriction of the problem to the linear vibronic coupling terms only, which "accidentally" causes a higher symmetry of the Hamiltonian than the reference symmetry (Sect. 3.2).

4.2.2 Quasi-Classical Approach to Jahn-Teller Problems

For highly excited states, satisfactory results can be obtained in the quasi-classical approximation. The energies of the vibronic states can be obtained from the Born-Sommerfeld condition of quantization [Ref. 4.17; §48], provided the nonadiabatic mixing is neglected

$$\oint \sqrt{2[E - E_\pm(\varrho)]}\, d\varrho = 2\pi\hbar(n + \tfrac{1}{2}) \ . \tag{4.2.18}$$

Here $E_\pm(\varrho)$ are given by (4.2.6). Passing to the dimensionless coordinate $x = \varrho(\omega_E/\hbar)^{1/2}$ and energy $\mathscr{E} = E/\hbar\omega_E$ one can obtain

$$\frac{1}{\pi\hbar} S_\pm(\mathscr{E}) = \int_{a_\pm}^{b_\pm} \sqrt{2\left[\mathscr{E} - \left(\frac{j^2}{2x^2} + \frac{1}{2}x^2 \pm k_E x\right)\right]}\, \frac{dx}{\pi} = n + \frac{1}{2} \ , \tag{4.2.19}$$

where $S_\pm(\mathscr{E})$ is the shortcut action for the upper and lower sheets of the adiabatic potential respectively, and $a_\pm$ and $b_\pm$ are the classical turning points at which the integrand in (4.2.19) vanishes.

The solution of the transcendental equation (4.2.19) is given graphically in Fig. 4.5. The almost linear function $S_-(\mathscr{E})$ shows that for the highly excited vibronic states of the lower sheet of the adiabatic potential, the virbronic energy spectrum corresponding to a certain value of j is almost equally spaced. It means that the result of (4.2.12) has in fact a larger range of validity than that determined by the criterion of the harmonic approximation used when it was deduced. The equally spaced energy levels of the lower sheet are, in accordance with (4.2.12), almost linearly decreasing with increasing V_E^2 (or k_E^2), while those of the upper sheet are increasing. For some of the k_E values some of the upper sheet energy levels

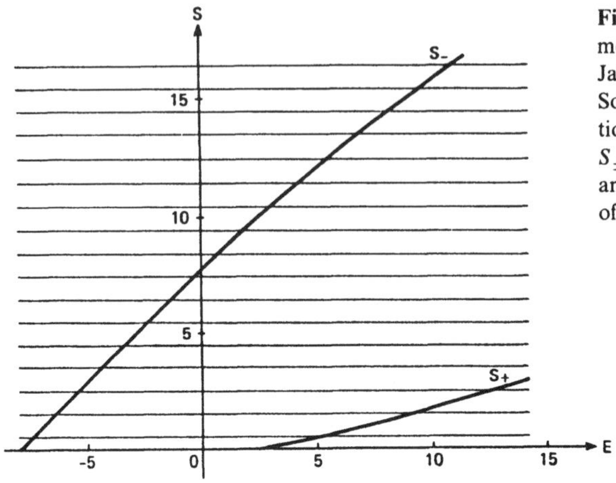

Fig. 4.5. Quantization of the radial motion in the linear ($E \otimes e$)-type Jahn-Teller effect after Bohr and Sommerfeld: the graphical solution of the transcendental equation $S_{\pm}(E) = 2\pi\hbar(n + \frac{1}{2})$. The energy E and the action $S_{\pm}$ are plotted in units of $\hbar\omega_E$ and $\pi\hbar$ respectively

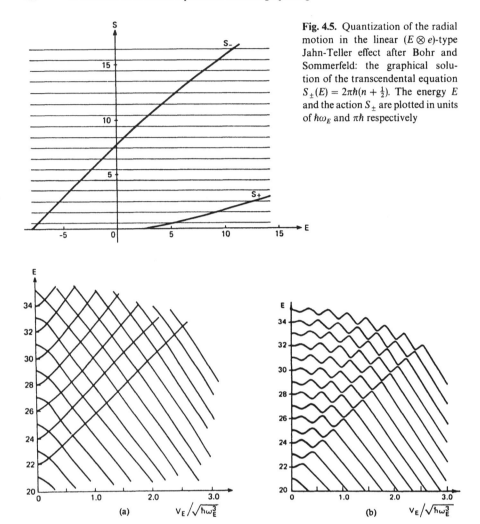

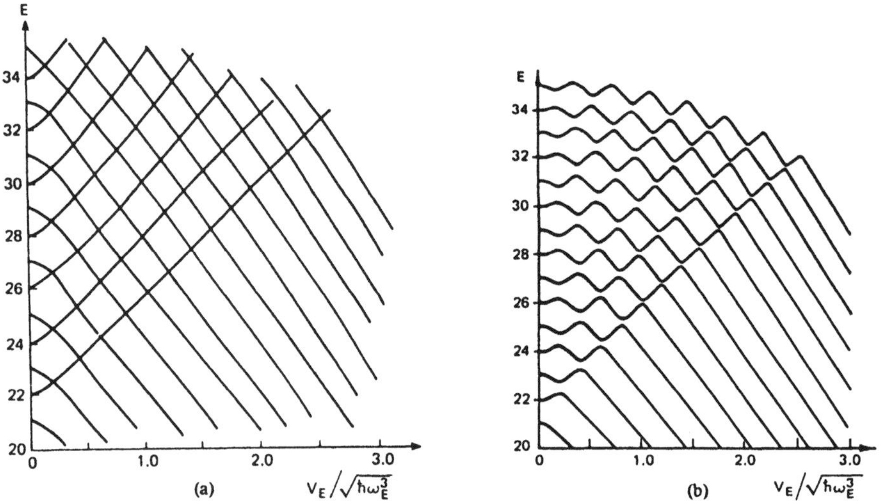

Fig. 4.6a, b. Vibronic energy levels (in units of $\hbar\omega_E$) versus the dimensionless vibronic constant in the $E \otimes e$ problem for $j = 41/2$ without (**a**) and with (**b**) including nonadiabaticity [4.19]

intersect with that of the lower sheet, and an accidental degeneracy arises. At these points the nonadiabaticity becomes important since it removes the degeneracies (Fig. 4.6).

The magnitude of the nonadiabatic splitting can be estimated in the perturbation theory approximation. It proves to be proportional to $|W_{+-}|$, the matrix element of the nonadiabaticity operator $-\hbar^2 j\sigma_x/(2\varrho^2)$ calculated with the wave functions of the excited degenerate states. If quasi-classical functions are used, the appropriate integral can be estimated by means of the bridge-wall method

[Ref. 4.17; §51]. It is obvious that the largest contribution to the integral arises from the region of small ϱ values where the phases of the rapidly oscillating wave functions of the upper and lower sheets are similar. By integrating along the path around the singular point $\varrho = 0$ it can be shown that the matrix element $|W_{+-}|$ is, as expected, exponentially small [4.19]:

$$W_{+-} \sim \exp(-\tau) \ . \tag{4.2.20}$$

Here

$$\tau = \int_0^{a_+} \sqrt{2\left[\frac{j^2}{2x^2} + \frac{1}{2}x^2 + k_E x - \mathscr{E}\right]}\,dx - \int_0^{a_-} \sqrt{2\left[\frac{j^2}{2x^2} + \frac{1}{2}x^2 - k_E x - \mathscr{E}\right]}\,dx \ . \tag{4.2.21}$$

In particular, for circular trajectories on the upper sheet, i.e., when the rotational quantum number is much larger than the quantum number of radial vibrations n,

$$\tau \approx (\pi - \ln 4)j/4 \approx 0.44j \ . \tag{4.2.22}$$

It follows from (4.2.20) that the criterion for applicability of the adiabatic approximation for the highly excited states under consideration can be expressed as

$$\tau \gg 1 \ .$$

For instance, for the lower vibrational states of the upper sheet ($n \sim 1$) the adiabatic approximation is valid only for rather large j values (see (4.2.22)).

A more rigorous treatment of the nonadiabatic mixing in the framework of the quasi-classical approximation is given in [4.19]. From this work it follows in particular that, allowing for the nonadiabaticity, the condition of quantization (4.2.19) has to be changed to

$$\cos\left(\frac{S_+}{\hbar}\right)\cos\left(\frac{S_-}{\hbar}\right) = \pm\frac{1}{4}W_{+-} \ . \tag{4.2.23}$$

The expression transforms to (4.2.19) when assuming that $W_{+-} \to 0$.

A somewhat different approach to the problem is developed in [4.20]. In this work the operator $-\hbar^2 j\sigma_x/2\varrho^2$ is included in the potential energy matrix. The adiabatic potentials are thus redetermined, resulting in, cf. (4.2.6),

$$\tilde{E}_{\pm}(\varrho) = \frac{\hbar^2 j^2}{2\varrho^2} + \frac{1}{2}\omega_E^2\varrho^2 \pm \sqrt{\frac{\hbar^2 j^2}{4\varrho^4} + V_E^2\varrho^2} \ .$$

The change of the potential energy due to this redetermination is most important for small ϱ values, especially for centrifugally stabilized states of the upper sheet localized at the minimum, cf. (4.2.10),

$$\tilde{\varrho}_0^{(+)} = \left[\frac{\hbar^4 j^2}{2V_E^2}\left(j^2 + 1 + \sqrt{j^4 + 3j^2}\right)\right]^{1/6} .$$

On the other and, for the lower sheet the result (4.2.9) remains valid and the lower vibronic states localized near the point ϱ_0 are not affected by the above redetermination of the adiabatic potential. This is due to the fact that for $\varrho \sim |V_E|$ the contribution of the nonadiabaticity is negligibly small.

4.3 Hindered Rotations and Pulse Motions of Distortions. Tunneling Splitting

Consider now the case when instead of a trough on the adiabatic potential energy surface there are several absolute minima divided by potential barriers. The second-order $(E \otimes e)$–type Jahn-Teller efffect with the Hamiltonian (2.2.2) and the potential energy matrix (3.1.16) seems to be a convenient subject to reveal the general features of this case. By changing the magnitude of the constant of quadratic vibronic coupling W_E from zero to $|W_E| \lesssim \omega_E^2$, one can pass gradually from the above case of a trough on the adiabatic potential energy surface to the case of several deep minima on this surface. For the Hamiltonian of the lower sheet in the adiabatic electronic basis neglecting the nonadiabatic mixing with the upper sheet and taking account of (3.1.19), we have

$$H = -\frac{\hbar^2}{2}\left[\frac{1}{\varrho}\frac{\partial}{\partial\varrho}\left(\varrho\frac{\partial}{\partial\varrho}\right) + \frac{1}{\varrho^2}\frac{\partial^2}{\partial\varphi^2}\right] + \frac{\hbar^2}{8}\left[\left(\frac{\partial\Omega}{\partial\varrho}\right)^2 + \frac{1}{\varrho^2}\left(\frac{\partial\Omega}{\partial\varphi}\right)^2\right]$$
$$+ \frac{1}{2}\omega_E^2\varrho^2 - \left(V_E^2\varrho^2 + 2V_E W_E\varrho^3\cos 3\varphi + W_E^2\varrho^4\right)^{1/2} , \qquad (4.3.1)$$

where Ω is determined by (3.1.18). As in the linear case, we take the wave function of this Hamiltonian in the form $\Psi = \Phi/\sqrt{\varrho}$, and obtain for the function $\Phi(\varrho, \varphi)$ the Hamiltonian

$$H = -\frac{\hbar^2}{2}\frac{\partial^2}{\partial\varrho^2} - \frac{\hbar^2}{2\varrho^2}\frac{\partial^2}{\partial\varphi^2} + \frac{1}{2}\omega_E^2\varrho^2 - \left(V_E^2\varrho^2 + 2V_E W_E\varrho^3\cos 3\varphi + W_E^2\varrho^4\right)^{1/2}$$
$$- \frac{\hbar^2 W_E}{8\varrho}\left[6V_E^3\cos 3\varphi + W_E V_E^2\varrho(5 + 4\cos^2 3\varphi) - 3W_E^3\varrho^3\right]$$
$$\times \left(V_E^2 + 2V_E W_E\varrho\cos 3\varphi + W_E^2\varrho^2\right)^{-2} . \qquad (4.3.2)$$

From this it is seen that, as distinguished from the linear $E \otimes e$ problem, the variables ϱ and φ are no longer separable.

4.3.1 Warping Terms in the $E \otimes e$ Problem as Small Perturbations

In order to simplify the following discussion, consider first the case of weak quadratic vibronic coupling

$$|W_E| \ll \omega_E^2 \ . \tag{4.3.3}$$

Taking into account that in the neighbourhood of the warped trough, i.e., along the line of steepest slope given by (3.1.22), $\varrho \sim |V_E|/\omega_E^2$, and expanding the Hamiltonian (4.3.2) in a power series with respect to W_E/ω_E^2, we obtain in the linear approximation

$$H = -\frac{\hbar^2}{2}\frac{\partial^2}{\partial\varrho^2} - \frac{\hbar^2}{2\varrho^2}\frac{\partial^2}{\partial\varphi^2} + \frac{1}{2}\omega_E^2\varrho^2 - |V_E|\varrho - W_E\varrho^2\cos 3\varphi \ . \tag{4.3.4}$$

If W_E is small enough, the last term in this equation can be taken into account by the perturbation theory. Then H_0 is the Hamiltonian of the lower sheet of the adiabatic potential of the linear $E \otimes e$ problem, its wave functions and eigenvalues being given by (4.2.8) and (4.2.12) respectively. As the perturbation operator is proportional to $\cos 3\varphi$, the selection rules $\Delta j = \pm 3$ are valid. This means that the ground state with $j = \pm\frac{1}{2}$ cannot be split by this perturbation (in any order with respect to W_E). The same condition is true for all the excited vibronic states except those for which $|j| = \frac{3}{2}(2i+1)$, $i = 0, 1, 2, \dots$.

It is obvious that the first excited state with $j = \pm\frac{3}{2}$ is split by the perturbation even in the first order, the fourth state with $j = \pm\frac{9}{2}$ is split in the third order of the perturbation theory, the seventh one ($j = \pm\frac{15}{2}$) is split in the fifth order and so on. The splitting of the first excited state is

$$\Delta E_{0,3/2} \approx |W_E|\varrho_0^2 = \frac{W_E V_E^2}{\omega_E^2} \ . \tag{4.3.5}$$

This result is meaningful only when the splitting $\Delta E_{0,3/2}$ is much smaller than the spacing of the rotational levels of the linear $E \otimes e$ case, $\hbar^2/\varrho_0^2$, i.e., when

$$\frac{|W_E|}{\omega_E^2} \ll 4\left(\frac{E_{\mathrm{JT}}}{\hbar\omega_E}\right)^2 \ . \tag{4.3.6}$$

A simple group-theoretical analysis shows that the vibronic states of the nonsplit levels, i.e., for $j \neq \pm\frac{3}{2}(2i+1)$, belong to the twofold irreducible representation of the point symmetry group of the molecule. On the other hand, the singlet states resulting from the split levels with $j = \pm\frac{3}{2}(2i+1)$ belong to the representations A_1 and A_2.

4.3.2 Hindered Rotations of Jahn-Teller Distortions

If the inequality (4.3.6) is not fulfilled, but inequality (4.3.3) is, resulting in

$$4\left(\frac{E_{JT}}{\hbar\omega_E}\right)^2 \lesssim \frac{|W_E|}{\omega_E^2} \ll 1 \ , \tag{4.3.7}$$

then the perturbation theory with respect to W_E becomes invalid, while the approximate Hamiltonian (4.3.4) still holds. In this case an appropriate separation of the variables ϱ and φ in (4.3.4) can be carried out in the adiabatic approximation. Indeed, as shown below, the vibrational quanta of the radial motion in the case of strong vibronic coupling are much larger than the energy spacing corresponding to the angular motion. This inequality is similar to (2.1.10) and provides the condition for applicability of the adiabatic approximation in which the radial motion is fast, while the angular motion is slow. The Hamiltonian of the fast subsystem,

$$H(\varrho) = -\frac{\hbar^2}{2}\frac{\partial^2}{\partial\varrho^2} + \frac{1}{2}\omega_E^2\rho^2 - |V_E|\varrho - W_E\varrho^2\left(1 + \frac{3\hbar^2}{V_E\varrho^3}\right)\cos 3\varphi \ , \tag{4.3.8}$$

depends parametrically on the angle φ, the coordinate of the slow subsystem. Allowing for the inequality (4.3.3), the last term in (4.3.8) can be ignored when looking for the wave functions of the fast subsystem. Excluding thus the W_E constant, one reduces the radial Schrödinger equation to that of the linear $E \otimes e$ case (the centrifugal energy term is not important here). The solutions of this equation are linear harmonic oscillations with the equilibrium point at $\varrho_0 = |V_E|/\omega_E^2$.

The Hamiltonian of the slow subsystem describing the rotational states corresponding to the ground vibrational state of radial motion can be obtained from (4.3.4) by averaging over the ground vibrational state $\chi_0(\varrho - \varrho_0)$, resulting in [4.21]

$$H(\varphi) = -\frac{\hbar^2}{2\varrho_0^2}\frac{\partial^2}{\partial\varphi^2} - W_E\varrho_0^2\cos 3\varphi - E_{JT} \ . \tag{4.3.9}$$

The numerical solution of the Schrödinger equation with this Hamiltonian can be obtained by passing to its matrix representation, taking the rotational states of the linear $E \otimes e$ case, $\exp(\mathrm{i}j\varphi)/\sqrt{2\pi}$, $j = \pm\frac{1}{2}, \pm\frac{3}{2}, \ldots$, as the basis of the calculation. With a finite but rather large basis set the calculations were carried out in [4.21] yielding the vibronic energy spectrum illustrated in Fig. 4.7 and corresponding rotational wave functions[2].

---

[2] In fact this was not necessary since (4.3.9) is the Mathieu equation with a special boundary condition $\Phi(\varphi + 2\pi) = -\Phi(\varphi)$, and its solutions are well known (e.g., see [4.22]).

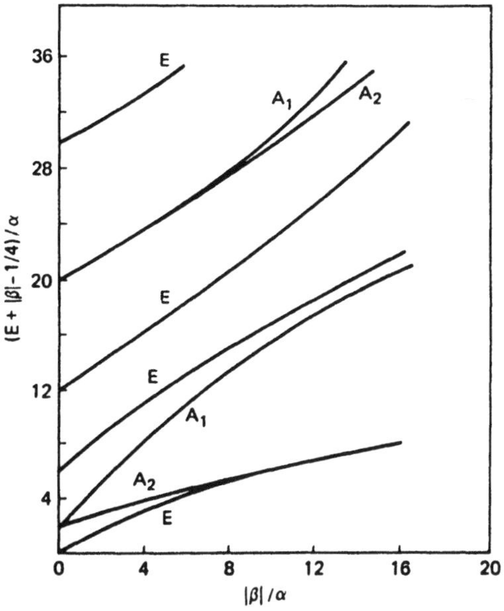

Fig. 4.7. Energy spectrum of hindered rotations in the quadratic $E \otimes e$ case; $\beta/\alpha = 2|W_E|V_E^4/\hbar^2\omega_E^8$ is the dimensionless quadratic vibronic constant, while $\varepsilon/\alpha = 2EV_E^2/\hbar^2\omega_E^2$ is the dimensionless energy [4.21]

The adiabatic wave functions of the quadratic $E \otimes e$ problem are

$$\Psi_{\Gamma\gamma}(\mathrm{r}, Q) = \psi_-(\mathrm{r}, Q)\chi_n(\varrho - \varrho_0)\Phi_{\Gamma\gamma}(\varphi)/\sqrt{\varrho}\ , \tag{4.3.10}$$

where $\chi_n(\varrho - \varrho_0)$ are the radial oscillator states of the linear $E \otimes e$ problem, and

$$\Phi_{\Gamma\gamma} = \frac{1}{\sqrt{2\pi}}\sum_j C_{\Gamma\gamma}^{(j)}\exp(\mathrm{i}j\varphi)\ , \qquad j = \pm\frac{1}{2}, \pm\frac{3}{2}, \ldots, \tag{4.3.11}$$

the coefficients of this expansion being determined from the numerical diagonalization of the Hamiltonian (4.3.9) by the basis set of rotational states $\exp(\mathrm{i}j\varphi)/\sqrt{2\pi}$.

The physical meaning of these results is as follows. In the case of very small W_E values, when inequality (4.3.6) is fulfilled, the angular motion suffers over-barrier reflections, the rotational states with $j = \pm\frac{3}{2}(2i + 1)$, $i = 0, 1, 2, \ldots$, interfere and the corresponding vibronic levels split. When the value of W_E increases, the angular motion becomes more and more hindered by the growing barriers, and when inequality (4.3.7) is satisfied (i.e., when the barriers become higher than the lowest excited rotational energy levels) the angular motion becomes localized in the three potential wells. This is the case of *hindered rotation of Jahn-Teller distortions*. The lowest vibronic states in this case are triply degenerate, provided the tunneling between the wells through the barriers is not taken into account (see below).

The spacing of the vibronic triplets is $\hbar\tilde{\omega}_B$, where $\tilde{\omega}_B$ is the frequency of harmonic vibrations near the bottom of the wells along the line of steepest slope, see (3.1.23). The above adiabatic separation of the variables ϱ and φ is valid if the quantum of radial vibrations $\hbar\tilde{\omega}_A$ is much larger than $\hbar\tilde{\omega}_B$. Taking $\tilde{\omega}_A$ and $\tilde{\omega}_B$ from (3.1.23), we find that the adiabatic approximation, resulting in the Hamiltonian (4.3.9), and the vibronic spectrum of Fig. 4.7 is valid if inequality (4.3.3) is fulfilled. In this case the radial vibrations are not affected by small barriers, and the nuclear motions, indeed, can be considered as internal hindered rotation of Jahn-Teller distortion.

In the case of larger barriers between the wells, the results obtained from the Hamiltonian (4.3.9) and represented by (4.3.10) and (4.3.11) and Fig. 4.7 are invalid since the adiabatic separation of the angular and the radial motions becomes unfounded. For this case the approach of tunneling splitting seems to be more adequate (see below).

Similar ideas for weak second-order interactions were also employed in the $T \otimes (e + t_2)$ case [4.14]. As mentioned in Sect. 3.3, if $E_{\mathrm{JT}}(E) = E_{\mathrm{JT}}(T)$, the lowest sheet of the adiabatic potential of the linear $T \otimes (e + t_2)$ problem has a two-dimensional trough. In the particular case of $\omega_E = \omega_T$, it is convenient to pass to five-dimensional spherical coordinates. Then the angular coordinates describing the motion along the trough can be separated from the "radial" ones [4.14, 4.15]. The free motion along the trough corresponds to the rotation of the Jahn-Teller distortion and, as mentioned in Sect. 4.2, it can be described by spherical functions $Y_{lm}(\theta, \varphi)$. If $E_{\mathrm{JT}}(E) \neq E_{\mathrm{JT}}(T)$, then additional humps and wells, corresponding to trigonal, tetragonal and orthorhombic extrema, occur along the bottom of the trough (Sect. 3.3). In other words, the two-dimensional trough is "warped", quite similarly to the case of the $E \otimes e$ problem, when the second-order terms of the vibronic interaction are included.

If the height of the barriers dividing the minima of the adiabatic potential is less than the magnitude of the quantum of radial vibrations, then the angular motion can be separated from the radial one by the adiabatic approximation, in the same way as in the $E \otimes e$ case. The Hamiltonian of the slow (angular) subsystem is

$$H = \frac{\hbar^2\omega^4}{6V^2}\left[-\frac{1}{\sin\theta}\frac{\partial}{\partial\theta}\left(\sin\theta\frac{\partial}{\partial\theta}\right) - \frac{1}{\sin^2\theta}\frac{\partial^2}{\partial\varphi^2}\right] + \varepsilon(\theta, \varphi) \ , \tag{4.3.12}$$

where $\varepsilon(\theta, \varphi)$ is the lowest sheet adiabatic potential energy averaged over the coordinates of the fast subsystem. This expression, since it corresponds to a singlet adiabatic electronic state, has to be a scalar of the cubic group G_0 of the undistorted polyatomic system. Therefore it can be expanded in a series with respect to cubic harmonics. Cutting off this expansion after the fourth-order harmonics, one obtains

$$\varepsilon(\theta, \varphi) \approx -E_{\mathrm{JT}} + \Delta\{Y_{4,0}(\theta, \varphi) + \sqrt{\tfrac{5}{14}}[Y_{4,4}(\theta, \varphi) + Y_{4,-4}(\theta, \varphi)]\} \ , \tag{4.3.13}$$

where E_{JT} is given by (3.3.7), while Δ is a coefficient independent of θ and φ (see also [4.23]).

It can be easily checked that the potential surface (4.3.13) has the same extrema as the exact surface $\varepsilon(\theta, \varphi)$. For instance, for negative Δ values the three points with the coordinates

$$(\theta, \varphi) = (0,0) \ , \quad \left(\frac{\pi}{2}, 0\right) \ , \quad \left(\frac{\pi}{2}, \frac{\pi}{2}\right) \tag{4.3.14}$$

are minima. Substituting these coordinates into (3.3.6), we find that they correspond to tetragonal minima (Table 3.3). For positive Δ values the minima are situated at four points with the coordinates

$$(\theta, \varphi) = \left(\theta_0, \frac{\pi n}{4}\right) \ , \qquad n = 1, 3, 5, 7 \ , \tag{4.3.15}$$

where $\cos\theta_0 = 1/\sqrt{3}$. These points prove to be of trigonal symmetry. At arbitrary Δ values the surface (4.3.13) has additional six extrema at

$$(\theta, \varphi) = \left(\frac{\pi}{4}, \frac{\pi n}{2}\right) \ , \quad \left(\frac{\pi}{2}, \frac{\pi}{4} + \frac{\pi n}{2}\right) ; \qquad n = 0, 1, 2, 3 \ , \tag{4.3.16}$$

which have orthorhombic symmetry and are always saddle points. All these results coincide with those obtained for the exact expression $\varepsilon(Q)$ in Sect. 3.3, but, of course, the coincidence is only qualitative, not quantitative, since it is impossible to reproduce all the Jahn-Teller stabilization energies for three different types of nonequivalent extrema points by means of only one parameter Δ. As follows from Sect. 3.3, this can be done by at least two parameters, $E_{JT}(E)$ and $E_{JT}(T)$. Nevertheless, one can hope that the main qualitative features of the vibronic energy spectrum and wave functions of the linear $T \otimes (e + t_2)$ problem can be satisfactorily reproduced by the eigenvalues of the Hamiltonian (4.3.12) with the potential energy taken as in (4.3.13).

The numerical diagonalization of the Hamiltonian (4.3.12) was carried out in the same way as for the $E \otimes e$ case. The rotational functions $Y_{lm}(\theta, \varphi)$ were taken as a basis set for the calculations, the total function being represented by the expansion [4.14, 4.23]

$$\Phi_{\Gamma\gamma}(\theta, \varphi) = \sum_{l,m} C_{lm}^{(\Gamma\gamma)} Y_{lm}(\theta, \varphi) \ . \tag{4.3.17}$$

The results obtained in [4.14] for the case when all the terms up to $l = 13$ are kept in the expansion (4.3.17) are illustrated in Fig. 4.8. The motion of the system in this case can be visualized as hindered rotations of the Jahn-Teller distortion, in much the same way as in the above case of the second-order $E \otimes e$ problem.

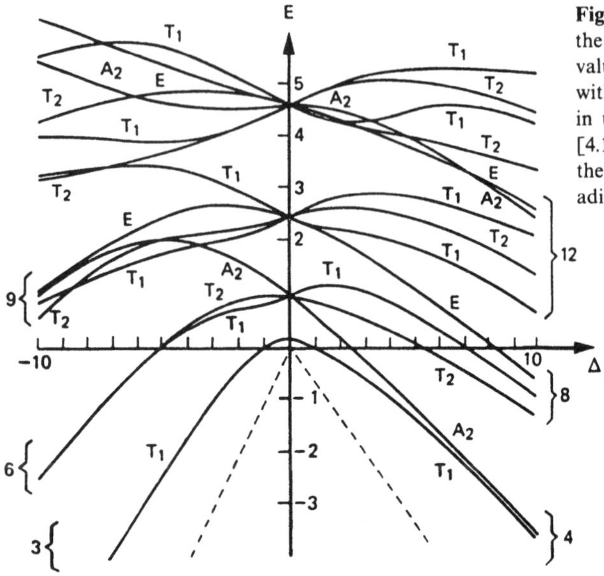

Fig. 4.8. Vibronic energy levels of the $T \otimes (e + t_2)$ problem. Eigenvalues of the Hamiltonian (4.3.12) with the potential energy (4.3.13) in units of $E_D = \hbar^2\omega^2/6V^2$ (after [4.14]). The dashed lines indicate the energy of the minima of the adiabatic potentials

4.3.3 Tunneling of Jahn-Teller Distortions Through Potential Barriers. Tunneling Splitting

In the case of high potential barriers between the minima the inequality (4.3.3) is obviously invalid. In this case the localized states in the minima seem to be a much more adequate basis set for calculations of the vibronic states. In the limiting case of very strong vibronic coupling, the above barriers may be considered to be infinitely large, and hence the vibronic states are localized within the wells. The matrix of the Hamiltonian calculated with this basis is diagonal, and each vibronic state is r-fold degenerate, where r is the number of equivalent minima. Since the barrier height (and width) is not infinite, the system can pass between the equivalent wells by tunneling. Because of this, the r-fold degeneracy of the vibronic levels is completely or partly removed and the resulting wave function is a linear combination of the wave functions localized in the equivalent minima. This *tunneling splitting* was originally called *inversion splitting* [4.24, 25][3] by analogy with the inversion splitting of the energy levels in ammonia (for a more detailed bibliography see [4.26]).

Let us assume that in the absence of tunneling the system performs small harmonic oscillations near the minimum, and for the lowest states localized in the ith well the wave function can be taken as

[3] Note that the presentation of the topic in this book does not follow the chronology of the events. In particular the papers [4.24, 25] devoted to tunneling splitting in Jahn-Teller systems were published before the work [4.21], in which the warping terms in the $E \otimes e$ problem are taken into account by using the rotational states as a basis set for calculations.

$$\Psi_n^{(i)}(\mathrm{r}, Q) = \psi_-(\mathrm{r}, Q)\Phi_n(Q - Q^{(i)}) \ , \tag{4.3.18}$$

where $\psi_-(\mathrm{r}, Q)$ is the electronic wave function for the lower sheet of the adiabatic potential, and $\Phi_n(Q - Q^{(i)})$ is the vibrational wave function in the ith well, $Q^{(i)}$ being the coordinate of its minimum point. In particular, for the quadratic $E \otimes e$ case, $Q^{(i)}$ for the three minima are given in polar coordinates by (3.1.20), and the frequency of the harmonic vibrations near these minima is determined by the curvature of the adiabatic potential (3.1.23).

For the sake of simplicity the Born-Oppenheimer functions (Sect. 2.1.2)

$$\Psi_n^{(i)} = \psi_-(\mathrm{r}, Q^{(i)})\Phi_n(Q - Q^{(i)}) \tag{4.3.19}$$

will be used hereafter instead of the adiabatic ones, given by (4.3.18). For definiteness we assume that $V_E > 0$ and $W_E > 0$. Then, substituting the minima coordinates (3.1.20) into the expression

$$\psi_-(\mathrm{r}, Q) = \frac{1}{\sqrt{2}}[\mathrm{e}^{-\mathrm{i}\Omega/2}\psi_+(\mathrm{r}) - \mathrm{e}^{\mathrm{i}\Omega/2}\psi_-(\mathrm{r})]$$

and allowing for (3.1.18), one finds the following functions for the ground state:

$$\Psi_0^{(1)} = \frac{1}{\sqrt{2}}[\psi_+(\mathrm{r}) - \psi_-(\mathrm{r})]\Phi_0(Q - Q^{(0)}) \ ,$$

$$\Psi_0^{(2)} = \frac{1}{\sqrt{2}}[\mathrm{e}^{2\pi\mathrm{i}/3}\psi_+(\mathrm{r}) - \mathrm{e}^{-2\pi\mathrm{i}/3}\psi_-(\mathrm{r})]\Phi_0(Q - Q^{(2)}) \ , \tag{4.3.20}$$

$$\Psi_0^{(3)} = \frac{1}{\sqrt{2}}[\mathrm{e}^{-2\pi\mathrm{i}/3}\psi_+(\mathrm{r}) - \mathrm{e}^{2\pi\mathrm{i}/3}\psi_-(\mathrm{r})]\Phi_0(Q - Q^{(4)}) \ ,$$

where the superscript n in $Q^{(n)}$ corresponds to the value in (3.1.20).

In the absence of tunneling transitions the states (4.3.20) localized in the wells, as mentioned above, are triply degenerate due to the equivalency of the minima. Assuming the tunneling to be weak (the criterion of weak tunneling is given below) one can use perturbation theory to diagonalize the full Hamiltonian using the functions (4.3.20) as a basis set.

The eigenfunction of the Hamiltonian (2.2.12) is sought as a linear combination of the functions (4.3.20)

$$\Psi = \sum_{i=1}^{3} C_i \Psi_0^{(i)}(\mathrm{r}, Q) \ . \tag{4.3.21}$$

Substituting this expression into the Schrödinger equation, multiplying it by $\Psi_0^{(k)*}(\mathrm{r}, Q)$ on the left side and integrating over the r and Q variables, we obtain the following system of equations for the energy E and the coefficients C_i

$$\sum_i H_{ki} C_i = E \sum_i S_{ki} C_i \ . \tag{4.3.22}$$

Here S_{ki} is the overlap integral for the basis functions (4.3.20) (which are not orthogonal) and H_{ki} are the Hamiltonian matrix elements. A nontrivial solution of the system of (4.3.22) is possible if

$$\det(H_{ik} - ES_{ik}) = 0 \ . \tag{4.3.23}$$

In the case of $E \otimes e$ problem by using symmetry considerations one can easily show that $H_{ik} = H_{12}$, $H_{ii} = H_{11}$ and $S_{ik} = S_{12} = S$, thereby reducing (4.3.23) to

$$\begin{vmatrix} H_{11} - E & H_{12} - ES & H_{12} - ES \\ H_{12} - ES & H_{11} - E & H_{12} - ES \\ H_{12} - ES & H_{12} - ES & H_{11} - E \end{vmatrix} = 0 \ . \tag{4.3.24}$$

The three localized states (4.3.20) belong to a threefold reducible representation of the symmetry point group of the molecule, which decomposes into the irreducible representations $A_1 + E$ (if $V_E < 0$ or $W_E < 0$ it will be $A_2 + E$). Since the latter does not contain repeating representations, the symmetry-adapted wave functions, being the correct zero-order functions, can be found directly using projection operators:

$$\begin{aligned} \Psi_A &= \frac{1}{\sqrt{3}}(\Psi_0^{(1)} + \Psi_0^{(2)} + \Psi_0^{(3)}) \ , \\ \Psi_{E+} &= \frac{1}{\sqrt{3}}(\Psi_0^{(1)} + \mathrm{e}^{2\pi \mathrm{i}/3}\Psi_0^{(2)} + \mathrm{e}^{4\pi \mathrm{i}/3}\Psi_0^{(3)}) \ , \\ \Psi_{E-} &= \frac{1}{\sqrt{3}}(\Psi_0^{(1)} + \mathrm{e}^{-2\pi \mathrm{i}/3}\Psi_0^{(2)} + \mathrm{e}^{-4\pi \mathrm{i}/3}\Psi_0^{(3)}) \ . \end{aligned} \tag{4.3.25}$$

The initial triply degenerate energy level is split by tunneling into a singlet and a doublet

$$E(A) = \frac{\langle \Psi_A | H | \Psi_A \rangle}{\langle \Psi_A | \Psi_A \rangle} = \frac{H_{11} + 2H_{12}}{1 + 2S} \ , \tag{4.3.26a}$$

$$E(E) = \frac{\langle \Psi_{E\pm} | H | \Psi_{E\pm} \rangle}{\langle \Psi_{E\pm} | \Psi_{E\pm} \rangle} = \frac{H_{11} - H_{12}}{1 - S} \ . \tag{4.3.26b}$$

The magnitude of the tunneling splitting of the ground state is thus

$$\delta = E(A) - E(E) = 3\Gamma \ , \qquad \Gamma = \frac{H_{12} - H_{11}S}{1 + S - 2S^2} \ . \tag{4.3.27}$$

The matrix element $H_{12} = H_{23}$ and the overlap integral $S = S_{23}$ can be calculated as follows. Due to the multiplicative form of the basis functions (4.3.20) the integration over the electronic variables can be carried out directly, whereas for the vibrational functions in the wells the following representation is useful

$$\Phi_0(Q - Q^{(2)}) = N\exp\left[-\frac{\tilde{\omega}_A}{2\hbar}\left(-\frac{1}{2}Q_\theta + \frac{\sqrt{3}}{2}Q_\varepsilon - \varrho_0\right)^2 - \frac{\tilde{\omega}_B}{2\hbar}\left(\frac{\sqrt{3}}{2}Q_\theta + \frac{1}{2}Q_\varepsilon\right)^2\right],$$

$$\Phi_0(Q - Q^{(4)}) = N\exp\left[-\frac{\tilde{\omega}_A}{2\hbar}\left(-\frac{1}{2}Q_\theta - \frac{\sqrt{3}}{2}Q_\varepsilon - \varrho_0\right)^2 - \frac{\tilde{\omega}_B}{2\hbar}\left(\frac{\sqrt{3}}{2}Q_\theta + \frac{1}{2}Q_\varepsilon\right)^2\right]. \quad (4.3.28)$$

Here the $Q - Q^{(i)}$ coordinates at the minimum points are taken as functions of the normal coordinates of the system as a whole, Q_θ and Q_ε (see e.g. [4.27]); $\tilde{\omega}_A$ and $\tilde{\omega}_B$ are the frequencies of the normal vibrations (in the minimum configuration) determined by (3.1.23), ϱ_0 is given by (3.1.20), and $N = (\tilde{\omega}_A\tilde{\omega}_B/\pi^2\hbar^2)^{1/2}$ is the normalization constant.

As a result of straightforward calculations, we obtain

$$S_{23} = -\tfrac{1}{2}\gamma\ , \qquad H_{ii} = -E_{\mathrm{JT}} + \tfrac{1}{2}\hbar(\tilde{\omega}_A + \tilde{\omega}_B)\ , \quad (4.3.29)$$

$$H_{23} = H_{11}S + \frac{3\gamma E_{\mathrm{JT}}(9 + 54\lambda - 6\lambda^3 - \lambda^4)}{2(9 - \lambda^2)(1 + 3\lambda)^2}\ , \quad (4.3.30)$$

where $\lambda = \tilde{\omega}_B/\tilde{\omega}_A = [9W_E/(\omega_E^2 - W_E)]^{1/2}$, E_{JT} is given by (3.1.21) and γ is the overlap integral for the vibrational functions (4.3.28)

$$\gamma = \left(\frac{16\lambda}{3\lambda^2 + 10\lambda + 3}\right)^{1/2}\exp\left(-6\frac{E_{\mathrm{JT}}}{\hbar\tilde{\omega}_A}\frac{\lambda}{(1 + 3\lambda)}\right). \quad (4.3.31)$$

Neglecting γ^2 compared with one, we find [4.24,25,27,28,29]:

$$\delta = 3\Gamma\ , \qquad \Gamma \cong 3\gamma E_{\mathrm{JT}}\frac{9 + 54\lambda - 6\lambda^3 - \lambda^4}{2(9 - \lambda^2)(1 + 3\lambda)^2}\ . \quad (4.3.32)$$

In particular, if $|W_E| \ll \omega_E^2$, i.e. when $\lambda \ll 1$, we have

$$\delta \approx 18E_{\mathrm{JT}}\sqrt{\frac{\lambda}{3}}\exp\left(-\frac{6\lambda E_{\mathrm{JT}}}{\hbar\omega_E}\right)\ ; \qquad \lambda^2 \approx 9\frac{W_E}{\omega_E^2}\ . \quad (4.3.33)$$

It follows from (4.3.32) and (4.3.33) that the criterion for applicability of the perturbation theory used above, and hence for the validity of the results obtained, is met if $E_{JT} \gtrsim \hbar\omega_E$, which means that there is at least one localized state in the well. Another point is that, since $E(A) - E(E) > 0$, the ground state of the system, as in the case of the linear $E \otimes e$ problem, remains doubly degenerate.

The picture of deep minima in the adiabatic potential separated by relatively high barriers is typical for Jahn-Teller polyatomic systems with strong vibronic coupling. Therefore *the tunneling phenomenon is a characteristic feature of Jahn-Teller systems*, and the method of calculation of the tunneling splitting given above is, in general, valid for all these cases.

The physical meaning of the tunneling splitting is as follows. As discussed above, in the case when the adiabatic potential energy surface has a continuum of equivalent minima (a trough), the orbital momentum (the sum of the momenta of electronic and nuclear motions) is preserved, and free rotations of Jahn-Teller distortions of the nuclear framework take place (Sect. 4.2). For a finite number of minima when the potential barriers between them are small compared with the quanta of radial vibrations, the rotations of the Jahn-Teller distortions are hindered. When the barrier height becomes comparable with or larger than the radial vibration energy quantum, the angular motion cannot be separated from the radial motion, and the idea of rotation of the Jahn-Teller distortion loses its physical sense. In this case the dynamics of the system is well described by localized vibrations at the minima accompanied by tunneling between neighbouring equivalent minima. Analogously to the previous cases, this phenomenon can be visualized as *pulse motions of the Jahn-Teller distortion*: the distorted nuclear configuration at the minimum periodically changes its orientation in space in accordance with the symmetries of the other minima equivalent to the first one. It does this with a frequency proportional to the tunneling splitting, which is much lower than the frequency of vibrations within the well. Note that in all the cases discussed above the term "motion of Jahn-Teller distortions" means the concerted motions of both the nuclear framework and the electrons.

The free and hindered rotations of Jahn-Teller distortions (which must not be confused with the usual rotation of the molecule or its separate parts) and their pulse motions form a new type of intramolecular motion which essentially distinguishes the Jahn-Teller systems from non–Jahn-Teller ones.

The tunneling splitting in the case of an electronic T term can be considered in a way similar to the one used above for the $E \otimes e$ case. In the case of the $T \otimes (e + t_2)$ problem (Sect. 3.3) without tunneling the ground vibronic states localized within each of the four trigonal minima of the adiabatic potential energy surface form a fourfold degenerate term. In the cubic symmetry groups there are no fourfold irreducible representations, and therefore the fourfold degeneracy is accidental. The tunneling removes the degeneracy resulting in a singlet A_1 (or A_2) and triplet T_2 (or T_1) state, $A_1 + T_2$ (or $A_2 + T_1$). The magnitude of the tunneling splitting is, cf. (4.3.27)

$$\delta = E(A) - E(T) = 4\Gamma \; ; \qquad \Gamma = \frac{H_{12} - H_{11}S}{1 + 2S - 3S^2} \; . \tag{4.3.34}$$

If, as above, the harmonic approximation is employed for the states localized in the wells, and taking into account the redetermined frequencies of vibrations near the bottom of the wells given by (3.3.15), for the $T \otimes t_2$ case we obtain [4.25–29]

$$\delta \approx 2E_{\mathrm{JT}}(T)\exp\left(-1.24\frac{E_{\mathrm{JT}}(T)}{\hbar\omega_T}\right), \tag{4.3.35}$$

where $E_{\mathrm{JT}}(T)$ is defined by (3.3.11). Note that $\delta = E(A) - E(T) > 0$, and hence the ground state is a vibronic triplet with the same symmetry properties as the initial electronic term.

The tunneling splitting in the case of six orthorhombic minima of the adiabatic potential in the quadratic $T \otimes (e + t_2)$ problem (Sect. 3.3) results in two lowest vibronic triplet states T_1 and T_2 with the splitting magnitude given by [4.28–31]

$$\delta = \frac{3}{2}E_{\mathrm{JT}}(OR)\frac{24 - 28B - 2B^2 + 9B^3}{(4 - 3B)(4 - 3B^2)}\frac{S}{1 - 4S^2}, \tag{4.3.36}$$

where the constant B is given by (3.3.21) and

$$S = \exp\left(-\frac{3}{2}\frac{E_{\mathrm{JT}}(OR)}{\hbar\omega}\frac{12 - 20B - 11B^2}{(4 - 3B)(4 - 3B^2)}\right). \tag{4.3.37}$$

In this case too, the ground state is a vibronic triplet with the same transformation properties as the initial electronic T term.

The magnitude of the tunneling splitting, as shown above, is determined by the overlap of the localized states of nearest-neighbor minima, i.e., by the behavior of the tails of the wave functions under the potential barriers. From this point of view the harmonic approximation used above with oscillator functions of the type (4.3.28) may lead to a rather crude approximation for the splitting, since these functions, although satisfactory near the minimum points, are decreasing too fast in the most important region under the barrier. This fault is partly removed by diagonalizing the exact Hamiltonian (which involves a variational procedure). However, the basis set of (4.3.20) and (4.3.28) is very limited, and therefore the results of (4.3.31–33), (4.3.35–37) should be considered as a semi-quantitative estimation of the tunneling splitting magnitude.

These results can be improved by two methods. First, one can increase the number of vibrational functions in the basis set and evaluate numerical results with a computer program. In this way the splitting can be obtained with any desired accuracy .

The second method is to improve the behavior of the basis functions in the barrier region. This can be done by an appropriate choice of parametrically dependent probe functions in a variational procedure [4.32].

In conclusion of this subsection, we note also that the oscillator wave functions in the wells of the type (4.3.28) can be regarded as coherent states of harmonic oscillators ([4.33], Chap. 2), their shift parameters being equal to the minimum point coordinates. From this point of view, [4.34] in which the tunneling splitting is evaluated by the method of coherent states, repeats in essence the first papers on tunneling splitting [4.25–29]. The fault of [4.34] (as well as [4.24, 25]) is neglecting the splitting of the vibrational frequencies in the minima. The curvature of the adiabatic potentials at the minimum points determines the behavior of the tails of the oscillator wave functions and affects strongly the magnitude of the tunneling splitting. The consideration of this circumstance leads to the much more accurate results given by (4.3.32) and (4.3.35).

A more accurate behavior of the wave functions in the region under the barriers can also be reached within the quasi-classical approximation. The problem can be reduced to one-dimension along the line of steepest slope from the barrier top to the minimum point (Fig. 3.3b). Postponing the discussion of this approximation to the end of this section, we present here some results obtained by the appropriate Wentzel-Kramers-Brillouin calculation of the tunneling splitting in the $E \otimes e$ and $T \otimes (e + t_2)$ problems [4.35].

In the case of three minima in the adiabatic potential energy surface in the $E \otimes e$ problem the quasi-classical tunneling splitting is

$$\delta = 3\Gamma\ ; \qquad \Gamma = \frac{\hbar\tilde{\omega}_B}{2\pi}\exp\left(-\frac{1}{\hbar}\int_a^b \sqrt{2m(\varepsilon_- - E_0)}\,dq\right)\ . \tag{4.3.38}$$

Here, as above, $\tilde{\omega}_B$ is given by (3.1.23), a and b are the turning points at the entrance to and exit from the forbidden region under the barrier, ε_- is given by (3.1.19), E_0 is the energy of the ground state in the minimum configuration, m is the effective mass of the "particle" corresponding to the generalized coordinate q,

$$m = \sum_{\Gamma\gamma}\left(\frac{\partial Q_{\Gamma\gamma}}{\partial q}\right)^2\ .$$

The integration is carried out along the line of steepest slope given by (3.1.22).

Equation (4.3.38) has a clear-cut physical meaning: it represents the probability of decay per second of the metastable state in the well. Indeed, Γ is proportional to the number of "particle" collisions with the barrier wall per second, $\tilde{\omega}_B/2\pi$, and to the probability of tunneling through the barrier at each of these collisions,

$$D = \exp\left(-\frac{1}{\hbar}\int_a^b \sqrt{2m(\varepsilon_- - E_0)}\,dq\right)\ .$$

The factor of three in (4.3.38) is related to the equal probability of the "particle" occurring at any of the three minimum configurations, i.e., to the existence of three equivalent metastable states.

Taking the integral in (4.3.38) along the path (3.1.22) and assuming that $|W_E| \ll \omega_E^2$, we obtain [4.35], cf. (4.3.33),

$$\delta \approx 2.46\hbar\omega_E\left(\lambda^3\frac{E_{\rm JT}}{\hbar\omega_E}\right)^{1/2}\exp\left(-\frac{16\lambda E_{\rm JT}}{9\hbar\omega_E}\right) . \tag{4.3.39}$$

Analogously, for the $T\otimes(e+t_2)$ case with four equivalent trigonal minima the tunneling splitting of the ground vibronic level is given by the expression

$$\delta = 4\frac{\hbar\tilde{\omega}_E}{2\pi}\exp\left(-\frac{1}{\hbar}\int_a^b\sqrt{2m(q)[\varepsilon(q)-E_0]}\,dq\right) . \tag{4.3.40}$$

Here $\tilde{\omega}_E \approx \omega_T\sqrt{2/3}$ is the frequency of the "particle" collisions with the barrier wall, $\varepsilon(q)$ is the lower sheet of the adiabatic potential, and the integral has to be taken again along the steepest slope path. The latter can be obtained from (3.3.9), if one takes into account that in the space of a_x, a_y and a_z the extreme points of the adiabatic potential are just the points piercing the unit sphere $a_x^2+a_y^2+a_z^2=1$ (see 3.1.28) on the symmetry axes of the cube, [111], [110], [100], and so on. Therefore the line of steepest slope is the one which links, say, the points [111], [110] and [11$\bar{1}$] by the shortest path on the surface of the unit sphere. Transforming to spherical coordinates,

$$a_x = \cos\varphi\cos\theta , \qquad a_y = \sin\varphi\cos\theta , \qquad a_z = \sin\theta , \tag{4.3.41}$$

and substituting them into (3.3.9) with the additional condition $\varphi = \pi/4$, we get a system of expressions determining parametrically the line of steepest slope in the five-dimensional Q space of E and T_2 nuclear coordinates [4.35]

$$\begin{aligned} Q_\theta &= -\frac{V_E}{4\omega_E^2}(3\cos 2\theta - 1) , \qquad Q_\varepsilon = 0 , \\ Q_\xi &= Q_\eta = \frac{V_T}{\omega_T^2\sqrt{2}}\sin 2\theta , \qquad Q_\zeta = \frac{V_T}{2\omega_T^2}(1+\cos 2\theta) . \end{aligned} \tag{4.3.42}$$

The values $\theta = 0$, $\theta = \pm\pi/2$ and $\theta = \pm\frac{1}{2}\arccos(\frac{1}{3})$ correspond to the orthorhombic saddle point, tetragonal extrema and trigonal minima, respectively. Taking the intergral in (4.3.40) along the path (4.3.42) and assuming for simplicity that $\omega_E = \omega_T = \omega$, $E_{\rm JT}(E) \approx E_{\rm JT}(T)$ (the d mode model) in the case of relatively weak warping of the two-dimensional trough [but holding $E_{\rm JT}^2(T) - E_{\rm JT}^2(E) \gg \hbar^2\omega^2$], we finally obtain $\delta = 4\Gamma$, where

$$\Gamma \approx 0.875\frac{[E_{\rm JT}(T)-E_{\rm JT}(E)]^{3/4}}{\sqrt{\hbar\omega}[E_{\rm JT}(T)+E_{\rm JT}(E)]^{1/4}}\exp\left(-0.976\frac{[E_{\rm JT}^2(T)-E_{\rm JT}^2(E)]^{1/2}}{\hbar\omega}\right) , \tag{4.3.43}$$

$E_{\rm JT}(T)$ and $E_{\rm JT}(E)$ are given by (3.3.11) and (3.3.10).

The quasi-classical approach to the tunneling splitting calculations discussed above allows one, in principle, to obtain asymptotically exact results in the case of rather strong vibronic coupling within the framework of the one-sheet adiabatic approximation. Indeed, the integral of action in (4.3.38) and (4.3.40) is second-order in the linear vibronic constant and therefore the increase of V_Γ improves the criterion for applicability of the quasi-classical approximation. Besides, large values of V_Γ justify neglecting nonadiabaticity along the whole line of steepest slope.[4]

Note that in the case of strong vibronic coupling to E vibrations and weak coupling to T_2 ones the adiabatic surfaces are very close in the region of the top of the barrier, and hence the adiabatic approximation and the above one-sheet quasi-classical approach become invalid.

Strictly speaking, in the multidimensional case under consideration the integral of action has to be taken along the classical path under the barrier [4.33]. Its substitution by the line of steepest slope is equivalent to neglecting the centrifugal energy which may be considered to be small, provided the V_Γ values are large enough.

The line of steepest slope differs from a circle (Fig. 3.3b), the difference being larger for larger W_E. The classical path is even more different from a circle because of centrifugal forces [4.37]. For large W_E values when inequality (4.3.3) is violated, the angular motion in the quadratic $E \otimes e$ case is fundamentally related to the radial one and cannot be reduced to the hindered rotations discussed above. The motion of the Jahn-Teller system in the case of deep wells with high barriers has the nature of pulsations. The system stays a relatively long time in a minimum configuration and then jumps into another one, equivalent to the first one, but differently oriented in space.

The tunneling splitting takes place for both the ground and the excited states. It is obvious that the splitting magnitude is larger when the energy of the local state in the well is higher, since for the states higher in energy the barrier height and width are smaller. The calculation scheme for the tunneling splitting in the excited state remains the same as for the ground state.

4.3.4 Symmetry Properties and Group-Theoretical Classification of Tunneling States

The group-theoretical classification of the tunneling states is much easier if a preliminary classification of the local states at the minima by the irreducible representations of the local symmetry group G_L is made. In this case the qualitative picture of the expected tunneling splitting can be predicted for any vibronic term using symmetry considerations only [4.38].

[4] The criterion for the validity of the adiabatic approximation in this case is somewhat different from (2.1.10) because the tunneling splitting is so small.

In the harmonic approximation employed above some of the vibrations in the wells may be degenerate. The appropriate normal coordinates transform after the irreducible representations of the subgroup G_L, corresponding to the lowered symmetry of the system in the configuration of the minimum. For instance, in the $T \otimes t_2$ problem the symmetry of an octahedral system is reduced to $G_L = D_{3d}$ in the configurations of trigonal minima. The T_{2g} vibrations at each of these minima are split into singlet A_{1g} and doubly degenerate E_g vibrations (Sect. 3.3). The degeneracy of the excited levels of E_g vibrations in the minima increases linearly with the number of the vibrational level n. (Sect. 4.1). This accidental degeneracy is removed by the local anharmonicity, each of the energy levels belonging to an irreducible representation of the local symmetry group G_L at the minimum.

Consider the symmetry operations of the reference symmetry group G_0 of the undistorted polyatomic system which transform the ith minimum of the adiabatic potential into itself, i.e., the identity elements for this nuclear configuration. These operations form the group $G_L^{(i)}$ (a subgroup of G_0), and the s manifolds $G_L^{(i)}$ (s is the number of equivalent minima) are the contiguous classes $G_L^{(1)}$, $G_L^{(2)} = g_2 G_L^{(1)}$, $G_L^{(3)} = g_3 G_L^{(1)}, \ldots$, where $g_2, g_3, \ldots$ are the elements of the G_0 group which transform the first minimum into others. Now consider the wave function $\Psi^{(i)}_{\Gamma_L \gamma_L}$ localized in the ith well and transforming as the γ_L row of the irreducible representation Γ_L of the symmetry group $G_L^{(i)}$. Neglecting the overlap of the wave functions localized in different wells for an arbitrary element G of the group G_0 the following equations hold:

$$\langle \Psi^{(j)}_{\Gamma_L \bar{\gamma}_L} | G | \Psi^{(i)}_{\Gamma_L \gamma_L} \rangle \neq 0 \ ,$$

if G transforms the ith minimum into the jth one, and

$$\langle \Psi^{(j)}_{\Gamma_L \bar{\gamma}_L} | G | \Psi^{(i)}_{\Gamma_L \gamma_L} \rangle = 0$$

in the opposite case. Considering the isomorphism of the groups $G_L^{(i)}$, these relations can be rewritten as

$$\langle \Psi^{(j)}_{\Gamma_L \bar{\gamma}_L} | G | \Psi_{\Gamma_L \gamma_L} \rangle = \langle \Psi_{\Gamma_L \bar{\gamma}_L} | G | \Psi_{\Gamma_L \gamma_L} \rangle \delta(j, Gi) \ ,$$

where $\delta(i,j)$ is the Kronecker delta. Then the character of the reducible representation realized by the state $\Psi^{(i)}_{\Gamma_L \gamma_L}$ localized in all the wells is

$$\begin{aligned} \chi(G) &= \sum_{i=1}^{s} \sum_{\gamma_L} \langle \Psi^{(i)}_{\Gamma_L \gamma_L} | G | \Psi^{(i)}_{\Gamma_L \gamma_L} \rangle = \sum_{i=1}^{s} \delta(i, Gi) \sum_{\gamma_L} \langle \Psi_{\Gamma_L \gamma_L} | G | \Psi_{\Gamma_L \gamma_L} \rangle \\ &= \chi_{\Gamma_L}(G) \sum_{i=1}^{s} \delta(i, Gi) \ , \end{aligned}$$

where $\chi_{\Gamma_L}(G)$ is the character of the irreducible representation Γ_L of the local group G_L. Using the characters of the reducible representation it is possible to find out the number of times $n(\Gamma)$ a given irreducible representation Γ of the group G_0, occurs in it:

$$n(\Gamma) = \frac{1}{g_0} \sum_{G_0} \chi(G)\chi_\Gamma^*(G) = \frac{1}{g_0} \sum_{G_0} \sum_{i=1}^{s} \delta(i, Gi)\chi_{\Gamma_L}(G)\chi_\Gamma^*(G)$$

$$= \frac{1}{g_0} \sum_{i=1}^{s} \sum_{G_L^{(i)}} \chi_{\Gamma_L}(G)\chi_\Gamma^*(G) \ .$$

Including the fact that the groups $G_L^{(i)}$ with different values of i are isomorphic and that the order of ach of these groups is equal to $g_L = g_0/s$, we obtain

$$n(\Gamma) = \frac{s}{g_0} \sum_{G} \chi_{\Gamma_L}(G)\chi_\Gamma^*(G) = \frac{1}{g_L} \sum_{G_L} \chi_{\Gamma_L}(G)\chi_\Gamma^*(G) \ . \quad (4.3.44)$$

This is the well-known formula of group theory for the number of times of occurrence of a given irreducible representation Γ in a reducible one, when the symmetry is reduced to G_L. It enables us to use the known group reduction correlation tables for the irreducible representations (see e.g. [4.39–41]) to find the symmetry labeling of the tunneling levels. Thus, the tunneling states belong to the irreducible representations Γ of the reference symmetry group G_0 which contain the representations Γ_L under consideration of the local group G_L.

In order to clarify this statement consider some examples. In case of the $E \otimes e$ problem for an octahedral system with, say, six equivalent absolute minima in which the symmetry O_h is reduced to D_{2h} (Sect. 3.2.2), the electronic E_g term is split into two A_g terms at the minimum points. The ground vibrational state $\Phi_0(Q - Q^{(i)})$ in the well is always an invariant of the group $G_L^{(i)}$, and therefore the lowest localized states $\psi_{A_g}(r, Q^{(i)})\Phi_0(Q - Q^{(i)})$ belong to the irreducible representation $\Gamma_L = A_g$ of the D_{2h} group. Using the character table one can calculate $n(\Gamma)$ from (4.3.44), where $G_L = D_{2h}$. It follows that $n(A_{1g}) = 1$, $n(A_{2g}) = 1$, $n(E_g) = 2$ and for other Γ's, $n(\Gamma)$ is zero. Thus, including the tunneling between six equivalent minima in the $E \otimes e$ case, we obtain four ground state tunneling levels, $A_{1g} + A_{2g} + 2E_g$.

In the same $E \otimes e$ problem, if there are only three absolute minima of D_{4h} symmetry (Fig. 3.5), the electronic E_g term is split at the minimum configuration into two singlets, $A_{1g} + B_{1g}$. The corresponding energy levels resulting from the tunneling of the ground vibrational states in the wells can be easily shown to be either $A_{1g} + E_g$ (when A_{1g} is the lowest electronic term at the minimum configuration), or $A_{2g} + E_g$ (when B_{1g} is lower).

In the case of the $T \otimes (e + t_2)$ problem with three tetragonal minima of $G_L = D_{4h}$ symmetry the triplet electronic state T_{1u} is split into $A_{2u} + E_u$, whereas T_{2u} results in $B_{2u} + E_u$. In both cases the E_u term is higher in energy, otherwise the instability will persist (in accordance with the Jahn-Teller theorem), and the symmetry will become lower than the tetragonal one assumed above. It follows from the character table that the only irreducible representation of the group O_h which by reduction to D_{4h} contains A_{2u} (or B_{2u}), is T_{1u} (or T_{2u}). This means that in the case under consideration with only tetragonal distortion, the triply degenerate vibronic ground level does not split by tunneling and remains a vibronic triplet T_{1u} (or T_{2u} in the case of B_{2u} electronic states at the minimum) formed by three equivalent A_{2u} (or B_{2u}) ground states in the wells, i.e., the same as the initial electronic term.

Note that all the above approaches to the problem are valid also in the corresponding cases of nondegenerate ground states, when the minima of the adiabatic potential energy surfaces are caused by the pseudo–Jahn-Teller effect. For instance, in the case of the off-center impurities in crystals (Sect. 3.5.3) within the cluster approximation (Sect. 4.5) the adiabatic separation of the slow angular motion from the fast radial one leads to the well-known *Devonshire model* of internal hindered rotations in cubic systems [4.42–44]. In the limiting case of a very strong pseudo–Jahn-Teller effect, with large potential barriers between the minima, the above approach based on localized vibrational states in the wells is known in the theory of low-symmetry crystal lattice defects as the tunneling model [4.45–48].

4.4 Numerical Solutions of Vibronic Equations

Analytical solutions of vibronic equations are possible only in the limiting cases of weak and strong vibronic coupling. However, estimates of the constants of vibronic interaction show that real polyatomic systems are, in most cases, in the regime known as intermediate vibronic coupling, where $E_{JT} \sim \hbar\omega$. In this section the methods most often used in numerical solutions of the Jahn-Teller vibronic equations are briefly considered, including the reduction of the problem to an algebraic eigenvalue one by choosing an appropriate basis for calculation, as well as other variational methods.

The simplest and most convenient basis for transforming to the matrix formulation of the problem is the so-called weak coupling basis (4.1.2). The exact eigenfunctions of the Hamiltonian can be represented by an expression in the form of the expansion

$$\Psi_{\Gamma'\gamma'}(\mathrm{r}, Q) = \sum_{\gamma} \sum_{\dots n_{\bar{\Gamma}\bar{\gamma}}\dots} C^{(\Gamma'\gamma')}_{\gamma,\dots n_{\bar{\Gamma}\bar{\gamma}}} \psi_{\Gamma\gamma}(\mathrm{r}) |\dots n_{\bar{\Gamma}\bar{\gamma}}\dots\rangle \ . \tag{4.4.1}$$

Substituting the expansion (4.4.1) into the Schrödinger equation with the Hamiltonian (2.2.12) one passes to the algebraic problem of the eigenvalues of an infinite matrix, representing the Hamiltonian (2.2.12) in the basis of the states (4.1.2). The eigenvectors of this matrix are columns formed by the coefficients of the expansion (4.4.1). The contribution of energy terms of the expansion (4.4.1) which lie higher than the upper sheets of the adiabatic potential, $\sum_{\bar{\Gamma}\bar{\gamma}} \hbar\omega_{\bar{\Gamma}} n_{\bar{\Gamma}\bar{\gamma}} \gg 4E_{JT}$ to the exact wave functions of the ground and the lowest excited states rapidly decreases with increasing quantum occupation numbers $n_{\Gamma\gamma}$ of the oscillator states. Therefore the expansion (4.4.1) can be cut off at a reasonable finite number of terms. In so doing, the matrix of the Hamiltonian becomes finite and its diagonalization can be realized by means of a computer, e.g., by the method of rotations.

This approach has some important limitations. First, although it gives good results for weak and intermediate coupling, it is practically unacceptable for the case of strong vibronic coupling, when the vibronic states are localized in the minima and are significantly different from the basis functions (4.1.2). It is obvious that the stronger the vibronic coupling, the more terms of the expansion (4.4.1) must be kept, and the greater the dimensionality of the Hamiltonian matrix. The algorithms of diagonalization of the matrix, usually used in numerical calculations, require the storage of the whole matrix in the operative memory. If the dimensionality of the basis space (i.e., the number of terms in the expansion) is M, then the number of the operative memory cells needed for computer storage is proportional to M^2. The restricted size of the operative memory of the computer thus determines the upper limit for the vibronic constant values for which the numerical diagonalization of the vibronic Hamiltonian in the above scheme gives reasonable results.

A second limitation is the fact that the finiteness of the basis has an influence on the accuracy of the results. The nearer the energy of the states under consideration is to the upper limit of the calculated energy spectrum, the less accurate are the results. Therefore reasonable accuracy can be obtained only for a relatively small number of vibronic states.

In order to weaken these constraints symmetry considerations are usually employed. In cases when the Jahn-Teller Hamiltonian possesses the symmetry of some continuous group which allows us to separate some of the dynamic variables (Sect. 3.2.1) there is no sense in numerical solution of the dynamic problem for all the degrees of freedom.

For instance, in the case of the linear $(E \otimes e)$–type Jahn-Teller effect, instead of looking for the eigenvalues of the Hamiltonian (3.2.1), it is convenient first to separate out the angular motion and to use numerical methods for determining the eigenvalues of the Hamiltonian (4.2.5) describing the radial motion only. The separation of the angular variables has to be performed not only in the Hamiltonian, but also in the basis functions. For this reason one can pass from the states $|n_\theta n_\varepsilon\rangle$ of the two-dimensional harmonic oscillator to linear combinations $\exp(il\varphi)\chi_{nl}(\varrho)$, $l = -n, -n+2, \ldots, n$ within the set of degenerate states with a fixed number of phonons $n = n_\theta + n_\varepsilon$. This can be done by means of the matrices of irreducible representations of the appropriate chain of subgroups, see the discussion after (4.1.13). Thus the radial functions $\chi_{nl}(\varrho)$ of the two-dimensional harmonic oscillator in the representation of the vibrational moment $L_z = -i\partial/\partial\varphi$ can be chosen as the basis states for the Hamiltonian (4.2.5). In this basis the matrix of the Hamiltonian (4.2.5) is [4.11]

$$\hat{H} = \begin{pmatrix} j+1 & k\sqrt{j+1} & 0 & 0 & 0 & \cdots \\ k\sqrt{j+1} & j+2 & k\sqrt{1} & 0 & 0 & \cdots \\ 0 & k\sqrt{1} & j+3 & k\sqrt{j+2} & 0 & \cdots \\ 0 & 0 & k\sqrt{j+2} & j+4 & k\sqrt{2} & \cdots \\ 0 & 0 & 0 & k\sqrt{2} & j+5 & \cdots \\ \cdots & \cdots & \cdots & \cdots & \cdots & \cdots \end{pmatrix}, \qquad (4.4.2)$$

where $j = l \pm 1/2$ are the quantum numbers of the conserved moment $\hat{J}_z = L_z + \frac{1}{2}\hat{\sigma}_z$, $k = |V_E|(\hbar\omega_E^3)^{-1/2}$ is the dimensionless vibronic coupling constant and the eigenvalues of the matrix (4.4.2) determine the energy values in units of $\hbar\omega_E$. The appropriate eigenvalue problem for the linear $E \otimes e$ case is solved in [4.5, 11, 49–52]. The resulting vibronic energies versus k are given in Fig. 4.9 (after [4.51]). For small values of k (less than 0.25) the energy levels are described well by (4.1.16) and (4.1.22) for weak coupling (Fig. 4.1). For large values of k (greater than 4) the vibronic levels have a vibrational-rotational structure described by (4.2.12) for strong coupling.

Note that the basis functions $\exp(il\varphi)\chi_{nl}(\varrho)$ used in the numerical calculation are the symmetry-adapted correct zero-order functions of the perturbation theory with the vibronic coupling constant considered as a small parameter (Sect. 4.1). This is the reason why these functions are used as the basis for weak coupling.

The separate solution of the eigenvalue problem of the Hamiltonian (4.4.2) with different j values means that the matrix of the total Hamiltonian is divided

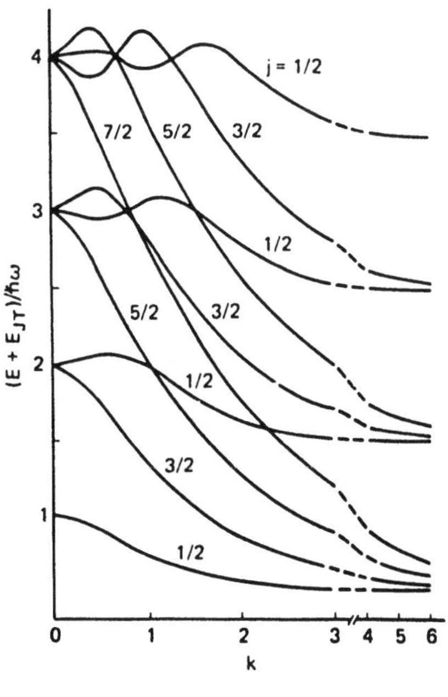

Fig. 4.9. Vibronic energy levels in the linear $E \otimes e$ case, obtained by numerical calculations, versus the dimensionless vibronic constant $k = |V_E|/\sqrt{\hbar\omega_E^3}$. (After [4.51])

into blocks numbered by different irreducible representations $E_{|j|}$ of the axial group. The diagonalization of the Hamiltonian by blocks considering only the mixing of repeating irreducible representations significantly enlarges the calculation possibilities of the method, allowing us to take into account a larger number of terms in the expansion (4.4.1). From this point of view the classification of the basis functions by irreducible representations of the chain of subgroups

$$U(f_\Gamma) \rightarrow SU(f_\Gamma) \rightarrow R(f_\Gamma) \rightarrow R(3) \rightarrow G_0 \ , \tag{4.4.3}$$

where f_Γ is the multiplicity of the degenerate Jahn-Teller vibrations, is very useful. It is important that the consequent reduction of the symmetry in the order (4.4.3) and appropriate reduction of irreducible representations do not as a rule result in repeating irreducible representations. This facilitates the construction of the basis functions of the required symmetry and allows us to provide them with genealogical quantum numbers. The Hamiltonian, being a scalar of the group G_0, transforms after the non-totally-symmetric representations of a higher symmetry group. Construction of the block matrix of the Hamiltonian becomes easier when one applies the Wigner-Eckart theorem to the non-totally-symmetric tensor operators of the Hamiltonian, and this also allows a deeper understanding of the nature of the mixing of different states classified by the same irreducible representations of the group G_0. The modern aspects of the Racah theory applied to Jahn-Teller problems are presented in more detail in comprehensive surveys [4.53, 54].

The *method of group genealogy* described above is most effective in cases when the group G_0 of the vibronic Hamiltonian is one of the continuous Lie groups, as, for instance, in the linear $E \otimes e$ case. Nevertheless, even when this is not so, the genealogical quantum numbers simplify greatly the solution of the problem. These considerations were used in [4.55] for the solution of the linear $T \otimes t_2$ problem. The wave functions $|n_\xi n_\eta n_\zeta\rangle$ of the three-dimensional isotropic oscillator of T_2 vibrations were employed in this work to construct linear combinations Φ_{lm}—eigenfunctions of the operator of the vibrational momentum (4.1.31), which transform as the irreducible representations $D^{(l)}$ of the group $O(3)$. Then allowing for the fact that the electronic states of the T_1 term transform as the irreducible representation $D^{(1)}$ of the group $O(3)$ (energy spin $S = 1$), and in accordance with the momentum composition rules, the states of the total momentum $\hat{\boldsymbol{J}} = \boldsymbol{L} + \hat{\boldsymbol{S}}$ (forming the basis of the weak coupling) were constructed:

$$\Psi_{JM_J}(\mathrm{r}, Q) = \sum_{m, M_S} \psi_{T_1 M_S}(\mathrm{r}) \Phi_{lm}(Q) \langle 1 M_S lm | JM_J \rangle \ , \tag{4.4.4}$$

where $\langle SM_S lm | JM_J\rangle$ are the Clebsch-Gordan coefficients of the group $O(3)$. Further reduction of the symmetry $O(3) \rightarrow O_\mathrm{h}$ results in repeating representations mixed by the operator of vibronic interaction. The energy levels, obtained by diagonalizing the appropriate matrix in the basis (4.4.4) cut off at the states with $J = 13$, are given in Fig. 4.10.

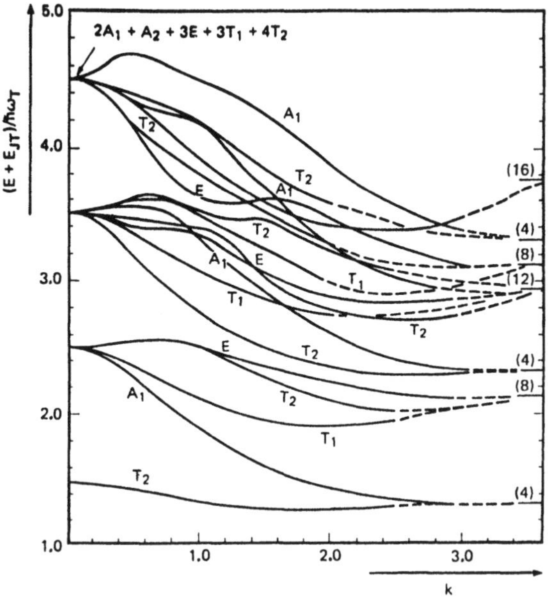

Fig. 4.10. Vibronic energy levels in the $T \otimes t_2$ case versus the dimensionless vibronic coupling constant. (After [4.55])

Note that for weak coupling the vibronic spectrum represents an equally spaced set of energy levels with a spacing $\Delta E = \hbar\omega_T$ weakly split in accordance with the results of perturbation theory. In the strong coupling limit [4.56] the energy levels are also assembled in equally spaced groups of levels with two spacings, $\Delta E = \hbar\omega_T$ and $\Delta E = \hbar\omega_T\sqrt{2/3}$, which corresponds to the harmonic vibrations of the system near the bottom of the trigonal minima, cf. (3.3.15) at $\eta = 0$. The small splitting of the levels in the equally spaced groups is determined by the tunneling between the minima (Sect. 4.3.3).

The same ideas were used in the papers [4.57, 58] for numerical solution of the quadratic $E \otimes e$ and $\Gamma_8 \otimes e$ cases. The vibronic energy levels as functions of the quadratic coupling constant are given in Fig. 4.11 for the particular case of $E_{\rm JT} = 4.6\hbar\omega_E$. Note the similarity of this exact result with the approximate one given in Fig. 4.7. In the limit of strong coupling, corresponding to deep adiabatic potential minima, the lowest energy levels are grouped in tunneling multiplets, forming almost equally spaced sets corresponding to small vibrations at the bottom of the minima. However, in spite of the relatively strong coupling, there is no quantitative agreement between the results of Figs. 4.7 and 4.11. The differences become more important as the vibronic constant decreases; at $E_{\rm JT}/\hbar\omega_E \lesssim 2.5$ only the lowest vibronic doublet and singlet may be satisfactorily described in terms of the results of Sect. 4.3.

The zero value of the linear vibronic constant reduces the Hamiltonian of the quadratic $E \otimes e$ problem to that of the Renner effect (Sect. 3.1.1). Figure 4.12

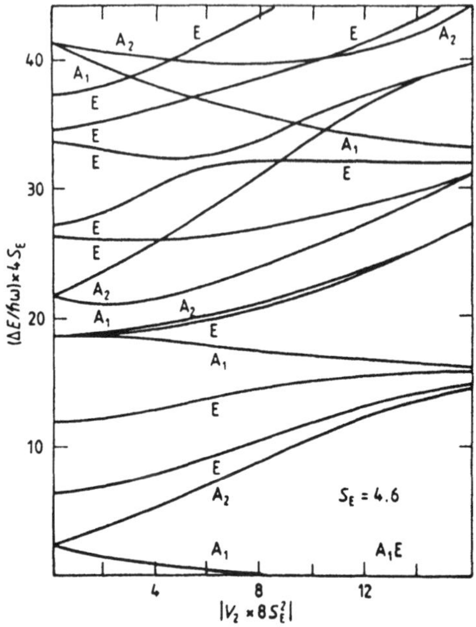

Fig. 4.11. Vibronic energy levels (in units of $\hbar\omega_E/4S_E$) versus the dimensionless vibronic constant $V_2 = W_E/\omega_E^2$ in the quadratic $E \otimes e$ case for $E_{JT}/\hbar\omega_E = S_E = 4.6$; cf. Fig. 4.7. (After [4.57])

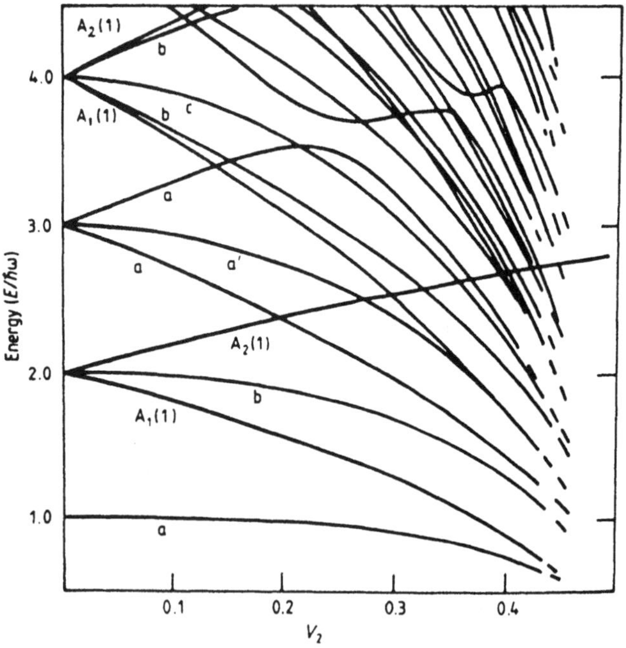

Fig. 4.12. Vibronic energy levels in the Renner $E \otimes e$ case versus the dimensionless quadratic vibronic constant. (After [4.57])

illustrates the vibronic energy levels for a Renner system obtained in [4.58] by numerical diagonalization of the Hamiltonian. In accordance with (3.1.24) the limiting value of the quadratic coupling constant W_E for which the lowest sheet of the adiabatic potential becomes a plane is $|W_E| = \omega_E^2/2$. As seen from Fig. 4.12, for this value W_E the vibronic levels condense into a continuous spectrum. For analogous calculations of the vibronic states of Renner type systems see, e.g., [4.59–61].

Similar ideas were used for numerical solutions of the $(A + B) \otimes b$ [4.62] and $(A + E) \otimes e$ [4.63, 64] cases, as well as the pseudo-Renner problem $(A + E) \otimes e$ [4.65]. More complicated cases of the pseudo–Jahn-Teller effect arising in the $^{2S+1}T \otimes (e + t_2)$ problem, taking account of the spin-orbital interaction, are considered below.

In more complicated cases when the symmetry of the vibronic Hamiltonian is not described by the Lie group but corresponds to the molecular symmetry of the nuclear framework, an increase in the possibilities of computer calculation is facilitated by the use of some special algorithms for matrix diagonalization best suited to the Jahn-Teller problems under consideration. Previously the method of minimal iterations, the *Lanczos method* [4.66], was suggested for this purpose (see also [Ref. 4.67; Sect. 50, 63]). This method is based on the construction of a new basis set by orthogonalizing the vectors obtained by the action of the Hamiltonian on a given initial function $|\psi_1\rangle$:

$$|\psi_2\rangle = H|\psi_1\rangle - |\psi_1\rangle \frac{\langle\psi_1|H|\psi_1\rangle}{\langle\psi_1|\psi_1\rangle} \; ;$$

$$|\psi_3\rangle = H|\psi_2\rangle - |\psi_1\rangle \frac{\langle\psi_1|H|\psi_1\rangle}{\langle\psi_1|\psi_1\rangle} - |\psi_2\rangle \frac{\langle\psi_2|H|\psi_2\rangle}{\langle\psi_2|\psi_2\rangle} \tag{4.4.5}$$

and so on. The process is continued according to the formula

$$|\psi_{n+1}\rangle = H|\psi_n\rangle + g_{n1}|\psi_1\rangle + g_{n2}|\psi_2\rangle + \dots + g_{nn}|\psi_n\rangle \; , \tag{4.4.6}$$

where

$$g_{ik} = -\frac{\langle\psi_k|H|\psi_i\rangle}{\langle\psi_k|\psi_k\rangle} \; , \qquad k = 1, 2, \dots, i \; , \tag{4.4.7}$$

until the zero vector is obtained. An interesting feature of the Lanczos method lies in the fact that when choosing the initial function which transforms after a certain line of an irreducible representation of the group G_0, we obtain by means of the recurrence formulae (4.4.6, 7), a new basis of states having the same transformation properties. This statement can be proved easily if one takes into account that the Hamiltonian is a scalar of the group G_0. It is obvious that the recurrent process results in the zero vector when the set of states of the given symmetry is exhausted and the dimensionality of the resulting basis space of states $\psi_i^{(\Gamma\gamma)}$ is the number of repeating irreducible representations Γ in the initial basis of the weak coupling.[5]

If the electronic degeneracy of the initial Jahn-Teller term is a non-Kramers one, then the Hamiltonian matrix is symmetric and the situation is significantly simplified, since in this case the Hamiltonian matrix in the basis of states (4.4.5), (4.4.6) is tridiagonal [4.67]. This circumstance is very important as it provides a considerable economy of computation time and operative memory. Approved algorithms with rapid convergence [4.68] are available for numerical diagonalization of tridiagonal matrices.

The Lanczos method is used in [4.51] for numerical solution of the $E \otimes (b_1 + b_2)$ case. The results of diagonalizing a 231×231 matrix corresponding to the mixing of vibrational states of the type (4.4.1) with quantum numbers $n_1 + n_2 \leq 20$ are given in Fig. 4.13. For $\omega_1 = \omega_2$ and $V_1 = V_2$ the Hamiltonian of the $E \otimes (b_1 + b_2)$ case coincides with the one of the linear $E \otimes e$ problem [cf. (3.1.7) with (3.1.25)]. Therefore as well as the $E \otimes (b_1 + b_2)$ problem, the linear $E \otimes e$ problem has also been solved to a higher accuracy. The diagonalization of the 200×200-matrix (4.4.2) in [4.51] at $j = 1/2$ corresponds to the mixing of all the states with quantum numbers $n_\theta + n_\varepsilon \leq 199$.

A remarkable regularity was revealed in this case. The vibronic energy levels of the excited states as functions of the vibronic constant oscillate about the mean values (base lines) $\hbar\omega_E(n + 1) - E_{JT}$ and intersect at the nodal points. After each intersection the level with the largest quantum number j smoothly goes down, taking its place among the levels of the rotational fine structure of the strong coupling energy level scheme (Fig. 4.14). There is a *multiple accidental degeneracy* at the points where the levels intersect. In fact this degeneracy may not be accidental. Some considerations about its nature are given in [4.69] (see below) but up to now no full group-theoretical analysis of the Schrödinger equation has been performed in order to reveal the internal dynamic symmetry, which determines the degeneracy described above.

Muramatsu and *Sakamoto* [4.51, 56, 57, 70–74] were the first to use the Lanczos algorithm for Jahn-Teller problems. *Haller*, *Cederbaum* and co-workers (see [4.75, 76] and references therein) also employed this method. Later, it was shown that the Lanczos method could be very useful without the re-orthogonalization procedure [4.66]. *O'Brien* and *Evangelou* [4.77] adopted this strategy for the $E \otimes (e_1 + e_2)$ case and for the $T \otimes (e + t_2)$ case with spin-orbit coupling. A similar technique was used in [4.78]. The Lanczos method was somewhat modified by *Evangelou* et al. [4.79] in order to obtain the vibronic ground state vector. For mathematical details, see the review paper by *Pooler* [4.80].

The number of mixing states which must be kept in the expansion (4.4.1) is determined not only by the magnitude of the constant of vibronic coupling but also by the number of Jahn-Teller active vibrational degrees of freedom. The

[5] In the case of accidental degeneracies the dimensionality of the basis may be lower.

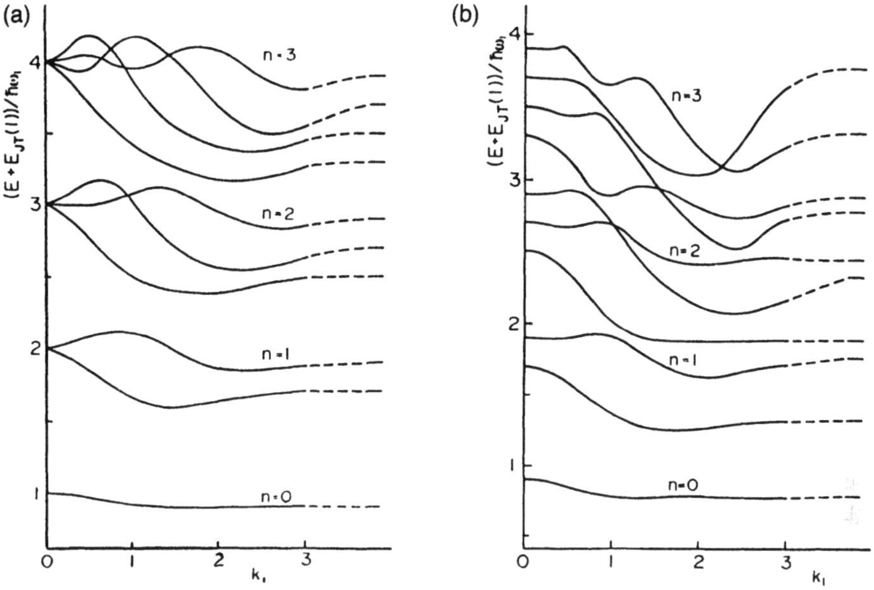

Fig. 4.13a, b. Vibronic energy levels in the $E \otimes (b_1 + b_2)$ case [4.51]. Here $k_i = |V_i|(\hbar\omega_i^3)^{-1/2}$ is the dimensionless coupling constant for (a) $\omega_1 = \omega_2$, $k_2 = 0.6k_1$ and (b) $\omega_2 = 0.8\omega_1$, $k_2 = 0.8k_1$. The number n denotes the energy levels arising from the vibrational states with $n = n_1 + n_2$

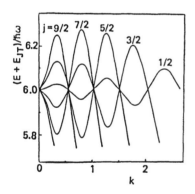

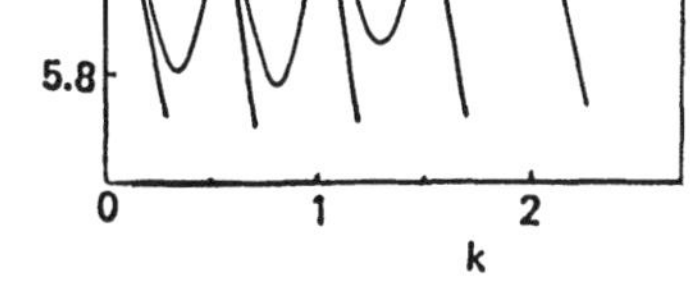

Fig. 4.14. Example of multiple accidental degeneracy of the excited vibronic states in the linear $E \otimes e$ case [4.51]. Here j is the quantum number of the angular motion, $k = |V_E|(\hbar\omega_E^3)^{-1/2}$ is the dimensionless vibronic coupling constant

degeneracy of the nth level of the s-dimensional isotropic oscillator is of the order of $n^{s-1}/(s-1)!$, and therefore, considering the mixing of N excited quantum states with the ground one, we must deal with a basis having a dimension of the order $N^s/(s-1)!$. If one takes into account that the functions of the weak coupling of the type (4.4.1) also have an electronic quantum number $\Gamma\gamma$, then the full dimension of the basis has to be multiplied by f_Γ, the degeneracy of the electronic term. For instance, for the $T \otimes (e + t_2)$ case ($f_\Gamma = 3, s = 5$), considering the admixture of the vibrational states with $n_\theta + n_\varepsilon + n_\xi + n_\eta + n_\zeta \leq 20$ ($N = 20$), a 59136×59136 matrix should be diagonalized. The classification of the states according to irreducible representations reduces the Hamiltonian matrix to a block form, the dimension of each block being less than that of the initial matrix by approximately an order of magnitude. But even after such a simplification of the problem the diagonalization of each block is at the limits of the possibilities of modern computers. Again, the cutoff of the basis by vibrational states with $N = 20$ allows us to obtain reliable results only for moderate vibronic coupling ($k \lesssim 4$), and only for the lowest vibronic states.

Thus the strong dependence of the basis dimension used in the diagonalization on s strongly constrains the computer possibilities when considering Jahn-Teller systems with a larger number of degrees of freedom. Considerable difficulties arise even in the solution of the linear $T \otimes (e + t_2)$ problem, therefore approximate approaches and models are usually employed. One of the models for particular cases is the so-called *d-mode approximation.* In this model it is assumed that $E_{\text{JT}}(E) = E_{\text{JT}}(T)$ and $\omega_E = \omega_T$, and consequently the Hamiltonian acquires the form (3.3.1) with the $O(3)$ symmetry. As mentioned above, this approximation is adequate to the situation in some F-centers in which the diffuse electronic orbitals are insensitive to the cubic symmetry of the local environment in the lattice [4.81]. The same vibronic Hamiltonian is suitable for the $T \otimes v$ case in icosahedral systems (Sect. 3.3).

The high symmetry of the Hamiltonian (3.3.1) allows us to use the powerful Racah method [4.15, 53, 54, 82] for its diagonalization or the symmetry advantages of the Lanczos method [4.71, 83].

The modern status of calculations of the vibronic spectrum for the linear $T \otimes (e + t_2)$ problem is illustrated schematically in Fig. 4.15. As can be seen from this figure, [4.3, 4, 14, 15, 55, 71, 84] in fact cover most of the domain of possible values of vibronic constants. Of course the borders of the regions in this figure are chosen by convention.

The solution of the $\Gamma_8 \otimes (e + t_2)$ problem is more complicated because there is a larger number of electronic states, but it becomes easier owing to additional integrals of motion (Sect. 3.3). The results of numerical calculations for the $\Gamma_8 \otimes t_2$ problem obtained in [4.5, 69, 85] are given in Fig. 4.16. Excited vibronic levels of the linear $\Gamma_8 \otimes t_2$ case oscillate as a function of vibronic constant about the mean values $\hbar\omega_T(n + \frac{3}{2}) - E_{\text{JT}}$, and at some vibronic coupling values they coincide at nodal points forming accidental degenerate multiplets, similar to that in the linear $E \otimes e$ problem (cf. Fig. 4.14 with Fig. 4.16).

The accidental degeneracy of the energy levels in the $\Gamma_8 \otimes t_2$ case was found by *Thorson* and *Moffit* [4.85] ten years before the similar degeneracy was discovered in the $E \otimes e$ problem [4.71]. Analysing the nature of the degeneracy in the $\Gamma_8 \otimes t_2$ problem *Judd* [4.69] came to the conclusion that a similar phenomenon should take place in the linear $E \otimes e$ problem. *Muramatsu* and *Sakamoto* [4.71] obtained this result by means of numerical calculations, probably independently of Judd. In order to explain this specific behavior of the vibronic levels, *Judd* [4.69] employs the similarity of the matrix (4.4.2) when

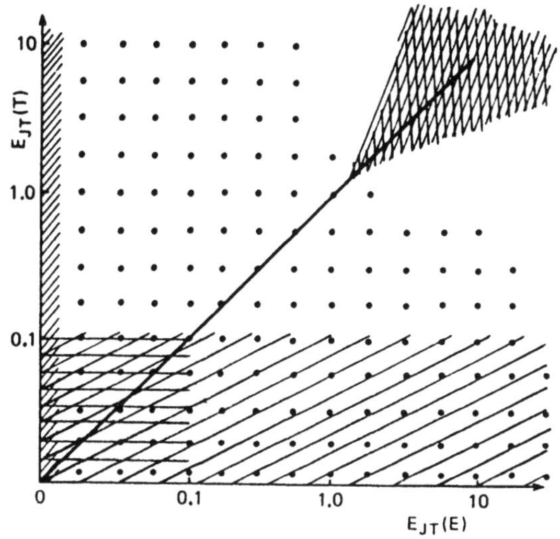

Fig. 4.15. Present status of the vibronic spectra calculations in the $T \otimes (e + t_2)$ case. The tetragonal $E_{JT}(E)$ and trigonal $E_{JT}(T)$ Jahn-Teller stabilization energies are given on the axes in units of $\hbar\omega_E = \hbar\omega_T = \hbar\omega$ in a logarithmic scale. The area of possible parameter values in this scheme is divided into domains of applicability of the perturbation theory of *Moffitt* and *Thorson* [4.3] (horizontal hatching), the numerical results of *Caner* and *Englman* [4.55] (narrow-spaced inclined hatching), the perturbation theory of *Bersuker* and *Polinger* [4.4] (widely spaced inclined hatching), the *d*-mode model, $E_{JT}(E) = E_{JT}(T)$, of *O'Brien* [4.15] (sloping line), the qualitative results of *O'Brien* [4.14] (cross hatching), the numerical calculations of *Sakamoto* and *Muramatsu* [4.71] and *Boldyrev* et al. [4.84] (dots)

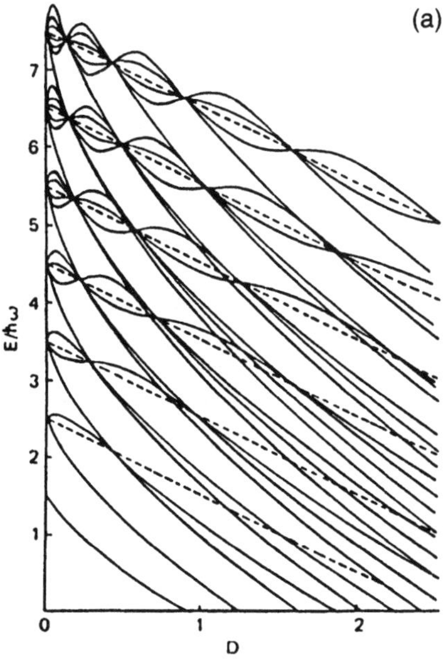

Fig. 4.16a, b. Vibronic energy levels (in units of $\hbar\omega$) in the $\Gamma_8 \otimes t_2$ problem versus $D = E_{JT}/\hbar\omega$ [4.5, 69, 85]. (**a**) General view. The dashed lines indicate the average values around which the oscillations of the energy take place. (**b**) Example of accidental degeneracy of the levels arising from the vibrational term with $n_\xi + n_\eta + n_\zeta = 6$

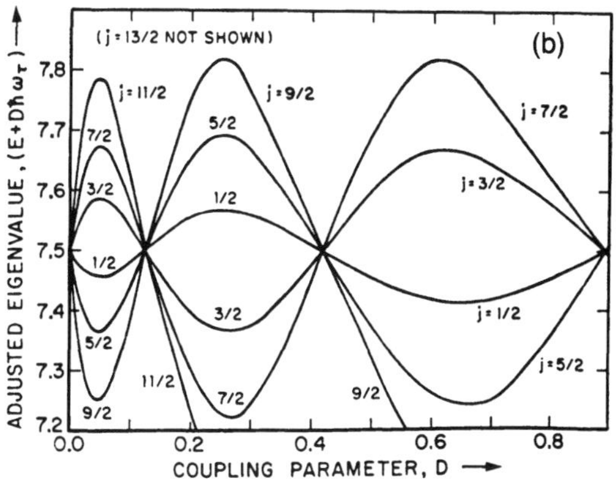

$j = -1/2$ to the matrix of the Hamiltonian $H_0 = \frac{1}{2}(p^2 + q^2) + kq$ in the basis of the functions of the undisplaced oscillator. Considering the difference between these matrices by perturbation theory he shows that the intersections of the lines occur at $J_1(2k\sqrt{2n+3}) = 0$, where $J_n(z)$ is the Bessel function, which is independent of the quantum number j.

The numerical solution of the linear $\Gamma_8 \otimes (e + t_2)$ case in the d-mode approximation, that is at $E_{\mathrm{JT}}(E) = E_{\mathrm{JT}}(T)$, $\omega_E = \omega_T$ is given in [4.86]. After the separation of the cyclic variables the secular matrix in the basis of weak coupling has the form (4.4.2), the same as in the linear $E \otimes e$ and $\Gamma_8 \otimes t_2$ cases, with the only difference being that the quantum number j has to be replaced by $2\mu + 3/2$, where μ is a half-integer. It is obvious that in this problem the excited vibronic levels oscillate about the mean values $\hbar\omega(n + 5/2) - E_{\mathrm{JT}}$ coinciding at nodal points at some values of the linear vibronic coupling constant.

The number of polyatomic systems with quasi-degenerate electronic levels (Sect. 3.4) is extremely large. In cases when the energy gap Δ between the electronic terms in the absence of vibronic interaction is much less than the quantum of Jahn-Teller active vibrations $\hbar\omega$, the low symmetry perturbation lifting the electronic degeneracy can be considered by perturbation theory (Sect. 4.7). In cases when $\Delta \gtrsim \hbar\omega$ the pseudo–Jahn-Teller problem has to be solved by a different method. The same is true when the group of neighboring electronic terms mixed by the vibrations arises because of spin-orbital splitting. If the spin-orbital interaction results in splitting even in first-order perturbation theory, then this splitting is, as a rule, not small and has to be taken into account along with the vibronic interaction. A typical example of this kind is the cubic polyatomic systems in a triply degenerate electronic state. The effective Hamiltonian of the ^{2S+1}T term including the second-order perturbation terms with respect to the spin-orbit interaction has the form

$$H_{SO} = \zeta\hat{\boldsymbol{L}}\hat{\boldsymbol{S}} + \mu(\hat{\boldsymbol{L}}\hat{\boldsymbol{S}})^2 + \varrho(\hat{L}_x^2\hat{S}_x^2 + \hat{L}_y^2\hat{S}_y^2 + \hat{L}_z^2 S_z^2) \ , \tag{4.4.8}$$

where $\hat{L}_x$, $\hat{L}_y$ and $\hat{L}_z$ are the matrices of the operator of the orbital momentum of the electrons determined in the basis of electronic states of the T term [they coincide with the matrices of the energy spin (3.3.3)], $\hat{S}_x$, $\hat{S}_y$ and $\hat{S}_z$ are the spin matrices, and ζ, μ and ϱ are the parameters of the effective Hamiltonian (in the approximation of the crystal field theory the constants ζ, μ and ϱ are simple functions of the crystal field parameter Dq [4.87]). If, besides the spin-orbital interaction, the low symmetry tetragonal or trigonal crystal field is also significant, the appropriate perturbation $\hat{V}_{\mathrm{tetr}} = \varepsilon\hat{C}_\theta$ or $\hat{V}_{\mathrm{trig}} = \tau(\hat{C}_\xi + \hat{C}_\eta + \hat{C}_\zeta)$, where $\hat{C}_{\Gamma\gamma}$ are the matrices determined by (3.3.2), and ε and τ are the reduced matrix elements (the parameters of the theory), must be added to the Hamiltonian (4.4.8).

The problem of the Jahn-Teller effect for the ^{2S+1}T term in the presence of spin-orbital interactions and low symmetry crystal fields is one of the most complicated in the theory of vibronic interactions. The difficulties are caused first by a large number of Jahn-Teller active vibrational degrees of freedom ($s \geq 5$) in the

$T \otimes (e + t_2)$ problem, secondly, by a large number of mixing states $f = 3(2S + 1)$, and thirdly, by the low symmetry of the problem, which prevents effective use of group-theoretical methods. The vibronic Hamiltonian in this case contains a large number of free parameters (two frequencies ω_E and ω_T, two vibronic constants V_E and V_T, three spin-orbital constants μ, ζ and ϱ, and a parameter of the low-symmetry crystal field ε or τ) thus requiring multiple calculations of the vibronic spectrum (as distinguished from, say, the linear $E \otimes e$ problem).

Using the above estimations of the number of basis states necessary to determine the vibronic spectrum, dependent on the numbers of vibrational degrees of freedom and mixing electronic states and the magnitude of the vibronic constants, it can be shown that even for moderate vibronic coupling the matrix which should be diagonalized is too large to be contained by on-line computer storage. For instance in the $^4T \otimes (e + t_2)$ problem with spin-orbital interaction and mixing of the vibrational states with $n_\theta + n_\varepsilon + n_\xi + n_\eta + n_\zeta \leq 20$ the appropriate matrix is of order 236544 × 236544. The classification of the states by the irreducible representations of the trigonal or tetragonal symmetry group allows us to reduce the dimension of the blocks to be diagonalized by 3–4 times.

These difficulties can be reduced if one looks for the solution of the so-called *partial eigenvalue problem*, i.e., when only one or a few of the most important eigenvalues and eigenfunctions are sought for. This partial solution of the problem has some advantages. First, some more powerful and rapidly converging methods have been worked out for solving the partial eigenvalue problem [4.67, 68]. Second, a considerable economy of the computer time and the on-line storage can be achieved.

The majority of the algorithms used in numerical methods for the solution of the partial eigenvalue problem are based on the variational principle. The lowest eigenvalue corresponds to the wave function which minimizes the Rayleigh quotient:

$$E\{\Psi\} = \frac{\langle \Psi | H | \Psi \rangle}{\langle \Psi | \Psi \rangle} \,. \tag{4.4.9}$$

In the matrix representation this quotient is a function of many variables, the coefficients of the expansion (4.4.1).

In [4.84] the *method of coordinate relaxation* ([4.67], Sect. 61) was suggested for determining the lowest vibronic eigenvalues and eigenfunctions of the $^3T \otimes (e + t_2)$ Jahn-Teller problem with the spin-orbital interaction. The method is based on the idea of the coordinate slope by successive minimization of the Rayleigh quotient along each of the coordinates in the multidimensional space in which the function $E\{\Psi\}$ is defined. In these calculations there is no need to keep the whole matrix to be diagonalized in the on-line storage of the computer. The simplicity of the matrix elements of the vibronic Hamiltonian evaluated by the vibrational oscillator functions allows us to calculate them each time they are needed and to use the on-line memory of the computer for storing the calculated eigenvectors only. If the dimension of the basis set is M, then the volume of the computer memory used is proportional to M in this method and not to M^2, as in the case of the full solution of the eigenvalue problem. This increases the computer possibilities (see also [4.83]).

It is important that in all the iteration methods based on the variational principle the transformation properties of the probe function are preserved during the minimization. If $\Psi_{\Gamma\gamma}^{(0)}$, transforming as the γ line of the irreducible representation Γ is chosen as the trial function, the iteration process results in the exact eigenfunction $\Psi_{\Gamma\gamma}$ with the same transformation properties for the lowest energy level.

Some other numerical methods for the diagonalization of the Jahn-Teller Hamiltonians using their specific forms (they are block-band and sparse matrices) have also been proposed by *Pooler* and *O'Brien* [4.86].

4.5 Multiparticle Methods in the Theory of the Jahn-Teller Effect

It is well known that the methods of multiparticle theory are very useful in quantum theory (see, e.g., [4.88]). It is natural to assume that these methods will yield good results for vibronic systems too. However, direct application of the standard procedures of quantum statistical physics to vibronic problems leads to incorrect results. For instance, in [4.89] the equation of motion for the quantum density matrix of a Jahn-Teller impurity center was employed; when approaching the limit $T = 0$ K in the absence of phonon dispersion the results of this paper do not give the correct splittings of the one-phonon states considered in Sect. 4.1. Ignoring the differences in the transformation properties of different vibronic states of the same vibronic system, some authors came to the conclusion that the usual Green's function methods were inadequate for the Jahn-Teller effect problems (see, e.g., [4.90]).

The essence of these difficulties lies in the fact that the ground state of the Jahn-Teller system is degenerate, the degeneracy and transformation properties of the appropriate vibronic states coinciding with that of the initial electronic states of the Jahn-Teller term (Figs. 4.7, 4.9, 4.10; discussion of this circumstance is given in more detail in Sect. 4.7). In such conditions the usual procedure of the Green's function method is invalid. As shown below, additional assumptions allowing the separation of any one of the degenerate states of the ground level are needed here [4.91, 92]. In the publications [4.93–95] devoted to the evaluation of the vibronic reduction factors characterizing the ground state of a Jahn-Teller system (Sect. 4.7), an external field of appropriate symmetry is introduced, removing the degeneracy of the ground state.

As mentioned in Chap. 1, there is a formal analogy between the Jahn-Teller problem and that of *pion-nucleon interaction* in the static model of the nucleon [4.12]. In [4.96–100] this analogy is used in order to transfer the ideas and methods of the scattering theory (developed in quantum field theory [4.101]) to the case of the Jahn-Teller effect. In [4.93–100] the Green's function method is used to solve the multimode problem. The discussion given below follows mainly the results of [4.96–100] reformulated for ideal vibronic systems. In conclusion (Sect. 4.5.4) the method of unitary transformations and its application to the theory of the Jahn-Teller effect are considered.

4.5.1 General Relationships

As is well known, several forms of Green's functions are used in quantum theory, differing in their explicit arguments, analytical properties and the procedure of averaging. Below we use the so-called *field Green's functions* which are determined by the matrix element of the commutator, anticommutator or chronologically ordered product of operators. Such Green's functions are usually employed in quantum field theory, while for a thermal quantum system they correspond to the case $T = 0$ K.

As already mentioned, the ground state of a one-center Jahn-Teller system is always degenerate, and therefore at $T = 0$ K there may be several matrix elements relating different states of the ground vibronic term. Combined, they form a matrix—the *matrix Green's function* [4.102].

Let us introduce the commutator retarded Green's function for the operators $A_{\Gamma_1\gamma_1}$ and $B_{\Gamma_2\gamma_2}$ ($\Gamma_1\gamma_1$ and $\Gamma_2\gamma_2$, as usual, denote the transformation properties of the operators in a given symmetry group):

$$\langle\langle\gamma|A_{\Gamma_1\gamma_1}, B_{\Gamma_2\gamma_2}|\bar{\gamma}\rangle\rangle = -\frac{i}{\hbar}\theta(t)\langle\bar{\Gamma}\gamma|[A_{\Gamma_1\gamma_1}(t), B_{\Gamma_2\gamma_2}]|\bar{\Gamma}\bar{\gamma}\rangle \;, \tag{4.5.1}$$

where $\bar{\Gamma}$ is the irreducible representation to which the exact states $|\bar{\Gamma}\gamma\rangle$ and $|\bar{\Gamma}\bar{\gamma}\rangle$ of the ground vibronic term belong, and the time dependence of the operator $A_{\Gamma_1\gamma_1}(t)$ is determined by the Heisenberg representation with the exact vibronic Hamiltonian. The trace with respect to the indices γ, $\bar{\gamma}$ in (4.5.1) divided by the multiplicity of the ground state $f_{\bar{\Gamma}}$ gives the usual retarded Green's function at $T = 0$ K:

$$\langle\langle A_{\Gamma_1\gamma_1}|B_{\Gamma_2\gamma_2}\rangle\rangle = f_{\bar{\Gamma}}^{-1}\sum_{\gamma}\langle\langle\gamma|A_{\Gamma_1\gamma_1}, B_{\Gamma_2\gamma_2}|\gamma\rangle\rangle \;. \tag{4.5.2}$$

From the point of view of transformation properties the operator $[A_{\Gamma_1\gamma_1}(t), B_{\Gamma_2\gamma_2}]$ is a tensor of the second rank. By using Clebsh-Gordan coefficients it can be represented in the form of a sum of irreducible tensor operators:

$$[A_{\Gamma_1\gamma_1}(t), B_{\Gamma_2\gamma_2}] = \sum_{\Gamma\mu}\langle\Gamma\mu|\Gamma_1\gamma_1\Gamma_2\gamma_2\rangle\hat{G}_{\Gamma\mu}(t) \;, \tag{4.5.3}$$

where

$$\hat{G}_{\Gamma\mu}(t) = \sum_{\gamma_1\gamma_2}\langle\Gamma_1\gamma_1\Gamma_2\gamma_2|\Gamma\mu\rangle[A_{\Gamma_1\gamma_1}(t), B_{\Gamma_2\gamma_2}] \;. \tag{4.5.4}$$

Substituting (4.5.3) into (4.5.1) and using the Wigner-Eckart theorem to express the matrix element $\langle \bar{\Gamma}\gamma | \hat{G}_{\Gamma\mu} | \bar{\Gamma}\bar{\gamma} \rangle$, we have

$$\langle\langle \gamma | A_{\Gamma_1\gamma_1}, B_{\Gamma_2\gamma_2} | \bar{\gamma} \rangle\rangle = \sum_{\Gamma\mu} \langle \Gamma\mu | \Gamma_1\gamma_1 \Gamma_2\gamma_2 \rangle \langle \Gamma\mu \bar{\Gamma}\bar{\gamma} | \bar{\Gamma}\gamma \rangle G_\Gamma(t) , \tag{4.5.5}$$

where the so-called *reduced Green's functions* $G_\Gamma(t)$ [reduced matrix elements of the operator (4.5.4)] are introduced:

$$G_\Gamma(t) = -\frac{i}{\hbar}\theta(t) \langle \Gamma\mu \bar{\Gamma}\bar{\gamma} | \bar{\Gamma}\gamma \rangle^{-1} \langle \bar{\Gamma}\gamma | \hat{G}_{\Gamma\mu}(t) | \bar{\Gamma}\bar{\gamma} \rangle . \tag{4.5.6}$$

The reduced Green's functions can be calculated from the usual thermodynamic Green's functions at $T = 0$ K by means of Bogoljubov's idea about quasi averages [4.91]. To do this, a low-symmetry perturbation $\lambda \hat{C}_{\Gamma_\gamma}$, removing the degeneracy of the ground vibronic term and shifting down (in energy) one of the vibronic states, has to be added to the Hamiltonian. When the limits $T \to 0$ K and then $\lambda \to 0$ are taken, the thermodynamic averaging procedure is reduced to the quantum-mechanical one over the state which occurs as the ground state. By introducing different $\lambda \hat{C}_{\Gamma_\gamma}$ operators one can find all the elements of the matrix Green's function (4.5.1) and hence the reduced Green's functions (4.5.4–6).

For instance, for the linear $E \otimes e$ problem having the symmetry of two-dimensional rotations (Sect. 3.2.1) the ground term vibronic states belong to the representation $D^{(1/2)}$. From this it follows that the matrix elements for ground state functions are nonzero only for those $\hat{G}_{\Gamma\mu}$ operators (4.5.4) for which Γ is contained in the product $D^{(1/2)} \times D^{(1/2)} = D^{(0)}_+ + D^{(0)}_- + D^{(1)}$. In the basis of the vibronic states $|\pm 1/2\rangle$ the appropriate matrix Green's functions are $G_0^{(+)}\hat{\sigma}_0$, $G_0^{(-)}\hat{\sigma}_z$, $G_1\hat{\sigma}_+$ and $G_1\hat{\sigma}_-$, see (3.1.1) and (3.1.4). The perturbation $\lambda\hat{\sigma}_z$, where $\lambda = +0$, lowers the energy of the state $|-1/2\rangle$, and taking the limits yields

$$\lim_{\lambda \to +0} \left(\lim_{T \to 0} \langle\langle A | B \rangle\rangle \right) = G_0^{(+)} - G_0^{(-)} . \tag{4.5.7}$$

But if $\lambda = -0$, then the same perturbation lowers the energy of the state $|+1/2\rangle$ and

$$\lim_{\lambda \to -0} \left(\lim_{T \to 0} \langle\langle A | B \rangle\rangle \right) = G_0^{(+)} + G_0^{(-)} . \tag{4.5.8}$$

Solving the systems (4.5.7) and (4.5.8) we find the reduced Green's functions $G_0^{(+)}$ and $G_0^{(-)}$. Similarly, introducing the perturbation $\lambda\hat{\sigma}_x$, one can separate one of the states $(|E\theta\rangle \pm |E\varepsilon\rangle)/\sqrt{2}$, and thus find the reduced function G_1.

The matrix Green's functions introduced here have all the analytical properties of the usual Green's functions. To reveal these properties we use the usual procedure based on the completeness of the set of exact vibronic states $|n\rangle$, and write the Fourier transform of (4.5.1) in the form

$$\langle\langle \gamma | A_{\Gamma_1\gamma_1}, B_{\Gamma_2\gamma_2} | \bar{\gamma} \rangle\rangle_\omega = \frac{1}{2\pi} \sum_n \left(\frac{\langle \bar{\Gamma}\gamma | A_{\Gamma_1\gamma_1} | n \rangle \langle n | B_{\Gamma_2\gamma_2} | \bar{\Gamma}\bar{\gamma} \rangle}{\hbar\omega - (E_n - E_{\bar{\Gamma}})} - \frac{\langle \bar{\Gamma}\gamma | B_{\Gamma_2\gamma_2} | n \rangle \langle n | A_{\Gamma_1\gamma_1} | \bar{\Gamma}\bar{\gamma} \rangle}{\hbar\omega + (E_n - E_{\bar{\Gamma}})} \right) , \tag{4.5.9}$$

where E_n and $E_{\bar{\Gamma}}$ are the exact energies of the vibronic states $|n\rangle$ and the ground state $|\bar{\Gamma}\gamma\rangle$, respectively. Performing as usual analytical continuation towards the region of complex values of ω (and hence introducing also advanced Green's functions), we come to the analytical function (4.5.9) having simple poles at $\omega = \pm(E_n - E_{\bar{\Gamma}})/\hbar$. From this all the known dispersion relationships for the field Green's functions can be obtained, as well as the relation of the so-called crossing symmetry:

$$\langle\langle \gamma | A_{\Gamma_1\gamma_1}, B_{\Gamma_2\gamma_2} | \bar{\gamma} \rangle\rangle_\omega = \langle\langle \gamma | B_{\Gamma_2\gamma_2}, A_{\Gamma_1\gamma_1} | \bar{\gamma} \rangle\rangle_{-\omega} . \tag{4.5.10}$$

It can be seen that the linear combination of the reduced Green's functions,

$$G^{(\Gamma_e)}_{\Gamma_1\Gamma_2} = \sum_\Gamma f_\Gamma^{1/2} \begin{Bmatrix} \bar{\Gamma} & \bar{\Gamma} & \Gamma \\ \Gamma_1 & \Gamma_2 & \Gamma_e \end{Bmatrix} G_{\bar{\Gamma}} \sum_{\gamma\bar{\gamma}} \sum_{\gamma_1\gamma_2} \langle \bar{\Gamma}\gamma \Gamma_1\gamma_1 | \Gamma_e\gamma_e \rangle \times \langle \Gamma_e\gamma_e | \Gamma_2\gamma_2 \bar{\Gamma}\bar{\gamma} \rangle \langle\langle \gamma | A_{\Gamma_1\gamma_1}, B_{\Gamma_2\gamma_2} | \bar{\gamma} \rangle\rangle , \tag{4.5.11}$$

where f_Γ is the dimensionality of the representation Γ and $\begin{Bmatrix} \cdot & \cdot & \cdot \\ \cdot & \cdot & \cdot \end{Bmatrix}$ is the so-called 6Γ symbol [4.103], accounts only for the excited states transforming as the representation Γ_e. In other words, the expression $\mathrm{Im}\{G^{(\Gamma_e)}_{\Gamma_1\Gamma_2}(\omega)\}$ reproduces the spectral density of the transitions from the ground state to the excited ones with Γ_e symmetry. To verify that this statement is valid one has to substitute the definition of G_Γ from (4.5.6) and (4.5.4) into (4.5.11), to use the expression (4.5.9) for $\langle\langle \gamma | A_{\Gamma_1\gamma_1}, B_{\Gamma_2\gamma_2} | \bar{\gamma} \rangle\rangle$, and then to employ the Wigner-Eckart theorem for the matrix elements of the operators $A_{\Gamma_1\gamma_1}$ and $B_{\Gamma_2\gamma_2}$. Curtailing the sum over the lines of the irreducible representations in the 6Γ symbol and using other properties of this symbol one can see that the Green's function $G^{(\Gamma_e)}_{\Gamma_1\Gamma_2}$ selects the contributions to the vibronic spectrum from the states of the Γ_e type, and therefore they are called partial Green's functions [4.96–100]. Note that the matrix of $G^{(\Gamma_e)}_{\Gamma_1\Gamma_2}$ magnitudes is nondiagonal with respect to the indices Γ_1 and Γ_2. For a full separation of partial contributions an additional diagonalization of this matrix is needed. This necessity usually occurs in Jahn-Teller problems with several types of active vibrations, e.g., in the $T \otimes (e + t_2)$ problem.

Let us start from the general Hamiltonian (3.1.29) of a Jahn-Teller system with linear vibronic coupling. In the representation of second quantization, using the operators (3.2.11), we have

$$\hat{H} = \sum_{\Gamma\gamma} \hbar\omega_\Gamma \left(b^\dagger_{\Gamma\gamma} b_{\Gamma\gamma} + \frac{1}{2} \right) + \sum_{\Gamma\gamma} V_\Gamma \sqrt{\frac{\hbar}{2\omega_\Gamma}} (b^\dagger_{\Gamma\gamma} + b_{\Gamma\gamma}) \hat{C}_{\Gamma\gamma} . \tag{4.5.12}$$

For the operators $b_{\Gamma\gamma}(t)$ in the Heisenberg representation the equations of motion are

$$\dot{b}_{\Gamma\gamma}(t) = \frac{\mathrm{i}}{\hbar}[H, b_{\Gamma\gamma}] = \mathrm{i}\omega_\Gamma b_{\Gamma\gamma} - \frac{\mathrm{i}V_\Gamma}{\sqrt{2\hbar\omega_\Gamma}} C_{\Gamma\gamma} \ . \tag{4.5.13}$$

Their solution can be presented in the integral form

$$\hat{b}_{\Gamma\gamma}(t) = \mathrm{e}^{-\mathrm{i}\omega_\Gamma t}\hat{B}_{\Gamma\gamma}(t_0) - \frac{\mathrm{i}V_\Gamma}{\sqrt{2\hbar\omega_\Gamma}} \int\limits_{t_0}^{t} d\tau\, \mathrm{e}^{-\mathrm{i}\omega_\Gamma(t-\tau)}\hat{C}_{\Gamma\gamma}(\tau) \ , \tag{4.5.14}$$

or, equivalently,

$$\hat{b}_{\Gamma\gamma}(t) = \mathrm{e}^{-\mathrm{i}\omega_\Gamma t}\hat{B}_{\Gamma\gamma}(t_0) + \frac{\mathrm{i}V_\Gamma}{\sqrt{2\hbar\omega_\Gamma}} \int\limits_{t}^{t_0} d\tau\, \mathrm{e}^{-\mathrm{i}\omega_\Gamma(t-\tau)}\hat{C}_{\Gamma\gamma}(\tau) \ . \tag{4.5.15}$$

Here $\hat{B}_{\Gamma\gamma}(t_0)$ are constants determined by the initial conditions at time t_0:

$$\hat{B}_{\Gamma\gamma}(t_0) = \exp\left(\frac{\mathrm{i}}{\hbar}\hat{H}t_0\right)\exp\left(-\frac{\mathrm{i}}{\hbar}\hat{H}_0 t_0\right) b_{\Gamma\gamma} \exp\left(\frac{\mathrm{i}}{\hbar}\hat{H}_0 t_0\right)\exp\left(-\frac{\mathrm{i}}{\hbar}\hat{H}t_0\right) , \tag{4.5.16}$$

where $\hat{H}_0$ is the Hamiltonian of the system in the absence of vibronic coupling.

Now, following the definition (4.5.1), we introduce the electronic and phonon Green's functions: $\langle\langle\gamma|\hat{C}_{\Gamma_1\gamma_1}, \hat{C}_{\Gamma_2\gamma_2}|\bar{\gamma}\rangle\rangle$ and $\langle\langle\gamma|b_{\Gamma_1\gamma_1}, b^\dagger_{\Gamma_2\gamma_2}|\bar{\gamma}\rangle\rangle$. For systems with the linear vibronic coupling described by the Hamiltonian (4.5.12) there is a simple relation between these functions. This relation can be obtained by means of (4.5.14) at $t_0 = 0$, if one substitutes the operator $b_{\Gamma_1\gamma_1}(t)$ from (4.5.14) into the phonon Green's function and then uses the symmetry relation (4.5.10) for the resulting Green's function $\langle\langle\gamma|\hat{C}_{\Gamma_1\gamma_1}, b^\dagger_{\Gamma_2\gamma_2}|\bar{\gamma}\rangle\rangle$, and after that, again uses (4.5.14) for the operator $b^\dagger_{\Gamma_2\gamma_2}(t)$. The result is

$$\begin{aligned}\langle\langle\gamma|b_{\Gamma_1\gamma_1}, b^\dagger_{\Gamma_2\gamma_2}|\bar{\gamma}\rangle\rangle_\omega &= \langle\langle\gamma|b_{\Gamma_1\gamma_1}, b^\dagger_{\Gamma_2\gamma_2}|\bar{\gamma}\rangle\rangle^{(0)}_\omega + 2\pi^2\hbar\frac{V_{\Gamma_1}V_{\Gamma_2}}{\sqrt{\omega_{\Gamma_1}\omega_{\Gamma_2}}}\\ &\times \langle\langle\gamma|b_{\Gamma_1\gamma_1}, b^\dagger_{\Gamma_1\gamma_1}|\gamma\rangle\rangle^{(0)}_\omega \langle\langle\gamma|\hat{C}_{\Gamma_1\gamma_1}\hat{C}_{\Gamma_2\gamma_2}|\bar{\gamma}\rangle\rangle_\omega\\ &\times \langle\langle\bar{\gamma}|b_{\Gamma_2\gamma_2}, b^\dagger_{\Gamma_2\gamma_2}|\bar{\gamma}\rangle\rangle^{(0)}_\omega \ , \qquad \text{where}\end{aligned} \tag{4.5.17}$$

$$\langle\langle\gamma|b_{\Gamma_1\gamma_1}, b^\dagger_{\Gamma_2\gamma_2}|\bar{\gamma}\rangle\rangle^{(0)}_\omega = \frac{\delta_{\gamma\bar{\gamma}}\delta_{\Gamma_1\Gamma_2}\delta_{\gamma_1\gamma_2}}{2\pi\hbar(\omega - \omega_{\Gamma_1})} \tag{4.5.18}$$

is the zeroth Green's function, i.e., the function $\langle\langle\gamma|b_{\Gamma_1\gamma_1}, b^\dagger_{\Gamma_2\gamma_2}|\bar{\gamma}\rangle\rangle$ in the absence of vibronic coupling. By comparing (4.5.17) with the known operator identity $\hat{G} = \hat{G}_0 + \hat{G}_0\hat{T}\hat{G}_0$, where $\hat{G} = (E - \hat{H})^{-1}$, $\hat{G}_0 = (E - \hat{H}_0)^{-1}$, $\hat{T} = \hat{V} + \hat{V}\hat{G}\hat{V}$, $\hat{V} = \hat{H} - \hat{H}_0$, one can see that the electronic Green's function is related to some of the matrix elements of the *transition matrix* T, namely to those which correspond to the processes of elastic scattering of the phonon by the Jahn-Teller center [4.98]:

$$T_{\gamma\Gamma_1\gamma_1,\bar{\gamma}\Gamma_2\gamma_2}(\omega) = 2\pi^2\hbar\frac{V_{\Gamma_1}V_{\Gamma_2}}{\sqrt{\omega_{\Gamma_1}\omega_{\Gamma_2}}}\langle\langle\gamma|\hat{C}_{\Gamma_1\gamma_1},\hat{C}_{\Gamma_2\gamma_2}|\bar{\gamma}\rangle\rangle_\omega \ . \quad (4.5.19)$$

This result acquires visible physical meaning in the theory of the multimode Jahn-Teller effect in impurity centers in crystals (see below, Sect. 4.6).

The phonon and electronic matrix Green's functions obey some more exact relationships. In particular, from (4.5.19) one can obtain the Lehmann dispersion relationship for the electronic Green's function [4.104]:

$$\begin{aligned}\langle\langle\gamma|\hat{C}_{\Gamma_1\gamma_1},\hat{C}_{\Gamma_2\gamma_2}|\bar{\gamma}\rangle\rangle_\omega &= \frac{K(\Gamma_1)K(\Gamma_2)}{2\pi\hbar\omega}\langle\psi_{\bar{\Gamma}\gamma}|[\hat{C}_{\Gamma_1\gamma_1},\hat{C}_{\Gamma_2\gamma_2}]|\psi_{\bar{\Gamma}\bar{\gamma}}\rangle - \frac{1}{\pi}\int_{+0}^{\infty} dv \\ &\times\left(\frac{\mathrm{Im}\{\langle\langle\gamma|\hat{C}_{\Gamma_1\gamma_1},\hat{C}_{\Gamma_2\gamma_2}|\bar{\gamma}\rangle\rangle_v\}}{\omega - v}\right. \\ &\left.+\frac{\mathrm{Im}\{\langle\langle\gamma|\hat{C}_{\Gamma_1\gamma_1},\hat{C}_{\Gamma_2\gamma_2}|\bar{\gamma}\rangle\rangle_{-v}\}}{\omega + v}\right) \ . \end{aligned} \quad (4.5.20)$$

Here $K(\Gamma_1)$ and $K(\Gamma_2)$ are the electronic reduction factors (Sect. 4.7), while $\psi_{\bar{\Gamma}\gamma}$ and $\psi_{\bar{\Gamma}\bar{\gamma}}$ are the electronic states of the degenerate term $\bar{\Gamma}$. Unlike the usual dispersion relationships, the contribution of transitions between the states of the ground vibronic term are separated explicitly in (4.5.20). The phonon Green's function $\langle\langle\gamma|b_{\Gamma_1\gamma_1}, b^\dagger_{\Gamma_2\gamma_2}|\bar{\gamma}\rangle\rangle$ obeys similar dispersion equations with the difference that in (4.5.20), instead of vibronic reduction factors, the so-called factors of vibronic amplification (Sect. 4.7) have to be inserted.

4.5.2 The Case of Weak Coupling. Perturbation Theory in the Green's Function Method

Perturbation theory in the Green's function method is of special importance in cases when the results depend on a small parameter in a nonanalytical way, as in the theory of superconductivity. For Jahn-Teller problems this is not the case (see Sect. 4.1). Therefore, as far as ideal Jahn-Teller problems are concerned, the perturbation theory presented below serves as one more way to obtain approximate results in the limit of weak coupling. However, the procedure developed in this section can almost be directly used in the theory of multimode vibronic systems (Sect. 4.6) where other methods become inapplicable, and this justifies the detailed presentation given below.

The perturbation theory for matrix Green's functions (4.5.1) is based on the same principles used for usual Green's functions. For instance, in [4.96–98] the perturbation theory is applied to the mass operator. We use here a simpler procedure, based on the idea of quasi-averages for the usual thermodynamic Green's functions which allows us, nevertheless, to obtain the same results as more complicated approaches.

Consider a Jahn-Teller system with linear vibronic coupling described by the Hamiltonian (2.2.13). As in Sect. 4.1, we present the latter in the form $\hat{H}_0 + \hat{V}$, where $\hat{V}$ is the operator of vibronic interaction (4.1.1), and H_0 is the Hamiltonian of free harmonic vibrations. As follows from (4.5.14) and (3.2.11), the following integral equation of motion holds ($t \geq 0; t_0 = 0$):

$$\hat{Q}_{\Gamma\gamma}(t) = Q^{(0)}_{\Gamma\gamma}(t) + \int\limits_0^t D^{(0)}_{\Gamma}(t-\tau)\frac{\partial \hat{V}(\tau)}{\partial Q_{\Gamma\gamma}}d\tau \ . \tag{4.5.21}$$

Here, as in (4.5.14), the time dependence of the operators $\hat{Q}_{\Gamma\gamma}(t)$ is determined by the Heisenberg representation with the exact Hamiltonian $\hat{H} = \hat{H}_0 + \hat{V}$ [as distinct from $Q^{(0)}_{\Gamma\gamma}(t)$ for which $\hat{H}$ is substituted by $\hat{H}_0$] and

$$D^{(0)}_{\Gamma}(t) = -\frac{\mathrm{i}}{\hbar}\theta(t)\langle[Q^{(0)}_{\Gamma\gamma}(t), Q_{\Gamma\gamma}]\rangle = -\theta(t)\frac{\sin(\omega_E t)}{\omega_E} \tag{4.5.22}$$

is the zeroth retarded Green's function. Similarly, the equation of motion of the electronic operator $\dot{C}_{\Gamma\gamma} = (\mathrm{i}/\hbar)[H, \hat{C}_{\Gamma\gamma}]$ can be presented in the integral form

$$\hat{C}_{\Gamma\gamma}(t) = \hat{C}_{\Gamma\gamma} + \frac{\mathrm{i}}{\hbar}\int\limits_0^t [\hat{V}(\tau), \hat{C}_{\Gamma\gamma}(\tau)]\,d\tau \ . \tag{4.5.23}$$

Substituting the expressions (4.5.21) and (4.5.23) into the retarded Green's functions $\langle\langle Q_{\Gamma_1\gamma_1}|Q_{\Gamma_2\gamma_2}\rangle\rangle$ and $\langle\langle \hat{C}_{\Gamma_1\gamma_1}|Q_{\Gamma_2\gamma_2}\rangle\rangle$ and performing the decoupling

$$\langle\langle[\hat{C}_{\Gamma\gamma}, \hat{C}_{\Gamma_1\gamma_1}]Q_{\Gamma\gamma}|Q_{\Gamma_2\gamma_2}\rangle\rangle \approx \langle[\hat{C}_{\Gamma\gamma}, \hat{C}_{\Gamma_1\gamma_1}]\rangle\langle\langle Q_{\Gamma\gamma}|Q_{\Gamma_2\gamma_2}\rangle\rangle \ , \tag{4.5.24}$$

we obtain the following system of equations:

$$\langle\langle Q_{\Gamma_1\gamma_1}|Q_{\Gamma_2\gamma_2}\rangle\rangle = D^{(0)}_{\Gamma_1}(\omega)\delta_{\Gamma_1\Gamma_2}\delta_{\gamma_1\gamma_2} + 2\pi V_{\Gamma_1}D^{(0)}_{\Gamma_1}(\omega)\langle\langle \hat{C}_{\Gamma_1\gamma_1}|Q_{\Gamma_2\gamma_2}\rangle\rangle_\omega \ ; \tag{4.5.25}$$

$$\langle\langle \hat{C}_{\Gamma_1\gamma_1}|Q_{\Gamma_2\gamma_2}\rangle\rangle_\omega = \frac{\mathrm{i}}{\hbar}\sum_{\Gamma\gamma} V_\Gamma\langle[\hat{C}_{\Gamma\gamma}, \hat{C}_{\Gamma_1\gamma_1}]\rangle\langle\langle Q_{\Gamma_1\gamma_1}|Q_{\Gamma_2\gamma_2}\rangle\rangle_\omega \ , \tag{4.5.26}$$

which can be solved easily for the Green's functions. In the deduction of (4.5.25) and (4.5.26) the representation of functions by Fourier integrals, as well as the theorem of convolution, were used.

For instance, for the linear $E \otimes e$ case for which the operator of vibronic interaction has the form $V_E\sqrt{2}(Q_+\hat{\sigma}_- + Q_-\hat{\sigma}_+)$ (Sect. 3.1.1), from (4.5.25) and (4.5.26) we find [4.102]

$$\langle\langle Q_+|Q_-\rangle\rangle_\omega = D^{(0)}_E(\omega)\left(1 - \frac{4\pi}{\hbar\omega}\langle\hat{\sigma}_z\rangle V_E^2 D^{(0)}_E(\omega)\right)^{-1} \ . \tag{4.5.27}$$

From this, using (4.5.7) and (4.5.8) and substituting ± 1 for $\langle\sigma_z\rangle$, we have

$$D_0^{(+)} = D_E^{(0)}(\omega)\left(1 - \frac{16\pi^2}{\hbar^2\omega^2} V_E^2 D_E^{(0)^2}(\omega)\right)^{-1} \tag{4.5.28}$$

$$D_0^{(-)} = \frac{4\pi}{\hbar\omega} V_E^2 D_E^{(0)^2}(\omega)\left(1 - \frac{16\pi}{\hbar^2\omega^2} V_E^2 D_E^{(0)^2}(\omega)\right)^{-1} . \tag{4.5.29}$$

Introducing the perturbation $\lambda\sigma_x$ and taking the limits $T \to 0$ K and $\lambda \to 0_+$ we find that $D_1 = 0$, and then from (4.5.11) one can evaluate the partial Green's functions $D^{(1/2)}_{(1/2)(1/2)}$ and $D^{(3/2)}_{(1/2)(1/2)}$

$$D^{(1/2)}_{(1/2)(1/2)} = \frac{1}{2}(D_0^{(+)} + D_0^{(-)}) = \frac{1}{2} D_E^{(0)}(\omega)\left[1 - \frac{4\pi^2}{\hbar\omega} V_E^2 D_E^{(0)}(\omega)\right]^{-1} , \tag{4.5.30}$$

$$D^{(3/2)}_{(1/2)(1/2)} = \frac{1}{2}(D_0^{(+)} - D_0^{(-)}) = \frac{1}{2} D_E^{(0)}(\omega)\left[1 + \frac{4\pi^2}{\hbar\omega} V_E^2 D_E^{(0)}(\omega)\right]^{-1} . \tag{4.5.31}$$

These functions, as mentioned above, reproduce the spectral density of excited one-phonon states corresponding to the momenta 1/2 and 3/2, respectively. This means that the spectrum of elementary excitations to the energy levels with the momenta 1/2 and 3/2 are determined by the roots of the transcendental equation

$$1 \pm \frac{4\pi}{\hbar\omega} V_E^2 D_E^{(0)}(\omega) = 0 . \tag{4.5.32}$$

Substituting $D_E^{(0)}(\omega) = [2\pi(\omega^2 - \omega_E^2)]^{-1}$ we find the energy gaps between the ground and "one-phonon" excited states (to an accuracy of $\sim V_E^2$)

$$\begin{aligned} \Delta E^{(1/2)} &\cong \hbar\omega_E + \frac{V_E^2}{\omega_E^2} = \hbar\omega_E + 2E_{JT} , \\ \Delta E^{(3/2)} &\cong \hbar\omega_E - \frac{V_E^2}{\omega_E^2} = \hbar\omega_E - 2E_{JT} , \end{aligned} \tag{4.5.33}$$

which correspond to the results (4.1.22) (Fig. 4.1).

This result can be improved if instead of decoupling (4.5.24) one substitutes the time-dependent operators $Q_{\Gamma\gamma}(t)$ and $\hat{C}_{\Gamma\gamma}(t)$ determined by the integral equations (4.5.21) and (4.5.23) into the Green's function on the left hand side, performing the decoupling at a later stage. The results take the form of a power series in the coupling constant which enters as a factor in the subintegral expressions in (4.5.21) and (4.5.23).

The integral equations (4.5.21) and (4.5.23) for the operators lead to multitime Green's functions. The same results can be obtained by double-time Green's functions, provided the usual technique based on differential equations for the operators (instead of the integral ones) is employed [4.105]. The perturbation method for the Jahn-Teller problems $T \otimes t_2$, $T \otimes d$ and $\Gamma_8 \otimes t_2$ is developed in [4.96–98].

4.5.3 The Method of Unitary Transformations

The method of unitary transformations proved to be rather useful, especially at the earlier stage of development of the multiparticle theory. It consists of a transformation of the Schrödinger equation to a new set of coordinates for which the operator of potential energy contains a small parameter, so that the usual perturbation theory may be used with respect to this small parameter. In the case of Jahn-Teller problems where the vibronic interaction results in localization of the nuclear motion in low symmetry minima of the adiabatic potential, the transformation to new coordinates can be reduced to a shift transformation of the type $\hat{U} = \exp(\lambda\hat{S})$, where $\hat{S}$ is given by, e.g., (4.1.7). As a result, the Hamiltonian acquires the form $\hat{U}\hat{H}\hat{U}^{-1} = \tilde{H}_0 + \tilde{V}$, for which the spectrum of H_0 is simple enough to find and the transformed operator of vibronic interaction is proportional to the renormalized constant of vibronic coupling, $V_\Gamma \exp(-\alpha E_{\mathrm{JT}}/\hbar\omega_\Gamma)$, where α is a numerical coefficient. The parameter λ is determined by minimizing the energy of the ground state of $\hat{H}_0$. It is clear that the renormalized coupling constant can serve as a small parameter of the perturbation theory both for $V_\Gamma \to 0$ and for $V_\Gamma \to \infty$. This allows us to hope that the perturbation theory with this parameter will yield reasonable results in the region of intermediate coupling, too [4.106–108].

The attempts to use the ideas of *coherent states* in Jahn-Teller problems [4.34, 4.109] are also related to the method of unitary (or canonical) transformations, since the transition to coherent states is performed by a shift transformation. The use of the so-called *para-Bose-operators* [4.110] can also be relevant here.

All these methods are based on the use of the variational principle. This statement becomes obvious if one applies the shift transformation not to the operators, but to the wave functions. Therefore one can suppose that these methods give satisfactory results for the ground state, but that they become less accurate when passing to excited states with higher energies. Nevertheless, good agreement with exact numerical results was also obtained for some excited states [4.111]. It is worth noting also the interesting attempts to find exact analytical solutions for some vibronic systems [4.112–117].

The principal shortcomings of the *method of unitary transformations* (as of other variational methods) consist in the fact that it is not so universally applicable as, for instance, the Green's function method. Another defect of the works mentioned above follows from neglecting the changes of the frequencies of vibrations in the minima of the adiabatic potentials from the initial frequencies; it results in inaccurate behavior of the "tools" of the wave functions and consequent inaccurate energy intervals in the spectrum. Besides, sometimes the transformation shifts the system towards one of the minima, thus lowering the symmetry of the Hamiltonian [4.118]. This complicates the calculation to restore the initial symmetry [4.118, 119] (see also [4.120]).

A comprehensive review of publications which use unitary transformations to solve vibronic problems is given by *Wagner* [4.108].

4.6 The Multimode Jahn-Teller Effect. Phonon Dispersion in Crystals

4.6.1 Qualitative Discussion

So far in this chapter only ideal vibronic systems have been considered for which the full vibrational representation contains only one set of normal coordinates for each Jahn-Teller active type of symmetry. As mentioned above, the main physical consequence of the Jahn-Teller effect in these simplest cases is that the rotation of the electron density around the Jahn-Teller center and the motion of the wave of distortion of the molecular framework are no longer independent. The vibronic coupling results in a coupled electron-vibrational system which performs free or hindered rotations around the Jahn-Teller center, and in the case of strong coupling the motion is reduced to pulsations of the molecule when it tunnels from one distorted nuclear configuration into another. This complicated dynamics is manifested in a radical reconstruction of the energy spectrum of the system, although the general symmetry properties of the states remain the same as in the undistorted molecule.

What kind of new properties are expected for multimode vibronic systems in this respect? First of all it is clear that all the Jahn-Teller active normal vibrations are involved in the coupled rotation of the electron density and the distortion of the nuclear framework. In the absence of the vibronic interaction the vibrations are independent and each of the waves of distortion of the framework, corresponding to the set of degenerate vibrations, has its own phase. Considering the vibronic coupling the normal coordinates become coupled via the electronic subsystem, and the motion of the waves of distortion becomes phase-correlated, resulting in a total wave covering the whole polyatomic system. This wave is linked to the electron density, resulting in an *electron-vibrational formation of a polaron type* which performs free or hindered rotations around the Jahn-Teller center. The vibronic spectrum of a multimode vibronic system is, in general, significantly different from the electron-vibrational spectrum of the system in the absence of the vibronic coupling.

So far the nature of the vibronic coupling and the dimensions of the polyatomic systems have not been considered. Two different cases are possible here. For relatively small polyatomic systems with a discrete set of vibrational frequencies, the vibronic coupling, removing the accidental degeneracy of the oscillator energy levels, results in a more dense, but still discrete energy spectrum. If the relaxation linewidth is much smaller than the energy level spacing, the vibronic spectrum remains discrete. The problem is similar to that considered above, but more complex from the quantitative point of view. Indeed, the Hamiltonian matrix to be diagonalized in the numerical solution has much larger dimensions than in case of an ideal vibronic system with the same value of Jahn-Teller stabilization energy [4.76–80].

Another case is provided by systems with a continuous energy spectrum, where a direct numerical diagonalization is, in principle, impossible. Point defects in crystals, having a continuous vibrational energy spectrum even in the absence of vibronic coupling, may serve as limiting examples of this kind. Under the effect of the vibronic coupling the spectral density of states becomes redistributed, its determination being one of the most difficult problems of the theory of vibronic interactions. Nevertheless, the problem can be greatly simplified, if one assumes that the electrons of the degenerate states of the Jahn-Teller center are rather localized in space and interact directly with the displacements of the atoms of the nearest-neighbor coordination spheres only (Sect. 3.5).

The amplitude of the wave of distortion covering the whole polyatomic system is therefore rapidly decreasing with distance from the Jahn-Teller center. Although the vibronic states are significantly delocalized compared with the geometric dimensions of the electronic distribution (this is manifest in the appropriate energy gaps), nevertheless the dimensions of the vibronic states are in many cases much smaller than the general dimensions of the polyatomic system. It follows that the vibronic states should manifest themselves in the observed spectra as pseudo-local resonances on the background of the weakly changed vibrational spectrum of a large number of peripheral degrees of freedom weakly involved in the vibronic interaction.

Let us illustrate this by an example of the multimode effect in a small-radius impurity center, the energy spectrum of which contains one discrete degenerate electronic term, well separated by a considerable energy gap from all the other electronic states of the crystal. The Hamiltonian of the system, considering only the linear vibronic interaction terms, has the form of (3.5.24). Let us pass from the normal coordinates q_κ to the symmetrized displacements of the atoms of all the coordination spheres

$$Q_{n\Gamma\gamma} = \sum_{\kappa} a_\kappa(n\Gamma\gamma) q_\kappa \; , \tag{4.6.1}$$

where n numerates the coordination spheres (details of the transformation (4.6.1) and the explicit form of the coefficients $a_\kappa(n\Gamma\gamma)$ for some special cases are presented in [4.121, 122]). Since, in accordance with the selection rules, the electronic states of the impurity under consideration do not interact with all the symmetrized displacements of the impurity, and the dynamic matrix of the elastic interaction is diagonal with respect to the indices of different irreducible representations, the symmetrized Jahn-Teller–nonactive displacements can be separated and the vibronic Hamiltonian of the active degrees of freedom acquires the form

$$\hat{H} = \frac{1}{2} \sum_{n\Gamma\gamma} P_{n\Gamma\gamma}^2 + \frac{1}{2} \sum_{\Gamma\gamma} \sum_{nm} Q_{n\Gamma\gamma}(\omega^2)_{nm}^{(\Gamma)} Q_{m\Gamma\gamma} + \sum_{\Gamma\gamma} V_\Gamma Q_{1\Gamma\gamma} \hat{C}_{\Gamma\gamma} , \tag{4.6.2}$$

where

$$(\omega^2)_{nm}^{(\Gamma)} = \sum_{\kappa} a_\kappa(n\Gamma\gamma) a_\kappa(m\Gamma\gamma) \omega_\kappa^2 .$$

The impurity center of small radius in the crystal can be represented as two interacting subsystems: one consists of the atoms of the first coordination sphere including the substituting impurity, the other contains the remainder of the crystal. Accordingly, the Hamiltonian (4.6.2) can be represented in the form $\hat{H}_{JT} + H'_{ph} + H_{int}$, where $\hat{H}_{JT}$ is the vibronic Hamiltonian of the first coordination sphere, which has the form of (2.2.13) with the force constants $\omega_\Gamma^2 = (\omega^2)_{11}^{(\Gamma)}$; H'_{ph} is the harmonic Hamiltonian of the remaining part of the crystal,

$$H'_{ph} = \frac{1}{2} \sum_{n\Gamma\gamma}{}' P_{n\Gamma\gamma}^2 + \frac{1}{2} \sum_{\Gamma\gamma} \sum_{n,m}{}' Q_{n\Gamma\gamma}(\omega^2)_{nm}^{(\Gamma)} Q_{m\Gamma\gamma} , \tag{4.6.3}$$

(the prime on summation sign means that the terms with $n = 1$, $m = 1$ are omitted) and H_{int} is the operator of interaction of the Jahn-Teller coordinates with the remainder of the crystal. For instance, in the multimode $E \otimes (e_1 + e_2 + \cdots)$ problem, where only E vibrations are Jahn-Teller active, for H_{int} we have

$$H_{int} = \sum_{n}{}' (\omega^2)_{n1}^{(E)} \varrho_n \varrho_1 \cos(\varphi_n - \varphi_1) . \tag{4.6.4}$$

Here the polar coordinates $Q_{n\theta} = \varrho_n \cos\varphi_n$, $Q_{n\varepsilon} = \varrho_n \sin\varphi_n$ are used.

Owing to the factor $\cos(\varphi_n - \varphi_1)$ the coupled rotation of electron density and the wave of distortion of the first coordination sphere involves the appropriate distortion of the other coordination spheres, resulting in a wave of distortion covering the whole crystal. Generally speaking, this collective rotation is stationary, since the appropriate momentum is an integral of motion (Sect. 3.2). Besides, the distortion of the first coordination sphere can be transferred to the next coordination spheres by the term $\varrho_n\varrho_1$ and propagate along the crystal in the form of waves diverging from the impurity center.

In spite of the obvious complexity of the Jahn-Teller effect in multimode vibronic systems, the physical ideas underlying the approximate methods of analytical solutions are the same as in the case of ideal vibronic systems. As in the latter case, the two limiting cases for which analytic calculations can be carried out, those of strong and weak vibronic coupling, must be distinguished. In the case of weak coupling there is a small parameter in the Hamiltonian, the vibronic constant, which allows us to use the methods of perturbation theory (Sect. 4.6.2). In the case of strong coupling the energy gap between the adiabatic potential sheets is much larger than the energy of the highest frequency phonons

$\hbar\omega_{max}$, and this allows to use the adiabatic approximation and separate the sheets of the adiabatic potential, considering the nuclear motion for each of them separately (Sect. 4.6.3). The vibrational sybsystem in the impurity crystal has a continuous spectrum, and therefore the vibronic energy spectrum should also be continuous. Thus in both the limiting cases the solution of the problem is conveniently carried out by means of the Green's function method.

In the low energy approximation the dispersion relationships for the Green function allow us to obtain some results which are valid for arbitrary values of the coupling constant (Sect. 4.6.4). In the conclusion of this section we discuss some model approaches used to simplify the multimode vibronic problem (Sect. 4.6.5).

4.6.2 The Case of Weak Coupling. Perturbation Theory

Consider first the case of weak coupling. The simultaneous change of the signs of $V_{n\Gamma}$ and all the $Q_{n\Gamma_\gamma}$ in (3.5.27) for given values of Γ and n does not change the explicit form of the Hamiltonian and hence of observables which do not contain $Q_{n\Gamma_\gamma}$ explicitly. Hence the power series, by which the observables are represented (if possible), do not contain odd powers of the vibronic constant. In particular, this means that the second-order contribution to the energy is additive with respect to the indices Γ and n. In its turn, this allows one to consider separately the multimode problems $E \otimes e$, $T \otimes e$, $T \otimes t_2$, $\Gamma_8 \otimes e$, $\Gamma_8 \otimes t_2$, $E \otimes b_1$, $E \otimes b_2$ in the second-order perturbation theory approximation. In the case of simultaneous vibronic coupling to several types of vibrations the second-order contributions for different Γ values should be summed. Similarly, the second-order contributions for a given type of vibrations are additive with respect to the mode number n.

In other words, in the second-order perturbation theory approximation the energy spectrum of an N-mode system with the linear Jahn-Teller effect is simply a superposition of N energy spectra of N one-mode systems obtained in the second order with respect to $V_{n\Gamma}$.

In the case of multimode systems with discrete energy spectra the perturbation theory has the usual form. For reasons given above all the odd-order corrections to the energy vanish, and mixed contributions of different modes being with the fourth order. For instance, the fourth-order correction to the energy of the ground state of the two-mode $E \otimes (e + e)$ problem is [4.123]

$$E_0^{(4)} = \frac{1}{2}\frac{V_{1E}^4}{\hbar\omega_{1E}^5} + \frac{1}{2}\frac{V_{2E}^4}{\hbar\omega_{2E}^5} + \frac{2V_{1E}^2 V_{1E}^2}{\hbar(\omega_{1E} + \omega_{2E})\omega_{1E}^2\omega_{2E}^2} . \tag{4.6.5}$$

The last term in this equation describes the effect of interaction of the two modes via the electronic subsystem. Within the same approximation the correction to the energy gap between the ground and first excited (one-phonon) states ($\hbar\omega_{1E}$

in the zeroth-order approximation) is [4.123]

$$\Delta E_{10}^{(4)} = -\frac{3}{2}\frac{V_{1E}^4}{\hbar\omega_{1E}^5} + 2V_{1E}^2 V_{2E}^2 \frac{3\omega_{2E}^2 - \omega_{1E}\omega_{2E} - \omega_{1E}^2}{\hbar\omega_{1E}\omega_{2E}^2(\omega_{1E}^2 - \omega_{2E}^2)} \ . \tag{4.6.6}$$

The energy gap to the next excited state (equal to $\hbar\omega_{2E}$ in the zeroth-order approximation) can be obtained by interchanging the indices 1 and 2 in (4.6.6). The second term in (4.6.6) diverges at $\omega_{1E} = \omega_{2E}$, where there is accidental degeneracy of the two one-phonon states. In this case degenerate perturbation theory has to be employed. If $\omega_{1E} < \omega_{2E}$, the second term in (4.6.6) is negative, which means that the mixing of the modes results in the reduction of the smaller energy gap. These results are confirmed also by numerical calculations for the $E \otimes (e + e)$ problem [4.76, 79, 124, 125].

In the case of multimode vibronic systems with a continuous energy spectrum, the vibronic interaction, as mentioned above, may give rise to pseudolocal resonances in the density of states. In order to reveal this effect the spectral density of the appropriate correlation functions has to be investigated. These functions can be obtained directly by means of the Green's function method. The perturbation theory for matrix Green's functions in the multimode case is quite similar to that for ideal vibronic systems (Sect. 4.5.2). Similarly to the procedure followed in Sect. 4.1 for ideal systems, one can use the perturbation theory for the mass operator [4.93–98] or the operator version of perturbation theory [4.102]. We use here a simpler version based on the idea of quasi-averages (Sect. 4.5.2).

Consider the multimode system described by the Hamiltonian (3.5.27). One can verify that for the interaction mode (3.5.15), the integral equation of motion (4.5.21) is also valid in the multimode case, provided

$$D_\Gamma^{(0)}(t) = -\frac{\mathrm{i}}{\hbar}\theta(t)\langle[q_{1\Gamma\gamma}^{(0)}(t), q_{1\Gamma\gamma}]\rangle = -\frac{\theta(t)}{\mathscr{V}_\Gamma^2}\sum_n \frac{V_{n\Gamma}^2}{\omega_{n\Gamma}^2}\sin(\omega_{n\Gamma}t) \ . \tag{4.6.7}$$

(for the notation used see Sect. 3.5). Repeating the arguments of Sect. 4.5.2, we come to the same system of equations (4.5.24) with respect to the Green's functions $\langle\langle q_{1\Gamma_1\gamma_1}|q_{1\Gamma_2\gamma_2}\rangle\rangle$ and $\langle\langle \hat{C}_{\Gamma_1\gamma_1}|q_{1\Gamma_2\gamma_2}\rangle\rangle$ which can be solved easily in each concrete case. The reduced and partial Green's functions can be derived from the thermodynamic double-time ones in the same way as in Sect. 4.5, i.e., by taking the limits $T \to 0$ K and $\lambda \to 0$, where λ is the parameter of the low symmetry field. For instance for the linear multimode $E \otimes e$ problem one obtains the same (4.5.30) and (4.5.31), but the zeroth-order Green's function $D_\Gamma^{(0)}(\omega)$ is now the Fourier transform of (4.6.7):

$$D_\Gamma^{(0)}(\omega) = \frac{1}{2\pi\mathscr{V}_\Gamma^2}\sum_n \frac{V_{n\Gamma}^2}{\omega^2 - \omega_{n\Gamma}^2} \ . \tag{4.6.8}$$

To reveal the local and pseudolocal resonances let us examine the spectral density $\varrho(\omega) = \mathrm{Im}\{D^{(j)}(\omega + i\varepsilon)\}$. From (4.5.30) and (4.5.31) we find

$$\varrho(\omega) = \frac{\varrho_E^{(0)}(\omega)}{\left(1 \pm \frac{4\pi}{\hbar\omega}\mathscr{V}_E^2 \mathrm{r}_E^{(0)}(\omega)\right)^2 + \left(\frac{4\pi}{\hbar\omega}\mathscr{V}_E^2 \varrho_E^{(0)}(\omega)\right)^2}, \quad \text{where} \tag{4.6.9}$$

$$\varrho_\Gamma^{(0)}(\omega) = 2\,\mathrm{Im}\{D_\Gamma^{(0)}(\omega + i\varepsilon)\} \; ; \qquad \mathrm{r}_\Gamma^{(0)}(\omega) = 2\,\mathrm{Re}\{D_\Gamma^{(0)}(\omega + i\varepsilon)\} \; . \tag{4.6.10}$$

In the particular case of an impurity center of small radius one can substitute (3.5.28) into (4.6.8), giving

$$\varrho_\Gamma^{(0)}(\omega) = \sum_\kappa a_\kappa^2(\Gamma\gamma)\delta(\omega_\kappa^2 - \omega^2) \; ; \qquad \mathrm{r}_\Gamma^{(0)}(\omega) = \frac{1}{\pi}\int_0^{\omega_{\max}} \frac{z\varrho_\Gamma^{(0)}(z)}{\omega^2 - z^2}dz \; , \tag{4.6.11}$$

where $\omega_{\max}$ is the largest lattice frequency. It follows from (4.6.9) that in the vicinity of the roots ω_α of the transcendental equation

$$1 \pm \frac{4\pi}{\hbar\omega}\mathscr{V}_E^2 \mathrm{r}_E^{(0)}(\omega) = 0 \; , \tag{4.6.12}$$

for which the spectral density $\varrho_E^{(0)}(\omega)$ is small enough, the redetermined density of states $\varrho(\omega)$ has a Lorentzian shape

$$\varrho(\omega) \approx \frac{1}{\pi}\frac{\gamma_\alpha}{(\omega - \omega_\alpha)^2 + \gamma_\alpha^2}, \quad \text{where} \tag{4.6.13}$$

$$\gamma_\alpha \approx \varrho_E^{(0)}(\omega_\alpha)/\mathrm{r}_E^{(0)}(\omega_\alpha) \; .$$

If some of the roots of (4.6.12) fall in the forbidden zone where $\varrho_E^{(0)}(\omega) = 0$, then the Lorentzian (4.6.13) becomes a δ-function and (4.6.12) using (4.6.11), transforms into (4.5.32), resulting in a local state. However, in the general case one has to use the known density of states $\varrho_E^{(0)}(\omega)$ in order to evaluate the integral in (4.6.11) and solve (4.6.12). Note that in the spectral regions near the roots of (4.6.12) where the initial density of E vibrations is not small, the redetermined density of states also differs significantly from $\varrho_E^{(0)}(\omega)$, although it does not have a Lorentzian shape. Note also that the roots of the dispersion equation (4.6.12) and the consequent significant redetermination of the spectrum are dependent on both the magnitude of the vibronic constant V_E and the phonon dispersion relation in the spectral regions where the Gilbert transform $\mathrm{r}_E^{(0)}(\omega)$ is great enough.

Thus even weak vibronic coupling may lead to a major redetermination of the spectrum of states, in particular, to the rise of local and pseudolocal states. Unlike the ideal $E \otimes e$ case where a weak Jahn-Teller influence results in the excited vibrational states splitting (Sect. 4.1), the occurrence of new peaks in the spectrum in the multimode case may not look like the old peaks splitting. The number and position of vibronic resonances are in large measure determined by the dispersion relation of the Jahn-Teller vibrations. The vibronic structure of the resulting density of states may be so complicated that there is no set of parameters for the ideal one-mode system with which one can obtain at least qualitative similarity between its vibronic spectrum and that of the multimode system with dispersion.

These results may be given a visual physical interpretation as follows. As in the ideal case, the wave function of the system does not have a multiplicative form. In the approximation of first-order perturbation theory the wave function of the ground state becomes mixed with the one-phonon excited state. Without vibronic coupling the continuous spectrum of the impurity-phonon system can be considered as a spectrum of free phonons. The discrete local energy levels which occur due to the vibronic coupling correspond to finite motions of the electron-phonon system, so the local peaks in the vibronic spectrum can be interpreted as *coupled states of the impurity center with a phonon.* In other words, the rotation of the low-symmetry electronic distribution around the Jahn-Teller impurity center involves, through the vibronic interaction, one phonon captured in a stationary orbital.

The qualitative conclusions about the vibronic structure of the density of states obtained above remain valid for more complicated cases of weak coupling. In particular, coupled states of the phonon with the impurity center also occur in the case of the pseudo-Jahn-Teller effect. Consider a simple example of two nondegenerate electronic states of different symmetry mixed by appropriate vibrations. The Hamiltonian of such a multimode system can be written as, cf. (3.4.1),

$$\hat{H} = \frac{1}{2}\sum_n (P_n^2 + \omega_n^2 Q_n^2) + \Delta\hat{\sigma}_z + \sum_n V_n Q_n \hat{\sigma}_x \ . \tag{4.6.14}$$

As above, we now use the integral equation (4.5.21) for the interacting mode (3.5.15) and the integral equation (4.5.23) for the electronic operators $\hat{\sigma}_x$, $\hat{\sigma}_y$ and $\hat{\sigma}_z$ (here the operator $\Delta\hat{\sigma}_z$ also has to be included in the perturbation). Substituting the appropriate time-dependent functions into the double-time thermodynamic Green's functions $\langle\langle q_1|q_1\rangle\rangle$, $\langle\langle\hat{\sigma}_x|q_1\rangle\rangle$ and $\langle\langle\hat{\sigma}_y|q_1\rangle\rangle$ and performing the decoupling $\langle\langle\hat{\sigma}_z q_1|q_1\rangle\rangle \approx \langle\hat{\sigma}_z\rangle\langle\langle q_1|q_1\rangle\rangle$ corresponding to the case of weak coupling, we come to a system of three linear algebraic equations for the Fourier transforms $\langle\langle q_1|q_1\rangle\rangle_\omega$, $\langle\langle\hat{\sigma}_x|q_1\rangle\rangle_\omega$ and $\langle\langle\hat{\sigma}_y|q_1\rangle\rangle_\omega$, which yield [4.126, 127]

$$\langle\langle q_1|q_1\rangle\rangle_\omega = \frac{(\hbar^2\omega^2 - 4\Delta^2)D_0(\omega)}{\hbar^2\omega^2 - 4\Delta^2 + 8\pi\mathscr{V}^2 D_0(\omega)\Delta\langle\sigma_z\rangle} \ , \tag{4.6.15}$$

where $D_0(\omega)$ is defined in (4.6.8).

The spectrum of elementary excitations is determined by the poles of the Green's function (4.6.15), i.e., by the roots of the transcendental equation

$$4\pi\mathscr{V}^2 D_0(\omega)\langle\sigma_z\rangle\Delta = 4\Delta^2 - \hbar^2\omega^2 . \tag{4.6.16}$$

Here $\langle\sigma_z\rangle = n_1 - n_0$ is the difference in the occupation number of the excited and ground electronic states, respectively. At $T = 0$ K $n_1 = 0$, $n_0 = 1$ and in the absence of phonon dispersion, (4.6.16) yields the results of Sect. 3.1. Note that it includes both the cases of presence and absence of resonance.

The local states split off from the continuous phonon bands were studied in [4.126] using the result (4.6.15), while pseudolocal states are considered in [4.127]. The theory of such coupled states of the impurity center with a phonon is presented in the review article [4.128] where they are called "hybrid states" and "dielectric modes".

The multimode vibronic problems $T \otimes t_2$ and $\Gamma_8 \otimes t_2$ in the case of weak coupling are considered in [4.96–98] where the general theory of perturbations for matrix Green's functions is also given.

4.6.3 Strong Coupling Case

Consider now multimode systems with strong vibronic coupling when in the lowest vibronic state the electronic and nuclear motions can be separated in the adiabatic approximation. As for ideal systems, two cases have to be distinguished: firstly, when the lowest sheet of the adiabatic potential has an equipotential continuum of minima (a trough) and secondly, when there are several absolute minima of the adiabatic potential divided by potential barriers.

Consider first the Jahn-Teller system related to the first case, the linear $E \otimes (e + e + \cdots)$ problem. In order to obtain qualitative ideas about the nuclear motion in this case and about the special features of the multimode problem, consider first the simplest example of this kind, the two-mode $E \otimes (e + e)$ problem. The adiabatic potentials, as eigenvalues of the potential energy matrix [see the Hamiltonian (3.2.30)], are here functions of four variables:

$$\begin{aligned}\varepsilon_\pm &= \tfrac{1}{2}\omega_1^2(Q_{1\theta}^2 + Q_{1\varepsilon}^2) + \tfrac{1}{2}\omega_2^2(Q_{2\theta}^2 + Q_{2\varepsilon}^2) \\ &\quad \pm \sqrt{(V_1 Q_{1\theta} + V_2 Q_{2\theta})^2 + (V_1 Q_{1\varepsilon} + V_2 Q_{2\varepsilon})^2} .\end{aligned} \tag{4.6.17}$$

The coordinates of the extreme points of the lowest sheet (with the minus sign before the square root) can be easily evaluated (see (3.5.23)):

$$\begin{aligned} Q_{1\theta}^{(0)} &= \frac{V_1}{\omega_1^2}\cos\varphi , \qquad Q_{2\theta}^{(0)} = \frac{V_2}{\omega_2^2}\cos\varphi , \\ Q_{1\varepsilon}^{(0)} &= \frac{V_1}{\omega_1^2}\sin\varphi , \qquad Q_{2\varepsilon}^{(0)} = \frac{V_2}{\omega_2^2}\sin\varphi , \end{aligned} \tag{4.6.18}$$

where φ is an arbitrary parameter, $0 \le \varphi \le 2\pi$. Their curvatures are

$$\omega_A^{(1)} = \omega_1 \ , \qquad \omega_A^{(2)} = \omega_2 \ ;$$
$$\omega_B^{(1)} = 0 \ , \qquad \omega_B^{(2)} = \sqrt{\frac{V_1^2\omega_2^4 + V_2^2\omega_1^4}{V_1^2\omega_2^2 + V_2^2\omega_1^2}} \ . \tag{4.6.19}$$

Here the subscripts correspond to the transformation properties of the vibrations near the bottom of the tetragonal minima with D_{4h} symmetry, cf. (3.1.23). The zero frequency $\omega_B^{(1)}$ corresponds to free motion along the trough. Hence the energy spectrum of the system consists of a superposition of the rotational spectrum $\hbar^2 j^2/2I$, where I is the moment of inertia (the masses are equal to 1),

$$I = \sum_n [Q_{n\theta}^{(0)^2} + Q_{n\varepsilon}^{(0)^2}] = \sum_n \frac{V_n^2}{\omega_n^4} \ , \tag{4.6.20}$$

and radial vibrations with frequencies ω_1, ω_2 and $\omega_B^{(2)}$. The equilibrium coordinates (4.6.18) can be somewhat improved by taking centrifugal energy into account, as in the single-mode case (4.2.9). Thus the lowest sheet of the adiabatic potential possesses a one-dimensional circular trough in the multimode problem as well, in full agreement with the results of Sect. 3.2 and 3.5. The rotation of the system along the trough determines the spectrum of the lowest vibronic states. As will be shown below, in the case of impurity centers in crystals the additional destabilizing centrifugal forces arising from these rotations can significantly redetermine the minima positions. For this reason the multimode impurity center problem differs significantly from the molecular vibronic ones.

In the adiabatic approximation the operator of potential energy of the crystal modes, including the centrifugal energy, has the form

$$E_\pm = \frac{\hbar^2 j^2}{2I} + \frac{1}{2}\sum_n \omega_n^2(Q_{n\theta}^2 + Q_{n\varepsilon}^2) \pm \left[\left(\sum_n V_n Q_{n\theta}\right)^2 + \left(\sum_n V_n Q_{n\varepsilon}\right)^2\right]^{1/2} , \tag{4.6.21}$$

where I is the moment of inertia of the collective rotation determined by (4.6.20) with all of the terms. By putting the derivatives $\partial E_\pm/\partial Q_{n\gamma}$ equal to zero we find

$$Q_{n\theta}^{(0)} = \mp\frac{V_n\cos\varphi}{\omega_n^2 - \Omega^2} \ , \qquad Q_{n\varepsilon}^{(0)} = \mp\frac{V_n\sin\varphi}{\omega_n^2 - \Omega^2} \ , \tag{4.6.22}$$

where $\Omega = \hbar j/I$ is the angular velocity of the collective rotation of the distortion and φ is the angle determining the orientation of the distortion. Substituting (4.6.22) into (4.6.21) we have

$$\Omega\sum_n \frac{V_n^2}{(\omega_n^2 - \Omega^2)^2} = \hbar j \ . \tag{4.6.23}$$

A graphical solution of this transcendental equation is shown schematically in

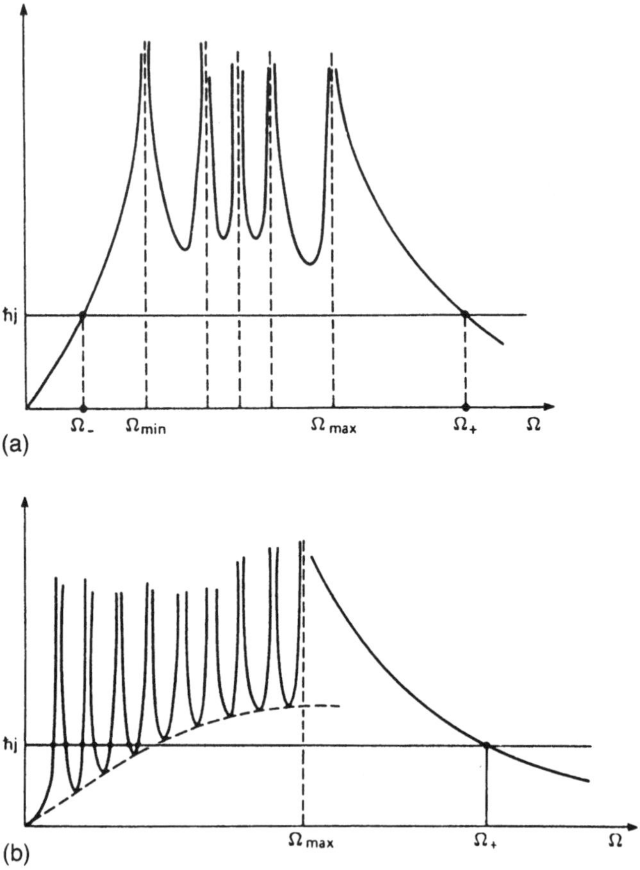

Fig. 4.17a, b. Graphical solution of the transcendental equation (4.6.23). (**a**) Molecular multimode case (vibronic interaction with one optical vibrational band). The dashed lines are the asymptotes of the left hand side of the equation crossing at the points $\Omega = \omega_n$. In the case of strong vibronic coupling only two solutions, Ω_+ and Ω_-, remain. (**b**) The case of a Jahn-Teller (impurity) center in a crystal (vibronic interaction with the acoustic vibrational band). The sloping dashed line is the envelope of the minimum points of the left side of (4.6.23) for low frequencies

Fig. 4.17. In the case with one or several optical phonon dispersion bands and rather strong vibronic coupling, all the minima of the left hand side of (4.6.23) at the singularities $\Omega = \omega_n$ lie above the straight line $\hbar j$. This means that (4.6.23) has only two solutions falling outside the boundaries of the phonon bands:

$$0 < \Omega_- < \omega_{\min} \; ; \qquad \Omega_+ > \omega_{\max} \; . \tag{4.6.24}$$

One can easily see that Ω_- determines the coordinates of the points of the trough in the lowest sheet of the adiabatic potential whereas Ω_+ is related to the upper sheet.

When the vibronic interaction with acoustic vibrations is included the singularities on the left side of (4.6.23) extend into the region of low frequencies down to $\Omega = 0$. The envelope of the minima lying between these singularities (the dashed line in Fig. 4.17) in the region of small Ω is described by the function $V_E^2 \overline{\omega^{-4}} \Omega$, directly proportional to Ω, where

$$\overline{\omega^k} = \sum_n \omega_n^k V_n^2 / V_E^2 \ ,$$
$$V_E^2 = \sum_n V_n^2 \ . \tag{4.6.25}$$

It is obvious that for any strong vibronic coupling (4.6.23) has an infinite number of roots in the region of small Ω, with the absolute minima corresponding to the value $\Omega_- = 0$. This result follows from inequality (4.6.24) at $\omega_{\min} \to 0$.

The angular velocity $\Omega_- = 0$ corresponds to an infinite moment of inertia $I_- = \hbar j / \Omega_-$. This result can be easily understood. As mentioned above, the minima of the lowest sheet of the adiabatic potential occur as a result of the compensation of the destabilizing Jahn-Teller and centrifugal forces by the stabilizing elastic ones $\sim \omega_n^2 Q_{n\gamma}$. Since for the limiting long wavelength acoustic phonons $\omega_\kappa \sim \kappa$ it is clear that if κ is reduced, the magnitude of the stabilizing elastic force decreases to zero and the appropriate $Q_{n\gamma}^{(0)}$ increase to infinity. This is the reason for the divergence of the integral (4.6.23) noted in [4.12].

Thus in the space of "radial" degrees of freedom orthogonal to the angular variable of the trough, the lowest sheet of the adiabatic potential has a ravine with a warped bottom going to infinity. The asymptotic value of the Jahn-Teller stabilization energy is $E_{JT} = V_E^2 \overline{\omega^{-2}} / 2$. Since the density of limiting long wavelength acoustic phonons is negligible the ravine contracts rapidly with the departure to infinity. This circumstance allows us to solve the nuclear dynamics despite the complicated form of the potential surface.

As was mentioned before, the coupled rotation of the electronic density and the wave of distortion of the nearest coordination spheres involves an appropriate distortion of the other coordination spheres. As a result a wave of distortions covering the whole crystal propagates around the impurity center. This collective rotation is generally stationary, since its appropriate momentum is an integral of motion.

If one considers the vibronic interaction with acoustic modes, the situation becomes somewhat complicated. The moment of inertia of such a collective rotation becomes infinite and thus it cannot be excited. On the other hand, as can be seen from (4.6.4), the interaction of the displacements of the atoms of the first coordination spheres with the most inert long wavelength acoustic phonons is of the order of $\omega_\kappa^2 a_\kappa(1\gamma) \sim \kappa^3$ and tends to zero with κ. This means that quasistationary rotations without the participation of long wavelength acoustic phonons are possible, the weak decay of these rotations being due to interactions with these acoustic phonons. Accordingly, there should be well-defined rotational resonances in the vibronic density of states.

The mathematical realization of these physical ideas in the case of a small radius impurity center can be performed in the same framework as in the case of the ideal $E \otimes e$ problem considered above. The transition to the adiabatic electronic basis is accomplished by means of the unitary transformation (3.1.9) resulting in a Hamiltonian of the first coordination sphere (4.2.3) and an integral of motion $\tilde{J}_z = L_z$, see (3.2.31). Now we pass to the reference system which rotates with the distortion of the first coordination sphere, which means that we will take φ_1 as being zero. Mathematically this is realized by means of a unitary transformation of a shift of φ_1 along φ_n: $S_1 = \exp(-\varphi_1 \sum_{n \neq 1} \partial/\partial\varphi_n)$. Neglecting the nondiagonal matrix σ_x (which mixes different sheets of the adiabatic potential) in the transformed Hamiltonian, taking the operator $L_z^{(1)} = -i\partial/\partial\varphi_1$ in the form $L_z - L_z'$ where $L_z' = \sum_{n \neq 1} L_z^{(n)}$ and replacing the integral of motion $\tilde{J}_z = L_z$ with the quantum numbers $j = \pm 1/2, \pm 3/2, \ldots$, we reduce the Hamiltonian of the system (4.6.2) to the form

$$H = \frac{1}{2}\sum_n P_{n\theta}^2 + \frac{1}{2}\sum_{nm} Q_{n\theta}(\omega^2)_{nm}Q_{m\theta} + \frac{1}{2}\sum_{n\neq 1} P_{n\varepsilon}^2 + \frac{1}{2}\sum_{n\neq 1}\sum_{m\neq 1} Q_{n\varepsilon}(\omega^2)_{nm}Q_{m\varepsilon}$$
$$+ \frac{\hbar^2}{2Q_{1\theta}^2}(j - L_z')^2 \pm V_E Q_{1\theta} , \tag{4.6.26}$$

where we have used the fact that in the rotating system of coordinates $\varphi_1 =$ const $= 0$, i.e., $Q_{1\theta} = \varrho_1$, $Q_{1\varepsilon} = 0$. To solve the problem in the harmonic approximation, the equilibrium values of the coordinates and momenta must be found. In the limit of leading terms in $1/V_E$, we obtain from the equations $\partial H/\partial Q_{n\gamma} = 0$ and $\partial H/\partial P_{n\gamma} = 0$, for the lowest sheet of the adiabatic potential

$$P_{n\theta}^{(0)} = Q_{n\varepsilon}^{(0)} = 0 , \qquad Q_{n\theta}^{(0)} = V_E \sum_\kappa \omega_\kappa^{-2} a_\kappa(n\theta) a_\kappa(1\theta) ,$$
$$P_{n\varepsilon}^{(0)} = \Omega^{(0)} Q_{n\theta}^{(0)} , \qquad \Omega_-^{(0)} = \hbar j/(\overline{\omega^{-4}} V_E^2) . \tag{4.6.27}$$

In the harmonic approximation in the limit of strong vibronic coupling the Hamiltonian (4.6.26) has the form $H = H_{\text{vib}} + H_{\text{rot}} + H_{\text{el}}$, where

$$H_{\text{vib}} \approx \tfrac{1}{2}\sum_n P_{n\theta}^2 + \tfrac{1}{2}\sum_{nm} Q_{n\theta}(\omega^2)_{nm}Q_{m\theta} + \tfrac{1}{2}\sum_{n\neq 1} P_{n\varepsilon}^2 + \tfrac{1}{2}\sum_{n\neq 1}\sum_{m\neq 1} Q_{n\varepsilon}(\omega^2)_{nm}Q_{m\varepsilon}$$
$$+ \tfrac{1}{2}(Q_{1\theta}^{(0)})^{-2}\left(\sum_{n\neq 1} Q_{n\theta}^{(0)} P_{n\varepsilon}\right)^2 ; \tag{4.6.28}$$
$$H_{\text{rot}} = \hbar^2 j^2/(2I^{(0)}) , \qquad H_{\text{el}} = -E_{\text{JT}} = -\tfrac{1}{2}V_E^2\overline{\omega^{-2}} .$$

Here $P_{n\gamma}$ and $Q_{n\gamma}$ are measured from the equilibrium values (4.6.27) and $I_-^{(0)} = \hbar j/\Omega_-^{(0)}$ is the moment of inertia of the collective rotation. Since the equilibrium values of the coordinates and momenta (4.6.27) are approximate, in the Hamiltonian (4.6.28) there should also be small terms, linear in $Q_{n\gamma}$ and $P_{n\gamma}$, which cause the relaxation of the rotational states:

$$V = \frac{\hbar j}{V_E(\overline{\omega^{-2}})^2} \sum_{n \neq 1} \omega_{1n}^{-2} P_{n\varepsilon} \ .$$

These can be neglected in a first approximation. In the Hamiltonian H_{vib} the variables $Q_{n\theta}$ can be separated out, and when returning to the q_κ coordinates they give the standard law of phonon dispersion.

In order to determine the spectrum of $Q_{n\varepsilon}$ vibrations one can use, e.g., the Green's function method. Note that the perturbation

$$\tfrac{1}{2}(Q_{1\theta}^{(0)})^{-2} \sum_{n \neq 1} \sum_{m \neq 1} Q_{n\theta}^{(0)} Q_{m\theta}^{(0)} P_{n\varepsilon} P_{m\varepsilon}$$

is local, since the appropriate correction to the dynamic matrix is multiplicative with respect to the indexes n and m. For the retarded Green's function $\langle\langle P_{n\varepsilon} | P_{m\varepsilon} \rangle\rangle$ we find [4.129]

$$\langle\langle P_{n\varepsilon} | P_{m\varepsilon} \rangle\rangle = \frac{1}{2\pi} \left[\omega^2 \left(D_{nm}^{(0)} - \frac{D_{n1}^{(0)} D_{1m}^{(0)}}{G_{11}^{(0)}} \right) + \omega^2 \overline{\omega^{-2}} \frac{D_{m1}^{(0)} D_{1n}^{(0)}}{D_{11}^{(0)}(D_{11}^{(0)} + \overline{\omega^{-2}})} \right.$$
$$\left. - \frac{\omega^2 \overline{\omega^{-2}} \delta_{n1} D_{1m}^{(0)}}{D_{11}^{(0)} + \overline{\omega^{-2}}} - \delta_{nm} + \delta_{n1} \delta_{m1} \right] \qquad (4.6.29)$$

where

$$D_{nm}^{(0)}(\omega) = \frac{1}{2\pi} \sum_\kappa \frac{a_\kappa(nE\gamma) a_\kappa(mE\gamma)}{\omega^2 - \omega_\kappa^2} \ ; \qquad D_{11}^{(0)} = D_E^{(0)} \ . \qquad (4.6.30)$$

The redefined spectrum of "radial" vibrations is determined by the poles of the Green's function (4.6.29), that is by the roots of the dispersion equation $2\pi D_E^{(0)}(\omega) + \overline{\omega^{-2}} = 0$ [4.129]. In order to determine the pseudolocal states the projected density of states

$$\varrho(\omega) = \sum_n \mathrm{Im}\{\langle\langle P_{n\varepsilon} | P_{n\varepsilon} \rangle\rangle_{\omega + i\varepsilon}\}$$

has to be investigated. Substituting (4.6.29) we find that the peaks in the density of states are located at frequencies which are the roots of the transcendental equation

$$2\pi \mathrm{r}_E^{(0)}(\omega) + \overline{\omega^{-2}} = 0 \ , \qquad (4.6.31)$$

where $\mathrm{r}_E^{(0)}(\omega)$ is determined by (4.6.11). The relaxation broadening of the rotational states caused by the terms linear in $P_{n\varepsilon}$, neglected in (4.6.28), was calculated in [4.130].

As mentioned in Sect. 4.3 the vibrational states centered at the equilibrium points of the trough with coordinates $Q_{n\gamma}^{(0)}$ can be expanded in a series of the full set of vibrational states of the Hamiltonian of the crystal without vibronic interaction. In the case of strong vibronic coupling the expansion contains mostly multiphonon states. From this point of view, the rotational states of the collective distortion of the coordination spheres describe a *multiphonon electron-vibrational system of a polaron type* localized on degenerate quasi-stationary orbitals at the impurity. By increasing the constant of vibronic coupling the lifetime of the phonons at this quasipolaron is increased and hence the width of the rotational reasonances decreases. Mathematically, as V_E increases the spacing in the rotational structure decreases and the excited rotational states fall into the spectral region with lower frequencies where the density of vibrations is lower and the probability of decay into the bottom of the ravine is also lower.

The vibronic spectrum of the upper sheet of the adiabatic potential can be obtained by considering the interaction of the first coordination sphere with the others, i.e., by considering H_{int} from (4.6.4) by means of perturbation theory. In the zeroth order with respect to the H_{int} the Jahn-Teller "interacting" coordinates $Q_{1\theta}$ and $Q_{1\varepsilon}$ are separated from the others, and in the centrifugally stabilized states of the upper sheet they lead to the discrete energy spectrum (4.2.14) superimposed on the continuous vibrational spectrum of the remaining degrees of freedom of the crystal. The spacing of the rotational states of the upper sheet $\Delta E_{\text{rot}}^{(+)}$ is proportional to $V_E^{2/3}$ as seen from (4.2.14), and in the case of rather strong vibronic coupling these states lie beyond the border of the band of the one-phonon states of the thermal system. Therefore direct decay processes are forbidden by the law of conservation of energy and the decay of rotational states takes place only in the second- and higher-order terms of the perturbation theory with respect to H_{int}. Bearing in mind that the probability of second-order processes is of the order of $H_{\text{int}}/\Delta E_{\text{rot}}^{(+)}$, the width of appropriate bursts in the vibronic density of states is less for stronger vibronic coupling. These quasi-stationary states of the upper sheet of the adiabatic potential are somtimes called the *Slonczewski resonances* [4.12].

In fact the width of the Slonczewski resonances is determined not so much by the processes of rotational-vibrational relaxation within the upper sheet of the adiabatic potential as by the nonadiabatic relaxation to the lower sheet of the adiabatic potential where the channel of coupling with the vibrational subsystem is wider [4.131].

For the states of the lower sheet of the adiabatic potential the treatment of H_{int} from (4.6.4) by means of perturbation theory in the limiting case of strong vibronic coupling generally leads to incorrect results, since in this case H_{int} is of the order of V_E. The perturbation theory with respect to H_{int} is applicable only in those cases where the coefficients $(\omega^2)_{n1}$ in (4.6.4) are small values of higher order in V_E^{-1}, i.e., when $(\omega^2)_{1n}/(\omega^2)_{11} \sim k_E^{-3}$, where k_E is the dimensionless constant of vibronic coupling. Since the same parameter $(\omega^2)_{1n}$ determines the width of the phonon bands, perturbation theory with respect to H_{int} is applicable only in those cases when the electrons of the impurity center interact mainly with one extremely narrow phonon band, for instance, with a well-defined pseudolocal vibration of the appropriate symmetry. In other words, this approach is valid when the elastic interaction of the displacements of the atoms of the first coordination sphere with the other ones is weak enough, and the coupled motion of the impurity electrons with the vibrations of the atoms of the first coordination sphere can be regarded in the first approximation as independent of the vibrations of the atoms of the remaining crystal lattice, that is as a "quasimolecule" (a cluster, Sect. 4.6.5).

The solution of the multimode $E \otimes e$ problem with strong linear vibronic coupling presented in this section follows closely [4.12, 129], and is not restricted by the framework of the cluster model. An analogous approach to the $\Gamma_8 \otimes (e + t_2)$ problem with strong linear vibronic coupling is used in [4.132, 133].[6]

Consider now another case when the lowest sheet of the adiabatic potential has several absolute minima divided by potential barriers. In the case of strong vibronic coupling, when the potential barriers are high enough, in the lowest vibronic states the motion of the system is localized near the minima and has the nature of small vibrations, provided tunneling is neglected. The quadratic $E \otimes e$ problem is a characteristic example of this kind. The Hamiltonian of the multimode $E \otimes e$ case with quadratic vibronic coupling differs from that of (3.5.24) by the additional term $W_E[(Q_\theta^2 - Q_\varepsilon^2)\sigma_x - 2Q_\theta Q_\varepsilon \sigma_y]$, where Q_θ and Q_ε, as above, are symmetrized displacements of the atoms of the first coordination spheres defined in (3.5.25). Considering only the lowest sheet of the adiabatic potential in the adiabatic approximation and performing some fairly simple transformations analogous to those which led to the Hamiltonian (4.6.26), we have

$$H = H_1 - W_E \varrho_1^2 \cos 3\varphi_1 \left(1 - \frac{3\hbar^2}{2V_E \varrho_1^3}\right) . \tag{4.6.32}$$

Here H_1 is the Hamiltonian (4.6.26) in which the expression $j - L_z'$ is preserved in the previous operator form $-i\partial/\partial\varphi_1$, and the coordinate $Q_{1\theta} \equiv \varrho_1$.

When the quadratic vibronic coupling is included, the symmetry reduces to the point group symmetry of the impurity center in the crystal (e.g., to cubic symmetry) and the motion along φ_1 can no longer be separated. However, if the spacing of the energy spectrum corresponding to rotations along φ_1 is small compared with the spacing of the vibrational spectrum, the above motions can be separated using the adiabatic approximation. We now assume that the criterion for such a separation of the motion is fulfilled.

The spectrum of the fast subsystem can be obtained by excluding the term with $\partial/\partial\varphi_1$ from (4.6.32). As above, in the case of not very high temperatures when only the lowest energy vibronic states are important, the motion of the nuclei along the remaining degrees of freedom can be reduced to harmonic oscillations near new equilibrium positions. Not repeating the deductions which are in essence analogous to those leading to the solution of the linear $E \otimes e$ case, we

[6] It is interesting to note that the interaction of a fixed nucleon with a field of mesons can be described by a Hamiltonian which is formally analogous to that of the multimode $\Gamma_8 \otimes t_2$ problem. Its first solution had already been obtained in the 1940s [4.101].

give the following results [4.134]. The vibrational Hamiltonian of the fast subsystem can be divided into two commutative parts, H_A and H_B, corresponding to the variables with the indexes θ and ε, respectively. This is due to the fact that the cubic symmetry of the system in the minima is lowered to D_{4h} and the E_g representation of the Jahn-Teller active vibrations in the O_h group becomes reducible, $E_g = A_1 + B_1$. The variables with the indexes θ and ε are separated because the minimum lying on the Q_θ axis is considered. The Hamiltonian H_B and its corresponding density of vibrations have the same form as when only the linear vibronic interaction terms are considered (see (4.6.28)–(4.6.31)).

The Hamiltonian H_A describes the radial vibrations in the minima weakly redetermined by the quadratic vibronic coupling. The nature of this redetermination is the same as when the change in force constants corresponding to the softening of the lattice in the vicinity of the impurity center is considered.

To consider the slow subsystem corresponding to rotations along the warped trough of the lowest sheet of the adiabatic potential, let us average the Hamiltonian (4.6.32) over the ground vibrational state of the fast subsystem. Keeping only the leading terms in V_E^{-1} and W_E, we have [4.134]

$$H(\varphi) = -\alpha \frac{\partial^2}{\partial \varphi_1^2} - \beta \cos 3\varphi - E_{\mathrm{JT}} \ , \quad \text{where} \tag{4.6.33}$$

$$\alpha = \tfrac{1}{2}\hbar^2 V_E^{-2} (\overline{\omega^{-2}})^{-2} \ ; \qquad \beta = V_E^2 W_E (\overline{\omega^{-2}})^2 \ . \tag{4.6.34}$$

The Hamiltonian (4.6.33), as expected, is formally identical to the corresponding Hamiltonian for ideal vibronic systems (Sect. 4.3), but unlike in the molecular case, the parameters α and β depend on the phonon density.

The solutions of the Schrödinger equation with the Hamiltonian (4.6.33) are discussed in detail in Sect. 4.3. They correspond to hindered rotations of the formation of the polaron type discussed above with multiple over-barrier reflections. For large β values the minima of the adiabatic potential are divided by rather high barriers. The motion of the polaron formation in this case is reduced to tunneling through the barriers: for most of the time the quasipolaron is centered at one of the fourth-order axes of the cubic system, jumping from time to time from one axis to another.

The energy spectrum of the problem under consideration consists of the superposition of the continuous spectrum of the fast subsystem (vibrations at the bottom of the minimum) and the discrete spectrum of the slow subsystem (hindered rotations along the trough or tunneling between the minima). Due to the strong vibronic coupling the positions of the rotational levels of the slow subsystem are shifted towards the region of low frequencies where the density of states of the fast subsystem is small. This causes the nonadiabatic corrections describing the broadening of the energy levels of the discrete spectrum (resulting from the interaction of the angular motion with the vibrations) to be very small; the broadening was estimated in [4.135]. The approximation of the adiabatic separation of the motions also constrains the magnitude of the constant of quadratic vibronic interaction: $W_E\langle\omega^{-2}\rangle_E \ll 1$. The same condition is used when the Hamiltonian (4.6.32) is deduced. The magnitude of the tunneling splitting can be calculated in principle by the method discussed in Sect. 4.3, which is valid for the multimode Jahn-Teller systems as well.

For instance, for the multimode $T \otimes (e + t_2)$ problem the quasi-classical treatment leads to the same result (4.3.43) as in the ideal case. Substituting the components of the vector $\hat{a}$ in spherical coordinates (4.3.41) into (3.5.30) with the additional condition $\varphi = \pi/4$, we obtain a system of equations which define parametrically the line of steepest slope in the multidimensional q space [(cf. (4.3.41)]:

$$q_\kappa(\theta) = \frac{V_T}{\omega_\kappa^2}\left(\frac{1}{\sqrt{2}}q_\kappa(\xi)\sin 2\theta + \frac{1}{\sqrt{2}}a_\kappa(\eta)\sin 2\theta + a_\kappa(\zeta)\cos^2\theta\right)$$

$$+ \frac{V_E}{4\omega_\kappa^2}a_\kappa(\theta)(1 - 3\cos 2\theta) \tag{4.6.35}$$

with the effective mass

$$m = \sum_\kappa \left(\frac{\partial q_\kappa}{\partial\theta}\right)^2 = V_T^2\overline{(\omega^{-4})}_T(4\cos^2 2\theta + \sin^2 2\theta) + \frac{9}{4}V_E^2\overline{(\omega^{-4})}_E\sin^2 2\theta\ . \tag{4.6.36}$$

The curvature of the adiabatic potential in the trigonal minima determined by the parameter values

$$\theta_a = -\tfrac{1}{2}\arccos\tfrac{1}{3}\ ; \qquad \theta_b = \tfrac{1}{2}\arccos\tfrac{1}{3} \tag{4.6.37}$$

along the line of steepest slope (4.6.35) determines the frequency of impacts with the barrier walls:

$$\tilde{\omega}_E = \left(\frac{6[E_{JT}(T) - E_{JT}(E)]}{2V_T^2\overline{(\omega^{-4})}_T + 3V_E^2\overline{(\omega^{-4})}_E}\right)^{1/2}\ . \tag{4.6.38}$$

Taking the integral in (4.3.40) along the contour (4.6.35) between the points (4.6.37) we obtain the magnitude of the tunneling splitting in the multimode $T \otimes (e + t_2)$ problem with strong vibronic coupling in the same approximation as (4.3.43):

$$\delta = 4\Gamma\ , \qquad \Gamma \approx 0.82\{\hbar\tilde{\omega}_E[E_{JT}(T) - E_{JT}(E)]\}^{1/2}$$

$$\times \exp\left(-1.127\frac{E_{JT}(T) - E_{JT}(E)}{\hbar\tilde{\omega}_E}\right)\ , \tag{4.6.39}$$

where $\tilde{\omega}_E$ is determined by (4.6.38), while $E_{JT}(T)$ and the frequency ω_T in its expression are given by (3.3.11) and (3.5.33) respectively. In a series of cases the tunneling between the minima of the adiabatic potential is forbidden by symmetry conditions (Sect. 4.3.4). A simple example of this kind is provided by the $E \otimes (b_1 + b_2)$ problem. In such cases the redetermination of the vibronic spectrum is caused by the change of the curvature of the minima for the lowest states, and by the vibronic anharmonicity of the adiabatic potential for higher states. Since the change of the minimum curvature is of local origin, the problem can be reduced to that of changes in the force constant and can be easily solved. A similar treatment for the multimode $T \otimes (e + t_2)$ problem with predominant coupling to E vibrations is given in [4.136].

In cases when the tunneling is not forbidden by symmetry (these cases are in the majority, see Sect. 4.3) the vibronic spectrum of local and pseudolocal vibrations in the minima is complicated by additional lines of tunneling splitting (see also Sect. 4.6.4).

The excited components of the ground vibronic term split by tunneling are superimposed on the continuous vibrational spectrum adjoint to the lowest tunneling level. The interaction of the isothermal vibrational system with the dynamic (tunneling) subsystem, not taken into account in the above treatments (e.g. anharmonicity), results in relaxational broadening of the discrete energy levels of the latter. The magnitude of this broadening γ is proportional to the projected density of vibrational states $\varrho(\omega)$ of the dissipative subsystem at the frequency $\omega_0 = \delta/\hbar$ of the tunneling splitting. For low frequency acoustic vibrations with a Debye dispersion law, $\varrho(\omega)$ is proportional to ω^4 and hence $\gamma \sim \delta^4$. Therefore the tunneling splitting can be observed when the broadening γ is less than the tunneling splitting. As seen from (4.6.39), in the case of strong vibronic coupling, δ may be very small, and since γ has a higher order of smallness, the above condition is fulfilled. As the vibronic coupling decreases, δ increases, but $\gamma \sim \delta^4$ increases faster and the effect of tunneling splitting disappears. Note, however, that due to the strong exponential dependence of δ on E_{JT} tunneling splitting takes place even for rather moderate intermediate vibronic couplings, and therefore the range of applicability of the tunneling splitting theory with respect to the vibronic coupling constant is very large.

The tunneling states occur as weakly broadened peaks (resonances) in the vibronic density of states which can be interpreted as coupled (with the impurity) multiphonon formations of a polaron type, i.e., as low symmetry inhomogeneities of the electronic density surrounded by a phonon furcoat. The initial high symmetry of the Jahn-Teller system is preserved. The ground state has the same transformation properties as the initial degenerate electronic ground state without the vibronic coupling, but it is now a hybrid state with vibronic character. In other words, the dressed quasiparticles localized at the impurity preserve the symmetry properties of the "nondressed" ones.

4.6.4 Intermediate Coupling. Low-Energy Approximation

As has already been mentioned, there is some analogy between the Jahn-Teller effect and pion-nucleon scattering problems in quantum field theory [4.12].

Using this analogy one can employ some ideas and methods developed in the theory of the pion-nucleon interaction [4.101] in the theory of the multimode Jahn-Teller effect [4.99, 100]. The results below are obtained by the so-called method of dispersion equations which is known to yield a series of very important results, not based on perturbation theory. The low energy approximation which underlies this method is applicable to vibronic states with low energies, i.e., for states in close proximity to the ground state. An important special feature of this method is that it is only applicable to systems with continuous energy spectra. This means that the results of this section are concerned only with Jahn-Teller impurity centers in crystals.

In order to obtain the main relationships of the method of dispersion equations it is convenient to introduce the assumption known as the adiabatic switch-on of the interaction (see, e.g., [4.88]). Suppose that at the time $t = \pm\infty$ the vibronic coupling is zero, and at the instant $t = 0$ it switches on infinitely slowly. To do this, it is necessary to make the substitution $V'_{n\Gamma} \to V_{n\Gamma}\exp(-\varepsilon|t|)$, where $\varepsilon = +0$, in the Hamiltonian (3.5.27). The evolution of the states is then described by the solutions of the time-dependent Schrödinger equation or, in the interaction representation, by the following evolution operator

$$\hat{S}(t,0) = \exp\left(\frac{\mathrm{i}}{\hbar}\hat{H}_0 t\right)\exp\left(-\frac{\mathrm{i}}{\hbar}\hat{H}t\right) = T\exp\left(-\frac{\mathrm{i}}{\hbar}\int\limits_0^t \hat{V}(\tau)d\tau\right) , \tag{4.6.40}$$

where $\hat{V} = \hat{H} - \hat{H}_0$ is the operator of vibronic interactions, T is the symbol of chronologic ordering of the operators, and the function $\hat{V}(\tau)$ is determined by the interaction representation. It is known [Ref. 4.17; §41] that if the system is in a nondegenerate state the adiabatic switch-on of the perturbation does not change this state. This statement remains valid also for degenerate states, provided the perturbation is a scalar of the symmetry group, and the matrix elements of transitions between degenerate states possessing different transformation properties are zero according to selection rules. This important circumstance allows us to apply the methods of scattering theory to Jahn-Teller problems.

As initial conditions for (4.5.14) and (4.5.15) the values $b_{\Gamma\gamma}(t_0)$ at the times $t = \pm\infty$ can be used, provided the vibronic coupling vanishes then. This results in the constants

$$\hat{B}_{\Gamma\gamma}(\pm\infty) = \lim_{t_0\to\pm\infty} \hat{B}_{\Gamma\gamma}(t_0) = \hat{S}^{+}(\pm\infty,0)b_{\Gamma\gamma}\hat{S}(\pm\infty,0) . \tag{4.6.41}$$

The operators $B_{\Gamma\gamma}(\pm\infty)$ introduced here are the known "in" and "out" operators of scattering theory [4.137]. They obey the usual (Bose) commutative conditions. In addition they have the properties

$$\hat{B}_{\Gamma\gamma}(\pm\infty)|\bar{\Gamma}\bar{\gamma}\rangle = 0 , \tag{4.6.42}$$

where $|\Gamma\gamma\rangle$ is the exact wave function of the ground state at the moment $t = 0$[7].

[7] At $t = 0$ the Schrödinger representation is the same as the interaction representation.

Equation (4.6.42) can be proved easily if one substitutes $\hat{B}_{\Gamma\gamma}(\pm\infty)$ from (4.6.41) and takes into account that $\hat{S}(\pm\infty, 0)|\Gamma\gamma\rangle$ is a zero-phonon wave function of the ground state in the absence of vibronic interactions. Analogously, it can be shown that

$$|\tilde{\Gamma}(\Gamma)\mu\rangle_{\pm} = \sum_{\gamma,\bar{\gamma}} \hat{B}^{+}_{\Gamma\gamma}(\pm\infty)|\bar{\Gamma}\bar{\gamma}\rangle\langle\Gamma\gamma\bar{\Gamma}\bar{\gamma}|\tilde{\Gamma}\mu\rangle \tag{4.6.43}$$

are exact vibronic states at the instant $t = 0$ originating from symmetrized one-phonon states weakly coupled with Γ vibrations. For instance, for the ideal $T_2 \otimes t_2$ problem these states are related to the vibronic levels A_1, T_1, T_2 and E situated directly above the ground vibronic level T_2 (Fig. 4.10).

Consider the dispersion relationship (4.5.9). For small ω the main contribution to the summation (4.5.9) is made by the terms with small energy denominators which correspond to transitions to low energy vibronic states. Limiting the summation in (4.5.9) to exact one-phonon states (4.6.43) only, one can write down the following approximate equation for the electronic Green's function:

$$\begin{aligned}
\mathrm{Im}\{\langle\langle\gamma|\hat{C}_{\Gamma_1\gamma_1}, \hat{C}_{\Gamma_2\gamma_2}|\bar{\gamma}\rangle\rangle_{\omega}\} &\approx -\frac{K(\Gamma_1)K(\Gamma_2)}{2\hbar}\delta(\omega) \\
&\times \langle\psi_{\bar{\Gamma}\bar{\gamma}}|[\hat{C}_{\Gamma_1\gamma_1}, \hat{C}_{\Gamma_2\gamma_2}]|\psi_{\bar{\Gamma}\bar{\gamma}}\rangle - \frac{1}{2\hbar}S\sum_{\tilde{\Gamma}\mu}\sum_{n\Gamma}[\langle\bar{\Gamma}\gamma|\hat{C}_{\Gamma_1\gamma_1}|\tilde{\Gamma}(n\Gamma)\mu\rangle_{+} \\
&\times {}_{+}\langle\tilde{\Gamma}(n\Gamma)\mu|\hat{C}_{\Gamma_2\gamma_2}|\bar{\Gamma}\bar{\gamma}\rangle\delta(\omega-\omega_{n\Gamma}) - \langle\bar{\Gamma}\gamma|\hat{C}_{\Gamma_2\gamma_2}|\tilde{\Gamma}(n\Gamma)\mu\rangle_{+} \\
&\times {}_{+}\langle\tilde{\Gamma}(n\Gamma)\mu|\hat{C}_{\Gamma_1\gamma_1}|\bar{\Gamma}\bar{\gamma}\rangle\delta(\omega+\omega_{n\Gamma})] ,
\end{aligned} \tag{4.6.44}$$

where, as in (4.5.20), the contribution due to transitions between the degenerate states of the ground vibronic level is separated. The constant S is introduced to effectively take into account the omitted part of the sum.

The relation (4.6.44) is essentially based on the assumption of a continuous energy spectrum of the system. For the latter the so-called flip-over relation holds [Ref. 4.137; Eq. (3.1)]:

$$\hat{H}\hat{S}(\pm\infty, 0) = \hat{S}(\pm\infty, 0)\hat{H}_0 .$$

Because of this circumstance the vibrational frequencies $\omega_{n\Gamma}$ [and not the exact energy intervals $(E_n - E_\Gamma)/\hbar$ as in (4.5.9)] appear in the arguments of the δ function in (4.6.44). Substituting the expression (4.6.43) into the matrix element ${}_{+}\langle\tilde{\Gamma}(n\Gamma)\mu|\hat{C}_{\Gamma_2\gamma_2}|\Gamma\gamma\rangle$, using the relation between $b_{n\Gamma\gamma}$ and $\hat{B}_{n\Gamma\gamma}(\infty)$ which follows from (4.5.15) at $t = 0$, $t_0 = \infty$, and taking into account that $b_{n\Gamma\gamma}$ commutates with $C_{\Gamma_2\gamma_2}$, by means of (4.6.42) one finds

$${}_{+}\langle\tilde{\Gamma}(n\Gamma)\mu|C_{\Gamma_2\gamma_2}|\bar{\Gamma}\bar{\gamma}\rangle = \pi\hat{V}_{n\Gamma}\sqrt{\frac{2\hbar}{\omega_{n\Gamma}}}\sum_{\lambda\bar{\lambda}}\langle\tilde{\Gamma}\mu|\Gamma\lambda\bar{\Gamma}\bar{\lambda}\rangle\langle\langle\bar{\lambda}|\hat{C}_{\Gamma\lambda}, \hat{C}_{\Gamma_2\gamma_2}|\bar{\gamma}\rangle\rangle_{\omega_{n\Gamma}} . \tag{4.6.45}$$

Then substituting this expression into (4.6.44) we finally obtain

$$\mathrm{Im}\{\langle\langle\gamma|C_{\Gamma_1\gamma_1},\hat{C}_{\Gamma_2\gamma_2}|\bar{\gamma}\rangle\rangle_\omega\} \approx -\frac{K(\Gamma_1)K(\Gamma_2)}{2\hbar}\delta(\omega)$$

$$\times \langle\psi_{\Gamma\gamma}|[\hat{C}_{\Gamma_1\gamma_1},\hat{C}_{\Gamma_2\gamma_2}]|\psi_{\bar{\Gamma}\bar{\gamma}}\rangle - \pi S\sum_{n\Gamma}\frac{V_{n\Gamma}^2}{\omega_{n\Gamma}}\sum_{\tilde{\Gamma}}\sum_{\mu_1\lambda_1}\sum_{\mu_2\lambda_2}\langle\bar{\Gamma}_{\mu_1}\Gamma\lambda_1|\tilde{\Gamma}\mu\rangle$$

$$\times \langle\tilde{\Gamma}\mu||\bar{\Gamma}_{\mu_2}\Gamma\lambda_2\rangle[\langle\langle\gamma|\hat{C}_{\Gamma_1\gamma_1},\hat{C}_{\Gamma\lambda_1}|\mu_1\rangle\rangle^*_{\omega_{n\Gamma}}\langle\langle\mu_2|\hat{C}_{\Gamma\lambda_2},\hat{C}_{\Gamma_2\gamma_2}|\bar{\gamma}\rangle\rangle_{\omega_{n\Gamma}}$$

$$\times \delta(\omega-\omega_{n\Gamma}) - \langle\langle\gamma|\hat{C}_{\Gamma_2\gamma_2},\hat{C}_{n\Gamma\lambda_1}|\mu_1\rangle\rangle^*_{\omega_{n\Gamma}}$$

$$\times \langle\langle\mu_2|\hat{C}_{\Gamma\lambda_2},\hat{C}_{\Gamma_1\gamma_1}|\bar{\gamma}\rangle\rangle_{\omega_{n\Gamma}}\delta(\omega+\omega_{n\Gamma})] \;. \tag{4.6.46}$$

Considering the relation between the transition matrix Γ and the electronic Green's function (4.5.19) one can easily see that (4.6.46) is an approximate representation of the optical theorem of the scattering theory which relates the imaginary part of the scattering amplitude to the full cross section of the elastic scattering of one phonon [4.137]. The proof of this theorem is based on the preservation of the number of scattered particles (in our case, the number of phonons). This reveals the physical meaning of the basic approximation of (4.6.46).

The optical theorem (4.6.46) can be rewritten for the partial electronic Green's functions (4.5.11) as

$$\mathrm{Im}\{G^{(\Gamma)}_{\Gamma_1\Gamma_2}(\omega)\} = -\frac{R^{(\Gamma)}_{\Gamma_1\Gamma_2}}{2\hbar}\delta(\omega) - S\pi\sum_{nL}\frac{V_{nL}^2}{\omega_{nL}}[G^{(\Gamma)*}_{\Gamma_1 L}(\omega_{nL})$$

$$\times G^{(\Gamma)}_{L\Gamma_2}(\omega_{nL})\delta(\omega-\omega_{nL}) - G^{(\Gamma)*}_{\Gamma_2 L}(\omega_{nL})G^{(\Gamma)}_{L\Gamma_2}(\omega_{nL})\delta(\omega+\omega_{nL})] \;. \tag{4.6.47}$$

where

$$R^{(\Gamma)}_{\Gamma_1\Gamma_2} = \mathrm{i}^{j(\Gamma_1)+j(\Gamma_2)}K(\Gamma_1)K(\Gamma_2)\Big((-1)^{j(\Gamma_1)}\delta_{\bar{\Gamma}\Gamma}\delta(\Gamma_2\bar{\Gamma}\bar{\Gamma})$$

$$\times \delta(\Gamma_2\bar{\Gamma}\bar{\Gamma})f_{\bar{\Gamma}}^{-1} - (-1)^{j(\Gamma_2)}\begin{Bmatrix}\Gamma_1 & \bar{\Gamma} & \Gamma\\ \Gamma_2 & \bar{\Gamma} & \bar{\Gamma}\end{Bmatrix}\Big) \;. \tag{4.6.48}$$

Here, as above, it is assumed that the Jahn-Teller electronic term of the vibronic problem is classified by the irreducible representation $\bar{\Gamma}$, $f_{\bar{\Gamma}}$ being its dimensionality. Here $\begin{Bmatrix}\cdots\\ \cdots\end{Bmatrix}$ is the 6Γ symbol, $K(\Gamma)$ is the vibronic reduction factor for electronic operators transforming as the representation Γ (Sect. 4.7), $j(\Gamma)$ is the fictitious moment of the representation Γ, and

$$\delta(\Gamma_1\Gamma_2\Gamma_3) = \begin{cases}1, & \Gamma_3 \in \Gamma_1\otimes\Gamma_2\\ 0, & \Gamma_3 \notin \Gamma_1\otimes\Gamma_2\end{cases} \;.$$

Using the fact that the system has a continuous spectrum of vibrations, the sum over n in (4.6.47) can be curtailed by means of the δ function:

$$\mathrm{Im}\{G^{(\Gamma)}_{\Gamma_1\Gamma_2}(\omega)\} = -\frac{R^{(\Gamma)}_{\Gamma_1\Gamma_2}}{2\hbar}\delta(\omega) - S\pi\sum_L \frac{V_L^2}{\omega}[\varrho_L(\omega)G^{(\Gamma)*}_{\Gamma_1 L}G^{(\Gamma)}_{L\Gamma_2}(\omega) + \varrho_L(-\omega)G^{(\Gamma)*}_{\Gamma_1 L}(-\omega)G^{(\Gamma)}_{L\Gamma_2}(-\omega)]\ , \quad \text{where} \tag{4.6.49}$$

$$\varrho_L(\omega) = \sum_\kappa a_\kappa^2(L\gamma)\delta(\omega-\omega_\kappa)\ .$$

In the particular case of the linear $E \otimes e$ problem, (4.6.49) acquires the simpler form [4.99, 100]

$$\mathrm{Im}\{G^{(J)}(\omega)\} = -\frac{R^{(J)}}{2\hbar}\delta(\omega) - \pi S V_E^2 \varrho_E^{(0)}(\omega)|G^{(J)}(\omega)|^2\ ; \qquad J = \frac{1}{2}, \frac{3}{2} \tag{4.6.50}$$

where $\varrho_E^{(0)}(\omega)$ is determined by (4.6.11). The subscripts $\Gamma_1 = \Gamma_2 = 1/2$ are omitted for the sake of simplicity.

The dispersion relationships (4.5.20) for partial electronic Green's functions (4.5.11) have the form [4.99, 100]

$$G^{(\Gamma)}_{\Gamma_1\Gamma_2}(\omega) = \frac{R^{(\Gamma)}_{\Gamma_1\Gamma_2}}{2\pi\hbar\omega} - \frac{1}{\pi}\int_{+0}^{\infty} d\nu\left(\frac{\mathrm{Im}\{G^{(\Gamma)}_{\Gamma_1\Gamma_2}(\nu)\}}{\omega-\nu} + \frac{\mathrm{Im}\{G^{(\Gamma)}_{\Gamma_1\Gamma_2}(-\nu)\}}{\omega+\nu}\right)\ . \tag{4.6.51}$$

In the particular case of the linear $E \otimes e$ problem, (4.6.51) is simpler:

$$G^{1/2(3/2)}(\omega) = \pm\frac{K^2(E)}{4\pi\hbar\omega} - \frac{1}{\pi}\int_{+0}^{\infty} d\nu\left(\frac{\mathrm{Im}\{G^{1/2(3/2)}(\nu)\}}{\omega-\nu} + \frac{\mathrm{Im}\{G^{3/2(1/2)}(\nu)\}}{\omega+\nu}\right)\ . \tag{4.6.52}$$

The lower sign in (4.6.52) corresponds to the values of the momentum indicated as a superscript to G in parenthesis.

In the deduction of this expression the crossing-symmetry relation (4.5.10) was employed, which for the partial Green's functions looks like

$$G^{(\Gamma)}_{\Gamma_1\Gamma_2}(-\nu) = \sum_L f_L \begin{Bmatrix} \bar\Gamma & \Gamma_1 & L \\ \bar\Gamma & \Gamma_2 & \Gamma \end{Bmatrix} G^{(L)}_{\Gamma_2\Gamma_1}(\nu)\ , \tag{4.6.53}$$

where f_L is the dimensionality of the representation L, and $\bar\Gamma$ is the representation of the electronic Jahn-Teller terms. In particular, for the linear $E \otimes e$ problem we have

$$G^{(1/2)}(-\nu) = G^{(3/2)}(\nu)\ . \tag{4.6.54}$$

Expressions (4.6.50), (4.6.52) and (4.6.54) form a system of nonlinear integral equations with respect to two functions of complex variables $G^{(1/2)}$ and $G^{(3/2)}$. Their solution may be sought in the form [4.99, 100]

$$G^{(J)}(z) = \frac{R^{(J)}}{2\pi\hbar z g^{(J)}(z)} \,. \tag{4.6.55}$$

The function $g^{(J)}$ has the same behavior as $G^{(J)}$ except for the pole at $z = 0$ where, as seen from (4.6.55), $g^{(J)}(0) = 1$. Therefore $g^{(J)}(z)$, being an analytic function, satisfies the dispersion relationship

$$g^{(J)}(z) = 1 - \frac{z}{\pi} \int\limits_{-0}^{\infty} \frac{dv}{v} \left(\frac{\mathrm{Im}\{g^{(J)}(v)\}}{z - v} + \frac{\mathrm{Im}\{g^{(J)}(-v)\}}{z + v} \right) . \tag{4.6.56}$$

From (4.6.50) by means of (4.6.55) we obtain

$$\begin{aligned} \mathrm{Im}\{g^{(J)}(\omega)\} &= -\frac{R^{(J)}}{2\pi\hbar\omega} \frac{\mathrm{Im}\{G^{(J)}(\omega)\}}{|G^{(J)}(\omega)|^2} \\ &= \frac{1}{\pi\omega} \frac{\delta(\omega)}{|G^{(J)}(0)|^2} + \frac{R^{(J)}}{2\hbar\omega} S V_E^2 \varrho_E^{(0)}(\omega) \,. \end{aligned} \tag{4.6.57}$$

Substituting (4.6.57) into (4.6.56) and taking into account (4.6.54) and (4.6.55), we finally obtain

$$g^{1/2(3/2)}(\omega) = 1 \mp \frac{\omega K(E)^2}{2\pi\hbar} S V_E^2 \int\limits_{-0}^{\infty} \frac{dv}{v} \varrho_E^{(0)}(v) \frac{1}{\omega^2 - v^2} \,. \tag{4.6.58}$$

Thus the solution of the dispersion equations is [4.99, 100]

$$G^{1/2(3/2)}(\omega) = \pm \frac{K(E)^2}{4\pi\hbar\omega} \left(1 \mp \frac{\omega K(E)^2}{2\pi\hbar} S V_E^2 \int\limits_{0}^{\infty} \frac{dv}{v} \varrho_E^{(0)}(v) \frac{1}{\omega^2 - v^2} \right)^{-1} \tag{4.6.59}$$

The unknown constant S can be evaluated by comparing (4.6.59) with the results obtained in the limits of strong and weak coupling. The spectrum of elementary excitations is determined by the poles of the Green's function (4.6.59):

$$1 \pm \frac{\omega K(E)^2 S V_E^2}{2\pi\hbar} \int\limits_{0}^{\infty} \frac{dv}{v} \varrho_E^{(0)}(v) \frac{1}{\omega^2 - v^2} = 0 \,. \tag{4.6.60}$$

By comparing (4.6.60) with (4.6.12) (in the case of weak coupling) and with (4.6.28) (in the case of strong coupling) one finds that the interpolation expression $S = 8\pi/K(E)$ is valid for both the weak and strong coupling cases. The final expression for the partial electronic Green's function $G^{1/2(3/2)}$ in the linear multimode $E \otimes e$ problem is

$$G^{1/2(3/2)}(\omega) = \pm\frac{K(E)^2}{4\pi\hbar\omega}\left(1 \pm \frac{4\omega K(E)V_E^2}{\hbar}\int_{+0}^{\infty}\frac{dv}{v}\varrho_E^{(0)}(v)\frac{1}{v^2-\omega^2}\right)^{-1} . \qquad (4.6.61)$$

Thus the low energy spectrum of the multimode $E \otimes e$ system with linear vibronic coupling is a superposition of two continuous spectra which correspond to the following values of the momentum: $J = \pm 1/2$ and $J = \pm 3/2$. For weak coupling this superposition corresponds to the Moffitt-Thorson splitting of one-quantum states (Sect. 4.1.2). In the case of strong vibronic coupling, as seen from (4.6.61), the spectral density for $G^{1/2}$ has only one singularity at the point $\omega = 0$, corresponding to transitions between the degenerate states of the ground vibronic level. The remainder of this function is a smooth curve with no singularities. The spectral density of $G^{3/2}$ in the strong coupling limit contains a resonance in the low energy region at the frequency [cf. (4.6.27) and (4.6.28)]

$$\omega_r = \frac{\hbar}{2K(E)V_E^2\overline{(\omega^{-4})}_E} \approx \frac{\hbar}{V_E^2\overline{(\omega^{-4})}_E} , \qquad (4.6.62)$$

where $\overline{(\omega^{-4})}_E$ is determined by (4.6.21). This resonance corresponds to the transition from the ground state to the rotational one with $J = \pm 3/2$ [4.138]. The broadening of this level due to processes of direct one-phonon decay is [4.99]

$$\gamma = \frac{V_E^2}{\hbar^2}\varrho_E^{(0)}(\omega_r) . \qquad (4.6.63)$$

As mentioned above, see the passages after (4.6.13) and after (4.6.31), the resonance between the vibrational and rotational states results in the decay of the latter, i.e., the rotational state is metastable. As seen from (4.6.62), as the coupling constant increases, the frequency of rotations decreases, leading to a weakening of the decay processes due to the small density of phonons $\varrho_E^{(0)}(\omega_r)$ in (4.6.63).

In spite of the fact that the results of this section were obtained for systems with a continuous spectrum, a limiting transition to the case of an infinitely narrow phonon band corresponding to the case of an ideal vibronic system (Sect. 3.5) was performed in [4.99, 100]. As a result, an analytical expression for the energy gap between the ground and first excited vibronic states was obtained:

$$\Delta E = E_{3/2} - E_{1/2} = \left(1 + \frac{q^2 V_E^4}{\hbar^4\omega_E^8}\right)^{1/2} - \frac{qV_E^2}{\hbar^2\omega_E^4} ,$$

where $q = K(E)$ is the vibronic reduction factor for the operators $\hat{C}_{E\gamma}$ (Sect. 4.7). If one uses q values obtained by numerical calculations of the wave functions for the $E \otimes e$ case, the ΔE values calculated by the above formula agree well with numerical calculations of the energy spectrum for any value of vibronic coupling. Similarly, an expression for the energy of the ground state was deduced in [4.99, 100].

As has already been mentioned, in a Jahn-Teller system with linear vibronic coupling there is a relation between the electronic Green's function and the T matrix of elastic scattering of a phonon by the Jahn-Teller center. This relation is given by (4.5.19). If one substitutes (4.5.20) into it, then the first term on the right describes the contribution of the Born approximation to the phonon scattering with a renormalized coupling constant, i.e., with a constant multiplied by the reduction factor (which thus plays here the role of a renormalization parameter). Using the Wigner-Eckart theorem one can separate out the reduced matrix elements and partial contributions to both the scattering matrix T and the electronic Green's function. The appropriate expressions are similar to (4.5.6) and (4.5.11). Performing such a transformation we obtain the following expression for the T matrix of phonon scattering by the multimode Jahn-Teller system in the Born approximation [4.99, 100]:

$$\frac{F^{(\Gamma)}_{\Gamma_1\Gamma_2}}{2\pi\hbar z} = \frac{\hbar}{2\omega^2\pi} V_{\Gamma_1} V_{\Gamma_2} [\varrho^{(0)}_{\Gamma_1}(\omega)\varrho^{(0)}_{\Gamma_2}(\omega)]^{1/2} R^{(\Gamma)}_{\Gamma_1\Gamma_2} . \tag{4.6.64}$$

The negative and positive signs of the eigenvalues of the matrix $F^{(\Gamma)}_{\Gamma_1\Gamma_2}$ mean respectively attraction and repulsion between the phonon and impurity center, while the absolute values characterize the magnitude of the effect. Therefore the occurrence of low energy resonances in the spectral density is expected for states with the largest negative eigenvalues of the matrix $F^{(\Gamma)}_{\Gamma_1\Gamma_2}$. For instance, for the $E \otimes e$ case $\Gamma_1 = \Gamma_2 = \Gamma = E$. Substituting the appropriate 6Γ symbols into (4.6.48) we obtain the eigenvalues $R^{(A_1)} = R^{(A_2)} = -\frac{1}{2}$, $R^{(E)} = \frac{1}{2}$. Hence the low energy resonances are expected for the states A_1 and A_2. The equal effective coupling of the phonon to these two states is due to the axial symmetry of the linear $E \otimes e$ problem (Sect. 3.2.1). They have the momenta $J = \pm 3/2$; the resonance mentioned above at the frequency ω_r, see (4.6.62), corresponds to these states.

In [4.99, 100], a similar analysis of the possible existence of low energy resonances in the spectral density of the other multimode cases, $E \otimes (b_1 + b_2)$, $T \otimes t$, $T \otimes (e + t_2)$, $T \otimes d$, is also given. It emerges that the occurrence of low energy resonances in multimode Jahn-Teller systems is expected in the same cases as in ideal vibronic systems. The physical sense of this result becomes clear when one considers the limit of strong vibronic coupling, where a separate investigation of the lower sheet of the adiabatic potential is valid. In this case there are some distinguished directions of transitions between the minima of the adiabatic potential along the line of steepest slope. The motion along these degrees of freedom can be reduced to either free (or hindered) rotations, or to tunneling transitions between the minima. The presence of the other degrees of freedom results in relaxational broadening of these low energy states, similar to the result given in (4.6.63). The occurrence of pseudolocal low energy resonances in the spectral density of states of multimode systems is thus one of the most characteristic demonstrations of the multimode Jahn-Teller effect (sse also [4.139]).

4.6.5 Other Models and Approximations

The theory of the multimode Jahn-Teller effect can be greatly simplified by means of the so-called *quasimolecular (cluster) model.* The main assumption of this model is that the Jahn-Teller center together with a small number of its nearest-neighbor atoms involved in the degenerate (or near-degenerate) electronic state may be considered as an isolated (or almost isolated) molecular group (cluster) [4.140–143], and instead of the normal modes of the crystal lattice the normal vibrations of this separate group can be used. For example, for the impurity center Tl^+ substituting the cation in the crystal NaCl, this cluster contains the Tl^+ ion plus six nearest neighbour Cl^- ions; the octahedral complex $[TlCl_6]^{5-}$ may serve as a cluster model of the impurity center in this case. In this model the real interaction of the electrons with the infinite number of crystal lattice vibrations of given symmetry is substituted by one or several vibrations of the cluster, and the problem is reduced to the solution of vibronic equations considered in Sects. 4.1–4.5.

Within the framework of the cluster model quite a number of vibronic properties of lattice defects can be satisfactorily described. They include, first of all, the ground state properties which are manifest in the low-temperature radio-spectroscopy region. As shown in Sects. 5.4 and 5.5, the spectral properties of defects in the radio-frequency region can be described by means of an effective spin Hamiltonian [4.87] with a matrix (operator) structure determined by symmetry conditions. The parameters of this Hamiltonian are the reduced matrix elements calculated by the ground state wave functions. As far as just the ground state is concerned, the operator structure of the effective Hamiltonian is the same with and without vibronic coupling, and only its parameters differ by the so-called vibronic reduction factors (Sect. 4.7). However, in the case of strong vibronic coupling the first excited vibronic (rotational or tunneling) state may be very close to the ground one, and then it must be included in the basis of the effective Hamiltonian (Sect. 5.5).

The results obtained in this way are in very good agreement with the experimental observations of ESR spectra of Jahn-Teller defects in crystals [4.144]. On the other hand, it is well known that a crystal with a paramagnetic defect, unlike molecular systems, has a continuous spectrum of phonon excitations among which there are some infinitesimally small in energy. The vibronic coupling involves all the "one-phonon", "two-phonon", etc., states in the structure of the magnetic sublevels, and at non-zero temperatures they should apparently be included in the basis of the effective Hamiltonian. This is very difficult to do, since the number of states to be considered is infinite, involving the whole low energy range of acoustic phonons.

The question is: how is it possible to describe correctly the ESR experiments on Jahn-Teller defects in crystals in the cluster model ignoring the contribution of low energy phonon states? The answer is that the density of low frequency acoustic phonons is very small and the vibronic coupling with them is negligible, even when the appropriate vibronic constant is large. Therefore the region of the phonon spectrum important in vibronic coupling is separated from the ground state by significant energy gap of $\sim 0.1\ \hbar\omega_D$, where ω_D is Debye frequency, and the acoustic phonons within this gap can be excluded. At low temperatures the region of the spectrum above $\sim 0.1\ \hbar\omega_D$ is not populated, and to a good approximation the treatment can be limited to the ground state. At higher temperatures, relaxation processes, in which the crystal plays the role of a dissipative system, become important. The cluster model describes satisfactorily only the dynamic subsystem, but its interaction with the rest of the system can be included in the theory in a phenomenological way [4.145].

Another example of the successful uses of the cluster model is provided by the explanation of the complicated shape of optical impurity absorption and luminescence of light corresponding to electronic transitions with the participation of degenerate electronic terms (Sect. 5.1.6). As can be seen from Figs 5.5 and 5.12, the envelope of the discrete spectrum evaluated in the cluster model reproduces well the shape of the band observed experimentally. This coincidence is due to the similarity of the adiabatic potential in the cluster model to that of the multimode impurity-phonon system. Indeed, the number of sheets of the surface, as well as the number of minima and their symmetry properties, are the same in these two cases. In order to fit the most characteristic features of the band envelope (the position of the maximum and the halfwidth) with the experimental ones, only two free parameters are needed, $E_{JT}/\hbar\omega$ and $\hbar\omega$. For the spectrum shown in Fig. 5.5 they are 2.5 and 0.049 eV respectively. However, the details of the fine structure of the spectrum, are not reproduced by the cluster model, as can be seen from the same Figs 5.5 and 5.12.

The following conclusions can be drawn from the above discussion: (1) The cluster model of defects in crystals is a convenient tool allowing us to obtain relatively simple solutions in limiting cases of weak and strong coupling and numerical results for intermediate coupling, as well as a very useful visual physical interpretation. (2) In the framework of the cluster model one can describe quite satisfactorily the properties of the system in its ground state and obtain good agreement with experimental data for band envelopes of electronic transitions by means of fitting free parameters. In the same way one can reproduce integral (with respect to the spectrum) characteristics of the system (susceptibility, heat capacity, etc.). (3) The cluster model is in principle invalid for the description of the vibronic spectrum, as well as for interpretation of relaxation processes for localized excitations of the defect and related temperature effects [4.145].

A simple way to extend the framework of the cluster model consists in considering the elastic interaction between the atoms of the first coordination spheres (forming the cluster) and the remaining crystal by means of perturbation theory. However, this method leads to bad results if the force constants of the cluster are taken as the constants of actual elastic interactions of nearest-neighbor atoms obtained from the solution of the lattice dynamics problem (and not as phenomenological parameters chosen to fit the experimental data). Therefore the division of the system into the cluster and the remainder of the crystal taking account of the interaction between them has to be carried out in a special way.

In [4.146] this division is performed by the method of linear canonical transformations in the space of the coordinates and momenta of the lattice (or, similarly, in the space of the operators of creation and annihilation of phonons). The coefficients of such a transformation are determined by the condition that the Jahn-Teller stabilization energy of the chosen cluster is minimized.

By means of fairly simple transformations one can verify that the interaction modes introduced in Sect. 3.5 satisfy this condition. Indeed, as shown in Sect. 3.5, the Jahn-Teller stabilization energy of a cluster is the same as the exact value of the Jahn-Teller energy of the crystal with the defect. It was also shown that in the case of one infinitely narrow phonon band the separation of the interaction mode can be carried out exactly.

If the dispersion cannot be ignored, then the canonical transformation (3.5.13), (3.5.15) is accompanied by the occurrence of a bilinear (with respect to the momenta) interaction between the cluster and the remainder of the crystal (the crystal with a cavity). This interaction results in a broadening of the vibronic levels of the cluster even in the first-order perturbation theory. Therefore this approximation can be called *the model of the relaxing cluster*. It is almost as simple as the cluster model. The Jahn-Teller Hamiltonian of the cluster can be diagonalized, e.g., by means of computer calculations. Therefore the model of the relaxing cluster allows us to consider the vibronic interaction even in the most difficult case of intermediate coupling. The calculation of the interaction of the cluster with the remainder of the crystal is also not difficult in principle. All this explains the attractiveness of the model of the relaxing cluster. A review of works in this area in the theory of the multimode Jahn-Teller effect is given in the review article [4.147].

The shortcomings of the model of the relaxing cluster are obvious. First, the application of perturbation theory requires the interaction between the cluster and the remainder of the crystal to be small compared with the average spacing of the vibronic spectrum of the cluster. On the other hand, the interaction alone leads to phonon band formation. This means that the model of the relaxing cluster is valid only for crystals with narrow phonon bands, their width being smaller than, for instance, the energy spacing in the rotational structure of the vibronic spectrum in the case of strong coupling.

Secondly, the use of a crystal with a cavity as a zeroth-order approximation means that there are artificial local and resonance states formed by the cavity. This artifact has to be diminished by considering the interaction between the crystal and the cluster exactly. It is clear that perturbation theory is not suitable for this purpose. It will be shown below that the occurrence of new local and resonance states, which are simultaneously both vibrational and electronic states (i.e., vibronic states), is the main feature of the multimode Jahn-Teller effect. Therefore the above artifact should be treated with care and excluded from the consideration by means of special procedures.

Note also that the method of cluster separation described above gives the best agreement of the ground state properties of the cluster with those of the real ground state of the multimode system. In this sense the above approach is a kind of variational method. The accuracy is known to fall off when passing to excited states requiring additional minimization and orthogonalization. For example, it is known that for reproduction of the envelope of the optical band for singlet-multiplet transitions, better results are given by another method of cluster separation based on a comparison of the first moments of the band of the model one-mode system with the exact moments of the multimode system (Sect. 5.1.6). A somewhat different approach to the problem using the ideas of renormalization group theory is suggested in [4.148].

One of the special features of vibronic systems is the chaotic (irregular) arrangement of the vibronic energy levels. This irregularity increases as the number of Jahn-Teller modes increases, and this suggests that a statistical investigation of the appropriate energy distribution is possible. There are grounds to assume that this energy distribution alone determines the spectral density of the observable features of the multimode system. Such a statistical analysis of the irregular vibronic spectra was undertaken in [4.149–151] where it is shown that the vibronic spectrum of a two-mode system with a strong pseudo-Jahn-Teller effect of the $E \otimes (b_1 + b_1)$ type can be described well by the Brody distribution. In [4.151] it is also shown that the form of the distribution depends essentially on the dimensions of the phase volume filled up with irregular trajectories, and that the transition from regularity to chaos in the spectrum can be reproduced well by a semiclassical approximation.

4.7 Factors of Vibronic Reduction

Some general properties of the ground state of the vibronic system can be analysed by means of group-theoretical considerations alone, provided the so-called factors of vibronic reduction are known. Consider, for example, the changes in the energy spectrum caused by the Jahn-Teller vibronic interaction. In the absence of vibronic coupling the energy spectrum consists of a superposition of equally spaced levels of harmonic oscillators. The degeneracy of the levels corresponds to the irreducible representations of the unitary group describing the symmetry of the zeroth-order Hamiltonian (Sect. 3.2).

Inclusion of the vibronic coupling reduces the high "accidental" symmetry of the electron-vibrational system to the group G_0, because the operator of vibronic interactions is a scalar of this group. All the excited vibronic terms which have a higher degeneracy than is allowed by the group G_0 are, generally speaking, split and classified now by the irreducible representations of the group (Sects. 3.2 and 4.1).

This conclusion also follows from other considerations. In the absence of vibronic coupling the eigenfunctions of the vibronic Hamiltonian have the form of the products (4.1.2). The wave functions related to the same excited vibrational term realize some (in general, reducible) representations of the group G_0. Indeed, if in the Jahn-Teller effect there is only one set of degenerate vibrations transforming after the irreducible representation Γ, then the oscillator states $|\ldots n_{\Gamma_1\gamma}\ldots\rangle$ transform after the irreducible representations contained in the symmetric product $[\Gamma_1^n]$, where $n = \sum_\gamma n_{\Gamma\gamma}$. The splitting of the excited vibronic terms arising when the vibronic interaction is considered is determined by the number of irreducible representations of the group G_0 contained in the product $\Gamma \times [\Gamma_1^n]$, where Γ is the representation of the wave functions $\psi_{\bar{\Gamma}\gamma}(\mathrm{r})$ of the Jahn-Teller term under consideration.

The ground vibronic term related to the given electronic term is an exception. This vibronic level is not split by the vibronic interaction in any order of the perturbation theory [4.152]. Indeed, the ground vibrational state $|\ldots 0\ldots\rangle$ transfoms as the totally symmetric representation $[\Gamma_1^0] = A_1$ of the group G_0. Therefore, in the absence of vibronic coupling, the electron-vibrational functions of the ground state $\psi_{\Gamma\gamma}(\mathrm{r})|\ldots 0\ldots\rangle$ transform after the representation $\Gamma \times A_1 = \Gamma$, which is irreducible in the group G_0. As mentioned above, the operator of vibronic interactions is a scalar of the group G_0 and therefore its operation preserves the irreducibility of the representation Γ, i.e., it does not cause any splitting of the ground vibronic term [4.152].

In order to be sure that this term will always be the ground one, and hence that the symmetry of the ground vibronic term of a polyatomic system will always be known a priori without calculation, we must prove that there are no values of coupling constants for which the ground term intersects with any of the excited ones.

In a general form this statement has not yet been proved, but the numerical calculations of concrete systems confirm it (Sect. 4.4). For some special cases this can be done analytically [4.153]. Meanwhile it allows one to draw a conclusion of extreme importance to the theory of the Jahn-Teller effect, that *the ground vibronic level of the system has the same symmetry as the initial degenerate electronic term* Γ. In many cases only the ground vibronic term determines the observable properties of the polyatomic system (see Chap. 5) and the knowledge of its symmetry allows their reliable interpretation without knowledge of the explicit form of the vibronic wave functions of the ground state.

For instance, the magnitude of the splitting of the ground vibronic term by some low symmetry perturbation is determined in accordance with degenerate perturbation theory by the eigenvalues of the matrix of the perturbation. If the perturbation is described by the operator $F_{\bar{\Gamma}\bar{\gamma}}$ transforming as the row $\bar{\gamma}$ of the representation $\bar{\Gamma}$, the perturbation matrix can be constructed by means of the Wigner-Eckart theorem. The situation is simplified if the operator $F_{\bar{\Gamma}\bar{\gamma}}$ acts in one of the subspaces only, either electronic or vibrational. In this case the reduced matrix element $\langle \Psi_\Gamma \| F_{\bar{\Gamma}} \| \Psi_\Gamma \rangle$ can be expressed by the reduced matrix element of the states of only one of the subsystems.

4.7.1 General Discussion

Consider first the matrix elements of the electronic operator $F_{\bar{\Gamma}\bar{\gamma}}(\mathrm{r})$. The vibronic wave function of the ground state can be written in the form of a tensor convolution:

$$\Psi_{\Gamma\gamma}(\mathrm{r},Q)=\sum_{\Gamma_1\gamma_1\gamma_2}\chi_{\Gamma_1\gamma_1}(Q)\psi_{\Gamma\gamma_2}(\mathrm{r})\langle\Gamma_1\gamma_1\Gamma\gamma_2|\Gamma\gamma\rangle\ , \tag{4.7.1}$$

where the summation is performed over the irreducible representation Γ_1 contained in the direct product $\Gamma\times\Gamma$. Calculating the matrix element of $F_{\bar{\Gamma}\bar{\gamma}}(\mathrm{r})$ with the wave functions (4.7.1) we find

$$\langle\Psi_{\Gamma\gamma_1}|F_{\bar{\Gamma}\bar{\gamma}}|\Psi_{\Gamma\gamma_2}\rangle=\langle\psi_\Gamma\|F_{\bar{\Gamma}}\|\psi_\Gamma\rangle\sum_{\Gamma_1}\langle\chi^2_{\Gamma_1}\rangle$$
$$\times\sum_{\lambda_1\lambda_2\lambda_3}\langle\Gamma\lambda_2\bar{\Gamma}\bar{\gamma}|\Gamma\lambda_3\rangle\langle\Gamma\lambda_2\Gamma_1\lambda_1|\Gamma\gamma_1\rangle\langle\Gamma\lambda_3\Gamma_1\lambda_1|\Gamma\gamma_2\rangle\ . \tag{4.7.2}$$

When deriving (4.7.2) we used the Wigner-Eckart theorem for the matrix elements $\langle\psi_{\Gamma\gamma_1}|F_{\bar{\Gamma}\bar{\gamma}}|\psi_{\Gamma\gamma_2}\rangle$ calculated with the function $\psi_{\Gamma\gamma}$ of the initial electronic term, and the orthogonality relation for the vibrational function $\chi_{\Gamma\gamma}(Q)$:

$$\langle\chi_{\Gamma_1\gamma_1}|\chi_{\Gamma_2\gamma_2}\rangle=\langle\chi^2_{\Gamma_1}\rangle\delta_{\Gamma_1\Gamma_2}\delta_{\gamma_1\gamma_2}\ , \tag{4.7.3}$$

where $\langle\chi^2_\Gamma\rangle=\langle\chi_{\Gamma\gamma}|\chi_{\Gamma\gamma}\rangle$ is the normalization integral. Reducing the sum over λ_1, λ_2, λ_3 in (4.7.2) by means of the Wigner-Eckart theorem we obtain [4.4, 36]

$$\langle\Psi_\Gamma\|F_{\bar{\Gamma}}\|\Psi_\Gamma\rangle=K(\bar{\Gamma})\langle\psi_\Gamma\|F_{\bar{\Gamma}}\|\psi_\Gamma\rangle\ ,\quad\text{where} \tag{4.7.4}$$

$$K(\bar{\Gamma})=f_\Gamma\sum_{\Gamma_1}(-1)^{j(\Gamma_1)}\begin{Bmatrix}\Gamma & \bar{\Gamma} & \Gamma\\ \Gamma & \Gamma_1 & \Gamma\end{Bmatrix}\langle\chi^2_{\Gamma_1}\rangle\ . \tag{4.7.5}$$

Here $\begin{Bmatrix}\cdots\\ \cdots\end{Bmatrix}$ is the 6Γ symbol for the G_0 group and $j(\Gamma_1)$ is the fictitious momentum of the representation Γ_1. For instance, for the E term interacting with E vibrations in trigonal or cubic group we obtain from (4.7.5)

$$\begin{aligned}K(A_1)&=1=\langle\chi^2_{A_1}\rangle+\langle\chi^2_{A_2}\rangle+\langle\chi^2_E\rangle\ ,\\ K(A_2)&=p=\langle\chi^2_{A_1}\rangle+\langle\chi^2_{A_2}\rangle-\langle\chi^2_E\rangle\ ,\\ K(E)&=q=\langle\chi^2_{A_1}\rangle-\langle\chi^2_{A_2}\rangle\ .\end{aligned} \tag{4.7.6}$$

Analogously, for the cubic T term interacting with E and T_2 vibrations from (4.7.5) we have

$$\begin{aligned}K(A_1)&=1=\langle\chi^2_{A_1}\rangle+\langle\chi^2_E\rangle+\langle\chi^2_{T_1}\rangle+\langle\chi^2_{T_2}\rangle\ ,\\ K(E)&=\langle\chi^2_{A_1}\rangle+\langle\chi^2_E\rangle-\tfrac{1}{2}\langle\chi^2_{T_1}\rangle-\tfrac{1}{2}\langle\chi^2_{T_2}\rangle\ ,\\ K(T_1)&=\langle\chi^2_{A_1}\rangle-\tfrac{1}{2}\langle\chi^2_E\rangle+\tfrac{1}{2}\langle\chi^2_{T_1}\rangle-\tfrac{1}{2}\langle\chi^2_{T_2}\rangle\ ,\\ K(T_2)&=\langle\chi^2_{A_1}\rangle-\tfrac{1}{2}\langle\chi^2_E\rangle-\tfrac{1}{2}\langle\chi^2_{T_1}\rangle+\tfrac{1}{2}\langle\chi^2_{T_2}\rangle\ .\end{aligned} \tag{4.7.7}$$

Since all the $\langle \chi_{\bar{\Gamma}}^2 \rangle \geq 0$ and their sum is 1 (4.7.6, 7), it is clear that for non-totally-symmetric representations $\bar{\Gamma} \neq A_1$, the factors $K(\bar{\Gamma}) \leq 1$. Thus for the ground vibronic term the Jahn-Teller effect results in a decrease of the matrix elements of the electronic operators $F_{\bar{\Gamma}\bar{\gamma}}(\mathrm{r})$ by a factor of $K(\bar{\Gamma})$. This effect is called the *vibronic reduction of electronic operators* and the factors $K(\bar{\Gamma})$ are called the *factors of vibronic reduction* [4.154].[8]

In cases when the symmetry of the problem is accidentally higher than the group G_0, some of the irreducible representations Γ_1 of the group G_0, over which the summation in (4.7.5) is performed, form irreducible representations of higher dimensionality related to the covering group. This means that some of the $\langle \chi_{\Gamma}^2 \rangle$ values cease to be independent and there may be some relation between the vibronic reduction factors $K(\Gamma)$.

Consider, for example, the $E \otimes e$ case. In the case of the trigonal symmetry D_{3h} the real electronic wave functions $\psi_\theta(\mathrm{r})$ and $\psi_\varepsilon(\mathrm{r})$ from (4.7.1), see (3.1.5), are

$$\left.\begin{aligned} \Psi_\theta &= \chi_{A_1}\psi_\theta + \chi_{A_2}\psi_\varepsilon + \frac{1}{\sqrt{2}}\chi_\theta\psi_\theta - \frac{1}{\sqrt{2}}\chi_\varepsilon\psi_\varepsilon \,, \\ \Psi_\varepsilon &= \chi_{A_1}\psi_\varepsilon - \chi_{A_2}\psi_\theta - \frac{1}{\sqrt{2}}\chi_\varepsilon\psi_\theta - \frac{1}{\sqrt{2}}\chi_\theta\psi_\varepsilon \,. \end{aligned}\right\} \tag{4.7.8}$$

Since the functions Ψ_θ, Ψ_ε, ψ_θ, ψ_ε and the normal coordinates Q_θ, Q_ε are real, the coefficients $\chi_{\Gamma\gamma}(Q)$ in (4.7.8) can be chosen to be real also.

As shown in Sect. 3.2, the restriction of the calculations to only linear vibronic interaction terms leads to a higher axial symmetry of the problem. Taking the wave functions of the reference electronic E term to be the states $\psi_{\pm 1/2}(\mathrm{r})$ of the energy spin 1/2, one can rewrite (4.7.1) in the form

$$\Psi_{1/2} = \chi_0\psi_{1/2} + \chi_1\psi_{-1/2} \,, \qquad \Psi_{-1/2} = \chi_1^*\psi_{1/2} + \chi_0^*\psi_{-1/2} \,, \tag{4.7.9}$$

where the index $j = 0, \pm 1/2, 1$ indicates the transformation properties of rotations around the symmetry axis of the axial group. Expressing $\Psi_{\pm 1/2}$ and $\psi_{\pm 1/2}$ in (4.7.9) in terms of the real functions Ψ_θ, Ψ_ε, ψ_θ, ψ_ε by means of the relations (3.1.5), and comparing with (4.7.8), we obtain the relations [4.155]

$$\begin{aligned} \chi_{A_1} &= \mathrm{Re}\{\chi_0\} \,, & \chi_{A_2} &= -\mathrm{Im}\{\chi_0\} \,, \\ \chi_\theta &= \sqrt{2}\,\mathrm{Re}\{\chi_1\} \,, & \chi_\varepsilon &= -\sqrt{2}\,\mathrm{Im}\{\chi_1\} \,. \end{aligned} \tag{4.7.10}$$

In the general $E \otimes e$ problem each pair of degenerate E vibrations transforms as the representation E_1 of the axial group, i.e., it possesses the transformation properties of the momentum $L_z = 1$. In accordance with the momentum

---

[8] In the literature the terms "Ham effect" and "Ham factors", named after the author who first suggested the general significance of the reduction effect [4.154], are often used.

composition rules

$$\chi_l(Q) = \sum_{\ldots l_n \ldots} f_{\ldots l_n \ldots}(\ldots \varrho_n \ldots) \exp\left\{ \mathrm{i} \sum_n l_n \varphi_n \right\} , \qquad (4.7.11)$$

where $\varrho_n = \sqrt{Q_{n\theta}^2 + Q_{n\varepsilon}^2}$, $\varphi_n = \arctan(Q_{n\varepsilon}/Q_{n\theta})$, and $f_{\ldots l_n \ldots}(\ldots \varrho_n \ldots)$ is a real function independent of φ_n, e.g., it is a scalar of the axial group. The summation in (4.7.11) is carried out for the values l_n which satisfy the requirement $\sum_n l_n = l$.

Since, from (4.7.10) and (4.7.11), we have

$$\chi_{A_2}(Q) = - \sum_{\ldots l_n \ldots} f_{\ldots l_n \ldots}(\ldots \varrho_n \ldots) \sin\left(\sum_n l_n \varphi_n \right) , \qquad \sum_n l_n = 0 , \qquad (4.7.12)$$

it is clear that in the ideal case $\chi_{A_2} = 0$. Excluding $\langle \chi_T^2 \rangle$ from (4.7.6) and for $\langle \chi_{A_2}^2 \rangle = 0$ we obtain $K(A_2) + 1 = 2K(E)$ or [4.156]

$$2q - p = 1 . \qquad (4.7.13)$$

It is clear also that this relationship holds only for the ideal $E \otimes e$ problem with linear vibronic coupling when the conditions of applicability for the perturbation theory with respect to $F_{\Gamma\gamma}(\mathrm{r})$ (in the framework of which the above results were obtained) are satisfied. Therefore there are quite a number of reasons why the relationship (4.7.13) may be violated, namely, the influence of the quadratic terms of the vibronic interaction, violation of the criterion of first-order perturbation theory with respect to $F_{\Gamma\gamma}(\mathrm{r})$, the multimode nature of the vibronic coupling and so on.

In the general case, if $\langle \chi_{A_2}^2 \rangle \neq 0$ we obtain from (4.7.6) the relation [4.87, 155] (see also [4.157])

$$2q - p = 1 - 4\langle \chi_{A_2}^2 \rangle \leq 1 . \qquad (4.7.14)$$

Similarly, for the linear $T \otimes (e + t_2)$ problem, if $E_{\mathrm{JT}}(E) = E_{\mathrm{JT}}(T)$, $\omega_E = \omega_T$ and the Hamiltonian "accidentally" has the higher symmetry of the group $SO(3)$, the irreducible representations E and T_2 of the cubic group are combined in one representation $D^{(2)}$ of the three-dimensional group of rotations. This means that $\chi_{E\theta}$, $\chi_{E\varepsilon}$, $\chi_{T_2\xi}$, $\chi_{T_2\eta}$, $\chi_{T_2\zeta}$ have the transformation properties of the lines of the irreducible representation $D^{(2)}$, $(3z^2 - \mathrm{r}^2)$, $\sqrt{3}(x^2 - y^2)$, $2\sqrt{3}yz$, $2\sqrt{3}xz$, $2\sqrt{3}$. From this it follows that $\langle \chi_{T_2}^2 \rangle / \langle \chi_E^2 \rangle = 3/2$ and in accordance with (4.7.7), $K(E) = K(T_2)$.

By means of arguments similar to those given above for the linear $E \otimes e$ problem it can be shown that in the case of the $T \otimes (e + t_2)$ problem, if $E_{\mathrm{JT}}(E) = E_{\mathrm{JT}}(T)$ and $\omega_E = \omega_T$, when only the linear terms of the vibronic coupling with one set of E vibrations and one set of T_2 vibrations are taken into account, the tensor convolution (4.7.1) does not contain the terms $\chi_{T_1\gamma}(Q)$. Accordingly, in (4.7.7), $\langle \chi_{T_1}^2 \rangle = 0$. Combining this condition with the equality $K(E) = K(T_2)$ one can easily obtain the relationship [4.15, 158]

$$K(E) = \tfrac{2}{5} + \tfrac{3}{5}K(T_1) \ . \tag{4.7.15}$$

If $\langle \chi_{T_1}^2 \rangle \neq 0$ one can find from (4.7.7) that

$$K(E) + \tfrac{3}{2}[K(T_2) - K(T_1)] = 1 - 3\langle \chi_{T_1}^2 \rangle \leq 1 \ . \tag{4.7.16}$$

In the general case it can be shown easily by means of (4.7.5) that [4.100]

$$\sum_{\bar{\Gamma}} (-1)^{j(\bar{\Gamma})} f_{\bar{\Gamma}} K(\bar{\Gamma}) = f_{\Gamma} \left\{ 1 - \sum_{\Gamma_0 \in \Gamma \otimes \Gamma} [1 - (-1)^{j(\Gamma_0)}] \langle \chi_{\Gamma_0}^2 \rangle \right\} \ . \tag{4.7.17}$$

The expression in square brackets on the right of (4.7.17) is nonzero if $j(\Gamma_0)$ is odd, i.e., for the anti-symmetric part of the direct product $\Gamma \otimes \Gamma$. The wave functions $\chi_{\Gamma_0\gamma_0}(Q)$ may be represented as a power series in Q, or as a series of powers of the operators of creation and annihilation of phonons. These operators possess Bose-type commutation properties, and therefore $\chi_{\Gamma_0\gamma_0}(Q)$ is invariant with respect to permutations of phonons. Being antisymmetric in its angular dependence, it must also be antisymmetric in its radial part. It follows that the radial dependence of $\chi_{\Gamma_0\gamma_0}(Q)$ has nodal points and $\langle \chi_{\Gamma_0}^2 \rangle$ *has to be small compared to* 1, $\langle \chi_{\Gamma_0}^2 \rangle \approx 0$ [4.100][9] and therefore (4.7.17) can be reduced to

$$\sum_{\bar{\Gamma}} (-1)^{j(\bar{\Gamma})} K(\bar{\Gamma}) f_{\bar{\Gamma}} \lessapprox f_{\Gamma} \ . \tag{4.7.18}$$

As mentioned above, for some cases of accidentally high symmetry of the vibronic Hamiltonian, $\langle \chi_{\Gamma_0}^2 \rangle = 0$ exactly. The multimode nature of the Jahn-Teller effect may lead to nonvanishing values of $\langle \chi_{\Gamma_0}^2 \rangle$, but as shown by numerical calculations, it remains small in accordance with the above arguments. For instance, for the two-mode $E \otimes (e + e)$ problem, $\langle \chi_{A_2}^2 \rangle$ is less than 0.07 or even 0.04 [4.79]. It is shown below (Sect. 4.7.2) that (4.7.18) becomes an exact equality in the limiting cases of weak and strong vibronic coupling.

For concrete calculations it is more convenient to use the simple definition of the vibronic reduction factors which follows from (4.7.4) and the Wigner-Eckart theorem [4.154]:

$$K(\Gamma) = \frac{\langle \Psi_{\Gamma\gamma_1}(\mathrm{r}, Q) | \hat{C}_{\bar{\Gamma}\bar{\gamma}} | \Psi_{\Gamma\gamma_2}(\mathrm{r}, Q) \rangle}{\langle \psi_{\Gamma\gamma_1}(\mathrm{r}) | \hat{C}_{\bar{\Gamma}\bar{\gamma}} | \psi_{\Gamma\gamma_2}(\mathrm{r}) \rangle} \ : \tag{4.7.19}$$

Taking into account the fact that the Pauli matrices σ_x, σ_z determined in the real basis ψ_θ and ψ_ε transform after the lines of the representation E, while σ_y transforms after the singlet representation A_2, we have

$$q = \langle \Psi_\theta | \hat{\sigma}_x | \Psi_\varepsilon \rangle \ , \qquad p = \mathrm{i} \langle \Psi_\theta | \hat{\sigma}_y | \Psi_\varepsilon \rangle \ . \tag{4.7.20}$$

---

[9] These arguments were first given by Pauli in his treatment of the pion-nucleon interaction in the so-called static model of the nucleon [4.101].

Analogously, in the Jahn-Teller effect for the T_1 term of cubic groups

$$K(E) = -\langle \Psi_z | \hat{C}_\theta | \Psi_z \rangle \; , \qquad K(T_1) = \mathrm{i}\langle \Psi_x | \hat{S}_z | \Psi_y \rangle \; ,$$
$$K(T_2) = -\langle \Psi_x | \hat{C}_\zeta | \Psi_y \rangle \; , \tag{4.7.21}$$

where the matrices $\hat{C}_{\Gamma\gamma}$ and $\hat{S}_\alpha$ are determined in (3.3.2) and (3.3.3).

The idea of vibronic reduction factors can be generalized to the case of an arbitrary matrix element between different vibronic states instead of the above one defined in the basis of the ground state terms.

The tensor convolution (4.7.1) does not represent the wave functions of the ground vibronic term only. It is clear that any matrix element of an electronic operator necessarily contains the reduced matrix element $\langle \Psi_\Gamma \| F_{\bar{\Gamma}} \| \Psi_\Gamma \rangle$ as a factor. Let us define the vibronic factor[10] by the ratio of reduced matrix elements of the electronic operator

$$K(\Gamma_1 | \bar{\Gamma} | \Gamma_2) = \frac{\langle \Psi_{\Gamma_1} \| F_{\bar{\Gamma}} \| \Psi_{\Gamma_2} \rangle}{\langle \psi_\Gamma \| F_{\bar{\Gamma}} \| \psi_\Gamma \rangle}$$
$$= \frac{\langle \Psi_{\Gamma_1\gamma_1}(\mathrm{r}, Q) | F_{\bar{\Gamma}\bar{\gamma}}(\mathrm{r}) | \Psi_{\Gamma_2\gamma_2}(\mathrm{r}, Q) \rangle}{\langle \psi_\Gamma \| F_{\bar{\Gamma}} \| \psi_\Gamma \rangle \langle \bar{\Gamma}\bar{\gamma}\Gamma_2\gamma_2 | \Gamma_1\gamma_1 \rangle} \; . \tag{4.7.22}$$

It is obvious that the vibronic reduction factors are particular cases of the vibronic factors [cf. (4.7.19)] $K(\Gamma|\bar{\Gamma}|\Gamma) = K(\bar{\Gamma})$. The vibronic factors introduced in this way may be very useful in situations when it is necessary to take into account the effect of an electronic operator which mixes a group of vibronic levels which are close in energy. Such cases arise, for instance, in the case of strong vibronic coupling when the main physical properties of the system are determined by the tunneling between the equivalent minima of the adiabatic potential, and the energy gap dividing the ground vibronic term from the first excited one is exponentially small (Sect. 4.3). The secular matrix of the first-order perturbation theory in this case contains the matrix elements of the perturbation between the vibronic states of the tunneling sublevels.

For instance, for the quadratic $E \otimes e$ problem the tunneling states transform as the irreducible representations $A_{1(2)}$ and E. The electronic operator mixing these states transforms as $A_{1(2)} \times E = E$. Accordingly, the matrix elements $\langle \Psi_A | F_{E\gamma} | \Psi_{E\gamma} \rangle$ which can be reduced to the vibronic factor $\mathrm{r} = K(A|E|E)$ have to be considered. In the limit of strong vibronic coupling the factor r has to be calculated by means of the tunneling functions (4.3.25). As a result, we obtain $\mathrm{r}_{1(2)} = K(A_{1(2)}|E|E) = \pm 1/\sqrt{2}$ [4.159] (see also [4.160, 161]).

[10] The term "reduction factor" is not appropriate here, since for an arbitrary matrix element the Jahn-Teller effect does not necessarily result in its reduction.

4.7.2 Concrete Analytical and Numerical Calculations of the Vibronic Reduction Factors

Analytical expressions for the vibronic reduction factors can be obtained only in the limiting cases of weak and strong vibronic coupling, that is, in the same approximations in which the wave functions of the ground state can be evaluated.

In the case of weak vibronic coupling the operator version of perturbation theory (Sect. 4.1) can be used. Performing a unitary transformation $\exp(\mathrm{i}\hat{S})$ of the operator $F_{\bar{\Gamma}\bar{\gamma}}(\mathrm{r})$, cf. (4.1.3), and taking into account the fact that in second-order perturbation theory

$$\tilde{F}_{\bar{\Gamma}\bar{\gamma}} = F_{\bar{\Gamma}\bar{\gamma}} + \mathrm{i}[\hat{S}, F_{\bar{\Gamma}\bar{\gamma}}] - \tfrac{1}{2}[[F_{\bar{\Gamma}\bar{\gamma}}, \hat{S}]\hat{S}] \ , \tag{4.7.23}$$

substituting in (4.7.23) the matrices $\hat{\sigma}_x$ and $\hat{\sigma}_y$ with (4.7.20) instead of the operators $F_{\bar{\Gamma}\bar{\gamma}}$, and (4.1.14) instead of the operator $\hat{S}$ we obtain for the ideal linear $E \otimes e$ problem

$$q = 1 - 2\frac{E_{\mathrm{JT}}}{\hbar\omega_E} \ , \qquad p = 1 - 4\frac{E_{\mathrm{JT}}}{\hbar\omega_E} \ . \tag{4.7.24}$$

In the case of strong vibronic coupling the adiabatic states (4.2.8) must be substituted into (4.7.20). Then we obtain

$$q = 1/2 \ , \qquad p = 0 \ . \tag{4.7.25}$$

The vibronic reduction factors q and p obtained in [4.156] by means of numerically evaluated wave functions of the $E \otimes e$ problem [4.162] are presented in Fig. 4.18. The dependence of the reduction factors on the constant of vibronic coupling can be approximated by the formulas

$$p = \exp\left(-\frac{4E_{\mathrm{JT}}}{\hbar\omega_E}\right) , \qquad q = \frac{1}{2}\left[1 + \exp\left(-\frac{4E_{\mathrm{JT}}}{\hbar\omega_E}\right)\right] , \tag{4.7.26}$$

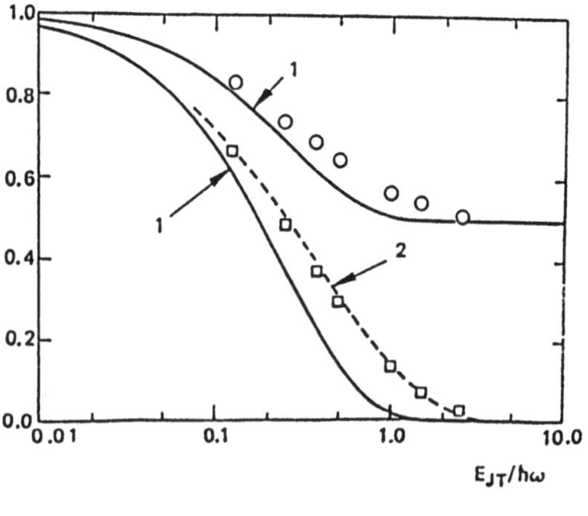

Fig. 4.18. Vibronic reduction factors in the linear $E \otimes e$ problem. (1) The curves obtained from (4.7.26); circles and squares indicate the results of numerical calculations [4.156]. (2) Approximation of (4.7.27)

which in the case of weak and strong coupling result in (4.7.24) and (4.7.25), respectively. In the region $0.1 \lesssim E_{JT}/\hbar\omega_E \lesssim 3$ a more exact interpolation has been suggested [4.156],

$$p = \exp\left[-1.974\left(\frac{E_{JT}}{\hbar\omega_E}\right)^{0.761}\right]. \tag{4.7.27}$$

Note that in the limiting case of strong vibronic coupling the asymptotic behavior of p is $p \sim (E_{JT}/\hbar\omega_E)^{-2}$, and is not exponential.

Equations (4.7.24–26) obey the relationship (4.7.13). In the multimode case this relationship is invalid, but in second-order perturbation theory it still holds. In fourth-order perturbation theory we have [4.93, 163]

$$\begin{aligned} p &= 1 - \frac{2}{\hbar}\sum_n \frac{V_n^2}{\omega_n^3} + 4\left(\sum_n \frac{V_n^2}{\hbar\omega_n^3}\right)^2 , \\ q &= \frac{1}{2}(1+p) - \frac{1}{2}\sum_{nm} \frac{V_n^2 V_m^2}{\hbar\omega_n^3 \hbar\omega_m^3}\left(\frac{\omega_n - \omega_m}{\omega_n + \omega_m}\right)^2 . \end{aligned} \tag{4.7.28}$$

In the case of strong vibronic coupling in the multimode $E \otimes e$ problem with a linear vibronic interaction, $p = 0$ (this is a result of the adiabatic approximation), whereas $q = \frac{1}{2}(1+p) = \frac{1}{2}$.

Including quadratic terms of the vibronic interaction also results in a violation of the relationship (4.7.13). In the limiting case of strong vibronic coupling using the tunneling states (4.3.26) to determine q and p we find [4.164, 165]

$$p = 0 , \quad q \lesssim \tfrac{1}{2} . \tag{4.7.29}$$

Analogously, for the $T \otimes (e + t_2)$ problem, substituting $\hat{S} = \hat{S}_E + \hat{S}_T$ in (4.7.23) (Sect. 4.1) and the operators $\hat{C}_\theta$, $\hat{S}_z$ and $\hat{C}_\zeta$ instead of $F_{\bar{\Gamma}\bar{\gamma}}$, see (4.7.21), in the limiting case of weak coupling gives

$$\begin{aligned} K(E) &= 1 - \frac{9}{4}\frac{E_{JT}(T)}{\hbar\omega_T} , \\ K(T_1) &= 1 - \frac{3}{2}\frac{E_{JT}(E)}{\hbar\omega_E} - \frac{9}{4}\frac{E_{JT}(T)}{\hbar\omega_T} , \\ K(T_2) &= 1 - \frac{3}{2}\frac{E_{JT}(E)}{\hbar\omega_E} - \frac{3}{4}\frac{E_{JT}(T)}{\hbar\omega_T} . \end{aligned} \tag{4.7.30}$$

In the absence of vibronic interactions with T_2 vibrations the $T \otimes e$ problem can be solved exactly. Simple calculations result in

$$K(E) = 1 , \quad K(T_1) = K(T_2) = \exp\left(-\frac{3}{2}\frac{E_{JT}(E)}{\hbar\omega_E}\right) . \tag{4.7.31}$$

In the limiting case of strong vibronic coupling with T_2 vibrations one can use the tunneling states (Sect. 4.3) to calculate the vibronic reduction factors. The limiting values of the reduction factors are

$$K(E) = 0 \ , \qquad K(T_1) = 0 \ , \qquad K(T_2) = 2/3 \ . \tag{4.7.32}$$

The following approximate expressions for the vibronic reduction factors in the $T \otimes t_2$ problem, found by means of an interpolation of the results of the strong and weak coupling cases were suggested in [4.154]:

$$\begin{aligned} K(E) &\approx K(T_1) = \exp\left(-\frac{9}{4}\frac{E_{\mathrm{JT}}(T)}{\hbar\omega_T}\right) , \\ K(T_2) &\approx \frac{1}{3}\left[2 + \exp\left(-\frac{9}{4}\frac{E_{\mathrm{JT}}(T)}{\hbar\omega_T}\right)\right] . \end{aligned} \tag{4.7.33}$$

The results of numerical calculations of the wave functions of the ground state in the $T \otimes t_2$ problem, and consequently the vibronic reduction factors [4.55], confirm the validity of the approximate formulas (4.7.33) (see however [4.56, 84, 166]).

It can be seen easily that in the case of weak vibronic coupling the reduction factors (4.7.30) obey the equality

$$K(E) + \tfrac{3}{2}[K(T_2) - K(T_1)] = 1 \ . \tag{4.7.34}$$

The exact results (4.7.31) obtained for the linear $T \otimes e$ problem obey the same equality. Comparing (4.7.34) with (4.7.16) one can find that for weak vibronic coupling with T_2 vibrations, $\langle \chi^2_{T_1} \rangle = 0$. A more accurate calculation by perturbation theory shows that the nonzero components $\chi_{T_1\gamma}$ occur in the wave function of the ground state in only the fifth order with respect to V_T. The zero value of $\chi_{T_1\gamma}$ in second-order perturbation theory follows from the "accidental" spherical symmetry of the $T \otimes t_2$ problem in this order (Sect. 4.1). It is likely that this high symmetry is preserved up to the fourth order with respect to V_T.

As follows from the approximate formulas (4.7.33), the equality (4.7.34) is also obeyed in the region of intermediate coupling, and in the case of strong vibronic coupling it is again an exact relationship [4.155]. This means that the contribution $\chi_{T_1\gamma}$ to the wave function of the ground state of the $T \otimes (e + t_2)$ problem is also small in the case of arbitrary values of the vibronic coupling constant V_T.

The strong dependence on $E_{\mathrm{JT}}/\hbar\omega$ is a general feature of all the reduction factors. Therefore even in the case of moderate vibronic coupling the reduction of the appropriate splitting in the vibronic spectrum can be very significant and reach several orders of magnitude. This was first discussed in [4.167] in which the vibronic reduction of the spin-orbital splitting of the $^2T_2$ term was evaluated. In [4.154, 159] the general nature of the effect was suggested and a generalization of the theory to the operators $F_{\bar{\Gamma}\bar{\gamma}}$ of arbitrary symmetry was given. In more

Table 4.1 References for calculations of the vibronic reduction factors

System	References
$E \otimes e$	
Linear coupling	[4.162]
Strong linear couling	[4.168]
Coupling to one and two modes	[4.79]
Strong linear and quadratic coupling	[4.56, 164, 169]
Intermediate linear and quadratic coupling	[4.57, 170]
Multimode coupling	[4.79, 95, 125]
$T \otimes e$	
Linear coupling, all strengths	[4.154]
$T \otimes t_2$	[4.55, 56]
$T \otimes (e + t_2)$	
Strong coupling	[4.14, 23]
Equal coupling, intermediate strength	[4.15]
Unequal coupling	[4.71, 73, 84]
$\Gamma_8 \otimes (e + t_2)$	[4.86]

complicated cases the vibronic reduction factors cannot be so easily evaluated analytically, and they must be computed numerically, but a great many of them have been worked out; the list of references is given in Table 4.1. The general result for any vibronic system (i.e., for any symmetry and any type of electronic degeneracy) was obtained only for the limiting case of strong vibronic coupling [4.100]. It can be shown easily that in this case $K(\bar{\Gamma}) \sim Q_0^2(\bar{\Gamma})$, where $Q_0(\Gamma)$ is the distance (in units of V_Γ/ω_Γ^2) from one of the equivalent minima of the adiabatic potential to the high symmetry point of the reference nuclear configuration (Sect. 2.2) in the subspace of nuclear displacements $Q_{\bar{\Gamma}\bar{\gamma}}$ with the given transformation properties $\bar{\Gamma}$. For instance, for the $T \otimes t_2$ problem the trigonal minima of the adiabatic potential are shifted from the high symmetry reference point in the subspace of the T_2 vibrations only (Sect. 3.3): $Q_0(T_2) = 2/\sqrt{3}$, whereas $Q_0(E) = 0$. It follows that the limiting values of $K(T_2)$ and $K(E)$ in the strong coupling case under consideration are $K(T_2) \neq 0$ and $K(E) = 0$. Similar conclusions about the magnitudes of reduction factors in the limiting case of strong vibronic coupling are summarized in [4.100] for all cases of the Jahn-Teller effect. Note that these results are also valid for the multimode cases (see Table 4.1).

4.7.3 Vibronic Reduction in Second-Order Perturbation Theory

In the case when it is necessary to consider the effect of the operators $F_{\bar{\Gamma}\bar{\gamma}}(\mathrm{r})$ in second-order perturbation theory, the effective operator can be written in the form of a tensor convolution

$$\mathscr{I}_{\Gamma\gamma}^{(2)}(\mathrm{r}, Q) = \sum_{\gamma_1\gamma_2} F_{\bar{\Gamma}\gamma_1}(\mathrm{r}) G(\mathrm{r}, Q) F_{\bar{\Gamma}\gamma_2}(\mathrm{r}) \langle \bar{\Gamma}\gamma_1 \bar{\Gamma}\gamma_2 | \Gamma\gamma \rangle \ , \qquad (4.7.35)$$

where

$$G(r, Q) = \sum_{nv}{}' \frac{|\Psi_{nv}(r, Q)\rangle\langle\Psi_{nv}(r, Q)|}{E_0 - E_{nv}} . \tag{4.7.36}$$

The summation in this equation is performed over all the excited states $\Psi_{nv}(r, Q)$ of the Jahn-Teller system, E_0 and E_{nv} being the energies of the ground and excited states, respectively. The operators introduced in this way are not pure electronic operators owing to the dependence of the Green's function $G(r, Q)$ on the Q coordinates in (4.7.35). In the absence of vibronic coupling the functions $\Psi_{nv}(r, Q)$ can be factorized, see (4.1.2), and considering the orthogonality of the vibrational states we have

$$\mathscr{I}^{(2)}_{\Gamma\gamma}(r) = \sum_{\gamma_1\gamma_2} F_{\bar{\Gamma}\gamma_1}(r)\left(\sum_{n}{}' \frac{|\psi_n(r)\rangle\langle\psi_n(r)|}{E_0 - E_n}\right) F_{\bar{\Gamma}\gamma_2}(r)\langle\bar{\Gamma}\gamma_1\bar{\Gamma}\gamma_2|\Gamma\gamma\rangle . \tag{4.7.37}$$

When the vibronic interaction is included, it results in two effects. First, due to the nonorthogonality of the nuclear functions the excited vibronic states of the electronic term under consideration can be taken as intermediate ones. This case was considered in [4.154] where in the framework of the simplest Jahn-Teller situation, the ideal linear $T \otimes e$ problem, second-order perturbation theory with respect to the spin-orbital interaction was used. Substituting the operator of orbital momentum L_α from $\lambda\hat{\boldsymbol{L}}\hat{\boldsymbol{S}}$ instead of $F_{\bar{\Gamma}\bar{\gamma}}(r)$ and the Green's function (4.7.36) with $n = T_1\gamma$ in (4.7.35) we find the second-order spin-orbital Hamiltonian

$$H^{(2)}_{SO} = -\frac{\lambda^2}{\hbar\omega_E}[f_a(\hat{\boldsymbol{L}}\hat{\boldsymbol{S}})^2 + (f_b - f_a)(\hat{L}_x^2\hat{S}_x^2 + \hat{L}_y^2\hat{S}_y^2 + \hat{L}_z^2\hat{S}_z^2)] , \tag{4.7.38}$$

where

$$\begin{aligned} f_a &= \exp\left(-\frac{3E_{JT}(E)}{\hbar\omega_E}\right) g\left(\frac{3E_{JT}(E)}{2\hbar\omega_E}\right) , \\ f_b &= \exp\left(-\frac{3E_{JT}(E)}{\hbar\omega_E}\right) g\left(\frac{3E_{JT}(E)}{\hbar\omega_E}\right) , \\ g(x) &= \int_0^x (e^t - 1)t^{-1}dt \approx x^{-1}e^{-x}\left(1 + \frac{1}{x} + \frac{2}{2!x^2} + \frac{2\cdot 3^2}{3!x^3} + \cdots\right) . \end{aligned} \tag{4.7.39}$$

The physical meaning of this result, as well as of the second-order effect in general, can be understood if one notes that in the case of strong vibronic coupling, $E_{JT} \gg \hbar\omega$, the following asymptotic formula is valid:

$$\frac{\lambda^2}{\hbar\omega_E}(f_b - f_a) \approx \frac{\lambda}{3E_{JT}(E)} .$$

The energy $3E_{JT}(E)$ is the gap between the lowest sheet of the adiabatic potential and the upper degenerate ones of the T term in the tetragonal distorted configuration of the nuclear framework at the minimum points. Thus in the case of strong vibronic coupling the Jahn-Teller distortion is equivalent to a static tetragonal crystal field, the upper sheets of the adiabatic potential playing the role of an excited level. In other words, in the case of strong vibronic coupling the correct result is given by (4.7.37) in which the summation can be limited by the upper sheets of the adiabatic potential in the configuration Q_0 of the minimum under consideration.

Thus the second-order perturbation theory splitting is usually proportional to E_{JT}^{-1} and decreases with E_{JT} slower than the first-order splitting (for which the vibronic reduction factor decreases more rapidly).

The other effect of vibronic coupling is due to the reduction of the contribution to the second-order splitting caused by the admixture of the excited electronic terms. Owing to the fact that the transformation properties of $\mathscr{I}_{\Gamma\gamma}^{(2)}(\mathrm{r}, Q)$ from (4.7.35) and of $\mathscr{I}_{\Gamma\gamma}^{(2)}(\mathrm{r})$ from (4.7.37) coincide, the differences in the reduced matrix elements of these operators can be considered by means of the introduction of the second-order vibronic reduction factors $K^{(2)}(\Gamma)$ [4.171]. If one takes into account only the nearest excited term and neglects the energy difference between the vibronic states compared with the energy gap Δ between the reference electronic terms, then, as can be shown easily, $K^{(2)}(\Gamma) \approx K(\Gamma)$.

However, this result is approximate and is based on the condition $\Delta_n \gg E_{nv} - E_{nv'}$. In the general case the reduction of the second- and higher-order effects is not determined by the factors $K(T)$. In order to describe the ground multiplet, the effective Hamiltonian written in the form of a scalar convolution of the orbital and spin operators is usually used. Different orbital irreducible tensor operators transforming as different irreducible representations result from the admixture of different excited electronic terms. The above condition $\Delta_n \gg E_{nv} - E_{nv'}$ may not be satisfied for all the admixed terms, and therefore not all the second-order reduction factors $K^{(2)}(\Gamma)$ can be reduced to the first-order factors $K(\Gamma)$.

4.7.4 Vibronic Amplification of Nuclear Displacement Operators

Consider now the matrix elements of the vibrational operator $F_{\bar{\Gamma}\bar{\gamma}}(Q)$ which is a function of the nuclear coordinates. The vibronic changes of splittings caused by perturbations linear in the nuclear coordinates,

$$F_{\bar{\Gamma}\bar{\gamma}} = A_{\bar{\Gamma}\bar{\gamma}} Q_{\bar{\Gamma}\bar{\gamma}} , \tag{4.7.40}$$

are of practical interest. Such a problem arises, for instance, in the analysis of quadrupole splitting in ESR spectra and in problems of acoustic nuclear resonance. In these cases, $A_{\bar{\Gamma}\bar{\gamma}}$ are functions of the operators of the nuclear spin of the Jahn-Teller center. In order to obtain the effect of vibronic changes of the reduced matrix element of such operators, consider the relation

$$[\hat{H}, P_{\Gamma\gamma}] = i\hbar\omega_\Gamma^2 Q_{\Gamma\gamma} + i\hbar V_\Gamma \hat{C}_{\Gamma\gamma} , \tag{4.7.41}$$

where H is the Hamiltonian (2.2.13) of an ideal vibronic system with linear vibronic coupling. Calculating the matrix element of this commutator with the exact vibronic functions of the ground state we have

$$\langle \Psi_{\bar{\Gamma}\gamma_1} | Q_{\Gamma\gamma} | \Psi_{\bar{\Gamma}\gamma_2} \rangle = -\frac{V_\Gamma}{\omega_\Gamma^2} \langle \Psi_{\bar{\Gamma}\gamma_1} | \hat{C}_{\Gamma\gamma} | \Psi_{\bar{\Gamma}\gamma_2} \rangle . \tag{4.7.42}$$

This exact relation is valid for the Jahn-Teller system with only linear vibronic coupling. From this it follows that the reduced matrix elements of $Q_{\Gamma\gamma}$ are proportional to the reduced matrix elements of $\hat{C}_{\Gamma\gamma}$, the latter being the vibronic reduction factors (4.7.20, 21) [4.159]

$$\langle \Psi_\Gamma \| Q_{\bar{\Gamma}} \| \Psi_\Gamma \rangle = -\frac{V_{\bar{\Gamma}}}{\omega_{\bar{\Gamma}}^2} K(\bar{\Gamma}) . \tag{4.7.43}$$

Similar consideration for the multimode Jahn-Teller system with linear vibronic coupling leads to the relation

$$\langle \Psi_\Gamma \| Q_{\bar{\Gamma}} \| \Psi_\Gamma \rangle = -V_{\bar{\Gamma}} \overline{(\omega^{-2})}_{\bar{\Gamma}} K(\bar{\Gamma}) , \tag{4.7.44}$$

where $Q_{\Gamma\gamma}$ are the symmetrized nuclear displacements of the first coordination sphere defined in (4.6.1), and $\langle \omega^{-2} \rangle$ is defined in (4.6.21).

As seen from these formulae the matrix elements of Jahn-Teller–active coordinates, which are zero in the absence of the Jahn-Teller effect, increase with the vibronic coupling constant $V_{\bar{\Gamma}}$. This *vibronic amplification* of splittings caused by operators of vibrational coordinates can be seen not only in the quadrupole splitting, but also in the Stark effect related to the dipole moment of the displacements $Q_{\Gamma\gamma}$.

Note that the nonzero values of the matrix elements of the operator $Q_{\Gamma\gamma}$ do not mean that the Jahn-Teller effect necessarily results in a distortion of the observable nuclear configuration. Indeed, the degeneracy of the ground vibronic state can be considered as the degeneracy of an energy spin (cf. the discussion in Sect. 3.2.1). *Such a pseudospin system can be completely or partially polarized by means of special preparation of the system in a mixed state determined by an appropriate density matrix.* Such a preparation can be performed, for instance, by means of an external low-symmetry field lifting the degeneracy, and by a consequent lowering of temperature down to $T = 0$ K. *Without this low-symmetry external perturbation the different components of the degenerate ground vibronic term are equally populated, and the cooling of the system leads to the zero value of the mean displacement*:

$$\langle Q_{\bar{\Gamma}\bar{\gamma}} \rangle = \sum_{\gamma_1} \langle \Psi_{\Gamma\gamma_1} | Q_{\bar{\Gamma}\bar{\gamma}} | \Psi_{\Gamma\gamma_1} \rangle = -\frac{V_\Gamma}{\omega_\Gamma^2} K(\bar{\Gamma}) \sum_{\gamma_1} \langle \bar{\Gamma}\bar{\gamma}\Gamma\gamma_1 | \Gamma\gamma_1 \rangle = 0 .$$

5. Spectroscopic Manifestations of Vibronic Effects

The essential changes of the energy spectrum and wave functions of molecules and crystals caused by vibronic interactions directly affect all the spectroscopic properties of the system. Even when the vibronic coupling is weak, the resulting changes in the spectra can be very significant, especially when the vibronic interaction leads to the appearance of additional spectral lines (for instance, in the rotational absorption spectrum of spherical-top molecules).

In this chapter the most-studied manifestations of vibronic effects over the full range of spectroscopy are considered. Attention is focused on changes in the shape and structure of the broad bands of electronic transitions (Sect. 5.1), splitting and temperature dependence of zero-phonon lines (Sect. 5.2), additional absorption in the IR region and the influence of vibronic effects on Raman spectra (Sect. 5.3), significant changes of the fine and hyperfine structure, line shape, temperature, frequency and angular dependence of EPR spectra, and related phenomena (Sects. 5.4 and 5.5). The strong influence of vibronic effects on spectral properties of polyatomic systems, especially on EPR spectra, played a decisive role in the development of a new field of the physics of molecules and crystals based on the concept of vibronic interactions.

5.1 The Shape of Optical Bands of Electron Transitions Between Degenerate Electronic Terms

The concept of vibrational bands of electron transitions in its modern formulation first arose in the theory of molecular spectra. In the case of diatomics the situation is relatively simple, since in such molecules, in the absence of accidental degeneracy of terms with identical symmetry, there are no vibronic effects. A correct understanding of the origin of the bands based on the Franck-Condon principle was first reached for these molecules, and the first quantum mechanical calculations of the distribution of the intensities over vibrational sublevels of the electron transition were carried out for them [5.1].

The same ideas applied to the similar situation of singlet-singlet transitions in impurity centers in solids allowed a corresponding theory of band shapes of impurity absorption to be constructed. The modern status of this theory is presented in [5.2–5].

However, in polyatomic molecules and in most cases of impurity centers, singlet-singlet transitions are the exception rather than the rule. Most often one or sometimes both the electronic levels participating in the optical transition are degenerate or quasi-degenerate. Electron transitions involving degenerate electronic levels also have possible applications. For example, the main working characteristics of quantum generators that determine the kinetics and efficiency of lasers are directly related to the parameters of the multiphonon bands of impurity absorption and luminescence in the activated crystals used as the working medium. This has stimulated a rapid development of the optics of impurity centers and coordination compounds in the last couple of decades, and, corresponding, the study of optical manifestations of the Jahn-Teller effect. The main results in this field are given in [5.4–12].

5.1.1 General Theory of Multiplet-Multiplet Optical Transitions

Consider a polyatomic system whose reference configuration Q_0 (Sect. 2.2.1) allows orbitally degenerate electronic states. Assume that the two electronic terms (or two groups of close-in-energy temrs) Γ_1 and Γ_2 participating in the optical transition are separated from one another and from the other electronic terms by rather large energy gaps, $|E(\Gamma_1) - E(\Gamma_2)| \gg \hbar\omega$, so that the modified Born-Oppenheimer approximation (Sect. 2.2.2) can be used for each of these terms separately. Assume also that the mean values of symmetrized nuclear displacements $Q_{\Gamma\gamma}$ in the electronic states Γ_1 and Γ_2 are relatively small, so that the second-order approximation (2.2.3) is valid for the vibronic Hamiltonians $\hat{H}(\Gamma_1)$ and $\hat{H}(\Gamma_2)$.

In a field of monochromatic electromagnetic radiation the polyatomic system undergoes transitions $\Gamma_1 n \rightleftarrows \Gamma_2 m$ between the vibronic states $\Psi_n(\Gamma_1)$ and $\Psi_m(\Gamma_2)$ for which the transition frequency $|E_m(\Gamma_2) - E_n(\Gamma_1)|/\hbar$ coincides with the frequency Ω of the electromagnetic wave. In these conditions absorption of radiation energy takes place. The absorption coefficient is given by

$$K(\Omega) = \frac{4\pi^2 N\Omega}{\hbar c n}(1 - \mathrm{e}^{-\hbar\Omega/kT})F_{\Gamma_1\to\Gamma_2}(\Omega) , \tag{5.1.1}$$

where n is the refractive index of the medium in which the absorption centers are placed, N is the concentration of these centers in the medium, c is the speed of light and $F_{\Gamma_1\to\Gamma_2}(\Omega)$ is the so-called form factor of the spectrum, determined by the expression

$$F_{\Gamma_1\to\Gamma_2}(\Omega) = \sum_{n,m} \varrho_n(\Gamma_1)|\langle\Psi_n(\Gamma_1)|d|\Psi_m(\Gamma_2)\rangle|^2\,\delta\left(\frac{E_m(\Gamma_2) - E_n(\Gamma_1)}{\hbar} - \Omega\right) . \tag{5.1.2}$$

Here $\varrho_n(\Gamma_1)$ is the Gibbs temperature occupation value for the vibronic level $E_n(\Gamma_1)$:

$$\varrho_n(\Gamma_1) = \exp\{[\mathscr{F}(\Gamma_1) - E_n(\Gamma_1)]/kT\} \ , \tag{5.1.3}$$

$\mathscr{F}(\Gamma_1)$ is the free energy of the system taking into account the occupations of the vibronic states of the ground electronic term (or group of terms) Γ_1 only, and d is the operator of interaction of the electromagnetic wave with the polyatomic system. In the particular case of electric dipole absorption, $d = \boldsymbol{e} \cdot \boldsymbol{D}$ is the projection of the dipole moment of the system $\boldsymbol{D}$ on the direction of the wave polarization given by the unit vector $\boldsymbol{e}$.

The spectral properties of the absorption coefficient (5.1.1) are determined mainly by the form factor of the spectrum which is the most rapidly changing function of frequency. Therefore, everywhere in this chapter, except for some special cases, the properties of the function $F_{\Gamma_1 \to \Gamma_2}$ are discussed.

Note that due to the δ-function one can assume that $E_n(\Gamma_1) = E_m(\Gamma_2) - \hbar\Omega$, and from (5.1.2) we find that the form factor of absorption $\Gamma_1 \to \Gamma_2$ is related to the form factor of absorption $\Gamma_2 \to \Gamma_1$ (for another system with the same terms Γ_1 and Γ_2 interchanged in energy) by the simple relationship [5.5]

$$F_{\Gamma_2 \to \Gamma_1}(\Omega) = F_{\Gamma_1 \to \Gamma_2}(\Omega) \exp\{[\mathscr{F}(\Gamma_2) - \mathscr{F}(\Gamma_1) - \hbar\Omega]/kT\} \ . \tag{5.1.4}$$

In cases when the time of both phase and energy relaxation within the group of excited vibronic states $\Psi_m(\Gamma_2)$ is much smaller than the lifetime with respect to the emission transition $\Gamma_2 \to \Gamma_1$ the form factor $F_{\Gamma_2 \to \Gamma_1}$ from (5.1.4) determines the band shape of the luminescence $\Gamma_2 \to \Gamma_1$. Note that (5.1.4) is an exact relationship and not based on any approximate calculations of the energy spectrum or transition probabilities. The only condition for its validity is the above-mentioned large energy separation of the energy terms under consideration, Γ_1 and Γ_2, from all other electronic terms of the system and from one another.

It is convenient to present the form factor $F(\Omega)$ in the form of a Fourier transform,

$$F(\Omega) = \frac{1}{2\pi} \int_{-\infty}^{\infty} I(t) \mathrm{e}^{-\mathrm{i}\Omega t} \, dt \ ; \qquad I(t) = \int_{-\infty}^{\infty} F(\Omega) \mathrm{e}^{\mathrm{i}\Omega t} \, d\Omega \ , \tag{5.1.5}$$

where $I(t)$ is known as the generating function of the transition band,

$$I(t) = \sum_{n,m} \varrho_n(\Gamma_1) \langle \Psi_n(\Gamma_1) | d | \Psi_m(\Gamma_2) \rangle \langle \Psi_m(\Gamma_2) | d | \Psi_n(\Gamma_1) \rangle \\ \times \exp\left(\frac{\mathrm{i}}{\hbar} [E_m(\Gamma_2) - E_n(\Gamma_1)] t\right) \ . \tag{5.1.6}$$

The frequencies of transitions $\Gamma_1 \to \Gamma_1$ and $\Gamma_2 \to \Gamma_2$ fall in regions far away from that of the $\Gamma_1 \to \Gamma_2$ transition under consideration in this section. Therefore we assume that all the matrix elements $\langle \Psi_m(\Gamma_1) | d | \Psi_n(\Gamma_1) \rangle$ and $\langle \Psi_m(\Gamma_2) | d | \Psi_n(\Gamma_2) \rangle$ are zero. Then the operator d can be represented by a rectangular matrix $f_2 \times f_1$, where f_i is the dimensionality of the space of the electronic states Γ_i. If one takes into account that $\Psi_n(\Gamma)$ are exact eigenfunctions of the vibronic Hamiltonian

$\hat{H}(\Gamma)$, so that

$$\exp[\mathrm{const}\cdot E_n(\Gamma)]|\Psi_n(\Gamma)\rangle = \exp[\mathrm{const}\cdot \hat{H}(\Gamma)]|\Psi_n(\Gamma)\rangle \tag{5.1.7}$$

then from (5.1.6) we obtain the expression

$$I(t) = \left\langle d^\dagger \exp\left(\frac{\mathrm{i}}{\hbar}\hat{H}(\Gamma_2)t\right) d \exp\left(-\frac{\mathrm{i}}{\hbar}\hat{H}(\Gamma_1)t\right)\right\rangle_{\Gamma_1} , \tag{5.1.8}$$

which can be considered as a generalization of the well-known Lax formula [5.5] for the case of multiplet-multiplet transitions. In (5.1.8), $\langle\cdots\rangle_\Gamma$ means the average over the vibronic states related to the electronic term Γ,

$$\langle A\rangle_\Gamma = \mathrm{Tr}\{\hat{\varrho}(\Gamma)A\} , \qquad \hat{\varrho}(\Gamma) = \exp\{[\mathscr{I}(\Gamma) - \hat{H}(\Gamma)]/kT\} . \tag{5.1.9}$$

The form of the generating function in (5.1.8) has the advantage that the trace is invariant with respect to any unitary transformation of the basis functions. In other words, any orthonormalized set of states that is complete in the same space as the eigenfunctions of the vibronic Hamiltonian can be taken as a basis for the calculation of the averaged values (5.1.8). In particular, it is convenient to use the Born-Oppenheimer multiplicative states (4.1.2) for this purpose.

The vibronic Hamiltonians $\hat{H}(\Gamma_1)$ and $\hat{H}(\Gamma_2)$ in principle also describe the motion of the polyatomic system in the space of the totally symmetric breathing coordinates Q_{A_1}. Within the limits of the linear vibronic interaction term, $\hat{H}(\Gamma_i)$ can be written in the form of two commutative terms $\hat{H}_{A_1}(\Gamma_i) + \hat{H}_{\mathrm{JT}}(\Gamma_i)$ and the variables can be separated (Sect. 2.2.3). The problem of electron transitions is trivial for the $\hat{H}_{A_1}(\Gamma_i)$ Hamiltonian and can be solved by an appropriate choice of the origin for the coordinate Q_{A_1} that results in linear harmonic oscillations of the system along Q_{A_1}.

In calculating the probabilities of the optical transition, it should be remembered that the interaction of the electrons with the A_1 vibrations in the Γ_1 and Γ_2 states is, generally speaking, different, and the harmonic oscillations along Q_{A_1} in these two states take place about different equilibrium positions. Therefore the matrix elements of the transition contain overlap integrals of the vibrational states of the breathing degrees of freedom, and hence the interaction with the A_1 vibrations influences the optical band shape.

Note that the Hamiltonians $\hat{H}_{A_1}(\Gamma_i)$ are multiples of the unit matrix of the dimension $f_i \times f_i$ (Sect. 2.2.3), while the rectangular matrix of the operator d commutes with the unit matrix, just changing its dimension. For example,

$$\begin{pmatrix} d_{11} & d_{12} & d_{13} \\ d_{21} & d_{22} & d_{23} \end{pmatrix}\begin{pmatrix} 1 & 0 & 0 \\ 0 & 1 & 0 \\ 0 & 0 & 1 \end{pmatrix} = \begin{pmatrix} 1 & 0 \\ 0 & 1 \end{pmatrix}\begin{pmatrix} d_{11} & d_{12} & d_{13} \\ d_{21} & d_{22} & d_{23} \end{pmatrix} . \tag{5.1.10}$$

Remembering that $H_A(\Gamma_i)$ commutes with $\hat{H}_{\mathrm{JT}}(\Gamma_i)$ and moving the value $\exp[\mathrm{i}H_{A_1}(\Gamma_2)t/\hbar]$ in (5.1.8) to the right of the matrix d (simultaneously changing the dimension of the unit matrix $f_2 \times f_2$ to $f_1 \times f_1$) and also using the invariance of the trace with respect to cyclic permutations of the operators, we obtain

$$I(t) = I_{A_1}(t) I_{\mathrm{JT}}(t) \ . \tag{5.1.11}$$

Here

$$I_{A_1}(t) = \mathrm{Tr}\left\{\hat{\varrho}_{A_1}(\Gamma_1) \exp\left(\frac{\mathrm{i}}{\hbar} \hat{H}_{A_1}(\Gamma_2) t\right) \exp\left(-\frac{\mathrm{i}}{\hbar} \hat{H}_{A_1}(\Gamma_1) t\right)\right\} , \tag{5.1.12}$$

$$I_{\mathrm{JT}}(t) = \mathrm{Tr}\left\{\varrho_{\mathrm{JT}}(\Gamma_1) d^+ \exp\left(\frac{\mathrm{i}}{\hbar} \hat{H}_{\mathrm{JT}}(\Gamma_2) t\right) d \exp\left(-\frac{\mathrm{i}}{\hbar} \hat{H}_{\mathrm{JT}}(\Gamma_1) t\right)\right\} . \tag{5.1.13}$$

Expression (5.1.12) is equivalent to the Lax formula for singlet-singlet transitions in the Condon approximation. Using the theorem of convolution for the Fourier transform (5.1.5) with the generating function (5.1.11), we find

$$F(\Omega) = \int_{-\infty}^{\infty} F_{A_1}(\Omega - \omega) F_{\mathrm{JT}}(\omega) \, d\omega \ , \quad \text{where} \tag{5.1.14}$$

$$F_{A_1}(\Omega) = \frac{1}{2\pi} \int_{-\infty}^{\infty} I_{A_1}(t) \mathrm{e}^{-\mathrm{i}\Omega t} \, dt \ ; \qquad F_{\mathrm{JT}}(\omega) = \frac{1}{2\pi} \int_{-\infty}^{\infty} I_{\mathrm{JT}}(t) \mathrm{e}^{-\mathrm{i}\omega t} \, dt \ . \tag{5.1.15}$$

The physical meaning of the convolution (5.1.14) can be easily understood if the form factor $F_{\mathrm{JT}}(\omega)$ is written as a sum of elementary δ-shaped contributions:

$$F_{\mathrm{JT}}(\omega) = \sum_n I_n \delta(\omega - \omega_n) \ , \tag{5.1.16}$$

where I_n is the intensity of a separate line and ω_n is its frequency in the spectrum. Substituting (5.1.16) into (5.1.14) and integrating over the δ function we obtain

$$F(\Omega) = \sum_n I_n F_{A_1}(\Omega - \omega_n) \ .$$

Thus the transition band consists of a superposition of elementary bands $F_{A_1}(\Omega - \omega_n)$ with relative intensities I_n. The form factor $F_A(\Omega)$ is well studied [5.13]. In particular, it is known that it has a bell-shaped asymmetric form with half-width

$$\Delta\Omega \sim |V_{A_1}(\Gamma_2) - V_{A_1}(\Gamma_1)| \sqrt{\frac{\coth(\hbar\omega_{A_1}/2kT)}{2\hbar\omega_{A_1}}} \ ,$$

where $V_{A_1}(\Gamma_i)$ is the constant of coupling of the electronic term Γ_i with A_1 vibrations, ω_{A_1} being its frequency. In the case of strong vibronic coupling with A_1 vibrations (and/or at high temperatures), the width of elementary bands $F_A(\Omega - \omega_n)$ is much larger than the spacing in the vibronic spectrum and hence the structure of the resulting absorption band is filled up. Considering the Jahn-

Teller features of the band we assume below that $V_{A_1}(\Gamma_i) = 0$, $H_{A_1}(\Gamma_1) = H_{A_2}(\Gamma_2)$ and, as follows from (5.1.12), $I_{A_1}(t) = 1$. In this case $F_{A_1}(\Omega) = \delta(\Omega)$ and interaction with the A_1 vibrations does not smearout the Jahn-Teller structure of the band.

From (5.1.2) it is seen that the envelope of the band shape and its structure contain information about both the electronic terms Γ_1 and Γ_2 between which the optical transition takes place, the optical spectrum being a superposition of the vibronic energy spectra of these terms. There are two ways to simplify the situation. First, one can consider singlet-multiplet transitions, for which the Jahn-Teller effect occurs for only one of the terms, Γ_1 or Γ_2. The same is true for multiplet-multiplet transitions when the electronic states of one of the terms are not mixed by the vibrations. The vibronic states $\Psi_n(\Gamma_1)$ or $\Psi_m(\Gamma_2)$ in this case have a simple Born-Oppenheimer form (2.1.14) and the appropriate vibrational spectrum is equidistant. Second, one can consider the case $T = 0$ K, when only the ground vibronic term is populated. Then only one term remains in the sum over n in (5.1.2):

$$F(\Omega) = \sum_{\gamma, n} |\langle \Psi_{\Gamma_1\gamma}(\Gamma_1)| d | \Psi_m(\Gamma_2)\rangle|^2 \, \delta\left(\frac{E_m(\Gamma_2) - E_0(\Gamma_1)}{\hbar} - \Omega\right) . \tag{5.1.17}$$

The structure of the band in this case reproduces the energy spectrum of the Jahn-Teller term Γ_2 for those vibronic states to which the optical transition is allowed.

Consider first singlet-multiplet transitions. The vibronic states of the A term, from which the optical transition starts, have the multiplicative from (2.1.14) and therefore the quantum-mechanical averaging over the electronic variables in (5.1.8) can be carried out independent of the statistical averaging over the vibrational variables:

$$I(t) = \sum_{\gamma\bar{\gamma}} \langle \psi_A | d^+ | \psi_{\Gamma\gamma}\rangle \langle \psi_{\Gamma\bar{\gamma}} | d | \psi_A\rangle$$
$$\times \left\langle\!\!\left\langle \psi_{\Gamma\gamma} \left| \exp\left(\frac{\mathrm{i}}{\hbar}\hat{H}(\Gamma)t\right) \right| \psi_{\Gamma\bar{\gamma}} \right\rangle \exp\left(-\frac{\mathrm{i}}{\hbar}H(A)t\right)\right\rangle_{\mathrm{vib}} . \tag{5.1.18}$$

This expression was deduced with the assumption that $\langle \psi_A | d | \psi_A\rangle = 0$, $\langle \psi_{\Gamma\gamma} | d | \psi_{\Gamma\bar{\gamma}}\rangle = 0$ [see the text after (5.1.6)], and $\langle \cdots \rangle_{\mathrm{vib}}$ means the average over the vibrational states of the A_1 term. Note that the Hamiltonians $\hat{H}(\Gamma)$ and $H(A)$ in (5.1.18) are scalars of the symmetry group G_0 of the reference nuclear configuration (Sect. 2.2.1). This averaging does not lower the symmetry and therefore, according to the Wigner-Eckart theorem,

$$\left\langle \exp\left(\frac{\mathrm{i}}{\hbar}\hat{H}(\Gamma)t\right) \exp\left(-\frac{\mathrm{i}}{\hbar}H(A)t\right)\right\rangle_{\mathrm{vib}} = I(t)\hat{C}_{A_1} , \tag{5.1.19}$$

where C_{A_1} is the unit matrix determined in the space of the electronic states of the degenerate term Γ, and $I(t)$ is the reduced matrix element. The matrix structure of (5.1.19) can also be obtained by expanding the left-hand side of (5.1.19) as a power series in t and by subsequent averaging of each of the series terms. Then (5.1.18) can be written as

$$I(t) = \frac{1}{f_\Gamma}\left\langle \mathrm{Tr}\left\{\exp\left(\frac{\mathrm{i}}{\hbar}\hat{H}(\Gamma)t\right)\right\}_\Gamma \exp\left(-\frac{\mathrm{i}}{\hbar}H(A)t\right)\right\rangle_{\mathrm{vib}} \sum_\gamma |\langle\psi_A|d|\psi_{\Gamma\gamma}\rangle|^2 , \tag{5.1.20}$$

where $\mathrm{Tr}\{\cdots\}$ means the trace operation over the electronic states of the Γ term only, and f_Γ is the multiplicity of the degeneracy of this term. From this it is seen that $\sum_\gamma |\langle\psi_A|d|\psi_{\Gamma\gamma}\rangle|^2$ determines the normalization of the band, while all spectral information is contained in $\langle\cdots\rangle_{\mathrm{vib}}$. Therefore, from now onwards we assume, for simplicity, that $\sum_\gamma |\langle\psi_A|d|\psi_{\Gamma\gamma}\rangle|^2 = 1$, i.e., that

$$I(t) = f_\Gamma^{-1}\left\langle \mathrm{Tr}\left\{\exp\left(\frac{\mathrm{i}}{\hbar}\hat{H}(\Gamma)t\right)\right\}_\Gamma \exp\left(-\frac{\mathrm{i}}{\hbar}H(A)t\right)\right\rangle_{\mathrm{vib}} . \tag{5.1.21}$$

Note that by means of (5.1.4) the optical band of the multiplet-singlet transition can be treated by the same formula (5.1.21).

5.1.2 The Jahn-Teller Effect in the Excitation Spectrum of the $A \to E$ Transition

As a first example consider the singlet-doublet transition in a cubic (or trigonal) system. The excited E term in this case interacts with the E vibrations (the $E \otimes e$ case, Sect. 3.1.1), and the Hamiltonian of the E term, considering just the linear terms of vibronic interaction, has axial symmetry (Sect. 3.2.1). The electronic states of the E term correspond to the energy spin values $\pm 1/2$, and therefore in order to make the transition $A \to E$ allowed the operator d should have the transformation properties of spin $\pm 1/2$. It is easily seen that in trigonal systems this condition is fulfilled even for electric dipole transitions. The selection rule $j \to j \pm 1/2$ follows. The ground vibrational state of the A term (zero quanta of E vibrations) corresponds to the total momentum $j = 0$, and therefore at $T = 0$ K transitions to the vibronic states with $j = \pm 1/2$ only are allowed.

The intensity of the vibronic components and the absorption band shape (the envelope) can be obtained for singlet-multiplet transitions by direct substitution of the results of numerical diagonalization of the vibronic Hamiltonian into (5.1.2) or, at $T = 0$ K, into the simpler (5.1.17). The results given in Figs. 5.1–5 based on the numerical solution of the linear $E \otimes e$ problem were obtained in this way [5.14–19]. Despite the essential irregularity of the vibronic spectrum of the linear $E \otimes e$ problem (Fig. 4.9) the optical excitation spectrum of the transition $A \to E$ at $T = 0$ K has almost equally spaced lines (Fig. 5.1). This is due to the equally spaced structure of the energy spectrum of states with a given value of the rotational quantum number but with different radial vibration quantum numbers (Sect. 4.2.2). Note that this apparent simplicity of the $E \otimes e$ Jahn-Teller

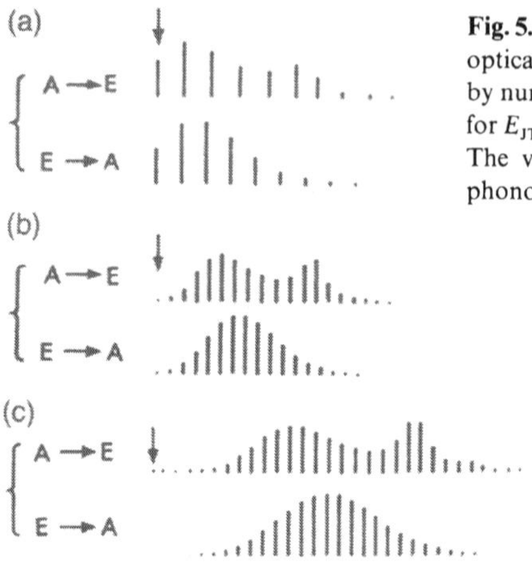

Fig. 5.1. Intensity of vibronic components of the bands of optical transitions $A \rightarrow E$ and $E \rightarrow A$ at $T = 0$ K obtained by numerical solution of the $E \otimes e$ problem (after [5.15]) for $E_{JT}/\hbar\omega_E$ equal to 2.5 (**a**), 7.5 (**b**), 15 (**c**) (see also Fig. 5.2). The vertical arrow indicates the position of the zero-phonon line

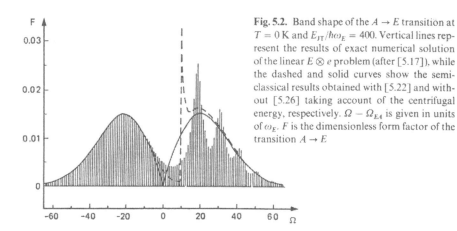

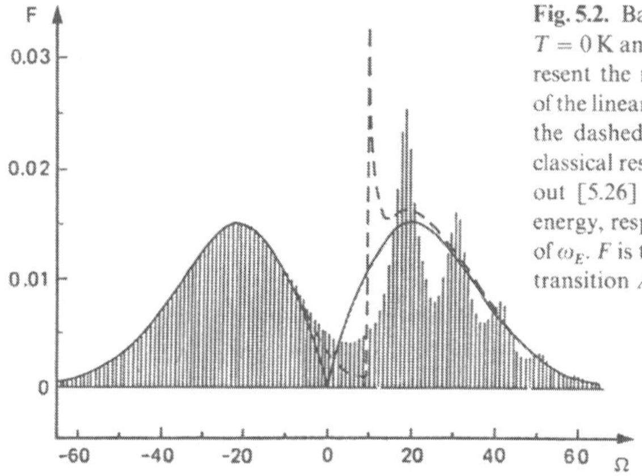

Fig. 5.2. Band shape of the $A \rightarrow E$ transition at $T = 0$ K and $E_{JT}/\hbar\omega_E = 400$. Vertical lines represent the results of exact numerical solution of the linear $E \otimes e$ problem (after [5.17]), while the dashed and solid curves show the semiclassical results obtained with [5.22] and without [5.26] taking account of the centrifugal energy, respectively. $\Omega - \Omega_{EA}$ is given in units of ω_E. F is the dimensionless form factor of the transition $A \rightarrow E$

excitation optical spectrum is a consequence of the high symmetry of the problem. A reduction of the symmetry, e.g. by taking into account higher-order terms of the vibronic interaction or external low-symmetry fields, increases the complexity of the spectrum enormously [5.20, 21].

From the results obtained it is seen that for strong vibronic coupling *the band of electronic transitions* $A \rightarrow E$ *is split into two*, and in the limit of very strong coupling (which has low probability) the envelope of the high frequency wing has additional structure (Fig. 5.2) which can be interpreted as the *Slonczewski resonances* [5.16, 18, 19]. The double-humped curve is obtained also for $T \neq 0$ K (Fig. 5.3), when in the band of $A \rightarrow E$ absorption there are contributions from transitions from the vibrational states of the term with $j \neq 0$ [5.23]. To obtain the envelope, each of the spectral lines is represented by a Gaussian band

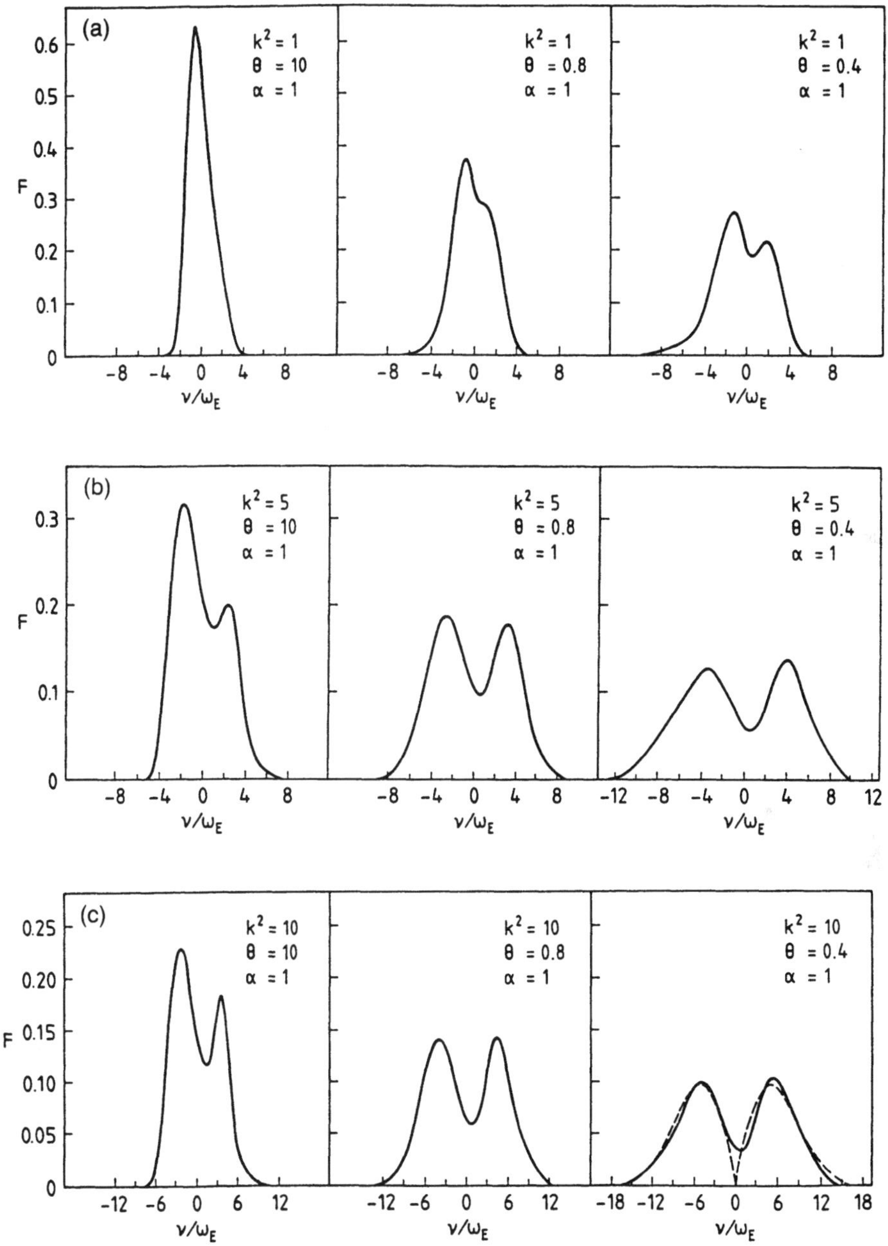

Fig. 5.3. Temperature dependence of the band shape for the $A \to E$ transition (dimensionless form factor F versus frequency) obtained by numerical solution of the linear $E \otimes e$ problem [5.23] for $E_{JT}/\hbar\omega_E$ equal to 0.5 (**a**), 2.5 (**b**) and 5 (**c**). $\nu = \Omega - \Omega_{EA}$ and $\theta = \hbar\omega_E/kT$. The dashed line illustrates the semiclassical result (5.1.26) for the same parameter values

$\exp[-(\Omega - \omega_n)^2/\sigma]$, where $\sigma = \alpha\omega_E^2$, normalized to the intensity I_n of the partial transition.

The double-humped structure of the band of the transition $A \to E$ in the case of strong vibronic coupling can be easily understood within the adiabatic approximation, if one assumes that the transitions to the two surfaces of the adiabatic potential of the E term take place independently, and the band of the $A \to E$ transition is a simple superposition of two singlet-singlet bands $A \to \varepsilon_+$ and $A \to \varepsilon_-$. Mathematically this can be realized by including the unitary transformation (3.1.9) in the electronic trace in (5.1.21) and by exclusion of the non-adiabaticity operator (Sect. 4.2) from $H(E)$. The trace over electronic states in (5.1.21) then decomposes into two terms $I_{A\to E} = I_+ + I_-$, each of which can be described by the Lax formula:

$$I_\pm(t) = \frac{1}{2}\left\langle \exp\left(\frac{\mathrm{i}}{\hbar}H_\pm(E)t\right)\exp\left(-\frac{\mathrm{i}}{\hbar}H(A)t\right)\right\rangle . \tag{5.1.22}$$

Here $H_\pm(E)$ are the Hamiltonians of the upper and lower surfaces of the adiabatic potential in the $E \otimes e$ problem.

At $T = 0$ K the statistical averaging is reduced to the quantum-mechanical one over the vibrational wave function of the ground state of the A term $(\omega_E/\pi\hbar)^{1/2}\exp[-\omega_E(Q_\theta^2 + Q_\varepsilon^2)/2\hbar]$, and after integration over the polar angle $\varphi = \arctan(Q_\varepsilon/Q_\theta)$ we obtain from (5.1.22)

$$I_\pm(t) = \frac{\omega_E}{\hbar}\int_0^\infty \exp\left(-\frac{\omega_E}{2\hbar}\varrho^2\right)\exp\left(\frac{\mathrm{i}}{\hbar}H_\pm(E)t\right) \times \exp\left(-\frac{\mathrm{i}}{\hbar}H(A)t\right)\exp\left(-\frac{\omega_E}{2\hbar}\varrho^2\right)\varrho\, d\varrho . \tag{5.1.23}$$

Here the operator $L_z = -\mathrm{i}\partial/\partial\varphi$ in the vibrational Hamiltonian $H(A)$ has to be replaced by the quantum number $l = 0$, whereas in the adiabatic Hamiltonian $H_\pm(E)$, in accordance with the selection rule $l \to l \pm 1/2$, it has to be replaced by $j = \pm 1/2$. In the semiclassical approach, neglecting the noncommutativity of ϱ with $\partial/\partial\varrho$, the expressions $\exp[\mathrm{i}H_\pm(E)t/\hbar]$ and $\exp[-\mathrm{i}H(A)t/\hbar]$ can be joined into a common exponential. Calculating the Fourier integral (5.1.6) with the generating function (5.1.23), for $F = F_+ + F_-$ we find

$$F(\Omega) = \omega_E\int_{-\infty}^{\infty} \exp\left(-\frac{\omega_E}{\hbar}\varrho^2\right)\delta\left(\frac{\hbar^2}{8\varrho^2} + V_E\varrho - \hbar\Omega\right)|\varrho|\, d\varrho , \tag{5.1.24}$$

where the frequency is with respect to the Franck-Condon value Ω_{EA}. The spectral curve (5.1.24) is shown in Fig. 5.2 as a dashed line [5.22]. Comparison with the envelope of the excitation spectrum obtained by exact numerical calculation (Fig. 5.2) shows that the semiclassical approximation provides a good description of the outline of the band of the singlet-multiplet transition.

An even simpler result can be obtained if one uses the semiclassical approximation at an earlier stage before performing the exact integration over φ, e.g. in (5.1.22). Neglecting the noncommutativity of ϱ with $\partial/\partial\varrho$ and φ with $\partial/\partial\varphi$ and joining together $\exp(\mathrm{i}H_{\pm}t/\hbar)$ and $\exp[-\mathrm{i}H(A)t/\hbar]$ into a common exponential, we obtain $I_{\pm}(t) = (1/2)\exp(\pm\mathrm{i}V_E\varrho t/\hbar)$. Calculating the Fourier integral (5.1.5) with this generating function, we have

$$F(\Omega) = \frac{1}{2}\sum_{+,-}\left\langle\delta\left(\Omega \pm V_E\frac{\varrho}{\hbar}\right)\right\rangle_{\mathrm{vib}}, \tag{5.1.25}$$

where, as above, Ω is with respect to Ω_{EA}. Performing the average in (5.1.25) with the quantum density matrix of harmonic oscillators Q_θ and Q_ε, we obtain

$$F(\Omega) = \frac{|\Omega|}{\langle\Omega^2\rangle}\exp\left(-\frac{\Omega^2}{\langle\Omega^2\rangle}\right), \tag{5.1.26}$$

where $\langle\Omega^2\rangle$ is the second moment of the exact distribution $F(\Omega)$:

$$\langle\Omega^2\rangle = \frac{V_E^2}{\hbar\omega_E}\coth\left(\frac{\hbar\omega_E}{2kT}\right). \tag{5.1.27}$$

Equation (5.1.26) in the high temperature limit $[kT \gg \hbar\omega_E, \langle\Omega^2\rangle \cong 2V_E^2kT/(\hbar\omega_E)^2]$ was obtained independently and almost simultaneously in [5.24–26] in the semiclassical adiabatic model with the classical density matrix. Averaging with the quantum density matrix was performed in [5.27], where the case of weak splitting of the E term by static fields was also considered.

The spectral distribution (5.1.26) gives a double-humped curve (Fig. 5.2) with maxima at $\Omega_m = \pm(\langle\Omega^2\rangle/2)^{1/2}$, the separation of which increases as $\sqrt{T}$. From comparison of the envelope of the $A \to E$ spectral curve (see Fig. 5.3c) obtained for the same parameter values by numerical diagonalization of the vibronic Hamiltonian [5.23] it follows that the semiclassical approximation performed at the earliest stage gives a satisfactory qualitative description of the band shape of singlet-multiplet transitions.

The semiclassical approximation is of course not the only possible approach to the evaluation of the curve of the optical singlet-multiplet transitions. The calculations can be performed in the above-mentioned adiabatic approximation without any additional simplifications (such as semiclassical consideration of nuclear dynamics on a given surface of the adiabatic potential). For instance, the nuclear eigenvalue problem for the spearate sheets of the adiabatic potential can be evaluated numerically. This procedure, though giving good results, especially for the low energy wing of the excitation spectrum in the case of strong vibronic coupling [5.12], needs considerable computational efforts, comparable with the ones for the exact numerical solution of the problem. This is caused by the strong anharmonicity of the Jahn-Teller adiabatic potentials. The Franck-Condon approximation, which is known to be very good for normal (non-Jahn-Teller) cases of singlet-singlet transitions, has been shown to be incorrect for Jahn-Teller excitation spectra [5.12] due to very strong borrowing effects. The adiabatic and Franck-Condon approximations in the Jahn-Teller and pseudo-Jahn-Teller electronic excitation spectra were compared numerically with the exact results in a series of papers by *Cederbaum* et al. (see [5.21] and the review article [5.12]).

In [5.27, 28] the so-called *independent ordering approximation* (IOA) was developed, as well as the iteration procedure subsequently used to improve it. Comparison of the approximate curves of the $A \rightarrow E$ absorption with the exact one is carried out in [5.27, 28] by means of the method of Gram-Sharlier expansion. In order to do this, the first nine momenta of all the spectral distributions, including the exact one, were calculated [5.27–29]. The same comparison carried out in [5.30] for the optical curve of the $A \rightarrow E$ transition constructed by means of IOA with the exact one obtained by numerical diagonalization of the vibronic Hamiltonian with the same parameter values shows that they agree well with one another.

Another approximate approach is based on the *method of canonical transformations.* The comparison of the results of the operator method, the method of canonical transformations, the IOA and the semiclassical approximation applied to the $A \rightarrow E$ absorption band is given in [5.31]. It can be concluded that the semiclassical approach, although being less accurate than other approximations mentioned above, gives a correct qualitative description of the envelope of the excitation spectrum, and it provides the simplest way to obtain the result. An approximate analytical treatment of the $E \otimes e$ Jahn-Teller effect applicable to all energy values and any coupling strengths, and giving quite a good reproduction of the exact numerical results, was proposed in [5.32]. See also [5.33].

In order to observe the vibronic structure of the bands of electronic transitions in molecules it is necessary that the relaxation broadening of the spectral lines is relatively small. This condition is fulfilled for a gas of Jahn-Teller molecules at low pressure and for the solid phase at low temperatures. Detailed studies of the vibronic structure of electronic spectra of Jahn-Teller molecules were started recently. One of the first investigations of this kind [5.34] was devoted to the forbidden electron transition ${}^1E'' \leftarrow \tilde{X}\,{}^1A_1$ in the *s*-triasine molecule (symmetry group D_{3h}). A systematic investigation of the luminescence band in cations $C_6H_6^+$ (symmetry group D_{6h}), 1,3,5-$C_6F_3H_3^+$ and 1,3,5-$C_6Cl_3H_3^+$ (symmetry group D_{3h}) is given in [5.34–42]. These cations have an E ground term, its vibronic states being manifest in the emission spectra ${}^2E \leftarrow {}^2A$ ($\tilde{X}\,{}^2E_{1g} \leftarrow \tilde{B}\,{}^2A_{2u}$ for $C_6H_6^+$ and $\tilde{X}\,{}^2E'' \leftarrow \tilde{B}^2A_2''$ for 1,3,5-$C_6F_3H_3^+$ and 1,3,5-$C_6Cl_3H_3^+$).

The above cations have four modes of the E type that are Jahn-Teller active, i.e., strictly speaking they are multimode systems (Sects. 3.5 and 4.6). Satisfactory agreement with the experimental data is achieved in [5.35] with the assumption that the vibronic interaction is essential only for one of these four modes. The best reproduction of the experimental data for these ions was achieved by *Sears* et al. [5.43] with the aid of Jahn-Teller calculations involving all four modes of the E type. See also [5.44].

Another example of a Jahn-Teller singlet-doublet excitation spectrum can be found in the outer-valence photoelectron spectrum of the BF_3 molecule [5.45].

In the region between 15 and 22 eV it contains six electron bands. In order of increasing energy the ionic final states of BF_3^+ corresponding to these bands are: $1A_2'$, $1E''$, $3E'$, $1A_2''$, $2E'$ and $2A_1'$ (symmetry group D_{3h}). Hence there are a variety of Jahn-Teller and pseudo-Jahn-Teller coupling mechanisms that can be operative between these ionic states [5.12, 45, 46]. Of the four vibrational modes transforming according to $A_1' + A_2'' + 2E'$, the coupling to the out-of-plane A_2'' bending mode turns out to be negligible. All the other vibrations are significant in the photoelectron spectrum, resulting in the two-mode Jahn-Teller effect for the ionic E terms. Figure 5.4 presents the results of [5.45] for the $2E'$ band. It is seen that the fine structure of this band is due to the superposition of the spectra of totally symmetric vibration A_1' and the two degenerate vibrations E' excited upon ionization. Note that the positions of the bands, as well as their vibrational structure, were obtained by numerical calculations based on first principles (not using any experimental information) [5.45]. See also [5.48].

The R' band in alkali halide crystals, which is related to the electron transition in the R' center in the system containing three halide vacancies with four electrons (symmetry C_{3v}), can be regarded as an example of the double-humped envelope band of the $A \to E$ transition in crystals [5.49] (Fig. 5.5).

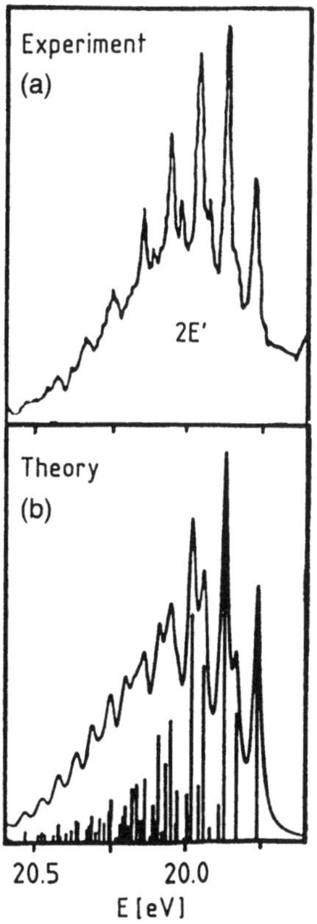

◀ **Fig. 5.4a, b.** Vibronic structure of the photoelectron excitation $2E'$ band of the BF_3 molecule. (**a**) Experimental curve (after [5.47]); (**b**) results of numerical solution of the multimode $E \otimes e$ problem taking account of totally symmetric vibrations (after [5.45])

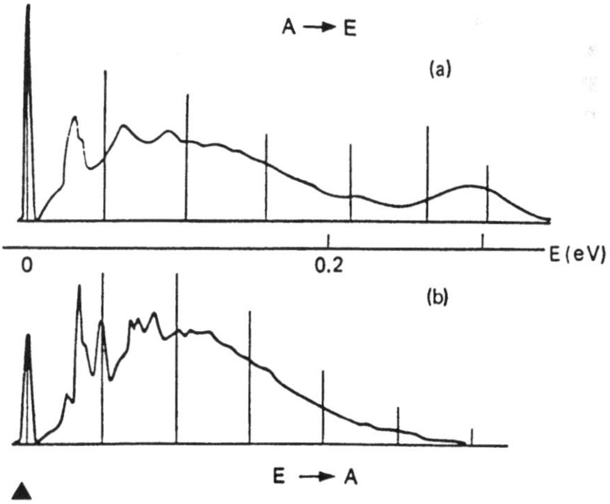

▲ **Fig. 5.5a, b.** R' band in KCl: comparison of the observed spectrum with that obtained by numerical solution of the linear $E \otimes e$ problem; the best agreement is obtained for $E_{JT}/\hbar\omega_E = 2.5$, $\hbar\omega_E = 395\ cm^{-1}$. (**a**) Absorption $A \to E$; (**b**) luminescence $E \to A$. (After [5.49])

The band shape of absorption and luminescence related to the $E \to A$ transition was also calculated in the semiclassical approximation [5.27]. However, it is simpler to use here the general relationship (5.1.4). The semiclassical calculation of the partial statistical sums results in the following expression for the factor $\exp\{[\mathscr{I}(E) - \mathscr{I}(A)]/kT\}$ in (5.1.4) [5.5]:

$$\exp\left(\frac{\mathscr{I}(E) - \mathscr{I}(A)}{kT}\right) = 1 + z\sqrt{\pi}\Phi(z)\exp(z^2) \; ,$$

where $z = |V_E|/\hbar\omega_E^2\sqrt{2kT}$, and $\Phi(z)$ is the error function. The Boltzmann factor $\exp(-\hbar\Omega/kT)$ in (5.1.4) describes the difference in population at a finite temperature of two sheets of the adiabatic potential of the E term, resulting in an additional asymmetry of the band of the $E \to A$ transition. An increase in temperature leads to the equalization of the population and to a decrease of the asymmetry of the band.

The optical absorption $^1A \to {}^2E$ band is considered in [5.50] for trigonal systems, where the spin-orbit interaction has to be included even in first-order perturbation theory.

The family of the cumulenes C_nH_4 with odd values of n (allene C_3H_4, pentatetraene C_5H_4, etc.) possess twofold degenerate electronic states, because

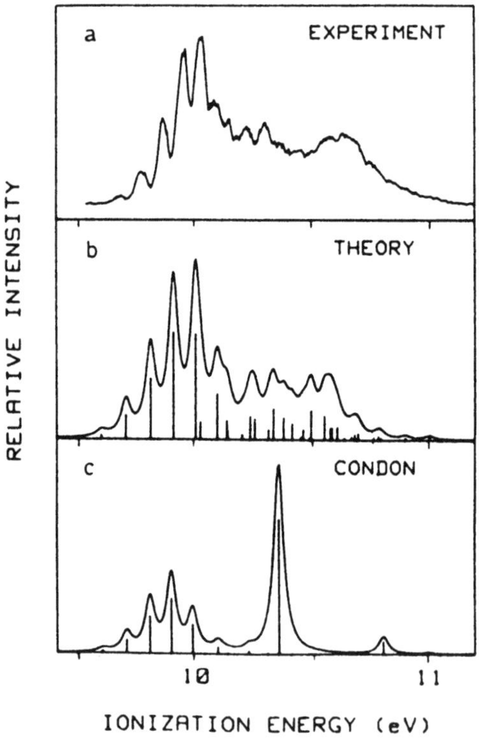

Fig. 5.6a–c. Vibronic structure of the photoelectron $A \to E$ excitation band in the spectrum of allene [$E \otimes (b_1 + b_2)$ case]. (**a**) Experimental spectrum [5.52]; (**b**) results of exact numerical solution of the vibronic $E \otimes (b_1 + b_2)$ problem; (**c**) results of Frank-Condon approximation. (After [5.51])

they have D_{2d} symmetry, and therefore they provided illustrations of the nature of Jahn-Teller excitation spectra of the $E \otimes (b_1 + b_2)$ type [5.12]. The photoelectron band of allene shown in Fig. 5.6a corresponds to excitation of the ionic E term at about 10 eV. An excellent reproduction of this experimental curve, see Fig. 5.6b, was obtained in [5.51] by retaining only one of the three Jahn-Teller–active B_2 modes and one B_1 mode and neglecting the totally symmetric modes. A similar result has also been obtained for pentatetraene [5.51].

Transitions of the type $A \to E$ in tetragonal systems with the Jahn-Teller effect of the type $E \otimes (b_1 + b_2)$ (Sect. 3.1.2) were considered in [5.50] in the framework of the semiclassical approximation, with the spin-orbital interaction for the ^2E term also being taken into account. The vibronic fine structure of the optical band in this case is considered in [5.53].

5.1.3 Manifestations of Vibronic Interactions in the Optical Spectra of the $A \to T$ Transition

In [5.54] the optical curve of the $A \to T$ transition at $T = 0$ K, considering the linear vibronic interaction with the T_2 vibrations only is given. It was obtained by numerical diagonalization of the $T \otimes t_2$ Hamiltonian. The spectral lines of the optical transitions to separate vibronic levels of the $T \otimes t_2$ system are replaced by Gaussian bands. In this way, in spite of the discrete nature of the energy spectrum, the optical curves of the $A \to T$ transition are represented by smooth envelopes (Fig. 5.7) each of which is an asymmetric *three-peaked curve.*

Some further calculations, which do not impose the limitation of zero coupling to one of the modes, are reported by *Sakamoto* [5.55]. The semiclassical calculation of the curve of the $A \to T$ transition, considering the linear vibronic coupling to the T_2 vibrations only, was carried out in [5.26]. It is based on the same approximations as the result (5.1.26) for the $A \to E$ transition.

If the curves given in Figs 5.7 and 5.8 are compared, one sees that the semiclassical approximation in this case gives qualitatively true pictures of the structure of the multiphonon band of the optical transition.

Note, however, that care is needed in interpreting the experimental data, since the situations in which an ideal vibronic $T \otimes t_2$ system is absorbing are seldom. For the T term additional effects may be important, namely, the spin-orbital interaction, simultaneous vibronic coupling to the E and T_2 vibrations, or Stark splitting of the T term by low-symmetry crystal fields. In the case of impurity centers in crystals, the multimode nature of the vibronic coupling has to be taken into account as well.

For instance, for the $^2T_1$ term the spin-orbit interaction, in general, results in the splitting $^2T_1 = \Gamma_6 + \Gamma_8$, where only Γ_8 is a Jahn-Teller term coupled to the E and T_2 vibrations. However, the Kramers doublet cannot be excluded from consideration, since the vibronic interaction with the above vibrations mixes the electronic states of the Γ_6 and Γ_8 terms. Thus the spin-orbit interaction results in a complicated combination of the Jahn-Teller and pseudo-Jahn-Teller effects.

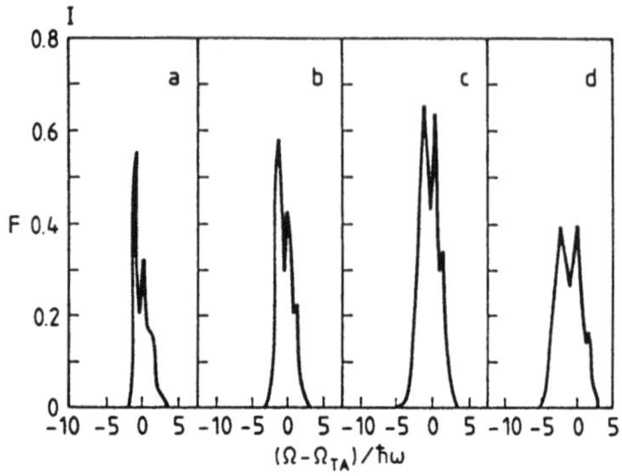

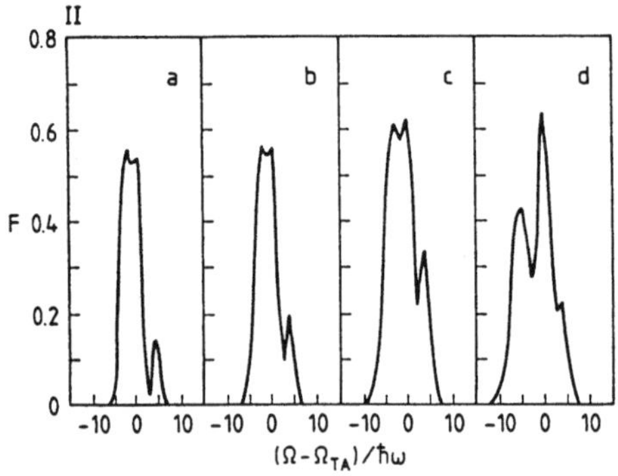

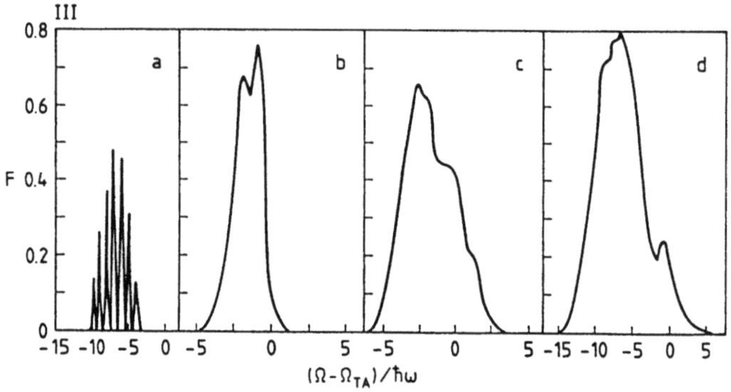

Fig. 5.7. Shape of the $A \to T$ (linear $T \otimes t_2$) transition band in absorption (**I, II**) and emission (**III**) at various temperatures: $kT/\hbar\omega = 0$ (**a**), 0.5 (**b**), 1 (**Ic, IIc**), 4 (**d**), and two values of the coupling constant: $E_{JT}/\hbar\omega = 2/3$ (**I, IIIb**), 3.527 (**II, IIIa, IIId**). (After [5.54])

The case of strong spin-orbit interaction is considered in [5.25], where the vibronic interaction, mixing the terms Γ_6 and Γ_8, is taken into account by means of perturbation theory (see also [5.56]).

A general discussion of the band shape for the ${}^2A \to {}^2T$ transition in the semiclassical approximation considering the spin-orbit interaction and the vibronic coupling to the A_1, E and T_2 vibrations is given in [5.57]. In this case the absorption curve is strongly dependent on the ratio between the vibronic coupling constants and the magnitude of the spin-orbit interaction. If the coupling to the

E vibrations is predominant (the $T \otimes e$ problem), but the spin-orbit interaction is zero, the absorption band of the $A \to T$ transition is not split, in spite of the significant splitting of the adiabatic potential (Fig. 3.6).

In the case of predominant coupling to the T_2 vibrations, the absorption band is split into three, and its envelope depends to a large extent on the magnitude of vibronic coupling to all the active vibrations, A_1, E and T_2. The special case of a ^2T term equally coupled to E and T_2 modes (the d mode model), with the spin-orbit interaction included, was considered by *O'Brien* [5.58]. Magnetic circular dichroism of the $^2A_1 \to {}^2T_1$ optical band was investigated in [5.59].

A more complicated case of the transition $^1A_{1g} \to (^1T_{1u} + {}^3T_{1u})$ for the so-called A and C absorption bands of thallium-like impurities in alkali halide crystals is considered in [5.60]. The A-band shape, taken separately and including vibronic coupling to the T_2 vibrations only, is discussed in [5.26], where the vibronic mixing of the $^1T_{1u}$ and $^3T_{1u}$ terms (the pseudo-Jahn-Teller effect) is also considered in the framework of perturbation theory (similarly to [5.25]). The B-band shape of the transition $^1A_{1g} \to (^3A_{1u} + {}^3T_{1u} + {}^3E_u + {}^3T_{2u} + {}^1T_{1u})$ considering all the mixing vibronic interactions in thallium-like centers was evaluated in [5.61] by means of semiclassical approximations. A detailed bibliography of studies of the shapes of the A, B and C absorption bands in thallium-like centers is given in [5.6, 10, 62, 63]. See also [5.64–68].

The case of equal coupling to the E and T_2 vibrations at $\omega_E = \omega_T$ (the d mode model) when the vibronic Hamiltonian possesses the high symmetry of three-dimensional rotations (Sect. 3.3.1) is considered in [5.69–71]. The absorption spectrum in this case is distinctly not equally spaced. The envelope is not Gaussian and has a weak three-humped structure. As mentioned in Sect. 3.3, the F^+ center in the crystal CaO may serve as a good example of a local center for which the d mode model is applicable. Analysis by the method of moments (Sect. 5.1.7) of the effects of polarization dichroism of the F^+ band in CaO under the action of the low-symmetry perturbations supports the approximate equality $E_{JT}(E) \approx E_{JT}(T)$, the contribution of the totally symmetric vibrations being negligible (see, however, [5.8]). The model of the d mode seems to be adequate for a series of cases Au^- : KCl, Fe^{2+} : MgO, Ca^+ : KBr, Tl^0 : KCl [5.72], etc. [5.56, 59]. In the work of *Nasu* and *Kojima* [5.73] the band shape of the $A \to T$ transition was considered in the IOA, while in [5.74] the $A \to T$ excitation curve was constructed by the cumulant expansion method. The optical response function from the Jahn-Teller system with nearly degenerate $(A_{1g} + T_{1u}) \otimes (a_{1g} + e_g + t_{1u} + t_{2g})$ excited electronic terms, as well as the IOA for this case, were investigated in [5.75, 76]. Operator methods (canonical transformations, the equations of motion, etc.) are used for the same purpose in [5.77]. In this paper a bibliography of previous publications is also given.

The band shape for the $T \to A$ transition can also be calculated in the semiclassical approximation [5.78]. As in the $E \to A$ case it is easier, however, to use the general relation (5.1.4). The Boltzmann factor $\exp(-\hbar\Omega/kT)$ explains why the symmetry of the band disappears at high temperatures (Fig. 5.8).

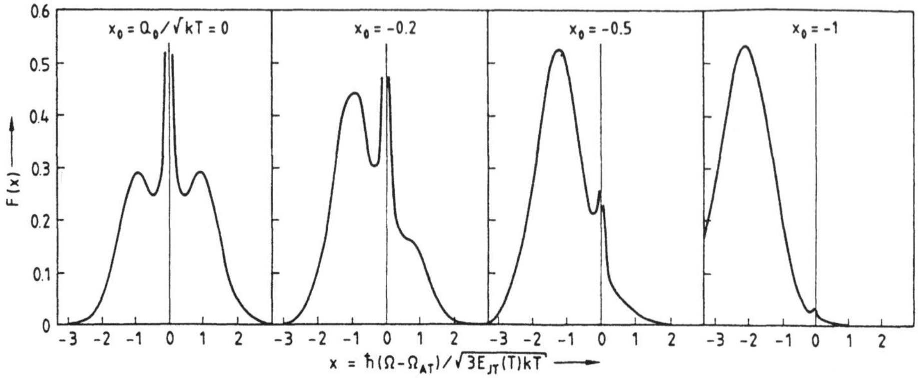

Fig. 5.8. Temperature dependence of the $T \rightarrow A$ transition band shape (linear $T \otimes t_2$ problem, $x_0 = Q_0/\sqrt{kT}$, $Q_0 = V_T\sqrt{2/3\omega_T}$), obtained in the semiclassical approximation [5.78]. In the high-temperature limit $x_0 = 0$ the bands $T \rightarrow A$ and $A \rightarrow T$ coincide

5.1.4 Other Vibronic Effects in Electron Excitation Spectroscopy

Optical transitions in a Jahn-Teller quadruplet Γ_8 coupled to the E and T_2 vibrations of a cubic polyatomic system are considered in [5.79]. The transition band $\Gamma_6(^2T_1) \rightarrow \Gamma_8(^2T_1)$ is constructed by means of numerical diagonalization of the vibronic Hamiltonian in the d mode model. As in the case of the $A \rightarrow E$ transition, the $\Gamma_6 \rightarrow \Gamma_8$ band has a double-humped structure, and for strong vibronic coupling a third peak is resolved at the high-frequency wing. This is interpreted in [5.79] as the optical manifestation of the first Slonczewski resonance (Sect. 4.2).

The pseudo-Jahn-Teller effect and its optical manifestations have received much attention since the foundation of the theory of vibronic interactions. In 1933 *Herzberg* and *Teller* [5.80] introduced the vibronic mixing of nondegenerate electronic states in order to explain the occurrence of optically forbidden bands in the electronic absorption spectrum. This *intensity borrowing effect*, obtained in [5.80] by a perturbation expansion of the electron wave functions in terms of the vibronic coupling constant, is known today as the Herzberg-Teller effect. A comprehensive review of the large number of publications in this area is outside the scope of this book, and we confine ourselves to citing [5.81] as a representative example of numerical calculations of the vibronic fine structure of the optical transition $A \rightarrow (A + B) \otimes b$, [5.82] as an example of semiclassical calculations of the envelope band shape, and [5.83–88] as examples of satisfactory explanations of different spectroscopic phenomena using the one-mode pseudo-Jahn-Teller Hamiltonian (3.4.1).

The pseudo-Jahn-Teller effect becomes more complicated if one takes into account the additional totally symmetric mode modulating the energy separation of the interacting states (Sect. 3.4.1). Since the coupling depends critically on the

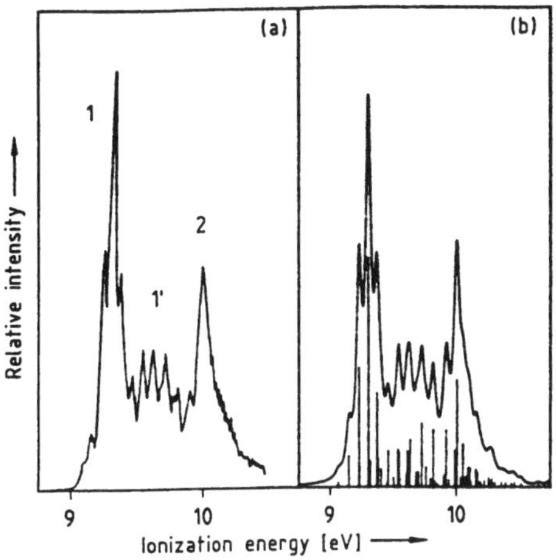

Fig. 5.9a, b. Vibronic structure of the photoelectron spectrum of butatriene. (**a**) Experimental spectrum (after [5.89]). Lines 1 and 2 are related to the ionization of the electron from the highest occupied molecular orbitals; without considering vibronic effects band 1′ cannot be explained, and is called [5.89] the "mystery band". (**b**) Results of numerical calculations taking account of vibronic coupling. (After [5.90])

energy gap between the mixing states, the vibronic interaction leads indirectly to an intricate coupling of the two modes. This model turned out to be able to explain the highly irregular and dense vibronic structure of the excitation spectra of some molecules and cation radicals [5.12]. As an example, Fig. 5.9 presents the low-energy photoelectron spectrum of butatriene from [5.89] and from the vibronic coupling calculation [5.90, 91]. The calculation takes into account the mixing of the $^2B_{3u}$ and $^2B_{3g}$ electronic terms of the $C_4H_4^+$ radical cation (symmetry group D_{2h}) by one stretching mode A_g and one torsional mode A_u. The central part of this band (in the range 9.4–10 eV) could not be assigned in [5.89] to any of the possible molecular orbitals, and was termed the "mystery band". However, the conical intersection of the two potential energy surfaces occurs at an energy of $\approx$9.7 eV. It can be seen that there is almost perfect agreement between the vibronic theory results and the experimental data. It follows that, as soon as the upper potential surface becomes energetically accessible to the electronic excitation, the vibrational motion is determined by the joint nuclear motion on both potential energy surfaces and it can no longer be attributed to the ionic states [5.91]. Other examples of the same kind are discussed in [5.12].

The low-energy $^3B_{1u}$ and $^3E_{1u}$ states of the benzene molecule provide a well-known spectroscopic example of the $(A + E) \otimes e$ pseudo-Jahn-Teller effect [5.92]. The same situation is found in the photoelectron spectrum of the BF_3 molecule. Some of its A_2' and E' terms are mixed by the E' vibrations of the symmetry group D_{3h} [5.45]. Analogous mixing of Σ and Π states determines the spectral properties of some linear molecules, such as C_2N_2 [5.93] or HCN [5.94]. In order to demonstrate the satisfactory quantitative agreement of the results of vibronic theory calculations [5.94] with the experimental data [5.95], the first band of the photoelectron spectrum of HCN is reproduced in Fig. 5.10. The irregularities in the spectrum at energies above 13.95 eV are of special interest. Since the first excited ionic state is of Σ symmetry and it has been calculated to lie in this energy

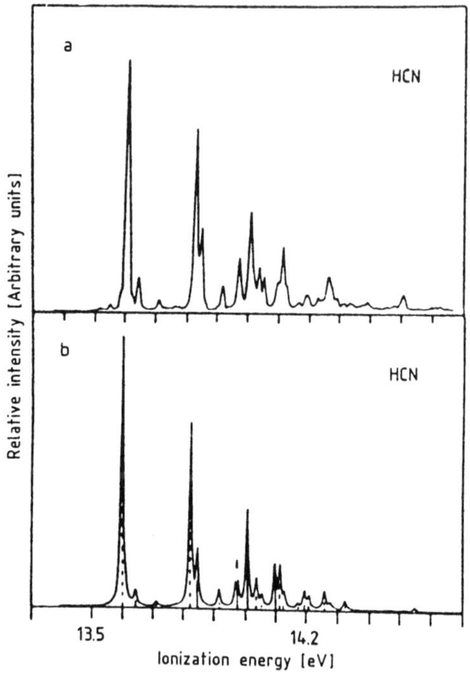

Fig. 5.10a, b. Vibronic structure of the photoelectron spectrum of the HCN molecule. (**a**) Experimental spectrum (after [5.95]); (**b**) numerical calculations. The vibronic states Π and Σ are shown by dashed and full lines, respectively. (After [5.94])

range, these spectral lines can be attributed to the $\Sigma - \Pi$ vibronic coupling [5.96].

The intrastate vibronic coupling $\Pi \otimes e$ in linear molecules (the Renner-Teller effect) and its manifestations in the electronic excitation spectra have been extensively discussed in the literature [5.97, 98]. The semiclassical envelope of the optical bands in Renner-Teller systems with strong vibronic coupling was evaluated in [5.99, 100]. The general theory of how vibronic coupling in linear molecules affects the spectral intensity distribution for transitions from a well-separated singlet state to the manifold of interacting states (obviously including the above case of $\Sigma - \Pi$ mixing, too), as well as appropriate numerical calculations of the spectral lines, is given in [5.101] (see also [5.12] and references therein).

The adiabatic separation of the sheets of the adiabatic potential of Jahn-Teller multiplets allows (at least in principle) the band structure of multiplet-multiplet transitions to be investigated. The optical curve, for instance, for the $E \to E$ transitions in the semiclassical approximation has a four-humped shape [5.27] corresponding to two transitions from the lower sheet of the ground E term to two sheets of the excited E term plus two other transitions from the upper sheet of the ground E term to two sheets of the excited one. The same approach was used recently [5.102] for the optical curve of the $^2E \to {}^2T_2$ transition in tetrahedral metal complexes.

The intensities of some low-frequency spectral lines for the $T \to E$ transition (*GR*1 band in diamond) are considered in [5.103–107]. Interesting experimental examples of the Jahn-Teller and pseudo-Jahn-Teller effects at deep impurity levels in diamond and silicon were reviewed by *Davies* [5.108]. See also [5.109, 110].

Band shapes of absorption and emission spectra for singlet-multiplet transitions with participation of the Jahn-Teller term $\Gamma_8 \otimes (e + t_2)$ were studied in [5.79] within the *d*-mode approximation. They are similar to the ones for the $E \otimes e$ term, except they are scooped out in the center because of the high dimensionality of the vibrational space.

5.1.5 The Jahn-Teller Effect in Radiative and Nonradiative Decay

Let us now consider one more spectral manifestation of the Jahn-Teller effect in the excited multiplet: the temperature dependence of the polarization of the luminescence [5.111]. The polarization of the exciting light can be chosen in such a way that the optical transition takes place to one of the minima of the adiabatic potential of the excited term. The tunneling relaxation of the excitation equalizes the population of equivalent minima resulting in a depolarization of the luminescence. At low temperatures, if the radiative lifetime is much shorter than the time of the tunneling relaxation, the polarization of the exciting light is predominant in the emission spectrum. An increase of the temperature leads consequently to:

a) population of the excited states in the minima for which the potential barrier is lower and hence the tunneling rate is higher;
b) temperature-induced jumps of the system to other minima through excited vibronic states of the lower sheet of the adiabatic potential which are higher in energy than the saddle points of the potential barriers;
c) population of the states of the higher sheets of the adiabatic potential.

All these processes, leading to *depolarization of the emission*, take place at a rather strong vibronic coupling, when the nuclear motions appropriate to different sheets of the adiabatic potential are independent. For weak and intermediate coupling the depolarization of the luminescence can be explained by dynamic and stochastic mechanisms of relaxation [5.112]. Experimental investigations of the polarized luminescence in alkali halide crystals activated by T1-like ions were also reported in [5.113, 114].

Studying the dependence of the polarization of luminescence on the polarization of the exciting light allows the symmetry of the Jahn-Teller minima of the excited multiplet to be evaluated [5.115, 116]. Consider, for example, the allowed magnetic dipole transitions in a cubic center with O_h symmetry. When the cubic system in the absolute minima of the adiabatic potential is tetragonally distorted the polarized absorption $\mathscr{H} \| C_4$ (where $\mathscr{H}$ is the magnetic field vector of the incident light) results in polarized luminescence with the same polarization. The polarization of the absorbed light $\mathscr{H} \| C_3$ in this case gives completely depolarized luminescence. In the case of trigonal distortions in the minima the luminescence is partly polarized for $\mathscr{H} \| C_3$ and depolarized for $\mathscr{H} \| C_4$ in the absorption light. It is assumed that before emission the system has time to relax to the minima states.

The dependence of the degree of polarization of luminescence light on the frequency of the exciting light can also be a characteristic feature of Jahn-Teller systems [5.115]. If the absorption center is selectively excited to the vibrational states near the bottom of the minimum of the adiabatic potential, the polarization of luminescence should be maximal and the depolarization increases with the increase in the frequency of excitation. The influence of the vibrational relaxation on the process of depolarization is discussed in [5.117, 118].

The problems of *nonradiative decay* are a field of special interest [5.119]. In almost all cases in which numerical results have been presented so far, the potential surfaces on which the nuclear motion takes place were assumed to have very simple forms: parabolic or paraboloidal [5.120]. The Jahn-Teller and pseudo-Jahn-Teller systems, however, possess potential surfaces which differ substantially from paraboloids, and hence for them the existing models are inadequate. The simple parabolic model can be used only when the relaxation occurs in the neighborhood of a single minimum, not when the relaxation or the excitation transfer takes place between different minima.

An alternative approach based on a quasi-classical consideration of the nuclear motion on separate sheets of the Jahn-Teller adiabatic potential was proposed in a series of papers by *Englman* and *Ranfagni* and co-workers [5.121–124] (see their review article [5.125]). Note also some archetypal cases worked out by *Wagner* ([5.126] and references therein). An interesting numerical calculation of the occupation probability of the upper electronic state $^2B_{2g}$ of the ethylene cation $C_2H_4^+$, taking into account the pseudo-Jahn-Teller mixing $(^2B_{2u} + {}^2B_{2g}) \otimes (a_g + a_u)$, was performed by *Köppel* [5.127] (see also [5.12]). It was shown that the calculated ultrafast decay may explain the absence of emission from the $\tilde{A}\ {}^2B_{2g}$ state in this molecule. More generally, in [5.12] it was proposed that the absence of detectable fluorescence may be expected whenever there exists a conical intersection close to the minimum of the corresponding adiabatic potential energy surface.

Another interesting effect confirmed by numerical calculations is the quenching of radiative decay rates for systems in individual vibronic states [5.12]. The fluorescence decay rate of any exact vibronic state written in the form (2.2.8) may easily be shown to gain contributions from the mixed electronic states [5.12]:

$$\Gamma = \sum_{k=1}^{f} \gamma_k \langle \chi_k | \chi_k \rangle \ , \tag{5.1.28}$$

where γ_k are the decay rates of the individual electronic states and χ_k satisfy the normalization condition

$$\sum_{k=1}^{f} \langle \chi_k | \chi_k \rangle = 1 \ .$$

The value of Γ is therefore smaller than the largest of the decay rates γ_k. The lifetime of the corresponding vibronic level, defined as the inverse decay rate Γ^{-1}, in the presence of vibronic coupling must always be longer than the shortest lifetime of the mixed electronic states. *Douglas* [5.128] was the first to state that the anomalously long radiative lifetimes observed for some molecules arise due to the mixing of different electronic states. Equation (5.1.28) thus provides a quantitative description of the *Douglas effect.* This result explains, for instance, the situation in the NO_2 molecule where the radiative decay rate of the excited electronic term $^2B_2$ has been found to be one or two orders of magnitude smaller than expected from the integrated absorption coefficient [5.129]. Theoretical calculations of *Haller* et al. [5.130] (see also [5.12]) confirmed the assumption of Douglas that the anomalously small fluorescence decay rates of the excited B_2 vibronic states of the NO_2 molecule are indeed caused by the strong nonadiabatic interaction between the $^2A_1$ and $^2B_2$ electronic states (the former having zero decay rate).

5.1.6 Multimode Vibronic Effects in Electronic Spectra

Both the exact results obtained by numerical solution of the vibronic equations and approximate analytic expressions for the curve of the optical absorption and luminescence, obtained above, are mainly relevant to ideal vibronic systems. The approximate methods developed for ideal systems are also valid in some cases for multimode Jahn-Teller systems. Indeed, the separation of the sheets of the adiabatic potential allows the problem to be reduced to a singlet-singlet transition with considerable anharmonicity of the potential surfaces. Note, however, that for systems with a conical intersection of the adiabatic potential the asymptotic behavior of the high energy wing of the optical band obtained within such an adiabatic approximation is wrong.

In the semiclassical approximation the difference between the ideal and multimode vibronic systems disappears. Indeed, neglecting the noncommutativity of $Q_{\Gamma\gamma}$ with $P_{\Gamma\gamma}$ in expressions of the type (5.1.22) allows the kinetic energy of the nuclei to be taken outside the averaging procedure. The substitution of variables in the multiple integral over the vibrational coordinates $\langle\cdots\rangle_{\text{vib}}$ corresponding to the scale transformation (3.5.13) with a subsequent rotation allows the "interaction" modes to be separated (3.5.17), that is, the multimode problem to be reduced to an ideal one. Therefore, all the results obtained above in the semiclassical approximation for the ideal vibronic systems are also relevant to multimode systems, provided the criteria of the applicability of the semiclassical approximation are fulfilled.

As discussed in Sect. 4.6, it is also possible to introduce the interaction mode in such a way that the multimode vibronic problem is reduced to a one-mode problem. This can be achieved by using the cluster model with a vibronic constant and a vibrational frequency chosen to reproduce the main features of the multimode problem being considered. *In the optical band investigation the centroid of the band, its width and other momenta of the spectrum have to be reproduced as well as possible.* We introduce here the centered moment of the nth order:

$$\langle \Omega^n \rangle = \langle \Omega^0 \rangle^{-1} \int_{-\infty}^{\infty} (\Omega - \langle \Omega \rangle)^n F(\Omega) d\Omega \ , \quad \text{where} \tag{5.1.29}$$

$$\langle \Omega^0 \rangle = \int_{-\infty}^{\infty} F(\Omega) d\Omega \ , \qquad \langle \Omega \rangle = \langle \Omega^0 \rangle^{-1} \int_{-\infty}^{\infty} \Omega F(\Omega) d\Omega$$

are the integral intensity and the centroid of the optical band $F(\Omega)$, respectively. The main advantage of the moments is that they can be calculated analytically for some cases of great interest, in particular, for transitions between electronic terms where one of them is vibronically uncoupled, irrespective of whether the system under consideration is a multimode one or not. For instance, in the case of the $A \to E$ transition the second moment can easily be shown to be equal to

$$\langle \Omega^2 \rangle = \sum_n \frac{V_n^2}{\hbar \omega_n} \coth \left(\frac{\hbar \omega_n}{2kT} \right) ,$$

whereas for the ideal vibronic system the corresponding expression is given by (5.1.27). The third moment is

$$\langle \Omega^3 \rangle = \frac{1}{\hbar} \sum_n V_n^2$$

and for the ideal vibronic system $\langle \Omega^3 \rangle = V_E^2/\hbar$.

Taking the above expressions for $\langle \Omega^2 \rangle$ and $\langle \Omega^3 \rangle$ in the multimode case equal to the ones for an ideal vibronic system, one obtains the effective coupling constant and effective frequency of an ideal system modeling the optical band of the $A \to E$ transition in the sense of the best reproduction of the first centered moments [5.131, 132]

$$V_{\text{eff}} = \left(\sum_n V_n^2 \right)^{1/2} ; \qquad \omega_{\text{eff}} = \sum_n V_n^2 \Big/ \sum_n \frac{V_n^2}{\omega_n} . \tag{5.1.30}$$

It can be shown [5.131] that the main contribution terms of all spectral moments in the case of strong coupling can be reproduced correctly, if one employs (5.1.30). Therefore one expects that the greater the value of the coupling constant, the more accurate the band shape curve that can be obtained by this approach.

In the case of weak coupling, i.e., when the main contribution to the optical absorption spectrum results from the transition to the ground vibronic state of the excited electronic term (the zero-phonon line, Sect. 5.2), the variational cluster model (Sect. 4.6), which is sufficient for the description of the lowest vibronic state, is the best one-mode model [5.75].

Another approximate approach to the multimode excitation spectrum problem was proposed in [5.133]. It is based on the fact that the Hamiltonian of the multimode system with linear vibronic coupling can be written as a sum $\sum_n \hat{H}_n$, where $\hat{H}_n$ is the vibronic Hamiltonian of the nth mode, cf. (3.5.27). It is worth remembering that different modes of a multimode vibronically coupled system cannot be considered separately because they interact through the electrons, and different $\hat{H}_n$ operators do not commute with one another. However, in the weak coupling limit one can use the fact that second-order corrections to the energies are additive with respect to the vibronic coupling contributions of different modes (Sect. 4.6). In other words, in the second-order approximation the modes are separable. This makes it possible to write the generating function (5.1.21) in the form of a product:

$$I(t) = \prod_n I_n(t) \ , \qquad I_n(t) = \frac{1}{f_\Gamma}\left\langle \mathrm{Tr}\left\{\exp\left(\frac{\mathrm{i}}{\hbar}\hat{H}_n(\Gamma)t\right)\right\}_\Gamma \exp\left(-\frac{\mathrm{i}}{\hbar}\hat{H}_n(A)t\right)\right\rangle_{\mathrm{vib}} \ .$$

Following the theorem of convolution for the Fourier transform, one can easily obtain the form factor of the excitation spectrum $F(\Omega)$ from (5.1.5) as a convolution of the partial excitation functions $F_n(\Omega)$,

$$F_n(\Omega) = \frac{1}{2\pi}\int_{-\infty}^{\infty} \mathrm{e}^{-\mathrm{i}\Omega t} I_n(t)dt \ .$$

That is why this approximation is called the *convolution approximation* [5.133]. It is understandable that the weaker the vibronic coupling, the better the structure of the optical excitation band and its shape are reproduced. Note that this approximation is applicable for multimode systems with a discrete vibronic spectrum only, since for systems with a continuous spectrum when the mode-mixing effects are considerable, even in the weak coupling case, normal perturbation theory is inapplicable (Sect. 4.6).

In Figs. 5.11 and 5.12 the results of numerical calculations of the vibronic structure and envelope of $A \to E$ type electronic bands for a two-mode $E \otimes (e + e)$ vibronic system are illustrated [5.12, 70, 133]. The first example (Fig. 5.11) belongs to the weak coupling regime. One can see that there is qualitative agreement of the exact spectrum with that obtained via the convolution approximation, at least as far as the positions of the first spectral lines are concerned. In Fig. 5.12 a case of strong Jahn-Teller coupling is presented. We can see that the band shapes (the envelopes) of the exact and effective single-mode spectra coincide [5.131, 132]. However, the band shape of the convoluted spectrum looks quite different and has a central maximum instead of the double peak. It can also be seen that the vibronic fine structure of the exact band in this case cannot be reproduced by the effective single-mode approximation in the same manner as by the convolution approximation. The increase in the line density of the exact spectrum relative to the convoluted one results from the loss of the artificial extra

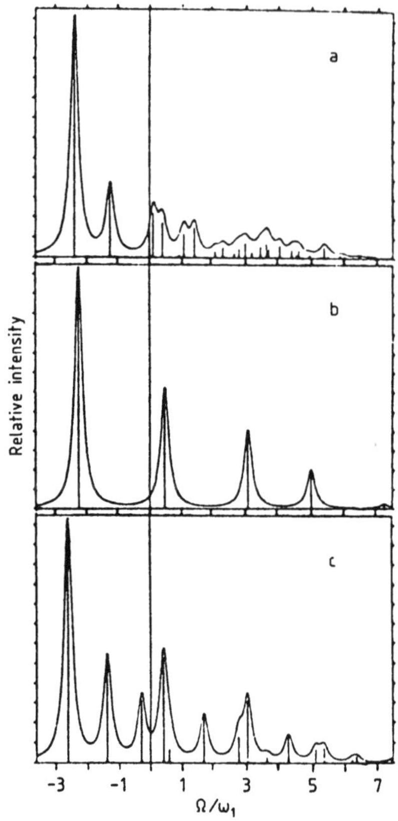

Fig. 5.11a–c. Band structure for the $A \to E$ transition in a two-mode Jahn-Teller $E \otimes (e + e)$ system $\omega_2 = 2.41\omega_1$; $V_1 = \sqrt{\hbar\omega_1^3}$; $V_2 = 2.41\sqrt{\hbar\omega_2^3}$. (**a**) Exact numerical solution; (**b**) one-mode cluster approximation; (**c**) approximation of independent modes (convolution approximation). (After [5.134])

Fig. 5.12a–d. The same as Fig. 5.11 for the case of strong vibronic coupling. $\omega_2 = 1.57\omega_1$; $V_1 = 2\sqrt{\hbar\omega_1^3}$; $V_2 = 3.13\sqrt{\hbar\omega_2^3}$. Part (**d**) illustrates the envelopes of the three bands in (**a–c**). The envelopes of the exact band (**a**) and the one-mode (cluster) approximation (**b**) merge into one double-humped curve [5.134]

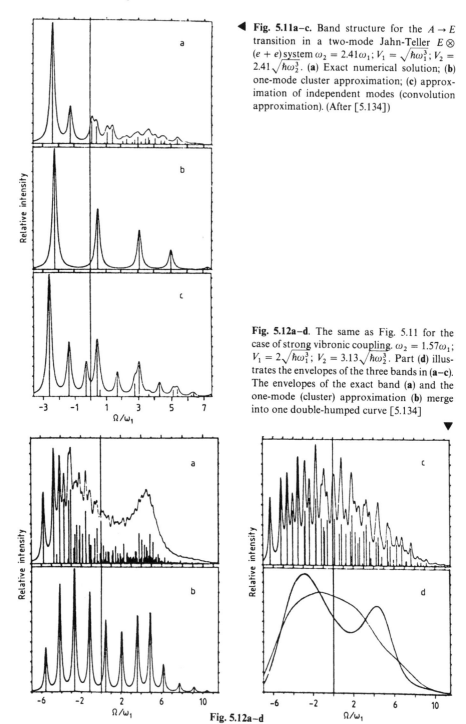

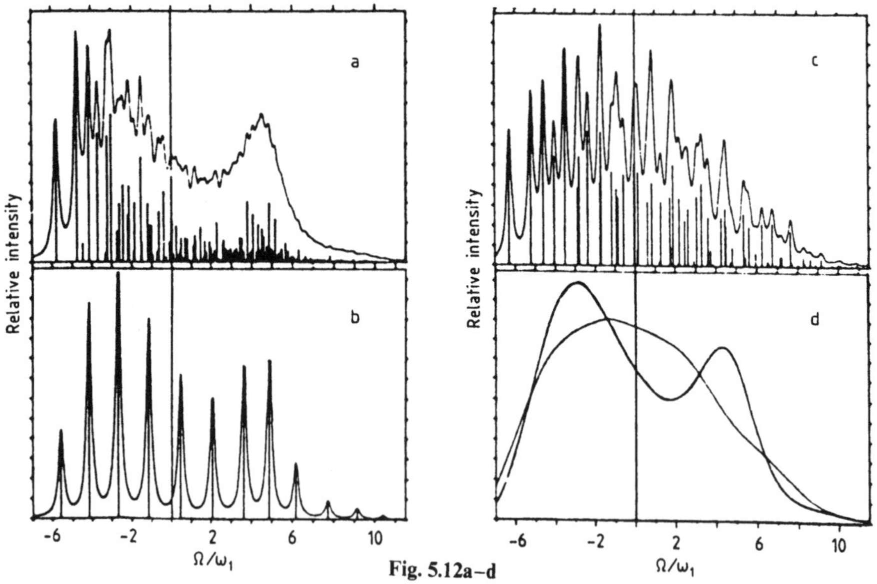

Fig. 5.12a–d

symmetry induced by the convolution approximation. Indeed, the number of integrals of motion for the separate-mode system equals the number of modes, whereas the exact ($E \otimes e$)-type multimode system possesses only one integral of motion (Sect. 3.2).

Some interesting experimental examples of multimode vibronic coupling involving electronic states of E symmetry, as well as their satisfactory theoretical reproduction are illustrated in Figs. 5.6, 5.9, and 5.10. Other examples are discussed in [5.12, 34–37, 135, 136].

The optical bands of transitions to and from degenerate electronic terms in multimode Jahn-Teller centers with continuous energy spectra (impurity centers in crystals) have some special features not reflected in the above approximations. For instance, it is known that the one-phonon satellites reproduce the structural features of the phonon density. As is shown in Sect. 4.6, the Jahn-Teller effect essentially changes the density of vibrations, resulting in the occurrence of local and pseudolocal resonances [5.136, 137]. At low temperatures, where the fine structure of the band is resolved, these resonances obviously determine the structure of the curve of optical absorption at the low-frequency wing of the band (Fig. 5.5).

To summarize all the foregoing, we note that the problem of determination of the band shape in the presence of the Jahn-Teller effect, at least in one of the multiplets between which the optical transition takes place, is very complicated. Practically, the most convenient way to calculate the band shape of the excitation spectrum with participation of degenerate states is the semiclassical approximation (which is simultaneously the adiabatic approximation). Even in this approximation, which has very strict criteria of applicability, investigation of the band shape is rather difficult due to the necessity of integrating in the multidimensional space of nuclear displacements.

5.1.7 The Method of Moments and Polarization Dichroism of Jahn-Teller Systems

In some cases when the Jahn-Teller effect is absent or negligibly small in the electronic states where the optical transition starts (or ends), exact calculation of the moments of the spectral distribution is possible, although the distribution itself remains unknown due to the Jahn-Teller effect in the final electronic state.

Applications of the *method of moments* to the optics of Jahn-Teller systems is developing in two directions. The first is devoted to the restoration of the band shape from the exactly calculated moments using the method of Gram-Charlier series [5.27, 28]. This method allows any trial distribution to be improved by expanding it in a series where the coefficients are the differences in the moment values of the exact distribution and the trial one. Naturally, the trial distribution should be close to the exact one, or it should at least preserve the most significant features.

The other direction starts with [5.138], where it is shown that an analysis of the changes of the moments in the conditions of polarized dichroism induced by external fields (electric, magnetic, or stress) provides information about the parameters of the Jahn-Teller system, orbital g-factors and constants of the spin-orbit interaction. In [5.139] a matrix formalism of the method of moments is developed, which allows the general solution of the problem and a group-theoretical analysis of the effects of dichroism. The status up to 1985 of optics of Jahn-Teller systems is reviewed in [5.8].

The application of the method of moments to the effects of polarization dichroism of broad optical bands of multiplet-multiplet transitions is based on the simple physical idea that under conditions of vibronic mixing of the final electronic states to which the transition takes place, the external perturbation changes both the centroid (the first moment) of the band and its shape.

In order to investigate the changes of the moments under the influence of low-symmetry electronic perturbations, we write the vibronic Hamiltonian of the term Γ_i (or of a group of close-in-energy terms) in the form

$$\hat{H}(\Gamma_i) = E(\Gamma_i)\hat{C}_A(\Gamma_i) + \hat{H}_{\mathrm{JT}}(\Gamma_i) + \hat{W}(\Gamma_i) \ , \tag{5.1.31}$$

where $E(\Gamma_i)$ is the energy of the degenerate term Γ_i (or the mean energy of the group of close-in-energy terms), $H_{\mathrm{JT}}(\Gamma_i)$ is the vibronic Hamiltonian of the electronic term Γ_i (Sect. 2.2), $\hat{C}_A(\Gamma_i)$ is the unit matrix, and $\hat{W}(\Gamma_i)$ is the matrix of the low-symmetry perturbation determined in the space of the electronic states Γ_i. The matrix $\hat{W}(\Gamma_i)$ can be chosen in such a way that $\mathrm{Tr}[\hat{W}(\Gamma_i)] = 0$. As mentioned above, for the calculation of moments it is important that the vibronic interaction is absent in the electronic states Γ_1 where the optical transition starts, that is, all the $V_\Gamma(\Gamma_1) = 0$.

Substituting (5.1.5), (5.1.8) and (5.1.31) into (5.1.29) and performing some not very complicated transformations we obtain [5.140]

$$\begin{aligned}\langle \Omega^n \rangle = [f_1\langle \Omega^0 \rangle]^{-1}\,\mathrm{Tr}\Bigg\{&\exp\left(-\frac{\hat{W}(\Gamma_1)}{kT}\right)\hat{d}^\dagger \sum_{m=0}^{n}\sum_{l=0}^{m}(-1)^{n-m} \\ &\times \binom{n}{m}\binom{m}{l}[\hat{W}(\Gamma_2) - \Delta\langle\Omega\rangle\hat{C}_A(\Gamma_2)]^l\hat{\sigma}_{m-l}\hat{d}\hat{W}^{n-m}(\Gamma_1)\Bigg\}_{\Gamma_1} \ ,\end{aligned} \tag{5.1.32}$$

where f_1 is the dimensionality of the space of electronic states Γ_1; $\mathrm{Tr}\{\cdots\}_\Gamma$, as previously, means the summation over the electronic states of the term Γ,

$$\Delta\langle\Omega\rangle = (\hbar\langle\Omega^0\rangle)^{-1}\langle \hat{d}^\dagger\hat{W}(\Gamma_2)\hat{d} - \hat{d}^\dagger\hat{d}\hat{W}(\Gamma_1)\rangle_{\Gamma_1} \ , \tag{5.1.33}$$

and the σ_n are known as the matrices of elementary moments [5.139],

$$\hat{\sigma}_0 = \hat{C}_A(\Gamma_2) \ , \qquad \hat{\sigma}_1 = 0 \ , \qquad \hat{\sigma}_2 = \sum_\Gamma \sigma_2(\Gamma) \ ,$$

$$\hat{\sigma}_2(\Gamma) = \frac{V_\Gamma^2}{2\hbar\omega_\Gamma}\coth\left(\frac{\hbar\omega_\Gamma}{2kT}\right)\sum_\gamma \hat{C}^\dagger_{\Gamma\gamma}(\Gamma_2)\hat{C}_{\Gamma\gamma}(\Gamma_2) \ ,$$

$$\hat{\sigma}_3 = \sum_{\Gamma} [\sigma_3^{(1)}(\Gamma) + \sigma_3^{(2)}(\Gamma)] \ ,$$

$$\hat{\sigma}_3^{(1)}(\Gamma) = \frac{V_\Gamma^2}{2\hbar} \sum_{\gamma} \hat{C}_{\Gamma\gamma}^{\dagger}(\Gamma_2)\hat{C}_{\Gamma\gamma}(\Gamma_2) \ ,$$

$$\hat{\sigma}_3^{(2)}(\Gamma) = \frac{V_\Gamma^2}{2\hbar\omega_\Gamma} \coth\left(\frac{\hbar\omega_\Gamma}{2kT}\right) \sum_{\gamma} \hat{C}_{\Gamma\gamma}^{\dagger}(\Gamma_2)[\hat{W}(\Gamma_2), \hat{C}_{\Gamma\gamma}(\Gamma_2)] - \frac{2}{\hbar}[\hat{W}(\Gamma_2), \hat{\sigma}_2(\Gamma)] \ . \tag{5.1.34}$$

In [5.139] the matrices $\hat{\sigma}_4$ are also given. Note that the component $\hat{\sigma}_3^{(2)}$ of the third moment also includes an explicit linear dependence on $\hat{W}$.

Returning to (5.1.32) we note that the moments are unitary invariants, since they can be expressed as traces of products of matrices acting in the electronic subsystem. This allows one to perform a unitary transformation on the system of zero-order functions, i.e., a transformation which transfers the perturbation from the wave functions to the Hamiltonian. In essence, this means the introduction of effective Hamiltonians acting in the space of zero-order wave functions. The use of the method of effective Hamiltonians in the theory of moments of spectral distributions of multiplet-multiplet transitions is developed in [5.140, 141].

Consider as an example the linear dichroism of the optical band of the $^1A \to {}^1T$ transition arising due to uniaxial stress [5.139]. The perturbation matrix $W(T)$ can be presented in the form

$$\hat{W}(T) = \sum_{\Gamma\gamma} P_\Gamma \varepsilon_{\Gamma\gamma} \hat{C}_{\Gamma\gamma}(T) \ , \tag{5.1.35}$$

where the matrices $\hat{C}_{\Gamma\gamma}(T)$ are given by (3.3.2) and $\varepsilon_{\Gamma\gamma}$ are the linear combinations of the components of the tensor of deformations transforming as the rows of the irreducible representation Γ. In the case of tetragonal stress (direction [001]), the perturbation matrix acquires the form $\hat{W}_{\text{tetr}} = P_E \varepsilon_\theta \hat{C}_\theta(T)$. For simplicity, we assume that the operator of optical transitions d and the operator of vibronic interactions are independent of the external perturbation. In other words, the intermultiplet mixing caused by the perturbation $\hat{W}$ is not considered. Under the influence of the tetragonal deformation the T term of the cubic system splits into the E and A terms of the group D_{4h}. Allowed transitions to these levels conform to the matrices $\hat{d}_\parallel^\dagger = (001)$, $\hat{d}_\perp^\dagger = (100)$ which are eigenvectors of the matrix $\hat{W}_{\text{tetr}}$ with the following eigenvalues [see (5.1.33) with $\hat{W}(\Gamma_1) = 0$, $\langle \Omega^0 \rangle = 1$]:

$$\Delta\langle\Omega\rangle_\parallel = -\frac{P_E}{\hbar}\varepsilon_\theta \ , \qquad \Delta\langle\Omega\rangle_\perp = \frac{P_E}{2\hbar}\varepsilon_\theta \ . \tag{5.1.36}$$

In the case being considered (5.1.32) can be written in the form $\langle \Omega^n \rangle = \hat{d}_\eta^\dagger \hat{\sigma}_n \hat{d}_\eta$, where the index η corresponds to different polarizations. For the second moment we find from (5.1.34) that

$$\langle \Omega^2 \rangle_\parallel = \langle \Omega^2 \rangle_\perp = \bar{\sigma}_2(A) + \bar{\sigma}_2(E) + 2\bar{\sigma}_2(T_2) \ , \tag{5.1.37}$$

where $\bar{\sigma}_2(\Gamma)$ are the contributions of different modes to the second moment,

$$\bar{\sigma}_2(\Gamma) = \frac{V_\Gamma^2}{2\hbar\omega_\Gamma} \coth\left(\frac{\hbar\omega_\Gamma}{2kT}\right) . \tag{5.1.38}$$

In particular, at $T = 0$ K one has $\bar{\sigma}_2(\Gamma) = V_\Gamma^2/2\hbar\omega_\Gamma$.

Similarly, for the third moment, calculating the commutators in the expression for $\sigma_3^{(2)}(\Gamma)$, we find

$$\langle \Omega^3 \rangle_\eta = \langle \Omega^3 \rangle_0 - 3\Delta\langle \Omega \rangle_\eta \bar{\sigma}_2(T_2) \ , \tag{5.1.39}$$

where $\langle \Omega^3 \rangle_0$ is the third moment of the band not perturbed by the stress.

For trigonal stress the perturbation matrix can be expressed in the form $\hat{W}_{\text{trig}} = P_T \varepsilon_T [\hat{C}_\xi(T) + \hat{C}_\eta(T) + \hat{C}_\zeta(T)]$. It is convenient to calculate the moments in the so-called trigonal basis [5.142] in which the matrix $\hat{W}_{\text{trig}}$ is diagonal. Omitting the details of the calculation, we have

$$\Delta\langle \Omega \rangle_\parallel = 2P_T\varepsilon_T/\hbar\sqrt{3} \ , \qquad \Delta\langle \Omega \rangle_\perp = -\tfrac{1}{2}\Delta\langle \Omega \rangle_\parallel \ . \tag{5.1.40}$$

The results (5.1.37, 38) hold also for trigonal stress, and instead of (5.1.39) we have

$$\langle \Omega^3 \rangle_\eta = \langle \Omega^3 \rangle_0 - \tfrac{1}{2}\Delta\langle \Omega \rangle_\eta [3\bar{\sigma}_2(E) + 2\bar{\sigma}_2(T_2)] \ . \tag{5.1.41}$$

From (5.1.39, 41) we obtain two expressions which can be used for the experimental determination of $\bar{\sigma}_2(E)$ and $\bar{\sigma}_2(T_2)$:

$$\begin{aligned} \bar{\sigma}_2(T) &= -\frac{\Delta\langle \Omega^3 \rangle_\eta^{\text{tetr}}}{3\Delta\langle \Omega \rangle_\eta^{\text{tetr}}} \ , \\ \bar{\sigma}_2(E) &= \frac{2\Delta\langle \Omega^3 \rangle_\eta^{\text{tetr}}}{9\Delta\langle \Omega \rangle_\eta^{\text{tetr}}} - \frac{2\Delta\langle \Omega^3 \rangle_\eta^{\text{trig}}}{3\Delta\langle \Omega \rangle_\eta^{\text{trig}}} \ , \end{aligned} \tag{5.1.42}$$

where $\Delta\langle \Omega^n \rangle = \langle \Omega^n \rangle - \langle \Omega^n \rangle_0$ is the change of the nth moment caused by the perturbation $\hat{W}$.

Thus the changes of the moments of the optical band due to the dichroism induced by external fields contain important information about the parameters of the vibronic coupling $V_\Gamma^2/\hbar\omega_\Gamma$. The accuracy of the method is determined by the accuracy of the moments measurement, since the relationships given in this section are based only on the symmetry of the Jahn-Teller center in question and not on its detailed structure [5.139]. It can be shown that the numerical coefficients in the expressions of the type (5.1.42) are the group-theoretical constants (6Γ, 9Γ, etc.) which are determined by the symmetry group of the center and by the nature of the low-symmetry perturbation, and also by the polarization of the light wave [5.143].

If the spin-orbital interaction is considered, new parameters [the spin-orbit constants $\lambda(S, \Gamma)$] appear in the effective Hamiltonian and the number of equations available for the determination of $\bar{\sigma}_2(\Gamma)$ and $\lambda(S, \Gamma)$ may be fewer than the number of unknowns. The missing equations can be obtained from the changes of moments due to the effects of magnetic circular dichroism [5.143, 144]. In [5.145] the theory of polarized dichroism of optical bands is generalized to the case of forbidden transitions, taking into account the dependence of the matrices of the effective dipole moment on the normal coordinates of vibrations. The spin-forbidden transitions and the effects of polarization dichroism, treated by the method of moments, are discussed in [5.146].

Note that the dispersion of crystalline vibrations (the multimode problem) can also be included in the exact calculation of moments. In particular, in all the publications on the method of moments cited above [5.139, 140, 144–146] appropriate expressions were obtained for the moments of optical bands of impurities in crystals taking the phonon dispersion into account. In the model of an impurity center of small radius (Sect. 3.5), instead of (5.1.38) we have

$$\bar{\sigma}_2(\Gamma) = \frac{V_\Gamma^2}{2\hbar} \sum_\kappa \frac{a_\kappa^2(\Gamma\gamma)}{\omega_\kappa} \coth\left(\frac{\hbar\omega_\kappa}{2kT}\right) . \tag{5.1.43}$$

Therefore, Jahn-Teller impurities which have broad multiphonon smoothed bands of impurity absorption and luminescence are the primary subjects of investigation by the method of polarized dichroism spectroscopy. The generalization of the matrix formulation of the method of moments, as applied to anisotropic centers in cubic crystals, is given in [5.147, 148].

The bands of exciton absorption of crystals with exciton-phonon interaction of the Jahn-Teller type are another interesting system that is treated by the method of moments in polarized dichroism spectroscopy. If one starts with the idea of localized Frenkel excitons, then (neglecting a resonance interaction) each elementary cell absorbs light independently and the problem is identical to that of impurity absorption. In addition to the complications caused by the Jahn-Teller effect in excited states, in exciton spectra new difficulties arise due to the resonant intercenter interaction resulting in exciton dispersion. So far, [5.149] is the only work in which the band shape for absorption of Jahn-Teller excitons in the case of the $A \to E$ and $A \to T$ transitions has been calculated. The above difficulties are overcome by means of the semiclassical approximation, as well as by the coherent potential approximation. However, not all Jahn-Teller active vibrational modes and not all types of resonance interactions are considered in [5.149].

An alternative approach to the problem is suggested in [5.150] in which the polarization dichroism of the exciton bands in external fields is employed. It is shown that the exciton dispersion does not prevent exact calculations of the moments of the spectral distribution of a Jahn-Teller exciton. The theory is applied to cation excitons in alkali halide crystals. For the exciton A band in

CsCl [the transition $^1A_{1g}(t_{1u}^6) \to {}^1T_{1u}(t_{1u}^5 a_{1g})$ in the ion Cs^+] a model microcalculation of the second and third moments that included their temperature dependence was carried out. It was shown that the dominant contribution to these moments comes from the vibronic coupling with the Jahn-Teller modes, causing the special structure of the A band. Note that the results [5.150] obtained by means of the method of moments are not constrained by the approximations used in [5.149], and in this sense they are exact.

5.2 Splitting of the Zero-Phonon Lines of the Optical Absorption and Luminescence

The pure electronic transition with no change of the vibrational state of the polyatomic system occupies a special place among the electron-vibrational transitions. It is obvious that for equal curvature of the adiabatic potentials $\varepsilon_1(Q)$ and $\varepsilon_2(Q)$ the energies of the transitions $0-0$, $1-1$, $2-2$, ... coincide, the corresponding lines in the spectrum of optical absorption are superimposed and at this frequency only one single line is observed. Transitions that do not involve a change of the vibrational state are usually called zero-phonon transitions, and the corresponding line in the optical spectrum is called the zero-phonon line [5.13]. It is significant that the idea that many zero-phonon transitions can add together and contribute to one narrow spectral line remains valid also for multimode systems, in particular, for crystals. Just for this reason, even in crystals the zero-phonon line is a sharp intense peak on the background of continuous multiphonon absorption.

The difference in the curvature of the adiabatic potentials $\varepsilon_1(Q)$ and $\varepsilon_2(Q)$, and hence in the frequencies of vibrations ω_1 and ω_2, causes noncoincidence of the energies of the zero-phonon transitions $0-0$, $1-1$, $2-2$, ... As a result, the zero-phonon line is slightly split into many lines, which in multimode systems merge into one broadened zero-phonon line. The influence of this so-called frequency effect (the Dushinski effect) on the width of the zero-phonon line of singlet-singlet optical transitions is well studied [5.5]. The anharmonicity of the potential curves $\varepsilon_1(Q)$ and $\varepsilon_2(Q)$ is another cause of broadening and shifts of the zero-phonon line.

The theory of zero-phonon lines has been developed for polyatomic systems with nondegenerate electronic states. Its generalization to the case of orbital degeneracy, when the Jahn-Teller effect occurs in at least one of the combining electronic terms, is not trivial. First, when the vibronic interaction is considered the oscillator occupation numbers of the Jahn-Teller–active normal vibrations cease to be good quantum numbers and the vibronic states of the Jahn-Teller systems are no longer characterized by a definite value of the phonon numbers. From this point of view, the "zero-phonon transitions", for which the "vibrational state" of the Jahn-Teller system does not change, have no meaning. Second, the vibronic energy levels are distinctly not equally spaced (see, e.g. Fig. 4.9), and therefore the coincidence of the energy of the allowed transitions, and thus the corresponding superposition of the spectral lines, is extremely improbable.

Nevertheless, the experimentally observed optical bands of absorption and luminescence of Jahn-Teller systems in some cases display a sharp peak: the zero-phonon line. A distinguishing feature of this line is that its position coincides in absorption and luminescence (Fig. 5.5). At $T = 0$ K it turns out to be the lowest-frequency line of the absorption spectra, i.e., it is located at the low-frequency edge of the band. With increasing temperature the zero-phonon lines of Jahn-Teller systems rapidly broaden, their intensities decreasing, and at any high temperatures they cease to be seen against the background of the multi-phonon band.

It is obvious that at $T = 0$ K only the ground vibronic state of the term from which the optical transition starts is populated, and therefore the lowest-frequency line of the absorption spectrum corresponds to the $0 \to 0$ transition, i.e., to the transition between the lowest vibronic states of the combining terms. (Note that the zeros in the expression "$0 \to 0$ transition" are no longer the vibrational occupation numbers as in the case of singlet-singlet transitions.)

In polyatomic systems with a discrete vibronic spectrum, the $0 \to 0$ transition line is similar to other δ-type spectral lines (Fig. 5.1). The situation is different in multimode vibronic systems with continuous spectra, in particular in crystals with Jahn-Teller impurity centers. As was stated in Sect. 4.6, the vibronic interaction localized on the center affects the electron-vibrational spectrum of the system, resulting in the occurrence of local and pseudolocal vibronic states (resonances). It is important that in the case of weak-to-intermediate vibronic coupling when the zero-phonon line is observable in the low-frequency region this reconstruction of the spectrum is related only to separate discrete regions of the spectrum determined by the appropriate transcendental equations, in other respects the density of states remains the same as without the impurity Jahn-Teller effect. If the vibronic resonances are neglected, one can see that the properties of the zero-phonon line of the Jahn-Teller systems are in many ways similar to the properties of this line in singlet-singlet transitions. In particular, the density of vibronic states close in energy to the ground state is near to zero. Therefore at $T = 0$ K there is a gap in the optical absorption band on the right of the $0 \to 0$ transition line, and this gap allows the zero-phonon line to be separated (Fig. 5.5).

Owing to the vibronic interaction localized on the center, the special Jahn-Teller dynamics of the electrons and nuclei takes place only for several collectivized degrees of freedom localized near the impurity. For most long wave vibrations ($\omega_\kappa \to 0$) of the crystal lattice the impurity Jahn-Teller effect hardly changes the oscillator nature of the motion. In this sense, besides the vibronic states for the Jahn-Teller degrees of freedom of the impurity center, one can also consider the vibrational states for most lattice modes. Therefore, as in the non-Jahn-Teller case, the zero-phonon line can be considered as resulting from electron transitions not altering vibrational quantum numbers, while the one-phonon, two-phonon, etc. side bands result from transitions involving one, two, etc. phonons.

The one-phonon transitions, comparable in their integral intensities with the zero-phonon transitions, have frequencies distributed over a spectral interval of the order of the width of the vibrational band, whereas the zero-phonon transitions are concentrated in one narrow line. This causes the zero-phonon line to be considerably more intense than the one-phonon satellite.

If the vibronic coupling in the initial and final electronic terms between which the optical transition takes place are significantly different, the overlap integral of the lowest vibronic states of these terms may be small, and the integral intensity of the zero-phonon line will be correspondingly small. In this case, in spite of the advantages mentioned above, the zero-phonon line will not be seen against the background of the broad multiphonon band. The consideration of the properties of the zero-phonon line (its shape, width, asymmetry and structure) has a physical meaning only for fairly weak vibronic coupling in both combining electronic terms, when the Debye-Waller factor determining the integral intensity of this line is not very small, and the line can be experimentally observed.

Note that the region of intermediate values of the vibronic coupling constants for which the zero-phonon line is still observable is superimposed on the region of large vibronic constants for which the approximation of strong coupling is already applicable (Chap. 4). This allows us, considering separately the cases of weak and strong coupling (from the standpoint of Chap. 4), to investigate the properties of the zero-phonon line using the analytical results obtained in Chap 4.

5.2.1 Vibronic Splitting and Broadening of the Zero-Phonon Line

Consider first the case of weak coupling. Suppose that for both electronic terms taking part in the optical transition the vibronic interaction is a small perturbation slightly splitting the equally spaced levels of the zero Hamiltonian (Sect. 4.1). In this case the oscillator occupation numbers may still be considered approximately as "good" quantum numbers, and one can characterize the vibronic states of the Jahn-Teller system with definite values of the phonon numbers. It is obvious that *the optical transitions* $0-0$, $1-1$, $2-2$, ..., *for which the "vibrational state" does not change, result in spectral lines occupying a narrow frequency interval with a small spread caused by the weak splitting of the excited oscillator levels* (Sect. 4.1).

Recall that the ground vibronic level of the system for both electronic terms is not split by the vibronic interaction in any order of the perturbation theory. Therefore at $T = 0$ K when the population of the excited vibronic states of the initial electronic term vanishes the vibronic structure and the width of the zero-phonon line are frozen out, and in the optical spectrum only the unsplit narrow peak of the zero-phonon transition $0-0$ remains. Many attempts [5.151–153] to obtain the splitting of the zero-phonon line at $T = 0$ K without considering low-symmetry crystal fields seem to be incorrect (note that the conclusion about the splitting of the zero-phonon line of the singlet-doublet optical transition made in [5.152] is a result of a misunderstanding: the direct substitution of the parameters of the $E \otimes e$ problem into Eq. (7) of this work taking into account the necessary symmetry conditions gives zero splitting).

As the temperature increases, the excited vibronic states of the initial electronic term are populated and the lines of the transitions $1-1$, $2-2$, etc., are added to the line of the $0-0$ transition at frequencies slightly different from that of the $0-0$ transition. This may lead to the temperature-dependent complexity of the zero-phonon line structure.

Let us consider as an example the zero-phonon line of the optical transition $A \to E$ with weak linear vibronic coupling of the E term with E vibrations (it will be recalled that the coupling constant of the A term with E vibrations is zero because of the symmetry selection rules). Substituting the equally spaced energy values of the E vibrations for the A term,

$$E_{m_1 m_2} = \hbar\omega_E(m_1 + m_2 + 1) \ ,$$

and for the E term from (4.1.22) into (5.1.2), separating the zero-phonon transitions $m_1 = n_1$, $m_2 = n_2$, from the sum (5.1.2), neglecting the small difference of the vibrational overlap integral from unity, and using the normalization condition $\sum_\gamma |\langle\psi_A|d|\psi_{E\gamma}\rangle|^2 = 1$, we obtain for the form function of the zero-phonon line of the $A \to E$ transition the expression [5.154]

$$F(\Omega) = 2\sinh^2\left(\frac{\beta_E}{2}\right) \sum_{n_1=0}^{\infty} \sum_{n_2=0}^{\infty} \sum_{+,-} \exp[-\beta_E(n_1 + n_2 + 1]$$
$$\times\, \delta\left[\Omega - \frac{1}{\hbar}(E_E - E_A - 2E_{JT}) + \omega_E(n_1 + n_2 + 1)\right.$$
$$\left. - \,\omega_\pm\left(n_1 + \frac{1}{2}\right) - \omega_\mp\left(n_2 + \frac{1}{2}\right)\right] , \tag{5.2.1}$$

where $\omega_\pm = \omega_E(1 \pm 2E_{JT}/\hbar\omega_E)$, $\beta_E = \hbar\omega_E/kT$. In (5.2.1) the frequency effect (i.e., the small difference in the frequencies of the E vibrations in the electronic states A and E) is neglected. It is seen that the zero-phonon line of the $A \to E$ transition is a superposition of two equally spaced spectra with frequency intervals $\Delta\Omega = 2E_{JT}/\hbar$ within each of them, the intensity of the spectral lines being determined by the Boltzmann exponents in (5.2.1). Thus in the case of the weak Jahn-Teller effect the zero-phonon line of an ideal vibronic system has a fine structure. Introducing the dimensionless parameters $D = E_{JT}/\hbar\omega_E$ and $x = [\Omega - (E_E - E_A - 2E_{JT})/\hbar]/\omega_E$, we rewrite (5.2.1) in the form

$$F(x) = \tanh\left(\frac{\beta_E}{2}\right) \exp\left(\frac{\beta_E|x|}{2D}\right) \sum_{n=-\infty}^{\infty} \delta(x - 2nD) \ . \tag{5.2.2}$$

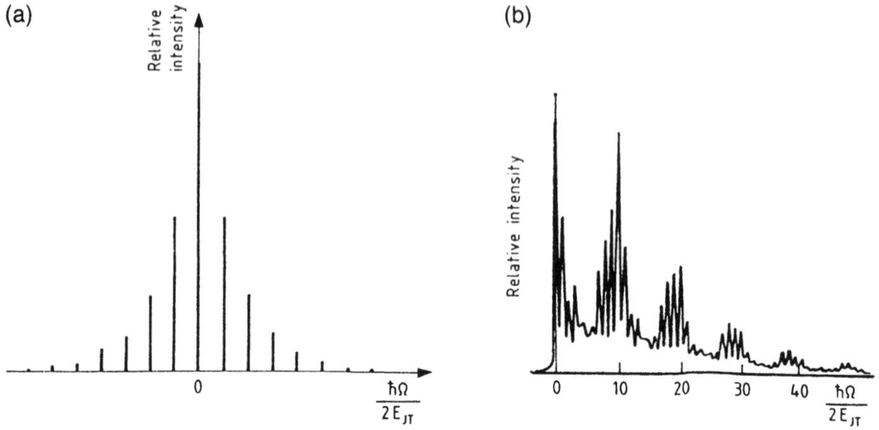

Fig. 5.13a, b. Vibronic structure of the zero-phonon line of the optical $A \to E$ transition. (**a**) In the absence of the frequency effect the equally spaced lines have a symmetric envelope which decreases exponentially towards the wings. (**b**) Taking into account the frequency effect, the line structure contains two sets of equally spaced lines, and the lines are asymmetric

The structure of the zero-phonon line described by (5.2.2) is shown in Fig. 5.13a.

For Jahn-Teller defects in crystals the vibronic structure is manifest in the zero-phonon line when the density of vibrations of the crystal contains a clearly defined peak corresponding to a local or pseudolocal vibration having predominant vibronic coupling. The decay broadening of the energy levels of these vibrations can be taken into account phenomenologically as shown in [5.155], by the introduction of the proper width $\bar{\gamma}$. If $\bar{\gamma} > \Delta\Omega$, the fine structure is not spectrally resolved and the effect under consideration is manifest in the form of a temperature broadening of the zero-phonon line. Its shape is described in this case by an envelope of the magnitudes of equally spaced lines (5.2.1) and represents a symmetric curve with exponentially falling wings:

$$F(x) \sim \tanh\left(\frac{\beta_E}{2}\right)\exp\left(-\frac{\beta_E|x|}{2D}\right). \tag{5.2.3}$$

The halfwidth of the zero-phonon line (5.2.3) $\delta\Omega = 4DkT\ln 2/\hbar$ increases linearly with temperature unlike, for example, Raman broadening, where $\delta\Omega \sim T^2$ [5.155].

In [5.154] the influence of the frequency effect (the Dushinski effect) on the fine structure of the zero-phonon line is also considered. It is shown that in cases when the frequencies of the E vibrations in the electronic states of the A and E terms do not coincide: $\Delta\omega = \omega_E(A) - \omega_E(E) \neq 0$, the fine structure of the zero-phonon line of the $A \to E$ transition can be described as before by a superposition of two spectra with constant spacings $\Delta\Omega_\pm = |\Delta\omega \pm 2E_{\rm JT}/\hbar|$. Here the zero-phonon line becomes asymmetric. Figure 5.13b shows the zero-phonon line of the $A \to E$ transition in the case of a strong frequency effect and weak vibronic coupling: $\Delta\omega/\omega_E(E) = -0.98 \times 10^{-2}$, $D = 0.25 \times 10^{-2}$, $\beta_E = 0.26$.

In the case of the electronic T term both the E and T_2 vibrations are Jahn-Teller active. In the second-order perturbation theory with vibronic interaction the Hamiltonian of the T term can be represented as the sum of two commutative terms $\hat{H}_E^{(2)} + \hat{H}_T^{(2)}$ (Sect. 4.1). Using the procedure which leads to (5.1.11) we obtain the following expression for the generating function of the singlet-triplet transition in second-order perturbation theory:

$$I(t) = I_E(t) I_T(t) \ , \tag{5.2.4}$$

where

$$I_E(t) = \exp\left(-\frac{it}{\hbar} E_{\rm JT}(E)\right) ,$$

$I_T(t)$ is the generating function for the $A \to T$ transition, provided the vibronic interaction with only T_2 vibrations is considered. It follows from (5.2.4) that the form factor of the $A \to T$ transition has the form of a convolution [of the type (5.1.14)].

Using

$$F_E(\Omega) = \frac{1}{2\pi} \int_{-\infty}^{\infty} e^{-i\Omega t} I_E(t)\, dt = \delta(\Omega + E_{\rm JT}(E)/\hbar) \tag{5.2.5}$$

we find that

$$F(\Omega) = F_T(\Omega + E_{\rm JT}(E)/\hbar) \ .$$

Thus in the second-order perturbation theory the interaction with E vibrations results in a hard shift of the band of the $E \to T$ transition by $E_{\rm JT}(E)/\hbar$ whereas the vibronic structure of the zero-phonon line is caused by the coupling with the T_2 vibrations only. Therefore we assume that there is no vibronic coupling with E vibrations, and we consider the fine structure of the zero-phonon line of the $A \to T$ transition as being due to the $(T \otimes t_2)$-type Jahn-Teller effect (Sect. 3.3).

Figure 5.14 shows the vibronic structure of the zero-phonon line of the $A \to T$ transition for three different temperatures $\beta_T = 2.74$, 1.45 and 0.37, obtained by means of summing the contributions to (5.1.2). As seen from Fig. 5.14, three qualitatively different types of zero-phonon line of the $A \to T$ transition are possible: low temperatures (Fig. 5.14a), when the intensity of the line of the $0 - 0$ transition is predominant; at intermediate temperatures (Fig. 5.14b), when the intensity of the line of the $0 - 0$ transition and the components of the fine structure nearest to it are comparable; and at high temperatures (Fig. 5.14c), when in the spectrum the line shifted towards the low-frequency region by $\Delta\Omega = \frac{3}{4}E_{\rm JT}(T)/\hbar$ from the $0 - 0$ transition is predominant. It is noteworthy that in this case the fine structure of the zero-phonon line is also equally spaced with a spacing $\Delta\Omega = \frac{3}{4}E_{\rm JT}(T)/\hbar$.

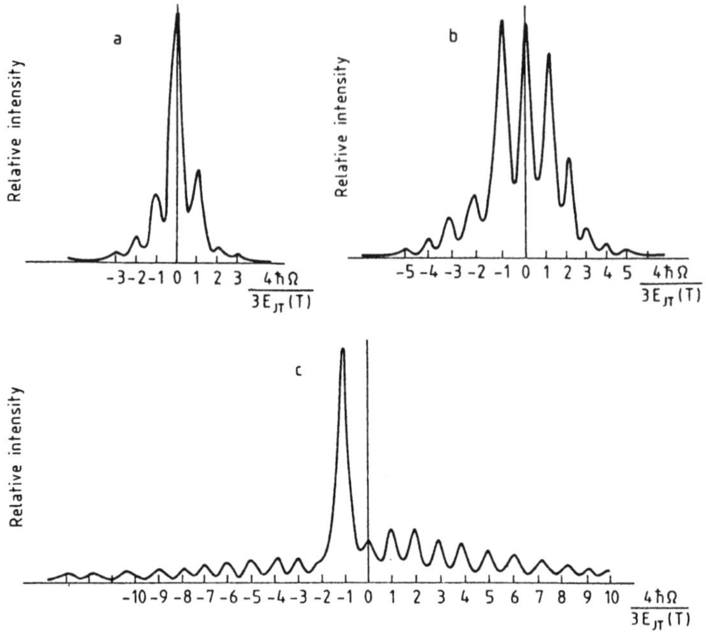

Fig. 5.14. Vibronic structure of the zero-phonon line of the optical $A \to T$ transition and its temperature variation: (**a**) $\hbar\omega/kT = 2.74$, (**b**) 1.45, (**c**) 0.37

In the case of high temperatures the width of the zero-phonon line wings increases as kT, their peak intensity decreases, and hence the zero-phonon line again becomes narrow, its position being shifted by $\frac{3}{4}E_{JT}(T)/\hbar$ in the long wavelength region with respect to the $0 - 0$ transition. The frequency effect $\omega_T(A) \neq \omega_T(T)$ can be considered in the same way as that for the $A \to E$ transition, the vibronic structure of the zero-phonon line becoming significantly more complicated. Giving no details here, we should note that the weak frequency effect causes a broadening and asymmetry of the most intense high temperature component of the zero-phonon line.

The above results are related to ideal vibronic systems and to impurity centers in crystals in the case of predominant coupling with one local vibration when the quasimolecular model is valid (Sect. 4.6.4). The contribution of the dispersion of the crystal lattice vibrations requires special consideration.

In the case of weak vibronic coupling under consideration, the generating function for the singlet-multiplet optical transition can be constructed by means of the cumulant expansions method [5.5]. Within this approach the information about the fine structure of the zero-phonon line arising due to Jahn-Teller local and pseudolocal vibronic states (Sect. 4.6) is lost, and therefore we shall discuss here only some general features of the zero-phonon lines in Jahn-Teller systems with continuous vibrational spectra, their shifts and broadening.

The Hamiltonian of the multimode $E \otimes e$ system in the second-order perturbation theory can be represented by a bilinear expression of phonon creation and annihilation operators. From this standpoint the Jahn-Teller effect for the excited E term results in a special frequency effect. Expressions for the shifts and broadening of the zero-phonon line including the quadratic terms in Bose operators in the Hamiltonian of the excited electronic term were obtained in [5.155, 156]. It turns out that for the $A \to E$ transition the temperature shift which is the first term of the cumulant expansion of the generating function is zero. An analogous result was obtained for the $A \to T$ transition as well [5.154]. The absence of a temperature shift of the zero-phonon line usually expected due to the interaction of impurity electrons with degenerate modes is an essential feature of the Jahn-Teller effect in the case of weak vibronic coupling. See also [5.157].

The cumulants of higher order result in a temperature broadening of the zero-phonon line, and this result can also be obtained directly from the expressions for the envelope of the zero-phonon line, e.g., (5.2.3). In both the case of the $A \to E$ transition and that of the $A \to T$ transition the mean-square dispersion $\langle \Delta\Omega^2 \rangle^{1/2} \sim E_{JT} kT/\hbar^2\omega$, i.e., it increases with temperature and with the vibronic coupling constant.

In the cases of intermediate and strong vibronic coupling, $E_{JT} \gtrsim \hbar\omega$, even a small increase of temperature results in a considerable broadening of the zero-phonon line, and the latter becomes invisible against the background of the wide band of the multiphonon absorption or luminescence. In these cases the zero-phonon line can be considered at rather low temperatures when it consists of practically just the one line of the $0-0$ transition.

5.2.2 Effect of External Field and Strain. Tunneling Splitting of the Zero-Phonon Line

Since the ground vibronic state of a Jahn-Teller system with arbitrary vibronic coupling has the same transformation properties as its electronic term (Sect. 4.7), all the foregoing about the absence of splitting of the $0-0$ transition line refers to the general case of an arbitrary value of the vibronic coupling constant.

The splitting of the $0-0$ transition line is possible only under the influence of low-symmetry perturbations removing the degeneracy of the ground vibronic state. When this perturbation is described by an electronic operator, i.e., when it is independent of nuclear coordinates, *the magnitude of the observed splitting is reduced compared with the primary one due to the vibronic reduction effects* (Sect. 4.7).

This can be evaluated experimentally, if the primary splitting is known. In the case of singlet-multiplet transitions it is manifest in the change of the first moment of the band (the centroid of the spectrum) under the influence of the perturbation (5.1.33). For instance, for the $A \to T$ transition under the influence of tetragonal strain the T term is split into the terms E and A of the group D_{4h}. In the parallel polarization the transition is allowed only to the electronic singlet term, whereas in the perpendicular polarization it is allowed to the E term. The change of the first moment of the band normalized to the oscillator strength is given in (5.1.36), from which we find the primary electronic splitting

$$\Delta\Omega_E = |\Delta\langle\Omega\rangle_{\parallel} - \Delta\langle\Omega\rangle_{\perp}| = \frac{3}{2\hbar}|P_E\varepsilon_\theta| \ .$$

Dividing the observable zero-phonon line splitting due to the above tetragonal deformation by $\Delta\Omega_E$ we obtain the vibronic reduction factor $K(E)$.

Similarly, in the case of trigonal deformation we find from (5.1.40)

$$\Delta\Omega_T = |\Delta\langle\Omega\rangle_{\parallel} - \Delta\langle\Omega\rangle_{\perp}| = \frac{\sqrt{3}}{\hbar}|P_T\varepsilon_T|$$

from which it follows that $\Delta\Omega_T$ describes the primary trigonal splitting. The ratio of the appropriate trigonal splitting of the zero-phonon line to $\Delta\Omega_T$ gives the vibronic reduction factor $K(T_2)$.

In this way experiments on polarization dichroism of singlet-multiplet optical transition bands and their zero-phonon lines allow the direct observation of the effect of vibronic reduction, unambiguously determined by the Jahn-Teller effect.

An experimental example of the effect of vibronic reduction of the zero-phonon line splitting is shown in Fig. 5.15 taken from [5.158] on V^{2+} : $KMgF_3$ $^4A_2 \rightarrow {}^4T_2$

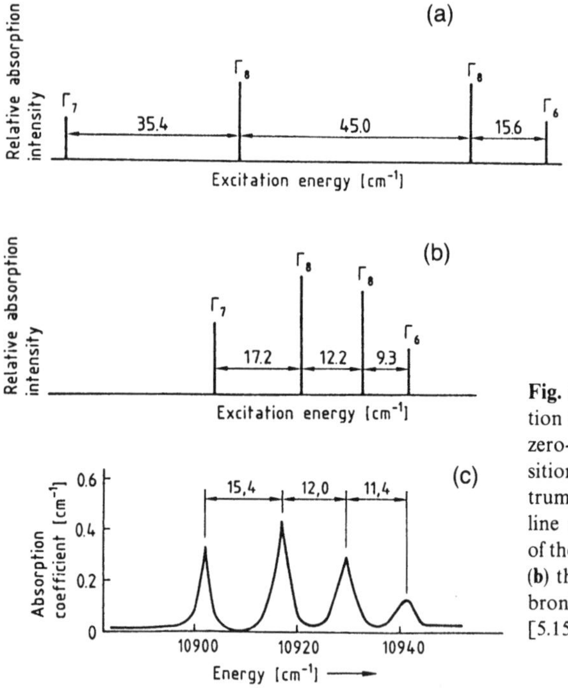

Fig. 5.15a–c. Effect of vibronic reduction of the spin-orbital splitting of the zero-phonon line of the $^4A_2 \rightarrow {}^4T_2$ transition in the optical absorption spectrum of V^{2+} : $KMgF_3$. (**a**) The expected line structure without taking account of the Jahn-Teller effect in the $^4T_2$ state; (**b**) the same, taking account of the vibronic interaction; (**c**) experiment (after [5.158])

impurity absorption. It is seen that the calculated multiplet structure with unreduced spin-orbital interaction does not correspond to the experimental zero-phonon line. The observable splitting of the latter is less than half the unquenched value of the splitting. Its interpretation in terms of the vibronic reduction, shown in Fig. 5.15b, looks quite satisfactory.

There are a great number of optical spectra data demonstrating the vibronic reduction of the zero-phonon line splitting. The analysis of the experimental data (up to 1974) is given in the monograph [5.5]. Some later works on this topic can be found in the review article [5.11]. See also [5.159, 160].

At $T = 0$ K the line of the $0 - 0$ transition is positioned at the low-frequency edge of the absorption band. On the high-frequency side of this line there are the lines of the allowed transitions $0 \rightarrow m$ from the ground vibronic state of the lower electronic term to the vibronic states of the upper electronic term. Among the latter there may also be excited components of the tunneling sublevels (Sect. 4.3), provided the appropriate transitions are allowed. If the vibronic coupling is strong enough and hence the tunneling splitting magnitude is relatively small, the spectral lines of the transition to a group of tunneling levels of similar energy are observed as one zero-phonon line split by the tunneling. In fact this line is composed of the unsplit line of the $0 \rightarrow 0$ transition accompanied by the components of the one-phonon satellite split by a strong vibronic interaction. See, for example, Fig. 4.7 and the passage after (4.3.11).

Thus the effect of tunneling splitting can be observed in a form of a corresponding splitting of the zero-phonon line.

Although the transition to some of the tunneling sublevels is often forbidden, there are always some low-symmetry perturbations which remove (partly or completely) this restriction. For impurity absorption in crystals this transition can be allowed due to random strain and inhomogeneities of the crystal structure, clustering of the impurities, the presence of other nearby defects, e.g., dislocations, and so on (Sect. 5.4).

In cases when the internal perturbations are not enough to allow the forbidden transition, one can use external low-symmetry perturbations for this purpose, e.g., uniaxial stress.

In the case of a cubic E term Jahn-Teller effect, the operator of a tetragonal deformation written in the real electronic basis $\psi_\theta(\mathrm{r})$, $\psi_\varepsilon(\mathrm{r})$ [cf. (3.1.5)], has the form

$$\hat{W} = P_E(-\varepsilon_\theta \hat{\sigma}_z + \varepsilon_\varepsilon \hat{\sigma}_x) \ , \tag{5.2.6}$$

where P_E is the parameter of the electron-deformation interaction (Sect. 5.1), $\hat{\sigma}_x$, $\hat{\sigma}_z$ are the Pauli matrices and $\varepsilon_{\Gamma_\gamma}$ are the symmetrized combinations of the components of the strain tensor.

In particular, for the deformation oriented along the [001] axis all the $\varepsilon_{\Gamma_\gamma}$ equal 0, except ε_θ. In the basis of the exact vibronic states of the tunneling E and A levels of the quadratic $E \otimes e$ problem (Sect. 4.3) we have

$$\hat{H}_0 = \begin{pmatrix} E(A) & 0 & 0 \\ 0 & E(E) & 0 \\ 0 & 0 & E(E) \end{pmatrix}, \qquad \hat{W} = P_E \varepsilon_\theta \begin{pmatrix} 0 & \mathrm{r} & 0 \\ \mathrm{r} & -q & 0 \\ 0 & 0 & q \end{pmatrix},$$

(rows and columns of both matrices: $|A\rangle$, $|E_\theta\rangle$, $|E_\varepsilon\rangle$)

where q and r are the vibronic reduction factors $K(E)$ and $K(A|E|E)$, respectively, and the values $E(E)$ and $E(A)$ in the case of strong vibronic coupling are determined by (4.3.26). The diagonalization of the Hamiltonian $\hat{H}_0 + \hat{W}$ results in

$$\begin{aligned} E_1 &= E(E) + qP_E\varepsilon_\theta\ , \\ E_{2,3} &= \tfrac{1}{2}[E(A) + E(E) - qP_E\varepsilon_\theta] \\ &\quad \pm \sqrt{[E(A) - E(E) + qP_E\varepsilon_\theta]^2 + (2\mathrm{r}P_E\varepsilon_\theta)^2}\ . \end{aligned} \tag{5.2.7}$$

In Fig. 5.16 the splitting of the zero-phonon line of the magnetic-dipole transition $^5T_{2g} \to {}^5E_g$ in a cubic Fe^{2+} : MgO impurity center observed at $T = 1.5$ K under uniaxial stress along the direction [001] is illustrated [5.161]. It is seen that by increasing the stress the forbidden line of the transition to the vibronic singlet emerges on the high-frequency side of the split line of the 0 – 0 transition. In Fig. 5.16b the full line shows the energy levels versus the stress described by (5.2.7); the points in this figure correspond to the experimental results and demonstrate excellent agreement with the predictions of the theory. It follows from this experiment that for the ^5E term of Fe^{2+} : MgO the magnitude of the tunneling splitting $\delta \approx 14\ \mathrm{cm}^{-1}$, $qP_E \approx 21 \times 10^3\ \mathrm{cm}^{-1}$, $|\mathrm{r}/q| \approx 1.2$.

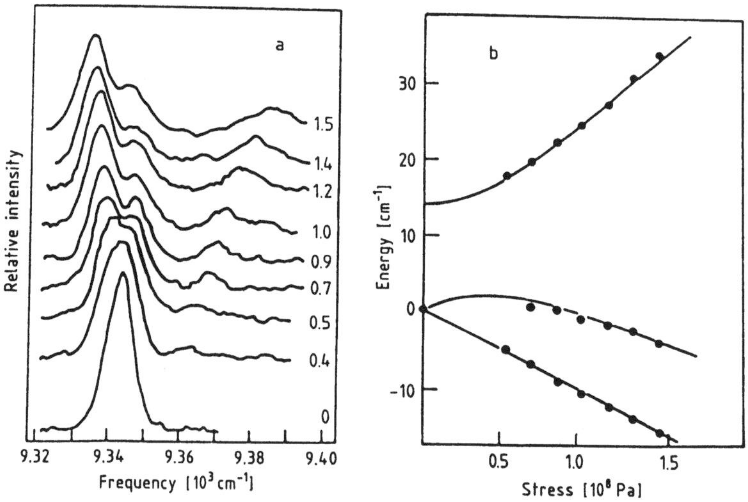

Fig. 5.16a, b. Splitting of the zero-phonon line of the optical $^5T_2 \to {}^5E$ transition in the impurity centre Fe^{2+} in the MgO crystal under uniaxial stress along the [001] axis. (**a**) Line shape at $T = 1.5$ K and different stress values shown at the right in 10^8 Pa; (**b**) observed (dots) and theoretically predicted (solid lines) absorption maximum positions. (After [5.161]

The first experiments in which the tunneling splitting of the zero-phonon line of the $A \rightarrow E$ transition was observed by the piezospectroscopic method were performed in 1965 and reported in [5.162]. For Eu^{2+} impurity centers in CaF_2 and SrF_2 crystals, the values $3\Gamma \approx 15.3\ cm^{-1}$ and $3\Gamma \approx 6.5\ cm^{-1}$, respectively, were obtained, and for Sm^{2+} impurities in CaF_2 and SrF_2 the magnitudes $27\ cm^{-1}$ and $26\ cm^{-1}$ were evaluated. The interpretation of these experiments in terms of tunneling splitting was given in [5.163].

The tunneling splitting of the zero-phonon line was observed directly also for the GR-1 line of absorption of the neutral vacancy in the diamond crystal [5.164, 165]. The magnitude of the tunneling splitting in this case is approximately $64\ cm^{-1}$. See also [5.103–108].

In conclusion to this section we note that in the limit of strong vibronic coupling the tunneling splitting may be comparable to or even smaller than that of the random deformation of the lattice. In this case the role of the latter can be reduced to a simple inhomogeneous broadening of the tunneling components of the spectrum. Averaging over the orientations of the random deformations, mixing the tunneling states, may result in a complicated spectral structure for the zero-phonon line [5.159, 166] (the role of random deformations in the EPR spectra is discussed in Sects. 5.4 and 5.5).

5.3 Vibronic Effects in Infrared Absorption and Raman Spectra

In molecular spectroscopy with nondegenerate electronic states (e.g., in diatomics) the division of the spectra into electronic, vibrational and rotational ones is widely accepted. This division is based on the difference between the energies of electronic, vibrational and rotational transitions [Ref. 5.167; Sect. 82] and corresponding differences in the frequency regions of absorbed electromagnetic radiation.

Even from the simplest qualitative considerations it follows that this relatively simple treatment of the spectra is invalid in the case of Jahn-Teller systems. For instance, in the case of strong vibronic coupling when adiabatic separation of the potential surface sheets can be accomplished, the Jahn-Teller effect results in the following two obvious consequences.

First, the shape of the adiabatic potentials is, as a rule, essentially different from paraboloid (Chap. 3). Because of this essential anharmonicity many overtones are allowed in the infrared spectrum, their spacing being not at all constant.

Second, along with the dipole moment of the nuclei $\boldsymbol{D}_n(Q)$ the operator of the electronic dipole moment $\boldsymbol{D}_e(r)$ contributes to the probability of the electric-dipole[1] infrared (IR) absorption resulting in transitions between different sheets of the adiabatic potential. Since the energy gap between the sheets is of the order of E_{JT} (Chap. 3) the frequencies of the electronic transitions between the sheets

[1] In this section electric-dipole transitions are considered, except where other cases are specified.

of the adiabatic potentials of the same Jahn-Teller term fall in the same frequency region as the intrasheet "vibrational" transitions do. This circumstance prevents the separation of the electronic IR spectra from the vibrational ones. The vibronic nature of the states participating in the transition makes the term "vibronic IR spectra" much more suitable.

5.3.1 Vibronic Effects in Vibrational Infrared Spectra

The operator of the electric-dipole transition is $\boldsymbol{e}\cdot\boldsymbol{D}$, where $\boldsymbol{e}$ is a unit vector of the electromagnetic wave polarization and

$$\boldsymbol{D} = \boldsymbol{D}_{\mathrm{e}}(\mathrm{r}) + \boldsymbol{D}_{\mathrm{n}}(Q) \tag{5.3.1}$$

is the vector of the dipole moment of the electrons and nuclei of the polyatomic system.

The dipole moment of the nuclear framework $\boldsymbol{D}_{\mathrm{n}}(Q)$ can be expressed as

$$\boldsymbol{D}_{\mathrm{n}}(Q) = \boldsymbol{D}_{\mathrm{n}}^{(0)} + \sum_{\bar{\Gamma}\bar{\gamma}} Z_{\bar{\Gamma}} Q_{\bar{\Gamma}\bar{\gamma}} \boldsymbol{n}_{\bar{\Gamma}\bar{\gamma}} \ ,$$

where the summation is performed over the irreducible representation $\bar{\Gamma}$ to which the components of the vector belong; $\boldsymbol{n}_{\bar{\Gamma}\bar{\gamma}}$ are the symmetrized combinations of the unit vectors of the coordinate system related to the high-symmetry configuration Q_0 of the polyatomic system (Sect. 2.2); $Z_{\bar{\Gamma}}$ is the effective charge of the appropriate vibrational mode arising as a result of passing from the Cartesian displacements of the atoms to the normal displacements and $\boldsymbol{D}_{\mathrm{n}}^{(0)}$ is the dipole moment of the nuclei fixed at the initial nuclear configuration Q_0. The operator $\boldsymbol{D}_{\mathrm{n}}^{(0)}$ obviously causes only pure rotational transitions between the states of a rigid top. Molecules for which $\boldsymbol{D}_{\mathrm{n}}^{(0)} \neq 0$ are called rigid-dipole molecules. For simplicity we restrict the consideration to high-symmetry molecules for which $\boldsymbol{D}_{\mathrm{n}}^{(0)} = 0$.

The scalar product $\boldsymbol{e}\cdot\boldsymbol{D}_{\mathrm{n}}$ can be written then in the form of a convolution of irreducible tensor operators

$$\boldsymbol{e}\cdot\boldsymbol{D}_{\mathrm{n}} = \sum_{\Gamma\gamma} Z_{\Gamma} e^{\dagger}_{\Gamma\gamma} Q_{\Gamma\gamma} \ . \tag{5.3.2}$$

The normal coordinates in the sum (5.3.2) are called *dipole-active.*

Of course, the vibronic effects in the IR spectra occur only in cases when the electromagnetic radiation interacts with the Jahn-Teller (pseudo-Jahn-Teller) degrees of freedom of the polyatomic system. If one assumes that $\boldsymbol{D}_{\mathrm{e}}(\mathrm{r}) = 0$ and the total interaction of the radiation with the molecule is described by (5.3.2) then the vibronic IR spectra are possible only for molecules for which among the dipole-active normal vibrations there are Jahn-Teller–active ones for the ground electronic term. In other words vibronic effects are possible in the IR spectra only for dipole-unstable systems (Sect. 3.4.2).

Two circumstances are of interest here. First, the ground electronic term can interact with Jahn-Teller–active vibrations absent in the sum (5.3.2). All the Jahn-Teller vibrations interacting with the same electrons interact indirectly with each other forming a coupled Jahn-Teller system. Therefore *the electromagnetic irradiation interacting only with dipole-active vibrations causes a response of the whole Jahn-Teller system, including those Jahn-Teller vibrations which are not dipole active.* In particular, weak Jahn-Teller interaction allows IR absorption in the region of nondipole (forbidden) Jahn-Teller vibrations. This circumstance was first noted in [5.168].

Second, in nondegenerate molecules in the harmonic approximation the operator (5.3.2) is subject to selection rules $v_{\Gamma\gamma} \to v_{\Gamma\gamma} \pm 1$, where $v_{\Gamma\gamma}$ is the vibrational quantum number of the oscillator $Q_{\Gamma\gamma}$. Because of the equal spacing of the energy spectrum of the normal vibrations, all the transitions $v_{\Gamma\gamma} \to v_{\Gamma\gamma} + 1$ correspond to the absorption of the same frequency $\Omega = \omega_\Gamma$. Therefore the temperature occupation of the excited levels does not cause the formation of new lines in the IR absorption while the probabilities of allowed transitions are redistributed between the partial contributions to each line of the fundamental tone, the total intensity remaining unchanged. From this it follows that in the harmonic approximation the IR spectra of nondegenerate molecules are independent of temperature. Small anharmonicity results in a temperature transfer of the intensity from the lines of the fundamental tone to that of the overtones, but this dependence is, as a rule, negligible.

In contrast to this the energy spectrum in the Jahn-Teller molecules is strongly unequally spaced (Chap. 4) and *the population at finite temperature of excited states results in the formation of new spectral lines.* This is manifest as a *strong temperature dependence of the vibronic* IR *spectra.*

Consider now the effects caused by the interaction of electromagnetic radiation with the dipole moment of electrons. As in (5.3.2) the scalar product $\boldsymbol{e}\cdot\boldsymbol{D}_e(\mathrm{r})$ can be written in the form of a tensor convolution

$$\boldsymbol{e}\cdot\boldsymbol{D}_e(\mathrm{r}) = \sum_{\bar{\Gamma}\bar{\gamma}} e^{\dagger}_{\bar{\Gamma}\bar{\gamma}} D_{\bar{\Gamma}\bar{\gamma}}(\mathrm{r}) \ , \tag{5.3.3}$$

where the summation is performed over the irreducible representations $\bar{\Gamma}$ of the vector representation. The matrix elements of the operator $D_{\bar{\Gamma}\bar{\gamma}}(\mathrm{r})$ calculated from the Jahn-Teller states of the electronic term Γ are nonzero if $\bar{\Gamma} \in [\Gamma^2]$. Therefore the interaction of the radiation with the dipole moment of the electrons causes a response of the Jahn-Teller system only in the case when the symmetrized square $[\Gamma^2]$ contains dipole-active representations. Since, on the other hand, the irreducible representations contained in $[\Gamma^2]$, except the trivial one A_1, determine the Jahn-Teller normal vibrations (Sect. 2.3) we come to the condition formulated above: the vibronic effects in the IR spectra occur only for dipole-unstable polyatomic systems.

As mentioned in Sect. 3.4 the occurrence of a dipole instability is also possible due to the vibronic mixing of a given term Γ with one of similar energy, Γ', i.e., due to the pseudo-Jahn-Teller effect on dipole-active vibrations, provided the product $\Gamma \times \Gamma'$ contains dipole-active representations. This case is more general and can be realized for both degenerate (but noninteracting with dipole-active vibrations) and nondegenerate electronic terms.

The coefficient of electric dipole IR absorption is described by the expression (5.1.1), where $d = \boldsymbol{e} \cdot \boldsymbol{D}$ with the operator $\boldsymbol{D}$ from (5.3.1) and $\Gamma_2 = \Gamma_1$ corresponds to the ground electronic term. It is convenient to write the form factor in the form of the Fourier integral (5.1.5) with the generating function (5.1.8). As a result we come to what is known as the Kubo formula

$$F(\Omega) = \frac{1}{2\pi} \int_{-\infty}^{\infty} \mathrm{e}^{-\mathrm{i}\Omega t} \langle d^\dagger d(t) \rangle_\Gamma \, dt \ , \tag{5.3.4}$$

where the function $d(t)$ is determined by the usual Heisenberg representation with the Hamiltonian $\hat{H}(\Gamma)$. Substituting $d = \boldsymbol{e} \cdot \boldsymbol{D}$ with $\boldsymbol{D}$ from (5.3.1) into (5.1.8) for $\Gamma_2 = \Gamma_1$ we obtain the generating function of the IR absorption containing the four terms

$$\begin{aligned} \langle d(t) d \rangle_\Gamma = \sum_{\alpha\beta} e_\alpha e_\beta \{ &\langle D_{\mathrm{e}\alpha}(t) D_{\mathrm{e}\beta} \rangle_\Gamma + \langle D_{\mathrm{n}\alpha}(t) D_{\mathrm{n}\beta} \rangle_\Gamma \\ &+ \langle D_{\mathrm{e}\alpha}(t) D_{\mathrm{n}\beta} \rangle_\Gamma + \langle D_{\mathrm{n}\alpha}(t) D_{\mathrm{e}\beta} \rangle_\Gamma \} \ , \qquad \alpha, \beta = x, y, z \ . \end{aligned}$$

The first one corresponds to the IR absorption by electrons. For a nondegenerate state this term is non-zero only in the case of rigid-dipole molecules, for which it causes the pure rotational IR spectrum of states nearest in energy to the ground vibrational state (Sect. 5.3.2). In the (pseudo-)Jahn-Teller case it gives rise to transitions between all the vibronic states. The second term in the above sum describes the IR absorption caused by the nuclear dipole moment. In the non-Jahn-Teller case it produces the transitions in the vibrational subsystem with a change in the quantum numbers of dipole-active vibrations. The last two terms in the above sum represent the interference effects which are absent in the non-degenerate case. In Jahn-Teller systems all four terms describe the transitions between all the vibronic states (with the restrictions imposed by the selection rules from group theory) and have the same order of magnitude throughout the spectral range.

In trigonal systems (e.g., in the triatomic molecule X_3) with a linear Jahn-Teller effect for the E term the actual symmetry is axial (Sect. 3.2.1) and it is convenient to write (5.3.3) in the form

$$\boldsymbol{e} \cdot \boldsymbol{D}_{\mathrm{e}}(\mathrm{r}) = e_z D_z(\mathrm{r}) + e_- D_+(\mathrm{r}) + e_+ D_-(\mathrm{r}) \ .$$

Here $e_\pm = (e_x \pm \mathrm{i} e_y)/\sqrt{2}$ and $D_\pm = (D_x \pm \mathrm{i} D_y)/\sqrt{2}$ have the transformation properties of the momentum $J_z = \pm 1$. In terms of matrices of the electronic operators $D_\pm(\mathrm{r})$ calculated using the electronic states of the E term $\psi_\pm(\mathrm{r})$, see (3.1.5), we have

$$e \cdot \boldsymbol{D}_e = M(e_+\hat{\sigma}_- + e_-\hat{\sigma}_+) \ , \qquad (5.3.5)$$

where M is the reduced matrix element and $\sigma_\pm$ are the matrices (3.1.1). The matrix elements $\langle\psi_\pm|D_z|\psi_\pm\rangle$ are zero owing to the symmetry with respect to reflection in the plane of the triangle.

For the example of the X_3 molecule under consideration the operator (5.3.2) can in principle be reduced to an analogous form

$$e \cdot \boldsymbol{D}_n = Z_E(e_+Q_- + e_-Q_+) \ . \qquad (5.3.6)$$

In more complicated systems for which the reflection symmetry in the plane σ_h is absent, the sum (5.3.2) contains terms of the type $Z_A e_z Q_z$ and the sum (5.3.3) contains terms like $M_z e_z \sigma_z$.

The operators $Q_\pm$ and $\hat{\sigma}_\pm$ causing transitions in the Jahn-Teller subsystem have the transformation properties of the momentum $J_z = \pm 1$. From this we can deduce simple selection rules for the quantum number j of the operator $\hat{J}_z$ for the IR transition in the linear $E \otimes e$ system: $j \to j \pm 1$. In particular at $T = 0$ K when only the ground vibronic state $|0, \pm 1/2\rangle$ is populated (Chap. 4), the following changes of j are allowed in transitions: $-\frac{1}{2} \to +\frac{1}{2}$, $-\frac{1}{2} \to -\frac{3}{2}$, $\frac{1}{2} \to \frac{3}{2}$, $\frac{1}{2} \to -\frac{1}{2}$.

The frequency structure is the same for electronic and vibrational dipole transitions because they have the same selection rules. The relative contribution to the intensity of the spectral lines is determined by the effective charge Z_E and the reduced matrix element M for each concrete system. Unlike in the nondegenerate case where the contributions of the electric dipole moment and nuclear dipole moment in the IR absorption can be separated since they have different frequency ranges, in the (pseudo-)Jahn-Teller case they fall in the same frequency range and in the general case cannot be separated from one another. Nevertheless in some special cases these contributions can be separated if the sums (5.3.2) and (5.3.3) contain terms having significantly different values of the reduced parameters Z_Γ and M_Γ for given Γ. Then these contributions may be separated by considering the IR absorption of polarized light and taking into account the appropriate selection rules from group theory.

The vibronic IR spectrum for arbitrary values of the vibronic coupling constant can be obtained from (5.1.2) using the results of the numerical solution of the vibronic equations (Sect. 4.4). Figure 5.17 illustrates the vibronic structure of the IR absorption spectra for the linear $E \otimes e$ system obtained by means of computer calculations [5.169]. The set of lines corresponding to transitions to the vibronic states with the same j value form an almost equally spaced paling (the spectral lines above the abscissa correspond to the values $|j| = \frac{1}{2}$, and those below the abscissa correspond to $|j| = \frac{3}{2}$). The nature of the accidental equal spacing of the lines is discussed in Sects. 4.2 and 4.4.

The numerical results given in Fig. 5.17 can be understood if one employs the approximate solutions of the vibronic equations obtained in Chap. 4. In the case of weak vibronic coupling the wave functions obtained from perturbation theory, the correct (symmetry-adapted) zeroth-order functions $|nlj\rangle$ can be used. In this

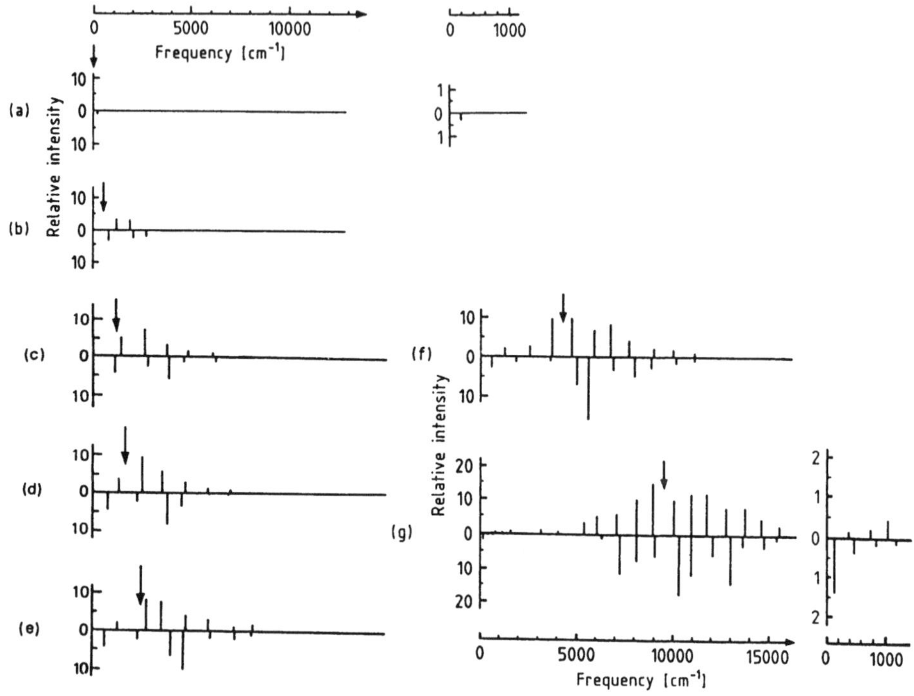

Fig. 5.17. Intensity of the vibronic components of the IR spectrum of a linear $E \otimes e$ system obtained by numerical calculations for **(a)** $E_{JT}/\hbar\omega_E = 0$, **(b)** 0.125, **(c)** 0.25, **(d)** 0.375, **(e)** 0.5, **(f)** 1.0 and **(g)** 2.5. The lines above and below the abscissa correspond to the transitions $|j| = 1/2 \rightarrow |j| = 1/2$ and $|j| = 1/2 \rightarrow |j| = 3/2$, respectively. The lowest frequency of intersheet transitions is shown by an arrow. (After [5.169])

approximation the vibrational quantum number n is still a good quantum number. The operators $Q_\pm$ change n by a unit, and therefore they cause transitions from the ground vibronic state (at $T = 0$ K) to only the first excited vibrational level weakly split into two sublevels by the vibronic interaction (Sect. 4.1). In this approximation the matrix elements of the transition are [5.169]

$$\left\langle 0, 0, \pm\frac{1}{2}\middle| Q_\pm \middle| 1, 1, \pm\frac{3}{2}\right\rangle = \left\langle 0, 0, \pm\frac{1}{2}\middle| Q_\pm \middle| 1, 1, \mp\frac{1}{2}\right\rangle = \sqrt{\frac{\hbar}{2\omega_E}} .$$

Accordingly the fundamental line of the E vibrations in the IR spectrum is slightly split by $\Delta\Omega = 2E_{JT}/\hbar$ (Fig. 5.17b).

The electronic dipole moment does not change the vibrational quantum number and at $T = 0$ K it just causes transitions between the degenerates states of the ground vibronic term. The appropriate matrix elements in the approximation of weak coupling are [5.169] $\langle 0, 0, \pm\frac{1}{2}| \hat{\sigma}_\pm |0, 0, \mp\frac{1}{2}\rangle = 1$. These transitions cannot be observed directly since the frequency is $\Omega = 0$. However, these matrix elements determine the intensity of the transitions between the rotational states of the free molecule related to the ground vibronic level (see below).

In the case of strong vibronic coupling the adiabatic states (4.2.8) can be used in (5.1.2). As mentioned above, all the IR transitions in this case can be divided into *intrasheet transitions* between the vibronic states of the same lower sheet of the adiabatic potential and *intersheet transitions* between the states of different sheets. At $T = 0$ K the lowest frequency line of the intersheet transitions has the frequency $\hbar\Omega_0 \cong \varepsilon_+(\varrho_0^{(+)}) - \varepsilon_-(\varrho_0)$ connecting the minima of the upper and lower sheets of the adiabatic potential (Sect. 4.2). Intrasheet transitions correspond to frequencies $\Omega < \Omega_0$.

It can be seen easily that $Q_\pm$ cause only the intrasheet transitions $|\pm, n, j\rangle \to |\pm, n', j \pm 1\rangle$. If one writes $Q_\pm$ in the form

$$Q_\pm = \frac{1}{\sqrt{2}}\varrho e^{\pm i\varphi} = \frac{1}{\sqrt{2}}(\varrho_0 + \mathrm{r})e^{\pm i\varphi} = \frac{1}{\sqrt{2}}\varrho_0 e^{\pm i\varphi} + \frac{1}{\sqrt{2}}\mathrm{r}e^{\pm i\varphi} \tag{5.3.7}$$

it can be seen that in the harmonic approximation the first term causes transitions with $n' = n$, while the second term causes transitions with $n' = n \pm 1$. At $T = 0$ K only transitions from the ground vibronic state remain, with matrix elements [5.169]

$$\left\langle -, 0, \pm\frac{1}{2}\middle| Q_\pm \middle| -, 0, \mp\frac{1}{2}\right\rangle = \left\langle -, 0, \pm\frac{3}{2}\middle| Q_\pm \middle| -, 0, \pm\frac{1}{2}\right\rangle \approx \frac{V_E}{\omega_E^2\sqrt{2}}, \tag{5.3.8}$$

$$\left\langle -, 0, \pm\frac{1}{2}\middle| Q_\pm \middle| -, 1, \mp\frac{1}{2}\right\rangle = \left\langle -, 0, \pm\frac{3}{2}\middle| Q_\pm \middle| -, 1, \pm\frac{1}{2}\right\rangle \approx \frac{1}{2}\sqrt{\frac{\hbar}{\omega_E}}. \tag{5.3.9}$$

The transitions (5.3.8) correspond to two intense lines at the frequencies $\Omega = 0$ and $\Omega = \hbar\varrho_0^{-2}$. Of course the line $\Omega = 0$ cannot be observed directly but it is related to the transitions between the rotational states of a free molecule (see below).

The transitions (5.3.9) correspond to two weak lines at $\Omega = \omega_E$ and $\Omega = \omega_E + \hbar/\varrho_0^2$, their intensities being $2E_{\mathrm{JT}}/\hbar\omega_E$ times weaker than those of the lines (5.3.8). The anharmonicity of the potential surface $\varepsilon_-(\varrho)$ also allows overtones with $n' > n + 1$. At first their intensities decrease with n in the region of weak anharmonicity (small n) and then when passing to the region of strong anharmonicity (large n) begin to increase (Fig. 5.17).

The electronic operators $\sigma_\pm$ cause both the intrasheet transitions $|\pm, n, j\rangle \to |\pm, n, j \pm 1\rangle$ and the intersheet ones $|\pm, n, j\rangle \to |\mp, n', j \pm 1\rangle$. The intensity of the latter is determined by the Franck-Condon overlap integrals $\langle \chi_n(\varrho - \varrho_0)| \chi_{n'}^{(+)}(\varrho - \varrho_0^{(+)})\rangle$. In this sense the intersheet transition is reduced to the usual electronic transition between electron-vibrational states of two singlet electronic terms with significantly anharmonic potential surfaces.

At $T = 0$ K when only the ground vibronic state is populated the following two types of intersheet transitions are possible: $|-,0,\pm\frac{1}{2}\rangle \to |+,n,\mp\frac{1}{2}\rangle$ and $|-,0,\pm\frac{1}{2}\rangle \to |+,n,\pm\frac{3}{2}\rangle$. As mentioned in Sects. 4.2 and 4.4 the energy spectrum of vibronic states of the linear $E \otimes e$ system with fixed j values is almost equally spaced and therefore each of the above two types of transition results in an almost equally spaced series of spectral lines with a bell-shaped envelope of the Pekar type. The frequency of the envelope maximum is that of the Franck-Condon transition between the sheets (Fig. 5.17):

$$\hbar\Omega_{\max} = \varepsilon_+(\varrho_0) - \varepsilon_-(\varrho_0) = 4E_{JT} \ .$$

Vibronic effects in the IR absorption in multimode Jahn-Teller systems can be described in the framework of the same approximations of the weak and strong coupling cases.

As mentioned above in the case of weak vibronic coupling, the transitions caused by the electronic dipole moment cannot be observed directly and therefore one can assume that $\boldsymbol{e}\cdot\boldsymbol{D}$ is determined by (5.3.2). In particular, for an uncharged impurity in a homopolar crystal with a trigonal E term in the ground state, the operator $\boldsymbol{e}\cdot\boldsymbol{D}$ acquires the form (5.3.6), where $Q_\pm$ are the symmetrized displacements of the atoms of the nearest coordination sphere. Substituting (5.3.6) into (5.3.4) we obtain

$$F(\Omega) = Z_E^2(\langle Q_\theta | Q_\theta \rangle_\Omega + \langle Q_\varepsilon | Q_\varepsilon \rangle_\Omega) \ ,$$

where $\langle Q_{\Gamma\gamma} | Q_{\Gamma\gamma} \rangle_\Omega$ is the spectral density of the appropriate correlation function. Thus the IR spectrum of the multimode Jahn-Teller system reproduces the density of vibronic states substantially redetermined by the vibronic coupling. In particular it contains local and pseudolocal resonances which are due to the Jahn-Teller effect [5.137].

In the case of strong vibronic coupling in the multimode system as in the ideal one considered above the approach based on separation of the sheets of the adiabatic potential is applicable. The broad band of the IR absorption is composed of intra- and intersheet transitions. The latter are due to the operator of the electronic dipole moment and differing slightly from singlet-singlet electronic transitions they result in a multiphonon band of the Pekar type. The zero-phonon line of this band is positioned at the frequency $\Omega_0 \approx E_{JT}$. The main contribution to the intrasheet transitions comes from $d = eD_n$, the operator of the interaction of light with the dipole moment of the nuclei which, considering (5.3.7), can be written in the form $d = d_0 + d_1$, where

$$d_0 = \frac{Z_E}{\sqrt{2}}\varrho_0(e_+\mathrm{e}^{-\mathrm{i}\varphi} + e_-\mathrm{e}^{\mathrm{i}\varphi}) \ , \qquad d_1 = \frac{Z_E}{\sqrt{2}}\mathrm{r}(e_+\mathrm{e}^{-\mathrm{i}\varphi} + e_-\mathrm{e}^{\mathrm{i}\varphi}) \ .$$

In the case of strong vibronic coupling the contribution of d_0 is obviously predominant since $\varrho_0 \sim V_E$.

The operator d_0 causes transitions between rotational states of the system with $j' = j \pm 1$. Since the states with different j values correspond to different equilibrium coordinates of the trough [see (4.2.9) and (4.6.27)], this transition is accompanied by a pulse of radial vibrations of the nuclei. In a sense the phenomenon described above is analogous to the Mössbauer effect, with the distinction that in the case under consideration the momentum is transferred, rather than the photon impulse. As in the case of the Mössbauer effect the IR absorption spectrum contains an intense line at the frequency $\Omega_-^{(0)}$, see (4.6.27), which corresponds to the process "without recoil", and weak phonon satellites reproducing the modified density of radial vibrations [5.170].

When considering the quadratic terms of the vibronic interaction in the case of the strong vibronic coupling, three deep minima occur at the bottom of the trough of the lowest sheet of the adiabatic potential (Sect 3.1) and the nuclei motion is localized at the bottom of these minima (Sect. 4.3). Neglecting tunneling, the system remains in the minima for an infinitely long time and this corresponds to a distorted nuclear configuration of the polyatomic system. For instance, the molecule X_3 has the form of an isosceles triangle (symmetry C_{2v}), and octahedral molecule ML_6 acquires the symmetry D_{4h}, etc. The degeneracy of the E vibrations at the minimum point is removed (Sect. 3.1) owing to the lowering of the symmetry. Therefore when tunneling is neglected the IR spectra of Jahn-Teller systems coincide with the usual IR spectra of normal molecules in a lower symmetry [5.168]. Here the lines of the fundamental tone are split in accordance with the splitting of the frequencies in the minima.

Allowing for tunneling the spectral lines of the IR absorption are split and the allowed transitions are determined by the usual selection rules. For instance, for a quadratic $E \otimes e$ system the ground vibronic level is a doublet. Since in the group D_{3h} we have $E' \times E' = A_1' + A_2' + E$ it is obvious that even in the dipole approximation the IR transitions from the ground vibronic E' state to all excited states are allowed. In particular at the frequency $\Omega = 3\Gamma/\hbar$, i.e., in the microwave range, an intense line corresponding to the transition between the tunneling sublevels of the ground state can be observed (Sect. 4.3).

Such transitions accompanied by an anomalous microwave absorption were predicted for the first time in [5.171]. The observed dipole microwave losses for Mn^{3+} ions (E term) in an yttrium iron garnet [5.172], which have a clear frequency dependence with a maximum, can be considered as those due to such transitions. This is also confirmed by the dependence of Ω_{max} on temperature ($\Omega_{max} = 15$ GHz at 37 K and $\Omega_{max} = 56$ GHz at 58 K) which agrees qualitatively with the theory of tunneling splitting. Indeed, going by the temperature population of the excited vibrational states in the minima for which the potential barrier is smaller, the tunneling splitting magnitude should increase. A relaxation mechanism of the temperature shift of Ω_{max} due to the interaction of the impurity with the lattice is also possible [5.173].

The special features of the vibronic IR spectra described above are general for all Jahn-Teller systems. Anomalous IR spectra for molecular systems with

$E \otimes (b_1 + b_2)$, $T \otimes (e + t_2)$ and $\Gamma_8 \otimes (e + t_2)$ type effects in the case of weak vibronic coupling are considered in [5.174]. Strong vibronic coupling with one of the two types of vibrations in the $\Gamma_8 \otimes (e + t_2)$ system and weak coupling with the other one are considered in [5.175]. In these works [5.169, 175] detailed tables of optically active vibrations for molecular systems with different symmetry are also given.

Vibronic IR spectra of systems with intermediate vibronic coupling can be obtained from numerical solutions of the vibronic equations (Sect. 4.4) in a way similar to that used to determine the spectral lines given in Fig. 5.17. For instance, in [5.176] the IR spectra of molecular systems with a ^3T term including the spin-orbital and linear vibronic interactions with E and T_2 vibrations were obtained.

A simple case of the pseudo-Jahn-Teller effect for two singlet electronic term of similar energy mixed by one vibration (Sect. 3.4) was considered in [5.177, 178]. As in the usual Jahn-Teller case the IR spectra of such systems have a complicated unequally spaced structure with a complicated temperature and polarization dependence.

5.3.2 Rotational Vibronic Infrared Spectra of Free Oriented Molecules

The discussion of the IR spectra of free molecules cannot be complete without considering transitions between the rotational levels.

Consider first a simple Jahn-Teller system, the X_3 molecule. As is well known, any molecule with D_{3h} symmetry in a nondegenerate electronic state has no dipole moment (i.e., it is not a rigid-dipole molecule) and therefore it does not display a purely rotational absorption of electro-magnetic radiation. However, in degenerate electronic states such a molecule may have a dipole moment linear in the field (a linear Stark effect) since the external electric field removes the electronic degeneracy. The Hamiltonian of the Stark interaction has the form of (5.3.5). This operator may cause purely rotational transitions too [5.169]. On the other hand, in the absence of vibronic coupling the electric dipole interaction with the nuclei (5.3.6) does not lead to purely rotational transitions since in this case the vibrational quantum number n is subject to the selection rule $n' = n \pm 1$. The vibronic interaction removes this restriction on n and allows purely rotational electric-dipole absorption based on the nuclear dipole moment.

This conclusion has a clear-cut physical sense. In systems without an inversion center, including the molecule X_3 under consideration, the dipole moment in the minima of the adiabatic potential can occur due to the Jahn-Teller effect with the participation of dipole-active vibrations. It is obvious that the contribution of the nuclear dipole moment to the electric-dipole rotational IR spectra is larger, the larger the magnitude of this moment in the minima of the adiabatic potential, i.e., the stronger the vibronic interaction. In the case of strong vibronic coupling the contribution of the nuclear dipole moment to the purely rotational transitions can be dominant.

The Jahn-Teller effect influences both the intensities and the positions of the spectral lines of the rotational IR absorption. As stated above (Sect. 3.2) in the degenerate electronic states there is in general an electronic momentum which in the basis of the orbital states $\psi_{\pm}(\mathrm{r})$, see (3.1.5), is given by the matrix

$$\hat{L}_z = \zeta_e \hat{\sigma}_z \ . \tag{5.3.10}$$

Here ζ_e is the reduced matrix element [do not confuse the operator of the electron's real rotation (5.3.10) with the infinitesimal operator $\hat{\sigma}_z$ from (3.2.9)]. The total momentum of the internal rotation in the molecule is composed of the electronic momentum (5.3.10) and vibrational momentum (3.2.25)

$$\hat{J}_z^{\text{intern}} = \zeta_e \hat{\sigma}_z + \frac{1}{\hbar}(Q_\theta P_\varepsilon - Q_\varepsilon P_\theta) \tag{5.3.11}$$

[do not confuse it with the electronic-vibrational momentum (3.2.24)].

Taking into account the total momentum of the internal rotation in the Coriolis interaction, the Hamiltonian of the rotational motion of a symmetric top-type molecule can be written in the form [Ref. 5.167; Sect. 104]

$$\hat{H}_{\text{rot}} = B(J - \hat{J}^{\text{intern}})^2 + (A - B)(J_z - \hat{J}_z^{\text{intern}}) \ , \tag{5.3.12}$$

where A and B are the rotational constants[2] and J is the preserved total momentum of the top. If the vibronic-rotational interaction is not considered, the total wave function of the molecule can be written in the form of a product of the vibronic function $|nj\rangle$ (n is the quantum number of the radial motion, $j = \pm\frac{1}{2}$, $\pm\frac{3}{2}, \dots$ is the quantum number of the electronic-vibrational momentum (3.2.24) in the linear $E \otimes e$ system, see Sects. 4.1 and 4.2) and the rotational wave function of a symmetric top $|JKM\rangle$:

$$|njJKM\rangle = |nj\rangle|JKM\rangle \ . \tag{5.3.13}$$

The required rotational energy is $\hat{H}_{\text{rot}}$ averaged over the states (5.3.13). Omitting the constants independent of the rotational quantum numbers we have [5.169, 179, 180]

$$E_{\text{rot}} = BJ(J + 1) + (A - B)K^2 - 2AK\zeta \ , \tag{5.3.14}$$

where the constant

$$\zeta = \langle nj|\hat{J}_z^{\text{intern}}|nj\rangle = (\zeta_e - \tfrac{1}{2})\langle nj|\hat{\sigma}_z|nj\rangle + j \tag{5.3.15}$$

describes the Coriolis interaction.

[2] For the sake of simplicity we neglect the dependence of the molecule's moment of inertia on nuclear displacements. Additional effects related to this dependence are considered in [5.179].

For the rotational levels accompanying the ground vibronic term ($n = 0$, $j = \pm 1/2$) the matrix element $\langle nj|\hat{\sigma}_z|nj\rangle$ is expressed by the vibronic reduction factor p from (4.7.20) and therefore

$$\zeta = p(\zeta_e - 1/2)\,\mathrm{sign}(j) + j \ . \tag{5.3.16}$$

In particular in the case of strong vibronic coupling when $p = 0$, ζ equals j. Thus the expression (5.3.14) for the energy of the rotational levels agrees with the usual result for nondegenerate molecules but the constant of the Coriolis interaction changes significantly from one vibronic level to another [5.179].

The selection rule for rotational transitions $\Delta K = \pm 1$ follows from the transformation properties of the operators $\hat{\sigma}_\pm$ and $Q_\pm$. For instance, for the Q transitions ($J = J'$) we obtain from (5.3.14) a set of lines with frequencies corresponding to the energy gaps [5.169]

$$\Delta E_K = [A(1 + 2\zeta) - B](2K + 1) \ . \tag{5.3.17}$$

With the wave functions (5.3.13) known one can determine the intensities of the lines of induced dipole transitions caused by the operators (5.3.5) and (5.3.6). As mentioned above, in the case of strong vibronic coupling the contribution of the operator (5.3.6) is dominant, and therefore the probability of rotational transitions between the states which accompany the ground vibronic doublet is determined by the matrix element (5.3.8). All the foregoing means that due to the dipole instability the Jahn-Teller systems which have no nuclear dipole moment show the properties of rigid-dipole systems with a dipole moment corresponding to a distorted molecule in the minimum of the adiabatic potential. From this it follows that *the division of all the molecules into those having a proper dipole moment (rigid-dipole molecules) and those without is, to a large degree, convention.*

This conclusion was first drawn in [5.181] on the basis of the temperature dependence of the averaged dipole moment of freely orienting Jahn-Teller molecules with a dipole instability. In accordance with the results obtained in [5.182] the difference between rigid-dipole and symmetric molecules depends on temperature. The approximate dependence of the averaged dipole moment on temperature is given by [5.182] $\bar{D} \sim \tanh(\delta/kT)$, where δ is the tunneling splitting magnitude (Sect. 4.3). In the limit of high temperature, $\delta \ll kT$, we have $\bar{D} \sim T^{-1}$, being the classical result for rigid-dipole molecules. In the other limiting case of low temperature $\bar{D}$ becomes a constant independent of temperature, which is characteristic of symmetric molecules not possessing a proper dipole moment.

The rotational IR spectra of Jahn-Teller spherical-top molecules having a dipole instability (symmetry group T_d, $T \otimes t_2$ system) were considered in [5.183]. Figure 5.18 gives an approximate scheme of the rotational energy levels which (in the absence of a rotational-vibronic interaction) adjoin the vibronic levels T_2 and A_1 arising in a tetrahedral molecule due to the tunneling in the case of strong vibronic coupling (Sect. 4.3)

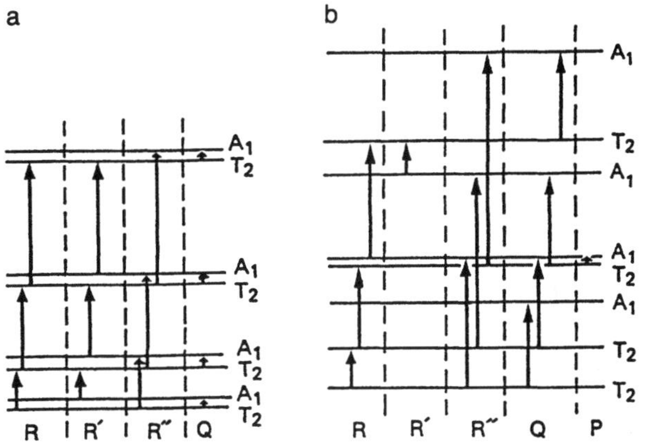

Fig. 5.18. Tunneling-rotational energy levels and allowed transitions for (**a**) $\delta < 2B$ and (**b**) $\delta > 2B$. The transitions related to different branches are divided by dashed lines [5.183]

$$E_{T_2 J} = BJ(J+1) \ , \qquad E_{A_1 J} = \delta + BJ(J+1) \ . \tag{5.3.18}$$

Here B is the rotational constant and δ is the tunneling splitting magnitude determined by (4.3.34).

For each rotational level the wave function can be written in the form of a product of the vibronic function $|\Gamma\gamma\rangle$ and the rotational function of a spherical top $|JKM\rangle$

$$|\Gamma\gamma JKM\rangle = |\Gamma\gamma\rangle |JKM\rangle \ , \qquad K, M = 0, \pm 1, \ldots, \pm J \ . \tag{5.3.19}$$

With the wave functions known one can determine the probabilities of transitions and the intensities of the induced dipole transitions $\Gamma J \rightarrow \Gamma' J'$ per unit of radiation density causing the transitions [5.183, 212]:

$$W_{\Gamma J \rightarrow \Gamma' J'} = \frac{8\pi^3 M_0^2}{9\hbar^2 c^2} \frac{N}{Z} (E_{\Gamma' J'} - E_{\Gamma J}) \times \left[\exp\left(-\frac{E_{\Gamma J}}{kT} \right) - \exp\left(-\frac{E_{\Gamma' J'}}{kT} \right) \right] C_{JJ'} g_{\Gamma J}(I) \ , \tag{5.3.20}$$

where Z is the statistical sum of tunneling-rotational levels (5.3.18),

$$C_{JJ'} = \begin{cases} (2J+1)(2J+3) \ , & J' = J+1 \\ (2J+1)^2 \ , & J' = J \\ (2J+1)(2J-1) \ , & J' = J-1 \ . \end{cases}$$

$g_{\Gamma J}(I)$ is the statistical weight depending on the nuclear spin I, M_0 is the modulus of the nuclear dipole moment vector of the system in the trigonal minimum and N is the number of absorbing centers per unit volume.

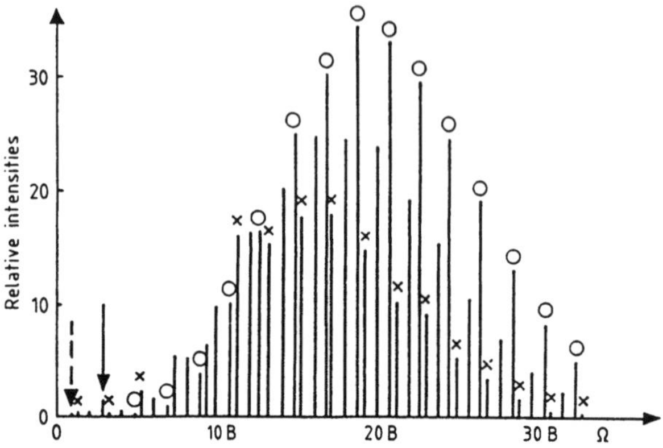

Fig. 5.19. Calculated line positions and intensities of the rotational IR spectrum for a system with a dipole instability at $\delta = 15\ \mathrm{cm}^{-1}$, $B = 5.24\ \mathrm{cm}^{-1}$, $kT = 200\ \mathrm{cm}^{-1}$ [5.183]. The R'', R' and R branches are labeled by rings, crosses, and no symbols, respectively. The Q and P transitions are shown by an arrow and a dashed arrow, respectively

From the relationship (5.3.20) it can be seen that three types of transitions with $J' = J + 1$ (R transitions) are possible: $T_2 J \to T_2(J + 1)$, labeled R, $A_1 J \to A_1(J + 1)$, labeled R', and $T_2 J \to A_1(J + 1)$, labeled R'', while the Q transitions ($J' = J$) and P transitions ($J' = J - 1$) are allowed only as $T_2 J \to A_1 J(Q)$ and $T_2 J \to A_1(J - 1)(P)$. If $\delta = 0$ (more exactly $\delta \ll B$) the transition frequencies in the three R-type bands coincide and the resulting spectrum consists of lines with a constant spacing of $2B$, without the P and Q branches. As δ increases, each of these lines splits and the Q transition occurs at a frequency δ, its intensity being small since the two tunneling levels have almost the same population.

In the case when $\delta > B$ the picture of the spectrum changes significantly (Fig. 5.19). Besides the increase of the frequency separation of the lines of the three branches, their intensities also change, increasing for the $T_2 \to A_1$ transitions and decreasing for the $A_1 \to T_2$ ones. Simultaneously the Q transition becomes stronger and the lines of the P transition appear. The number and the intensity of the latter increase as the inequality $\delta > B$ strengthens. Note that in the usual purely rotational spectra only the R-type transitions are possible whereas all the branches of R, P and Q transitions can be observed simultaneously in the rotational structure of the vibrational band [5.184]. In the predicted spectrum in the region of purely rotational transitions all the branches can be observed, with the distinction that there will be three R branches while the P and Q branches occur only for large δ.

At lower temperatures the intensity of the Q transition increases more rapidly than that of the other lines and hence at some specific temperature the Q line becomes distinct from the background of the band (Fig. 5.20). This line corresponds to the transition with no change of the rotational quantum number, and by its shape and temperature features it is quite analogous to the zero-phonon

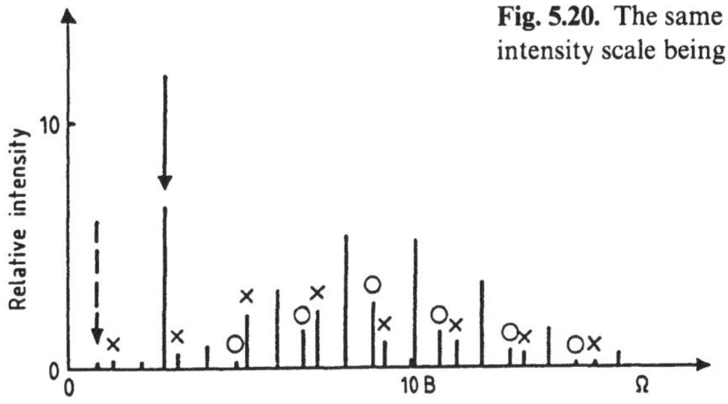

Fig. 5.20. The same as in Fig. 5.19 for $kT = 50\ \mathrm{cm}^{-1}$, the intensity scale being reduced by a factor of 100 [5.183]

line in the optical spectra (Sect. 5.2). Therefore it can be called the "zero-rotational" line. This line has already been discussed as a transition between the tunneling states.

Unlike in systems with a dipole instability, the spherical top molecules for which the dipole-active vibrations are not simultaneously Jahn-Teller active, have no pure rotational spectra. Examples of this kind can be found, in particular, among tetrahedral molecules with a Jahn-Teller E term in the ground state. For doubly degenerate electronic states of spherical top molecules there is no rotational fine structure in the excited vibronic state as well, since, as shown in Chap. 4, all the Jahn-Teller states transform as the representations E, A_1 and A_2, the squares of which do not contain the triplet representation of the operator of the dipole moment inducing the transition.

Nevertheless, as shown in [5.185], the spin-orbital interaction, by mixing in the excited T term (second-order perturbation theory), removes the prohibition, allowing the rotational structure in the infrared (microwave) spectrum of the ground E term. For instance, under the influence of the spin-orbital interaction the electronic ^2E term of a tetrahedral molecule becomes a cubic Kramers quadruplet Γ_8, its symmetrized square containing the T_2 representation of the dipole moment. This explains the origin of the expected new type of rotational spectrum.

The spin-vibronic-rotational interaction for a C_{3v} molecule in an electronic ^2E state and the hyperfine Hamiltonian for the appropriate nuclear-spin-electron-spin interaction is investigated in [5.180]. An analogous treatment of the vibronic effects in linear molecules in orbitally degenerate electronic states (the Renner effect) is given in [5.186–190]. The general theory of these effects is reviewed in [5.97, 98].

The presence of rigid-dipole properties in some Jahn-Teller molecules having no proper dipole moment can be manifest also in the collision-induced absorption of light by spherical top molecules in degenerate orbital states [5.191]. It is known that the collision interaction of two molecules induces a dipole moment which interacts with the electromagnetic wave, resulting in its absorption [5.192]. In the absence of electronic degeneracy the first nonzero multipole moments of spherical top molecules determining their interaction at large distances, are the octapole and hexadecapole (O_h symmetry) moments. But if the ground electronic term of the molecule is degenerate, then the first nonzero multipole moment is the quadrupole or even the dipole (in cases with a dipole instability). The estimates given in [5.191] show that the effective quadrupole moments under consideration may reach magnitudes comparable with the quadrupole moments of symmetric top molecules.

5.3.3 Vibronic Nonresonant Raman Spectra

As was mentioned above, the vibronic effects in the IR spectra take place only for Jahn-Teller systems with a dipole instability. Analogous spectra for systems having no dipole instability can be obtained by means of Raman spectroscopy. As is well known, the Raman spectra are determined by other selection rules, and they are in a sense complementary to the IR spectra.

The form function of the Raman spectrum accompanied by a transition of the Jahn-Teller system from the vibronic state $\Psi_n(\Gamma_1)$ to the state $\Psi_m(\Gamma_1)$ is determined by (5.1.2) with $\Gamma_2 = \Gamma_1$, where $\Omega = \Omega_s - \Omega_i$ is the frequency shift of the incident (Ω_i) and scattered (Ω_s) light, and d is the two-photon transition operator determined by the tensor of electronic polarizability [5.193]:

$$d = \sum_{\alpha,\beta} n_{i\alpha} n_{s\beta} P_{\alpha\beta}(\mathrm{r}, Q) \ , \qquad \alpha, \beta, = x, y, z \ . \tag{5.3.21}$$

Here n_i and n_s are the unit vectors of polarization of the incident and scattered light respectively.

The scalar convolution (5.3.21) can be written in the form of a convolution of irreducible tensor operators,

$$d = \sum_{\Gamma\gamma} N^+_{\Gamma\gamma} P_{\Gamma\gamma}(\mathrm{r}, Q) \ , \tag{5.3.22}$$

where Γ are the irreducible representations of the symmetry group of the system which are contained in the square of the vector representation, and $N_{\Gamma\gamma}$ and $P_{\Gamma\gamma}$ are linear combinations of the components of the second-rank tensors $n_{i\alpha} n_{s\beta}$ and $P_{\alpha\beta}$ respectively, having the transformation properties of $\Gamma\gamma$. For instance, for a cubic system of O_h symmetry the vectors transform as T_{1u} and therefore $\Gamma = T_{1u} \times T_{1u} = A_{1g} + E_g + T_{1g} + T_{2g}$,

$$N_{\Gamma\gamma} = \sum_{\alpha\beta} n_{i\alpha} n_{s\beta} \langle T_{1u}\alpha T_{1u}\beta | \Gamma\gamma \rangle \ ; \qquad P_{\Gamma\gamma} = \sum_{\alpha\beta} P_{\alpha\beta} \langle T_{1u}\alpha T_{1u}\beta | \Gamma\gamma \rangle \ . \tag{5.3.23}$$

Thus by choosing different polarizations for the incident and the scattered light one can separate the scattering effects determined by polarizabilities of different symmetry. For instance, for the O_h group one can separate A_{1g} scattering, E_g scattering, etc.

Among the irreducible representations in the sum (5.3.22) one can always separate symmetric and antisymmetric ones, i.e., those contained in respectively the symmetric and antisymmetric squares of the vector representation.

The selection rules for the symmetric scattering are the same as the ones for the electric quadrupole moment and therefore the spectral and the temperature dependence of the corresponding components of scattering are the same as those for the quadrupole IR absorption. For the antisymmetric scattering the selection rules are the same as those for the operator of the magnetic dipole moment and therefore the spectral and temperature dependence of the antisymmetric components of scattering is the same as for magnetic dipole IR absorption. All that was said above about the vibronic effects in the IR absorption is equally valid for Raman spectra with the distinction that instead of the frequency Ω of the absorbed IR irradiation one deals here with the frequency shift of the light when it is scattered.

In particular, for systems without a center of symmetry (i.e., for systems which are not invariant with respect to inversion of the coordinates) among the irreducible representations Γ in the sum (5.3.22) there are representations to which components of a vector belong. The appropriate components of the tensor $P_{\Gamma\gamma}$ cause the same transitions as the components of the dipole moment vector do in IR absorption. The remaining components of the tensor $P_{\Gamma\gamma}$ cause transitions which do not appear in the IR spectra. Therefore the Raman spectra have a richer vibronic structure. For instance, the rotational spectra of free Jahn-Teller molecules of the spherical top type contain O and S branches for which $\Delta J = \pm 2$ in addition to the P, Q and R lines in the IR spectra discussed above [5.194].

Now consider the fact that in nonresonant Raman scattering the operator of polarizability $P_{\Gamma\gamma}(\mathrm{r}, Q)$ is weakly dependent on the nuclear coordinates Q [5.166]. This dependence originates from the vibronic interaction (Jahn-Teller or pseudo-Jahn-Teller) in the excited electronic states which serve as the intermediate ones in the two-photon Raman process. We can expand this operator in a power series:

$$P_{\Gamma\gamma}(\mathrm{r}, Q) = P^{(0)}_{\Gamma\gamma}(\mathrm{r}) + \sum_{\Gamma_1\gamma_1} P^{(1)}_{\Gamma\gamma\Gamma_1\gamma_1}(\mathrm{r})Q_{\Gamma_1\gamma_1} + \cdots , \tag{5.3.24}$$

where

$$P^{(1)}_{\Gamma\gamma\Gamma_1\gamma_1}(\mathrm{r}) = \left.\frac{\partial P_{\Gamma\gamma}(\mathrm{r}, Q)}{\partial Q_{\Gamma_1\gamma_1}}\right|_{Q=Q_0} .$$

The first term in (5.3.24) is the electronic polarizability of the polyatomic system in the nuclear configuration Q_0 (Sect. 2.2). This operator in the sense discussed above is analogous to the operator of the electronic quadrupole (or magnetic dipole) moment. For systems with no center of inversion there are some components of this operator which have the transformation properties of a vector and therefore in the limited basis of the electronic states of the Jahn-Teller term Γ_1 they are represented by the same matrices as the operator of the electronic dipole moment. For instance, for the E term of a tetrahedral molecule, taking only the $P^{(0)}_{\Gamma\gamma}$ from (5.3.24), the expression (5.3.22) becomes

$$\hat{d} = P^{(0)}_{A_1} N_{A_1} \hat{\sigma}_0 + P^{(0)}_{E}(N_+\hat{\sigma}_- + N_-\hat{\sigma}_+) ,$$

which is quite analogous to (5.3.5). Here $P_{\Gamma}^{(0)}$ are the reduced matrix elements. Obviously the Raman transitions caused by the terms containing $P_{E}^{(0)}$ do not differ from the IR transitions caused by the operator (5.3.5).

Note that in molecules with no degeneracy the zero term of the expansion (5.3.24) $P_{\Gamma\gamma}^{(0)}(\mathrm{r})$ only causes Rayleigh scattering. The essential feature of Jahn-Teller systems is the possibility of observing Raman scattering caused by this operator.

The second term in (5.3.24) describes what is known as Raman scattering of the first order. As said above it is weak compared with the zeroth-order Raman scattering. It can be estimated to be of the order of $P^{(1)} \sim V_{\mathrm{e}}/(\omega_{\mathrm{eg}} - \Omega_{\mathrm{i}})$, where V_{e} is the vibronic coupling constant in the excited electronic states serving as intermediate ones for the Raman process, $\omega_{\mathrm{eg}} = (E_{\mathrm{e}} - E_{\mathrm{g}})/\hbar$ is the Bohr frequency of the dipole transition from the ground state to the excited intermediate one with the energies E_{g} and E_{e} respectively and Ω_{i} is the frequency of the incident light. In the case of nonresonant Raman scattering this parameter was estimated in [5.169] to be relatively small, $P^{(1)} \lesssim 0.1$ (see however [5.195]).

Let us regroup the terms in (5.3.24) so as to get the irreducible tensor operators

$$P_{\Gamma\gamma}(\mathrm{r}, Q) = P_{\Gamma\gamma}^{(0)}(\mathrm{r}) + \sum_{\Gamma_1\gamma_1}\sum_{\Gamma_2\gamma_2} P_{\Gamma_1\gamma_1}^{(1)}(\mathrm{r}) Q_{\Gamma_2\gamma_2} \langle \Gamma_1\gamma_1\Gamma_2\gamma_2|\Gamma\gamma\rangle \ , \tag{5.3.25}$$

where

$$P_{\Gamma_1\gamma}^{(1)}(\mathrm{r}) = \sum_{\gamma_1\gamma_2} P_{\Gamma_1\gamma_1\Gamma_2\gamma_2}^{(1)}(\mathrm{r}) \langle \Gamma_1\gamma_1\Gamma_2\gamma_2|\Gamma\gamma\rangle \ . \tag{5.3.26}$$

Passing to the matrix representation of the electronic operators by means of the Wigner-Eckart theorem we have

$$\hat{d} = \sum_{\Gamma\gamma} N_{\Gamma\gamma}^{+}\left(P_{\Gamma}^{(0)}\hat{C}_{\Gamma\gamma} + \sum_{\Gamma_1\gamma_1}\sum_{\Gamma_2\gamma_2} P_{\Gamma_1}^{(1)}\hat{C}_{\Gamma_1\gamma_1} Q_{\Gamma_2\gamma_2}\langle \Gamma_1\gamma_1\Gamma_2\gamma_2|\Gamma\gamma\rangle\right) , \tag{5.3.27}$$

where $P_{\Gamma}^{(0)}$ and $P_{\Gamma}^{(1)}$ are the appropriate reduced matrix elements.

Note that the terms of the operator of polarizability which are linear in $Q_{\Gamma_1\gamma_1}$ have the same transformation properties $\Gamma\gamma$ as the zeroth-order terms $P_{\Gamma\gamma}^{(0)}(\mathrm{r})$ and therefore they are subject to the same selection rules. Accordingly the first-order Raman scattering contains the same spectral lines as the Raman scattering caused by the zeroth-order terms. Therefore the separation of the contribution to the Raman scattering spectra is, in general, impossible. However, this can be done in limiting cases of weak and strong vibronic coupling. For instance, in the case of weak coupling when the vibrational occupation numbers $n_{\Gamma\gamma}$ are still good quantum numbers, the term linear in $Q_{\Gamma\gamma}$ cause the Raman transitions $n_{\Gamma\gamma} \to n_{\Gamma\gamma} \pm 1$. At $T = 0$ K this means that the appropriate spectral lines are at the frequencies $\Omega \approx \omega_{\Gamma}$. On the other hand the rotational Raman spectra related to the ground vibronic state are caused only by the operator $P_{\Gamma\gamma}^{(0)}(\mathrm{r})$ and are situated in the immediate neighborhood of the Rayleigh line, i.e., they lie in a different range of frequencies.

In the case of strong coupling for low energy vibronic states approximately reduced to the harmonic vibrations at the minima of the adiabatic potential (Chap. 4), $Q_{\Gamma\gamma}$ in (5.3.27) has to be replaced by $Q^{(0)}_{\Gamma\gamma} + q_{\Gamma\gamma}$, where $Q^{(0)}_{\Gamma\gamma}$ are the minima coordinates and $q_{\Gamma\gamma}$ are small displacements from these minima. The terms containing $Q^{(0)}_{\Gamma\gamma}$ merge with those containing $P^{(0)}_{\Gamma\gamma}(\mathrm{r})$ while the terms linear in $q_{\Gamma\gamma}$ lead to the first-order Raman scattering modified by the vibronic interaction. Since the occupation numbers at the bottom of the minimum are good quantum numbers, the operators $q_{\Gamma\gamma}$ at $T = 0$ K cause Raman scattering at the frequencies $\Omega \approx \tilde{\omega}_\Gamma$, where $\tilde{\omega}_\Gamma$ are the modified frequencies of normal vibrations at the bottom of the minimum (Chap. 3). The remaining terms of the polarizability operator result in Raman transitions at the frequencies $|\Omega| \lesssim \hbar\omega_\Gamma^2/V_\Gamma^2$ between the rotational vibronic states (Sect. 4.2) and between the tunneling states (Sect. 4.3) for even stronger coupling. In multimode Jahn-Teller systems the first-order Raman scattering reproduces the vibrational density of states modified by the vibronic interaction (Sect. 4.6).

If the distance from resonance $|\omega_{eg} - \Omega|$ in the nonresonant Raman scattering is large enough, the first-order scattering is negligible and the main contribution originates from the first term in (5.3.24). The intensity of the Rayleigh line is determined in this case by the square of the matrix element of the electronic polarizability operator $P^{(0)}_{\Gamma\gamma}(\mathrm{r})$, calculated with the exact vibronic wave functions of the ground state. This matrix element is obviously proportional to the square of an appropriate vibronic reduction factor if the polarization of the incident and the scattered light is taken so as to separate out the contribution to the scattering of the polarizability operator component transforming as a given irreducible representation Γ. The total value of the intensity is equal to the integral intensity of the Raman scattering, i.e., to the zeroth-order moment of the Raman band. From this it follows that it is possible to estimate the values of the vibronic reduction factors from the Raman scattering experiments by taking different polarizations of the incident and the scattered light [5.196, 197].

The vibronic spectrum of the Raman scattering of Cu^{2+} impurity centres in CaO crystals observed at $T = 4.2$ K [5.166, 198] is illustrated in Fig. 5.21. In the upper right corner the low-frequency part of the spectrum containing the line at the frequency $\hbar\Omega \approx 4\ \mathrm{cm}^{-1}$ is given on a large scale. This line is interpreted as the $E \rightarrow A$ transition between the tunneling states of the ground electronic E term of the Cu^{2+} ion.

Substituting the operator $\hat{d}$ from (5.3.22) into the Kubo formula (5.3.4), (5.1.1) we obtain the intensity of the Raman scattering,

$$I(\Omega) \sim \sum_{\Gamma\gamma}\sum_{\bar{\Gamma}\bar{\gamma}} N_{\Gamma\gamma} N^{+}_{\bar{\Gamma}\bar{\gamma}} \int_{-\infty}^{\infty} \mathrm{e}^{-\mathrm{i}\Omega t} \langle [P^{\dagger}_{\Gamma\gamma}, P_{\bar{\Gamma}\bar{\gamma}}(t)] \rangle_{\Gamma_1}\, dt\ . \qquad (5.3.28)$$

Assuming (as is common in most cases) that the expansion (5.3.22) does not

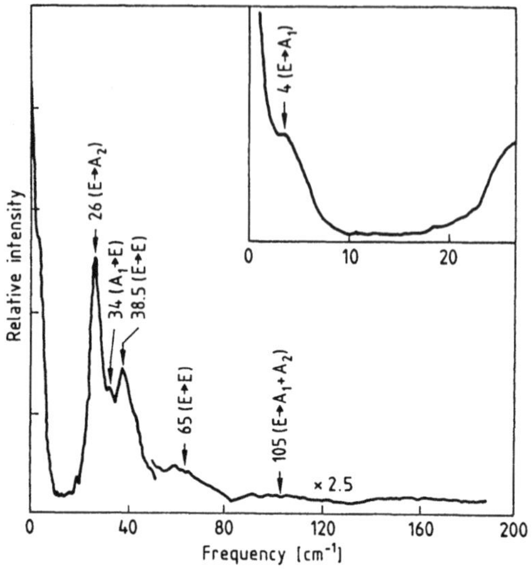

Fig. 5.21. Vibronic E-type Raman spectrum for the CaO : Cu^{2+} system at $T = 4.2$ K (frequencies in cm^{-1}). The inset shows (on an enlarged scale) the part of the spectrum containing the line of the $E \rightarrow A$ transition between the tunneling levels at a frequency of 4 cm^{-1}. (After [5.166])

contain terms transforming as the same irreducible representations, and taking into account that the average $\langle \cdots \rangle_{\Gamma_1}$ is not zero only for totally symmetric combinations of the operators, we obtain from (5.3.28)

$$I(\Omega) \sim \sum_{\Gamma\gamma} |N_{\Gamma\gamma}|^2 \int_{-\infty}^{\infty} \mathrm{e}^{-\mathrm{i}\Omega t} \langle [P_{\Gamma\gamma}^{\dagger}, P_{\Gamma\gamma}(t)] \rangle_{\Gamma_1} dt \ . \tag{5.3.29}$$

From symmetry considerations it is clear that the average $\langle P_{\Gamma\gamma} | P_{\Gamma\gamma} \rangle_{\Gamma_1}$ does not depend on γ. Introducing the notation

$$G_{\Gamma}(\Omega) = \frac{f_{\Gamma}}{2\pi} \int_{-\infty}^{\infty} \mathrm{e}^{-\mathrm{i}\Omega t} \langle [P_{\Gamma\gamma}^{\dagger}, P_{\Gamma\gamma}(t)] \rangle_{\Gamma_1} dt \ , \tag{5.3.30}$$

where f_{Γ} is the dimensionality of the representation Γ, we obtain from (5.3.29)

$$I(\Omega) \sim \sum_{\Gamma} G_{\Gamma}(\Omega) \frac{1}{f_{\Gamma}} \sum_{\gamma} |N_{\Gamma\gamma}|^2 \ . \tag{5.3.31}$$

Since the magnitudes $|N_{\Gamma\gamma}|^2$ are composed of the components of the vectors of the incident and the scattered light polarization with respect to the symmetry axes of the scattering system, this formula determines explicitly the angular dependence and polarization properties of the scattering.

For randomly oriented systems (molecules in the gas phase and solutions, polycrystals and so on) the magnitudes $|N_{\Gamma\gamma}|^2$ should be averaged over the orientations of the scattering systems. A direct calculation shows that the orientational averages $\overline{|N_{\Gamma\gamma}|^2/f_{\Gamma}}$ are the same for different irreducible representations $\Gamma_s \neq A_1$ contained in the symmetric square of the vector representation (i.e., for symmetric scattering) and separately for the representations Γ_a contained in the antisymmetric square (i.e., for antisymmetric scattering [Ref. 5.199; Sect. 2]. Denoting

$$G_s = \sum_{\Gamma_s \neq A_1} G_{\Gamma_s} \, , \qquad G_a = \sum_{\Gamma_a} G_{\Gamma_a} \tag{5.3.32}$$

we obtain from (5.3.31) [Ref. 5.193; Sect. 61]

$$I(\Omega) \sim [G_{A_1}(\Omega)|\boldsymbol{n}_s^* \boldsymbol{n}_i|^2 + \tfrac{1}{10}G_s(\Omega)(1 + |\boldsymbol{n}_s \boldsymbol{n}_i|^2 - \tfrac{2}{3}|\boldsymbol{n}_s^* \boldsymbol{n}_i|^2) + \tfrac{1}{6}G_a(\Omega)(1 - |\boldsymbol{n}_s \boldsymbol{n}_i|^2)] \tag{5.3.33}$$

Let us denote the angle between the direction of the scattering $\boldsymbol{k}_s$ and the direction of the polarization of the incident light $\boldsymbol{n}_i$ by θ. The scattering light contains two independent components, polarized in the plane $(\boldsymbol{k}_s, \boldsymbol{n}_i)$ (intensity I_1) and perpendicular to the plane (intensity I_2). The ratio $I_2/I_1 = \varrho_l$ is called the depolarization ratio [Ref. 5.199; Sect. 2]. For the scattering of linear polarized light we obtain from (5.3.33)

$$\varrho_l(\Omega) = \frac{3G_s(\Omega) + 5G_a(\Omega)}{30G_{A_1}(\Omega) + G_s(\Omega)(3 + \sin^2\theta) + 5G_a(\Omega)\cos^2\theta} \, . \tag{5.3.34}$$

Specifically, at $\theta = \pi/2$ (when the electric vector of the excited light is perpendicular to the plane containing the beams of the incident and the scattered light) we have

$$\varrho_l(\Omega) = \frac{3G_s(\Omega) + 5G_a(\Omega)}{30G_{A_1}(\Omega) + 4G_s(\Omega)} \, . \tag{5.3.35}$$

In the case when the incident light is natural (nonpolarized) the expression (5.3.33) has to be averaged over the directions of polarizations $\boldsymbol{n}_i$ for a given direction of the incident light $\boldsymbol{k}_i$. In the case of scattering to an angle of $\pi/2$ the expression for the depolarization ratio is different from (5.3.35), and is related to ϱ_l by the simple relationship

$$\varrho_n(\Omega) = \frac{2\varrho_l(\Omega)}{1 + \varrho_l(\Omega)} = \frac{6G_s + 10G_a}{30G_{A_1} + 7G_s + 5G_a} \, . \tag{5.3.36}$$

An important feature of molecules in nondegenerate electronic states is that the tensor of polarizability $P_{\alpha\beta}$ is real and symmetric [Ref. 5.193; Sect. 62]. From this it follows that there is no antisymmetric scattering in such molecules, i.e., $G_a = 0$. From (5.3.36) we obtain in this case $\varrho_n \leq \frac{6}{7}$ and from (5.3.35) we obtain $\varrho_l \leq \frac{3}{4}$. An essential feature of Jahn-Teller molecules is the possibility of the antisymmetric scattering resulting in $\varrho_n > \frac{6}{7}$ and $\varrho_l > \frac{3}{4}$[5.169].

The first results obtained for Raman scattering in Jahn-Teller systems are given in [5.169] for the $E \otimes e$ system in the X_3 molecule by means of a direct substitution of the vibronic states obtained by numerical calculations (Sect. 4.4) into (5.1.2). The experimental results obtained by 1965 for Raman scattering related mainly to hexafluorines of transition metals with the Jahn-Teller effect are given in the review paper [5.200]. From more recent results we should note the works [5.166, 198] in which experimental data and a theoretical interpretation of the Raman spectra for the ground E term (quadratic $E \otimes e$ system) of the Cu^{2+} impurity centre in CaO crystals are given (See also [5.201]). The same system was discussed in [5.202] with allowance for the dispersion of the phonons within the framework of the relaxing cluster (Sect. 4.6.5). Approximate weak and strong coupling results as well as the intermediate coupling case in the multimode Jahn-Teller effect were investigated in [5.137, 170, 196, 203–206]. Selection rules for first-order Raman scattering in Jahn-Teller systems were obtained in [5.207]. Rotational Raman spectra of free spherical-top-type molecules around the Rayleigh line for the Jahn-Teller E and T terms were considered in [5.194].

As is well known, the tensor of Rayleigh scattering of light by a molecule is closely related to its polarizability in the ground state [5.193]. The traditional description of the electric properties of a molecule is based on the assumption that the charge distribution (either classical or quantum-theoretical) is totally symmetric with respect to the equilibrium nuclear configuration. For instance, in the case of a spherical-top-type molecule in a nondegenerate state it follows that only the scalar part of the polarizability is not zero.

In degenerate states the situation is significantly different. Using the group-theory selection rules for the matrix elements of the polarizability operator it can easily be seen that there may also be nonzero matrix elements for components of the irreducible tensor operators of the polarizability and multipole moments which are not totally symmetric. If the vibronic ground state transforms as the irreducible representation Γ, then the nonzero contribution to the polarizability arises from the components of the operator of polarizability transforming as the irreducible representations containing in the symmetric square $[\Gamma^2]$ (or the antisymmetric square $\{\Gamma^2\}$ if Γ is a two-valued representation of a double group).

Thus in degenerate states the anisotropic components have to be taken into account in addition to the totally symmetric components of the tensors of polarizability and multipole moments, i.e., the point symmetry of the charge distribution becomes lower than in the nondegenerate case. Note that in nondegenerate states the diagonal matrix element of an operator is the same as the mean value of the corresponding quantity (polarizability, multipole moments, etc.), whereas in the case of degeneracy there is no such direct relation between the matrix elements and observables, since in the basis of the degenerate states each physical quantity corresponds not to a matrix element but to a matrix. Therefore the correlation between the matrix elements and the observables has to be carried out for each distinct experimental situation separately. Such an analysis of birefringence in gases of spherical top molecules, in particular of the Kerr and Cotton-Mouton effects, is performed in Sect. 5.3.5. The depolarization of light in Rayleigh, hyper-Rayleigh and purely rotational Raman scattering by degenerate molecules is considered in [5.210]. A similar investigation of the temperature-dependent optical activity of symmetric degenerate and pseudo-Jahn-Teller molecules in a magnetic field is presented in [5.211]. All these problems as well as some other manifestations of the anomalous electric properties of Jahn-Teller and pseudo-Jahn-Teller molecules are reviewed in [5.212].

5.3.4 Vibronic Effects in Resonant Raman Scattering

Note that the spectral density of nonresonant Raman scattering is determined by the vibronic properties of the ground electronic term only and does not depend on the properties of the virtual excited states through which the non-elastic scattering of light takes place. The situation is different in resonant Raman scattering when the frequency of the incident light Ω_i lies in the absorption band of the Jahn-Teller system. In this case the operator $P_{\alpha\beta}$ causing the two-photon transition (5.3.21) cannot be reduced to the electronic polarizability and depends strongly on the nuclear coordinates [5.213],

$$P_{\alpha\beta}(\mathrm{r}, Q) = \sum_k D_\alpha(\mathrm{r}) \frac{|\Psi_k^{(\bar{\Gamma})}(\mathrm{r}, Q)\rangle\langle\Psi_k^{(\bar{\Gamma})}(\mathrm{r}, Q)|}{\omega_{kn} - \Omega_i - \mathrm{i}\gamma_k} D_\beta(\mathrm{r}) \ . \tag{5.3.37}$$

Here $D_\alpha(\mathrm{r})$ are the Cartesian components of the vector of the dipole moment of the electrons, $\bar{\Gamma}$ is the index of the excited electronic term (or of a group of terms of similar energy) which is in resonance with the scattered light, $\omega_{kh} = (E_{\bar{\Gamma}k} - E_{\Gamma_1 n})/\hbar$ is the Bohr frequency of the transition, γ_k is the natural linewidth. In (5.3.37) as usual the terms containing nonresonant denominators are omitted.

As seen from (5.3.37) the intensity of the scattered light in resonant Raman scattering depends strongly on the properties of the excited states through which the scattering takes place. Of special interest is the case when the excited states are Jahn-Teller states. In the simplest situation when both the initial and final states of the scattering are that of the singlet electronic term $\Gamma_1 = A_1$, the spectrum of the shifted scattering reproduces the equally spaced vibrational energy spectrum of the A_1 term, while the intensity of the lines depends on the frequency Ω_i determining the resonant excited state.

Figure 5.22 shows the dependence of the intensity of the resonant Raman scattering on the frequency of the incident light Ω_i in the case when the scattering takes place through the excited E term having a $(E \otimes e)$-type Jahn-Teller effect while the ground term is an A_1 one. This result is obtained in [5.214] by means of a numerical calculation of the vibronic states of the linear and the quadratic $E \otimes e$ system and substitution of the results into (5.3.37).

The dashed line in Fig. 5.22 illustrates the dispersion of the depolarization ratio of natural light in this case. It is seen that the depolarization ratio of the overtones depends on the frequency of the incident light Ω_i, this dependence being stronger for larger values of the second-order vibronic coupling constant W_E, and this can be used for an experimental estimation of W_E. Analogous results

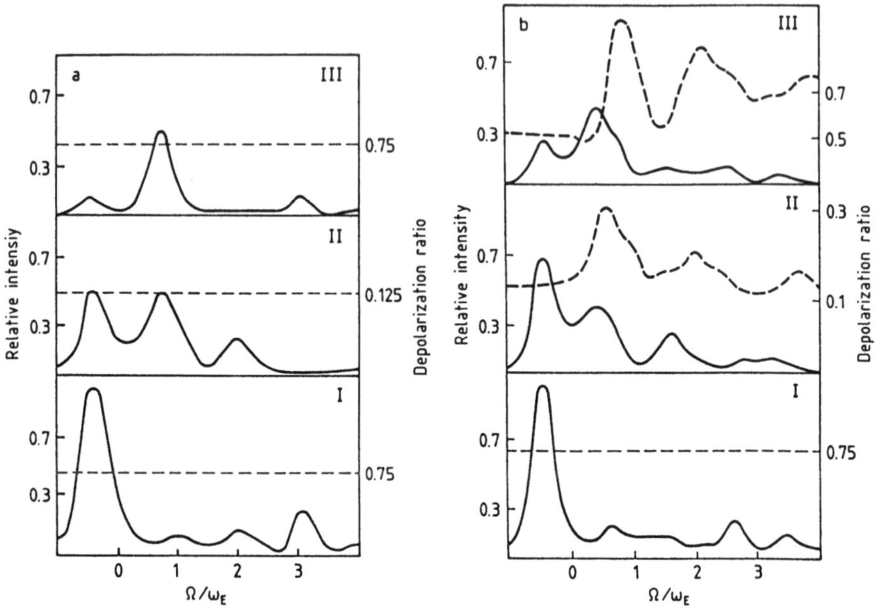

Fig. 5.22a, b. Excitation profiles (solid lines) and depolarization ratios (broken lines) for the resonance Raman scattering via the excited E term ($E_{JT} = \hbar\omega_E$) as functions of the incident light frequency Ω (in units of ω_E; the energy zero corresponds to the position of the degenerate E term) for (**a**) $W_E = 0$ and (**b**) $W_E = 0.1\omega_E^2$. **I**: fundamental tone; **II**: first overtone; **III**: second overtone. (After [5.214])

for the case when the excited resonance term is a Jahn-Teller T one (the $T \otimes t_2$ problem) were obtained in another work by the same authors [5.215]. The multimode aspects of resonant Raman scattering in Jahn-Teller systems is discussed in [5.216, 217]. The semiclassical approach to the problem developed in [5.217] gives satisfactory agreement with the numerical quantum-mechanical calculations in the limiting case of strong vibronic coupling.

5.3.5 Birefringence in Jahn-Teller Spherical Top Molecules

Another phenomenon in which Jahn-Teller spherical top molecules manifest anisotropic properties is birefringence in external electric fields, the electro-optical Kerr effect. The origin of this phenomenon in polar liquids and gases containing rigid-dipole molecules can be explained by the mechanism of Langevin and Born. In accordance with this theory the chaotic distribution of the rigid-dipole molecules in space results in the macroscopic isotropy of the matter. Under the influence of an external electric field the molecules orient themselves, causing the optical anisotropy of the medium, and hence the birefringence. A wave of light in this medium splits into two components with reciprocal perpendicular planes of polarization. One of these components, the ordinary ray, has a plane of polarization perpendicular to the external electric field, whereas the other one, the extraordinary ray, is polarized along the electric field. The difference in the pathlengths of the two rays in units of the wavelength is $Bl\mathscr{E}^2$, where l is the pathlength of the light ray as a whole in the medium, $\mathscr{E}$ is the intensity of the electric field, and B is the Kerr constant characterizing the difference between the refractive indices of the ordinary and extraordinary rays. The orientation of the molecules in the external electric field is hindered by their thermal motions, and therefore B is inversely proportional to the square of the temperature [5.218]. In addition, the Kerr constant depends on the magnitude of the dipole moment of the molecule and the wavelength of the light.

Apart from the orientational Langevin-Born mechanism, the anisotropy of the medium may be caused by the deformation of the molecular electronic shell induced by the electric field, i.e., by the anisotropy of the hyperpolarizability (the Voight mechanism). The contribution of this effect to the Kerr constant is relatively small and independent of temperature.

At first sight it seems that for spherical top molecules having no anisotropy in the absence of external fields, the birefringence may occur as a result of the Voight mechanism only. However, this conclusion implies that the ground state of the molecule is nondegenerate and transforms according to the totally symmetric representation. As mentioned above, the molecules with a dipole instability having no proper dipole moment manifest rigid-dipole properties due to the Jahn-Teller (or pseudo JT) effect; thus for these molecules a birefringence of the Langevin-Born orientational mechanism should occur. The temperature dependence of the Kerr constant here may be different from that indicated above, since the dipole moment itself depends on temperature. It emerges from the theory given below that even in the absence of a dipole instability (say, when the vibronic coupling constant is negligibly small) the presence of electronic degeneracy determines the anisotropy of electronic polarizability causing a birefringence in external electric fields of the Langevin-Born type [5.219, 220].

Consider, for instance, a molecule with T_d or O_h symmetry having a doubly degenerate ground E term. Neglecting rotational quantization, the molar Kerr constant ${}_{\mathrm{m}}B$ is related to the components of the tensor of polarizability of the molecule in a constant homogeneous electric field $\mathscr{E}$ paralel to the z axis, as follows [5.221]:

$$ {}_{m}B = 2\pi N_A\{\mathscr{E}^{-2}[\langle P_{zz}(\Omega_{\mathrm{i}},\mathscr{E})\rangle - \langle P_{xx}(\Omega_{\mathrm{i}},\mathscr{E})\rangle]\}_{\mathscr{E}\to 0} \ , \tag{5.3.38} $$

where N_A is the Avogadro number, $P_{\alpha\beta}$ is the operator of electric polarizability determined by (5.3.37), and Ω_{i} is the frequency of incident light propagating in the direction perpendicular to the electric field. In the latter all the states $\Psi_k^{(\Gamma)}$ and corresponding transition frequencies ω_{kn} in (5.3.37) depend on the intensity $\mathscr{E}$, resulting in the dependence of the components of the tensor of polarizability on $\mathscr{E}$ in (5.3.38). The angular brackets in (5.3.38) mean the average over all the vibronic states of the ground E term, and over all the orientations of the molecules. In order to calculate $\langle P_{\alpha\beta}\rangle$ one can use the relation between the polarizability and the Fourier transform of the correlation function,

$$\langle P_{\alpha\beta}(\omega,\mathscr{E})\rangle = \frac{1}{2\pi}\int_{-\infty}^{\infty}(1-\mathrm{e}^{\hbar\omega'/kT})\frac{I_{\alpha\beta}(\omega')}{\omega-\omega'}d\omega' \ , \tag{5.3.39}$$

where

$$I_{\alpha\beta}(\omega) = \int_{-\infty}^{\infty}\langle D_\alpha(t)D_\beta\rangle \mathrm{e}^{-\mathrm{i}\omega t}dt \tag{5.3.40}$$

Here D_α, as above, is a component of the operator of the dipole moment, and $D_\alpha(t)$ means the Heisenberg representation of this operator with the Hamiltonian of the molecule in the external field,

$$H_{\mathscr{E}} = H - D_z\mathscr{E} \ .$$

Assuming that the field $\mathscr{E}$ is small enough and the orientational saturation is far from being reached, one can expand the difference $I_{zz} - I_{xx}$ into a series in $\mathscr{E}$ and cut off the series at the second-order terms. The zeroth-order (in $\mathscr{E}$) terms are zero owing to the assumed isotropy of the system in absence of the electric field. The linear terms become zero after averaging over all the orientations of the individual molecule. Among the nonzero second-order (with respect to $\mathscr{E}$) terms the following are related to the orientational mechanism [5.220]:

$$I_{zz}^{(\mathrm{or})} - I_{xx}^{(\mathrm{or})} = \mathscr{E}^2\sum_{\varepsilon\varrho\sigma\delta} W^{\varepsilon\varrho\sigma\delta}J_{\varepsilon\varrho\sigma\delta} \ , \tag{5.3.41}$$

where

$$\begin{aligned} W^{\varepsilon\varrho\sigma\delta} &= \tfrac{1}{5}\sum_k(-1)^k C^{2,k}_{1\varepsilon 1\varrho}C^{2,-k}_{1\sigma 1\delta} \ , \\ J_{\varepsilon\varrho\sigma\delta} &= \int_0^{1/kT} d\beta_1\int_0^{\beta_1} d\beta_2\langle D_\varepsilon(-\mathrm{i}\beta_1\hbar)D_\varrho(-\mathrm{i}\beta_2\hbar)D_\sigma(t)D_\delta\rangle_E \ . \end{aligned} \tag{5.3.42}$$

Here $C^{2,k}_{1\varepsilon 1\varrho}$ are the Clebsch-Gordan coefficients of the full spherical group, the symbol $\langle\ldots\rangle_E$ means an average over the vibronic states of the ground electronic E term, and the Heisenberg representation of the operators is realized by the Hamiltonian H that depends only on the internal degrees of freedom of the molecule and does not take into account the external electric field.

The average $\langle\ldots\rangle_E$ in (5.3.42) is a sum of the diagonal matrix elements multiplied by the appropriate Boltzmann factors. Let us represent each of the matrix elements of the product of operators in (5.3.42) as a sum of products of matrix elements using the full orthonormalized set of exact vibronic states of the ground and excited electronic terms. In the cubic and icosahedral groups, to which spherical top molecules belong, the dipole moment transforms after one of the triplet states, T_1 or T_2, not contained in the symmetric square $[E^2]$. Therefore the matrix elements of the type $\langle Ek|D_\varepsilon|Em\rangle$, which intermix the vibronic states of the ground electronic E term, are zero, whereas the nonzero matrix elements mix the vibronic states of the excited electronic terms with that of the ground term. The Heisenberg representation of the operator D_α determines the dependence of the matrix elements on β_1, β_2 and t in the form of exponents with the appropriate transition frequencies in the index. Performing the integration in (5.3.42) over β_2 and over t [by means of formulas (5.3.39) and (5.3.40)], replacing the vibronic energy difference by the energy of pure electronic excitation, $E_{\Gamma n} - E_{Em} \approx E_\Gamma - E_E$, and assuming that for not very high temperatures the main contribution to the integral over β_1 is given by the region of relatively small β_1, so that the occupation of the excited electronic term may be neglected, $\exp[-\beta_1(E_\Gamma - E_E)] \approx 0$, we come to the following expression for (5.3.38) [5.220]:

$$_mB^{(\mathrm{or})} = 2\pi N_A \sum_{\varepsilon Q\sigma\delta} W^{\varepsilon Q\sigma\delta} \int_0^\beta d\beta_1 \langle e^{\beta_1 H} P_{\varepsilon Q}(0) e^{-\beta_1 H} P_{\sigma\delta}(\Omega_\mathrm{i})\rangle_E \ . \tag{5.3.43}$$

Here $\beta = 1/kT$; $P_{\varepsilon Q}(0)$ is the static electronic polarizability of the molecule determined by (5.3.37) at $\Omega_\mathrm{i} = 0$, in which, owing to the condition mentioned above, $\exp[-\beta_1(E_{\Gamma m} - E_{En})] \approx 0$ for $\Gamma \neq E$, only those matrix elements of the operator of polarizability which do not mix in the excited electronic states have to be taken into account. If the frequency Ω_i is far from the resonance frequencies Ω_{kn}, see (5.3.37), then for the operator $P_{\sigma\delta}(\Omega_\mathrm{i})$ as well as for $P_{\varepsilon Q}(0)$, the dependence on nuclear coordinates can be neglected. Together with everything said above, this means that $P_{\sigma\delta}(\Omega_\mathrm{i})$ and $P_{\varepsilon Q}(0)$ can be represented by electronic 2×2 matrices within the states of the ground electronic E term.

In the absence of vibronic coupling, $\exp(\pm\beta_1 \hat{H})$ commutes with $\hat{P}_{\varepsilon Q}(0)$ and these terms cancel each other. The integral over β_1 in (5.3.43) yields the factor $1/kT$. As a result we obtain [5.219]:

$$_mB^{(\mathrm{or})} = \frac{6\pi N_A}{5kT} P_E^{(0)}(0) P_E(\Omega_\mathrm{i}) \ , \tag{5.3.44}$$

where $P_E^{(0)}(0)$ and $P_E^{(0)}(\Omega_\mathrm{i})$ are the reduced matrix elements of the static and dynamic electronic polarizability, respectively, cf. (5.3.27). Thus, even without taking into account the vibronic coupling, the electronic degeneracy of the ground state results in an anomalous Kerr effect in spherical top molecules with the Kerr constant proportional to T^{-1}. The estimates in [5.219] show that the appropriate contribution to $_\mathrm{m}B$ is comparable with the Kerr constant of anisotropically polarizable molecules. When considering the vibronic coupling, the result (5.3.44) changes. The appropriate changes can be taken into account by the factor $Q_E(T)$,

$$_\mathrm{m}B^{(\mathrm{or})} = \frac{6\pi N_A}{5kT} P_E^{(0)}(0) P_E^{(0)}(\Omega_\mathrm{i}) Q_E(T) \ , \tag{5.3.45}$$

where [5.220]

$$Q_E(T) = kT \int_0^{1/kT} d\beta_1 [\langle \hat{\sigma}_+(-i\beta_1 \hbar)\hat{\sigma}_- \rangle_E + \langle \hat{\sigma}_-(-i\beta_1 \hbar)\hat{\sigma}_+ \rangle_E] \ . \tag{5.3.46}$$

Here $\hat{\sigma}_+$ and $\hat{\sigma}_-$ are the Pauli matrices (3.1.1) in the space of the ground E term states. The quantity $Q_E(T)$ may be called the temperature-dependent vibronic reduction factor in the Kerr effect. The index E in $Q_E(T)$ indicates that the operator containing the E components of the polarizability is reduced. In the absence of vibronic coupling $Q_E(T) = 1$. In the limiting case of low temperatures when only the ground vibronic state is populated, the averaging in (5.3.45) can be reduced to the calculation of the matrix elements of the operators $\hat{\sigma}_+$ and $\hat{\sigma}_-$ by the wave functions of the ground vibronic state. One can easily see that in this case $Q_E = K^2(E)$, where $K(E)$ is the usual vibronic reduction factor introduced in Sect. 4.7.1 As the constant of vibronic coupling increases from zero to infinity, $K(E)$ lowers from 1 to 1/2, and hence Q_E changes from 1 to 1/4.

At high enough temperatures $Q_E(T)$ can be represented by the first terms of the expansion in $\beta = 1/kT$:

$$Q_E(T) \approx 1 - \frac{2E_{\mathrm{JT}}}{3kT} \ . \tag{5.3.47}$$

Substituting this expression into (5.3.45) we come to an interesting result:

$$_{\mathrm{m}}B^{(\mathrm{or})} \approx \frac{A}{kT} + \frac{B}{(kT)^2} \ , \quad \text{where} \tag{5.3.48}$$

$$A = \frac{6\pi N_A}{5} P_E^{(0)}(0) P_E^{(0)}(\Omega_{\mathrm{i}}) \ ; \qquad B = -\frac{2}{3} A E_{\mathrm{JT}} \ .$$

Thus in addition to the dependence on T^{-1}, a term is present characteristic of rigid-dipole molecules. In accordance with the estimates in [5.220], the corresponding contribution to the Kerr constant at room temperature may be more than 20% of the pure electronic one.

In [5.220] the electro-optical Kerr effect for the electronic Γ_8 term in cubic molecules of O_{h} symmetry was also discussed. The contribution to the Kerr constant in this case is given by the electronic polarizability of the E_g and T_{2g} types. If the vibronic coupling is taken into account, both these contributions are reduced by $Q_E(T)$ and $Q_{T_2}(T)$, for which approximate expressions are found for the limiting cases of high and low temperatures, respectively.

The analysis of the temperature-dependent reduction factors in the Kerr effect performed in [5.222] is based on the approximation employing the coordinate representation of the temperature dependent density matrix of the Jahn-Teller system. The results of this approximation agree well with different simple limiting cases of weak and strong vibronic coupling, and of high and low temperatures.

As mentioned in Sect. 5.3.2, the spherical top molecules with a degenerate ground state possess a quadrupole (or even a dipole, in cases with a dipole instability) moment. It is clear that such quadrupole molecules should orient themselves in homogeneous external electric fields resulting in optical anisotropy of the medium and birefringence. The additional contribution to the difference between the refractive indices of the ordinary and extraordinary rays $\Delta n^{(or)}$ which is due to the orientating action of the gradient of the electric field on molecules of cubic symmetry O_h or T_d in degenerate electronic states E or Γ_8 in the absence of a dipole instability is considered in [5.223]. It emerges that $\Delta n^{(or)}$ is proportional to the absolute value of the gradient of the electric field, and in the case of the ground E term, is described by an expression like (5.3.45), with the distinction that instead of $P_E^{(0)}(0)$ (the static polarizability) the reduced matrix element of the operator of the quadrupole moment Θ_E is introduced. The temperature dependence of $\Delta n^{(or)}$ is determined by the term proportional to T^{-1} and by the factor $Q_E(T)$, and is therefore rather complicated. Similar results were obtained in [5.223] for cubic molecules of O_h symmetry in the ground electronic Γ_8 state.

Similar to the Stark splitting of the vibronic and rotational energy levels determining the Kerr effect, the Zeeman splitting of these levels in an external magnetic field also results in optical anisotropy and birefringence. This is known as the Cotton-Mouton effect, whose theory is similar to that of the Kerr effect. The difference in the path length of the ordinary and extraordinary rays is $Cl\mathscr{H}^2$, where l is the optical path length of the ray of light in the medium, and $\mathscr{H}$ is the intensity of the magnetic field, C being the Cotton-Mouton constant, which depends on the composition of the matter, the wavelength of the light and the temperature. According to widespread opinion, spherical top molecules do not orient under the external magnetic field, and hence the only reason for their optical anisotropy is the influence of the magnetic field on the molecular electronic shell, which gives no temperature dependence to the Cotton-Mouton constant. As shown in [5.224] this conclusion is based on the implicit assumption of the absence of electronic degeneracy, and hence it is invalid in the case of electronic degeneracy. Neglecting rotational quantization, the molar Cotton-Mouton constant is given by the following expression [cf. (5.3.38)]:

$$_{\mathrm{m}}C = 2\pi N_A\{\mathscr{H}^{-2}[\langle P_{zz}(\Omega_{\mathrm{i}},\mathscr{H})\rangle - \langle P_{xx}(\Omega_{\mathrm{i}},\mathscr{H})\rangle]\}_{\mathscr{H}\to 0} , \tag{5.3.49}$$

where $P_{\alpha\beta}(\Omega_{\mathrm{i}},\mathscr{H})$ is the polarizability of the system in the magnetic field $\mathscr{H}$ parallel to the z axis, the notation otherwise being the same as in (5.3.38). By means of arguments similar to those yielding (5.3.41, 42), one can separate the orientational contribution to the Cotton-Mouton constant [cf. (5.3.41)] [5.224]

$$_{\mathrm{m}}C^{(or)} = 2\pi N_A \sum_{iklm} W^{iklm}\mathscr{F}_{iklm} , \quad \text{where} \tag{5.3.50}$$

$$W^{iklm} = \tfrac{1}{30}(3\delta_{il}\delta_{km} + 3\delta_{im}\delta_{kl} - 2\delta_{ik}\delta_{lm}) ,$$

while the tensor $\mathscr{F}_{iklm}$ consists of three terms:

$$\mathscr{F}^{(1)}_{iklm} = \int_0^{\beta} d\beta_1 \int_0^{\beta_1} d\beta_2 \langle \mu_i(-i\beta_1\hbar)\mu_k(-i\beta_2\hbar)P_{lm}(\Omega_i)\rangle_E \; ; \tag{5.3.51}$$

$$\mathscr{F}^{(2)}_{iklm} = \tfrac{1}{2}\int_0^{\beta} d\beta_1 \langle \kappa_{ik}(-i\beta_1\hbar)P_{lm}(\Omega_i)\rangle_E \; ; \tag{5.3.52}$$

$$\mathscr{F}^{(3)}_{iklm} = \int_0^{\beta} d\beta_1 \langle \mu_i(-i\beta_1\hbar)\chi_{k,lm}(\Omega_i)\rangle_E \; . \tag{5.3.53}$$

Here $\beta = 1/kT$, μ_i are the components of the vector of the magnetic moment, κ_{ik} is the tensor of magnetic susceptibility $\kappa_{ik} = 2\mu_i G \mu_k$, where $G = (H - E)^{-1}$ at $E = 0$ takes into account all the excited states,

$$G' = \sum_{k \neq 0} \frac{|k\rangle\langle k|}{E_k}$$

and $\chi_{k,lm}$ is the operator of magneto-electric hyperpolarizability,

$$\chi_{k,lm}(t) = \mu_k G D_m(t) D_l + D_m(t) D_l G \mu_k - \frac{i}{\hbar}\int_0^t dt_1 [\mu_k(t_1), D_m(t)] D_l \; .$$

Accordingly, the orientational contribution to the Cotton-Mouton constant also has three components. In the absence of vibronic coupling the electronic operators μ_i, μ_k and κ_{ik} in the expressions (5.3.51–53) commute with the Heisenberg exponents and do not depend on β_1 and β_2. Performing the integration, we come to the following temperature dependence. The first contribution to the Cotton-Mouton constant, emerging from (5.3.51), is quadratic in T^{-1}. The second and third contributions, due to (5.3.52) and (5.3.53) respectively, are linear in T^{-1}. Usually, for paramagnetic molecules the first contribution is dominant, and the temperature dependence as a whole is ${}_{\mathrm{m}}C \sim T^{-2}$. For diamagnetic molecules the first and the third contributions are zero (since the reduced matrix element of the operator μ_i is zero), and hence ${}_{\mathrm{m}}C \sim T^{-1}$.

If the vibronic coupling is taken into account, all three contributions are reduced by temperature-dependent reduction factors, similar to $Q_\Gamma(T)$. Therefore the resulting temperature dependence of the Cotton-Mouton constant is rather complicated. Nevertheless, in the limiting case of very low temperatures when the population of excited vibronic states may be neglected, the temperature-dependent reduction factors transform to a combination of the usual vibronic reduction factors of electronic operators independent of temperature. In this case the temperature dependence is qualitatively the same as without the vibronic coupling. At higher temperatures the temperature dependence becomes more complicated, and unambiguous conclusions about the paramagnetic properties of molecules from the dependence of ${}_{\mathrm{m}}C$ on T^{-1} become impossible.

In [5.224] the general theory of the anomalous Cotton-Mouton effect for Jahn-Teller molecules including the analysis of the temperature behavior of the vibronic reduction factors at high and low temperatures is given, and examples of birefringence by cubic molecules in the Γ_8 ground state are considered.

The problems considered in this section are also related to the optical activity of Jahn-Teller molecules in Rayleigh and hyper-Rayleigh light scattering induced by external magnetic fields. This topic is considered in detail in [5.225].

5.4 The Dynamic Jahn-Teller Effect in Magnetic Resonance Spectra

The method of electronic paramagnetic resonance (EPR) played a primary role in the development of the ideas about the nature of the Jahn-Teller effect. Even the first experiments of *Bleaney* and co-workers [5.226, 227] on the temperature dependence of the EPR spectra of the Jahn-Teller ion Cu^{2+} performed in the early 1950s demonstrated the naiveté of the classical ideas about the static nature of the instability of the nuclear framework resulting from the Jahn-Teller effect. Qualitatively correct ideas about the essence of the phenomenon were given in [5.228] with an interpretation of Bleaney's experiments at high temperatures. The progress in this area achieved since then is presented in quite a number of papers on the Jahn-Teller effect in magnetic resonance spectra. A detailed bibliography can be found in the review articles [5.229–231] and monographs [5.232, 233] (see also [5.234, 235]) together with a presentation of the problem as a whole.

In general it is quite clear that the complicated nuclear dynamics due to the vibronic interaction should strongly influence the EPR spectra. Without considering this dynamics the expected EPR spectra are formed only by electronic states, the interaction with nuclear motion being taken into account only in the relaxation processes. If the vibronic interaction is taken into account the electronic and nuclear motion cannot be separated, the latter thus taking part directly in the formation of the spectrum itself (the number, position and shape of the lines, their intensity and temperature dependence). This influence is rather important in the case of strong vibronic coupling when the value of the tunneling splitting is of the same order of magnitude as the Zeeman splitting.

As already mentioned in Sect. 4.7 the ground vibronic term arising if the vibronic interaction is taken into account has the same transformation properties as the initial Jahn-Teller electronic term. In cases when the excited vibronic states are not populated, and in any case when their mixing with the ground vibronic multiplet by low-symmetry perturbations can be neglected, the spectrum of magnetic resonance is determined by the properties of the lowest vibronic level only. Low-symmetry perturbations here are the Zeeman interaction with the external homogeneous magnetic field as well as the interaction with the field of random strain of the crystal lattice (Sect. 5.4.2). The splittings of the vibronic levels due to these perturbations are usually less than or of the order of 5 cm^{-1}.

The contribution of the lowest excited vibronic level to the EPR spectrum at low temperatures ($T \sim 4$ K) can therefore be neglected if $\delta \gg 5\ \mathrm{cm}^{-1}$, where δ is the magnitude of the tunneling splitting (Sect. 4.3). Just this situation will be considered in the present section. The opposite case of small δ values will be discussed in Sect. 5.5.

5.4.1 Effect of Vibronic Reduction of the Angular Momentum of Electrons in EPR Spectra. Two-State Model

The magnetic resonance spectrum of a Jahn-Teller system is usually described by an effective Hamiltonian. In general it can be written as a tensor convolution of the form

$$\hat{H} = \sum_{\Gamma\gamma} G_{\Gamma\gamma} \hat{C}_{\Gamma\gamma} , \tag{5.4.1}$$

where, as before, $\hat{C}_{\Gamma\gamma}$ are the matrices of the Clebsch-Gordan coefficients determined in the space of the orbital states of the Jahn-Teller electronic term and $G_{\Gamma\gamma}$ are symmetrized combinations of the components of the vectors $\boldsymbol{\mathscr{H}}$ (magnetic field), $\boldsymbol{S}$ (electronic spin operator) and $\boldsymbol{I}$ (nuclear spin operator). For instance, for a cubic system

$$G_{A_1} = g_1 \beta \boldsymbol{\mathscr{H}S} + A_1 \boldsymbol{IS} + \gamma_1 \beta_{\mathrm{N}} \boldsymbol{\mathscr{H}I} , \tag{5.4.2a}$$

$$\begin{aligned} G_{E\theta} = {} & \tfrac{1}{2} g_2 \beta (3\mathscr{H}_z S_z - \boldsymbol{\mathscr{H}S}) + \tfrac{1}{2} A_2 (3 I_z S_z - \boldsymbol{IS}) \\ & + \tfrac{1}{2} \gamma_2 \beta_{\mathrm{N}} (3\mathscr{H}_z I_z - \boldsymbol{\mathscr{H}I}) + \tfrac{1}{2} P_2 (3 I_z^2 - \boldsymbol{I}^2) , \end{aligned} \tag{5.4.2b}$$

$$\begin{aligned} G_{T_2\xi} = {} & g_3 \beta (\mathscr{H}_y S_z + \mathscr{H}_z S_y) + A_3 (I_y S_z + I_z S_y) \\ & + \gamma_3 \beta_{\mathrm{N}} (\mathscr{H}_y I_z + \mathscr{H}_z I_y) + P_3 (I_y I_z + I_z I_y) , \end{aligned} \tag{5.4.2c}$$

$$G_{T_1 x} = g_L \beta \mathscr{H}_x + \zeta S_x + a I_x . \tag{5.4.2d}$$

The expressions for the remaining $G_{\Gamma\gamma}$ values can be obtained from (5.4.2) by means of symmetry transformations. Here β and β_{N} are the usual and the nuclear Bohr magnetons, respectively, g_L is the orbital g factor[3], the constant a determines the magnitude of the interaction between the nuclear magnetic moment and the magnetic field of the electrons. The physical meaning of the separate terms in the expressions (5.4.2) is well known and the constants of the Hamiltonian can be calculated, say in the crystal field approximation. For instance in second-order perturbation theory for cubic systems with the electronic configu-

---

[3] The g_L value introduced here differs from the true gyromagnetic ratio by the genealogic coefficient [5.232]. The genealogic factor is the ratio of the effective operator of angular momentum (effective spin) acting in the space of degenerate electronic states of a given electronic term Γ to the operator of the real angular momentum (spin operator) with appropriate quantum number L. This question is discussed in various books on EPR, e.g. [5.232].

ration d^n, we have

$$g_1 = g_1 - \frac{4\lambda}{10Dq}\,, \qquad g_2 = -\frac{4\lambda}{10Dq}\,, \qquad \gamma_1 = g_N\,,$$
$$A_1 = \mathscr{P}\left(\kappa + \frac{4\lambda}{10Dq}\right), \qquad A_2 = -\mathscr{P}\left(6\xi + \frac{4\lambda + g\lambda\xi}{10Dq}\right), \qquad \zeta = \lambda\,, \tag{5.4.3}$$

where λ is the constant of spin-orbital interaction, $10Dq$ is the crystal field parameter, $g_s = 2.0023$ is the spin g factor and g_N is the nuclear one, $\mathscr{P} = 2g_N\beta\beta_N\langle r^{-3}\rangle$ where $\langle r^{-3}\rangle$ is the one-electron averaged value of r^{-3}, κ is the constant of contact Fermi interaction of electrons with a nucleus, and ξ is determined by the electronic configuration. For a ^2D term, for example, $\xi = 2/21$.

Passing to the matrix representation of the Hamiltonian (5.4.1) in the basis of the exact vibronic states of the ground vibronic term we have (Sect. 4.7)

$$\hat{H} = \sum_{\Gamma\gamma} \tilde{G}_{\Gamma\gamma}\hat{C}^{(g)}_{\Gamma\gamma}\,, \tag{5.4.4}$$

where the $\hat{C}^{(g)}_{\Gamma\gamma}$ matrices have the same form as in (5.4.1) but they act in the space of the vibronic states of the ground vibronic term only, and $\tilde{G}_{\Gamma\gamma}$ differs from $G_{\Gamma\gamma}$ by the vibronic reduction factors. In some of the terms of the expressions (5.4.3) obtained in second-order perturbation theory second-order vibronic reduction (Sect. 4.7) has to be introduced: $\tilde{g}_1 = g_s - K^{(2)}(A_1)4\lambda/10Dq$, $\tilde{g}_2 = -K^2(E)4\lambda/10Dq$, etc.

It is worthwhile recalling that for large $10Dq$ values second-order vibronic reduction factors can be reduced to the first-order reduction factors, resulting in

$$\tilde{g}_1 = g_1\,, \qquad \tilde{g}_2 = K(E)g_2\,, \qquad \tilde{A}_1 = A_1\,, \qquad \tilde{A}_2 = K(E)A_2\,,$$
$$\tilde{g}_L = K(T_1)g_L\,, \qquad \tilde{\zeta} = K(T_1)\zeta\,. \tag{5.4.5}$$

Let us consider the structure of the magnetic resonance spectrum described by the Hamiltonian (5.4.4). To be specific we will deal with EPR for a cubic ^2E term. The generalization for other cases is rather trivial. Neglecting the hyperfine interaction for simplicity, using (5.4.4), (5.4.2) and (5.4.5) we can write the Zeeman Hamiltonian of the ground vibronic ^2E term as

$$\hat{H} = g_1\beta\mathscr{H}S\hat{\sigma}_0 + qg_2\beta\left(\frac{1}{2}(\mathscr{H}S - 3\mathscr{H}_zS_z)\hat{\sigma}_z + \frac{\sqrt{3}}{2}(\mathscr{H}_xS_x - \mathscr{H}_yS_y)\hat{\sigma}_x\right) \tag{5.4.6}$$

where the $\hat{\sigma}_i$ matrices have the same form as in Sect. 3.1 but act in the space of two vibronic states of the ground E term.[4] This is known as the two-state model.

[4] See the discussion after (3.1.6) concerning the unimportance of the difference in the sign of $\hat{C}_{E\theta} = \pm\hat{\sigma}_z$ for cases of axial and cubic symmetry groups.

When considering the two-fold spin degeneracy, the Hamiltonian (5.4.6) is represented by a 4 × 4 matrix. Its diagonalization can be performed by means of perturbation theory methods using the fact that $g_1 \gg g_2$. The zeroth-order Hamiltonian $H_0 = g_1 \beta \mathscr{H} S \hat{\sigma}_0$ can be diagonalized by means of passing to a new coordinate system with the unit vectors ξ, η, ζ oriented in such a way so that the ζ axis coincides with the axis of quantization determined by the vector $\mathscr{H}$, i.e., $\zeta \| \mathscr{H}$. Then $\hat{H}_0$ is given by a diagonal matrix resulting in two doubly degenerate levels with energies $E_{\pm}^{(0)} = \pm g_1 \beta \mathscr{H}/2$.

The perturbation matrix in the basis of these orbitally degenerate levels has the form

$$\hat{V} = \tfrac{1}{2} q g_2 \beta \mathscr{H} S_\zeta [(\zeta_x^2 + \zeta_y^2 - 2\zeta_z^2)\hat{\sigma}_z + \sqrt{30}(\zeta_x^2 - \zeta_y^2)\hat{\sigma}_x] , \tag{5.4.7}$$

where ζ_x, ζ_y, ζ_z are the direction cosines of the vector $\mathscr{H}$ with respect to the symmetry axes of the cubic polyatomic system. The diagonalization of the perturbation matrix (5.4.7) can be performed directly. As a result, we obtain the first-order corrections to the energies:

$$E_{1\pm} = E_{2\pm} = \pm \tfrac{1}{2} q g_2 \beta \mathscr{H} \sqrt{1 - 3(\zeta_y^2\zeta_z^2 + \zeta_x^2\zeta_z^2 + \zeta_x^2\zeta_y^2)} \ . \tag{5.4.8}$$

Here the quantum numbers 1, 2 identify the orbital states while $\pm$ denotes the spin states α and β with different values of the projection of the spin on the axis $\zeta \| \mathscr{H}$. Accordingly the wave functions have a multiplicative form $\psi_{1,2}\alpha$ and $\psi_{1,2}\beta$. Since the transition operator is $g_1 \beta(\mathscr{H}_+ S_- + \mathscr{H}_- S_+)$, the Zeeman interaction of the spin with the transverse wave, only two transitions $\psi_1\beta \rightleftarrows \psi_1\alpha$, $\psi_2\beta \rightleftarrows \psi_2\alpha$ are allowed. The other transitions are forbidden by the selection rules with respect to the orbital quantum numbers 1 and 2. The appropriate g factors are

$$g_\pm = g_1 \pm q g_2 f \ , \qquad f = \sqrt{1 - 3(\zeta_y^2\zeta_z^2 + \zeta_x^2\zeta_z^2 + \zeta_x^2\zeta_y^2)} \ . \tag{5.4.9}$$

Note that when the Jahn-Teller effect is not taken into account the angular dependence of the g factors is quite different. Figure 5.23 gives these two dependences with ($q = 1/2$, Sect. 4.7) and without allowance for the Jahn-Teller effect

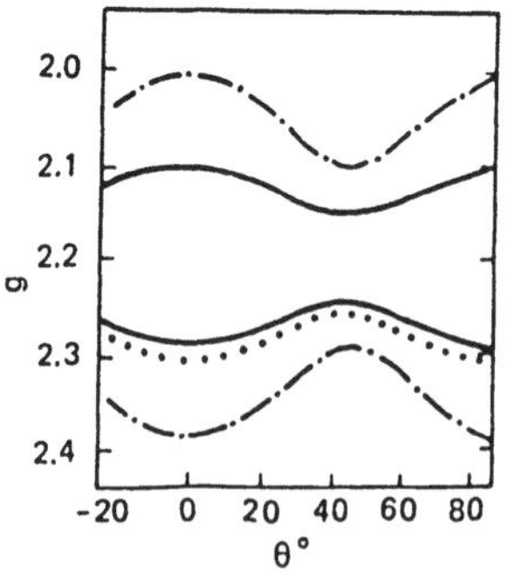

Fig. 5.23. Two limiting cases of angular dependence of the g-factor for the linear $E \otimes e$ problem with strong ($q = 1/2$, full line) and without ($q = 1$, dot-dash line) vibronic coupling. The experimental data are shown by points (after [5.236])

together with the experimental data marked by points obtained in [5.236, 237] in the case of Cu^{2+} : MgO.

The vibronic reduction of the orbital contribution to the g factor was subject to many experimental checks [5.229–231] confirming the main conclusions of the theory. The investigation of the EPR spectra and the acoustic paramagnetic resonance (APR) of the Ni^{3+} ion in corundum, where a trigonal crystal field exists and the ground Jahn-Teller state is ^4E, enabled the determination of both the reduction factor $K(E) = q$ and reduction factor $K(A_2) = p$ [5.238]. The relationship (4.7.9) turned out to be invalid for these experimentally found values of q and p. From this the authors of [5.238] concluded that in this case the Jahn-Teller effect has a significantly multimode nature. However, the reduction factors obtained in [5.238] describe the reduction of splittings in different (including higher) orders of the perturbation theory and therefore, in general, the relationship should not be valid [5.239]. Some new detailed experimental studies of the Ni^{3+} : Al_2O_3 system by EPR, by thermal conductivity and by Raman spectroscopy became available recently [5.231, 240]. However, one has to obtain still more experimental results to be able to test with certainty whether the present interpretation of the experimental data on Ni^{3+} : Al_2O_3 is good enough in practice. Other experimental examples of the EPR manifestations of the vibronic reduction can be found in the review article [5.231].

5.4.2 Effect of Random Strain

The result obtained above is related to the Jahn-Teller system with an ideal cubic reference configuration (Sect. 2.2.1). However, if the Jahn-Teller paramagnetic center is placed in a crystal of, say, cubic symmetry, it has to be taken into account that the crystalline environment of the center is never ideally cubic [5.229, 241]. Such imperfection in the crystal matrix has different causes: mosaic structure arising inevitably in the process of crystal growth, dislocations and other defects in the neighborhood of the paramagnetic center, local uncompensated charges, etc. As a result, near each paramagnetic center some local, low-symmetry field arises which removes the orbital degeneracy of the ground vibronic ^2E term. It can be described as a static low-symmetry strain $\hat{W}$. Considering the effect of the vibronic reduction for the ground vibronic ^2E term we have, cf. (5.2.6),

$$\hat{W} = qP_E(-e_\theta\hat{\sigma}_z + e_\varepsilon\hat{\sigma}_x) \ . \tag{5.4.10}$$

Here the fact that the orbital degeneracy of the ^2E term can be removed by tetragonal strain alone is taken into account.

It can easily be seen that the constants of electron-strain interaction P_Γ (5.1.35) are related to the constants of the linear vibronic coupling [5.229, 241]. Indeed the strain results in a change in the configuration of the atoms nearest to the paramagnetic center, i.e., in a nonzero symmetrized nuclear displacement, $Q_{\Gamma\gamma} \neq 0$. The latter are linked to the electrons by the vibronic interaction operator and this leads to the removal of the electronic degeneracy. It is obvious that $P_\Gamma \sim V_\Gamma$.

From this an important conclusion follows: the role of random local strain is more important for stronger vibronic coupling. For systems with a strong Jahn-Teller effect the interpretation of the magnetic resonance spectra has to be carried out allowing for the effect of random strain [5.229, 241].

In the case of large strain the hierarchy of perturbations is somewhat different from that considered above. In the first place the strain removes the orbital degeneracy. As a result of the diagonalization of the matrix (5.4.10) we obtain two Kramers doublets:

$$\begin{aligned} E_1 &= qeP_E\ , \qquad \psi_1 = \frac{1}{\sqrt{2}}(\sqrt{1-\cos\varphi}\,\psi_\theta + \sqrt{1+\cos\varphi}\,\psi_\varepsilon)\ , \\ E_2 &= -qeP_E\ , \qquad \psi_2 = \frac{1}{\sqrt{2}}(-\sqrt{1+\cos\varphi}\,\psi_\theta + \sqrt{1-\cos\varphi}\,\psi_\varepsilon)\ , \end{aligned} \tag{5.4.11}$$

where e is the strain and φ is its orientation angle:

$$e = \sqrt{e_\theta^2 + e_\varepsilon^2}\ , \qquad \varphi = \arctan\left(\frac{e_\varepsilon}{e_\theta}\right)\ . \tag{5.4.12}$$

The spin Hamiltonian of each of these two Kramers doublets can be obtained by averaging the Hamiltonian (5.4.6) over the orbital states (5.4.11):

$$\begin{aligned} \bar{H}_{1,2} &= g_1\beta\mathscr{H}\boldsymbol{S} \pm \tfrac{1}{2}qg_2\beta[(\mathscr{H}\boldsymbol{S} - 3\mathscr{H}_zS_z)\cos\varphi \\ &\quad + \sqrt{3}(\mathscr{H}_xS_x - \mathscr{H}_yS_y)\sin\varphi]\ . \end{aligned} \tag{5.4.13}$$

The diagonalization of the Hamiltonian (5.4.13) can be performed as in the above case using the perturbation theory with the small parameter g_2. In order to diagonalize the zeroth-order Hamiltonian $g_1\beta\mathscr{H}\boldsymbol{S}$ let us pass to the coordinate system with the unit vectors $\boldsymbol{\xi}$, $\boldsymbol{\eta}$ and $\boldsymbol{\zeta}$ oriented in such a way that $\boldsymbol{\zeta}\,\|\,\mathscr{H}$. Then $g_1\beta\mathscr{H}\boldsymbol{S} = g_1\beta\mathscr{H}S_\zeta$, that is, in the zeroth order both Kramers doublets are split by the same magnitude $g_1\beta\mathscr{H}$. The contribution to the diagonal matrix elements of the perturbation is made only by the terms in (5.4.13) proportional to S_ζ, that is,

$$V = \pm\tfrac{1}{2}qg_2\beta\mathscr{H}S_\zeta[(\zeta_x^2 + \zeta_y^2 - 2\zeta_z^2)\cos\varphi + \sqrt{3}(\zeta_x^2 - \zeta_y^2)\sin\varphi]\ . \tag{5.4.14}$$

From this we obtain the energies of the split Kramers doublets:

$$\begin{aligned} E_{1\pm} &= geP_E \pm \tfrac{1}{2}\beta\mathscr{H}[g_1 + qg_2 f\cos(\varphi - \alpha)]\ , \\ E_{2\pm} &= geP_E \pm \tfrac{1}{2}\beta\mathscr{H}[g_1 - qg_2 f\cos(\varphi - \alpha)]\ , \end{aligned} \tag{5.4.15}$$

Here the angle α is introduced which gives the orientation of the vector $\mathscr{H}$ with respect to the symmetry axes of the cubic system,

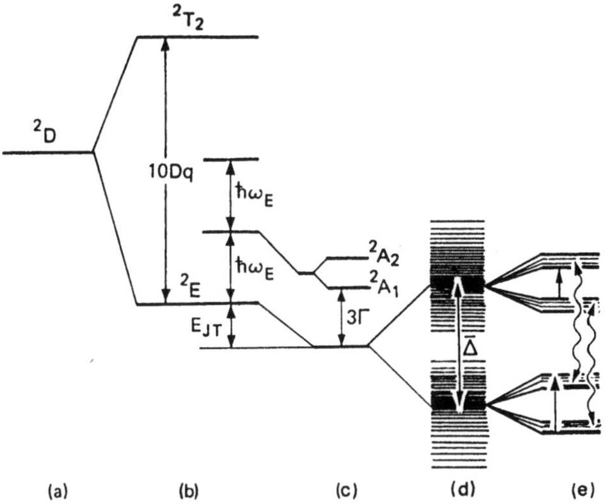

Fig. 5.24a–e. Lowest energy level scheme for an octahedral system in the ^2E state: (**a**) free ion term; (**b**) crystal field splitting; (**c**) tunneling splitting; (**d**) the splitting of the ground vibronic ^2E level into two Kramers doublets by random strain; (**e**) Zeeman splitting by magnetic fields. Allowed transitions in the radio frequency field are shown by straight arrows, whereas the relaxation transitions without spin reversal are displayed by wavy arrows. (After [5.242])

$$\tan\alpha = \frac{\sqrt{3}(\zeta_x^2 - \zeta_y^2)}{2\zeta_z^2 - \zeta_x^2 - \zeta_y^2} \ . \tag{5.4.16}$$

The scheme of energy levels (5.4.15) is given in Fig. 5.24.

The operator of radio-frequency transition $\beta(\mathscr{H}_+ S_- + \mathscr{H}_- S_+)$ is diagonal in the space of the orbital states (5.4.11) and therefore transitions between the states of different Kramers doublets are forbidden. As follows from (5.4.15) the frequencies of allowed transitions correspond to the g factors

$$g^{(1,2)} = g_1 \pm \tfrac{1}{2} q g_2 f \cos(\varphi - \alpha) \ . \tag{5.4.17}$$

In the framework of the perturbation theory used in the deduction of this result the wave functions of the system can be written as the products of the zeroth-order functions $|im_s\rangle = |i\rangle|m_s\rangle$, where m_s is the quantum number of the operator S_ζ of the projection of the spin on the axis of quantization.

The operator which describes the interaction of the transverse radio wave with the spin has the form

$$H_{\text{int}} = \beta(\mathscr{H}_+^{(0)} S_- e^{2\pi i \nu t} + \mathscr{H}_-^{(0)} S_+ e^{-2\pi i \nu t}) \ , \tag{5.4.18}$$

where $\mathscr{H}_{+(-)}^{(0)} = \mathscr{H}_\xi^{(0)} + (-)\, i\mathscr{H}_\eta^{(0)}$ is the amplitude of the right (left) circularly polarized transverse radio wave, ν is its frequency, $S_\pm = S_\xi \pm iS_\eta$.

The probability of the transition,

$$W_{\mathrm{i},m_S\to \mathrm{i},m_S\pm 1} = \frac{2\pi}{\hbar}|\langle \mathrm{i}, m_S \pm 1|\beta\mathscr{H}_{\mp}^{(0)}S_{\pm}|\mathrm{i}, m_S\rangle|^2\delta(h\nu - h\nu_{\mathrm{i}}) \ , \tag{5.4.19}$$

is independent of the orientation of the strain or the magnetic field. However, the transition frequency, that is, the EPR line position, is dependent on both of these. Due to the random distribution of the strain the frequency of the EPR signal for different centers is somewhat different, resulting overall in an inhomogeneous broadening of the EPR line.

Considering the temperature population of the Kramers doublets one can obtain the following expression from (5.1.2) and (5.4.19) for the form function of the radio-frequency absorption:

$$F = n_1\delta(h\nu - g^{(1)}\beta\mathscr{H}) + n_2\delta(h\nu - g^{(2)}\beta\mathscr{H}) \ , \tag{5.4.20}$$

where n_i are the Boltzmann occupation numbers,

$$n_i = \frac{1}{Z}\mathrm{e}^{-E_i/kT} \ , \qquad Z = \sum_i \mathrm{e}^{-E_i/kT} \ . \tag{5.4.21}$$

Here E_i are the energies of the Kramers doublets (5.4.11) and $g^{(1)}$ and $g^{(2)}$ are determined by (5.4.17). When writing (5.4.21) and hereafter for simplicity we neglect the difference in the temperature population of the Zeeman sublevels of each of the Kramers doublets.

Since the resonance value of the magnetic field for different centers is somewhat different owing to the different orientations of the random strain, the form function of the line should be averaged over the angle φ. On assuming that different φ values are equally probable, one can reduce the averaging to integration:

$$\overline{\delta(\hbar\nu - g^{(i)}\beta\mathscr{H})} = \frac{1}{2\pi}\int_0^{2\pi}\delta(h\nu - g^{(i)}\beta\mathscr{H})\,d\varphi \ . \tag{5.4.22}$$

Substituting (5.4.17) into (5.4.22) and performing the integration one can easily obtain [5.242]

$$\overline{\delta(h\nu - g^{(i)}\beta\mathscr{H})} = \frac{1}{\beta\mathscr{H}}\theta[qg_2\beta\mathscr{H}f)^2 - (h\nu - g_1\beta\mathscr{H})^2]\left.\frac{\partial g^{(i)}}{\partial\varphi}\right|_{\varphi=\varphi_i}^{-1} \ , \tag{5.4.23}$$

where $\theta(z)$ is a step function ($\theta = 1$ if $z > 0$, and $\theta = 0$ if $z < 0$) and φ_i is the solution of the transcendental equation $q^{(i)}\beta\mathscr{H} = h\nu$,

$$\varphi_{1,2} = \alpha + \arccos\left(\mp\frac{2(h\nu - g_1\beta\mathscr{H})}{qg_2\beta\mathscr{H}f}\right) \ . \tag{5.4.24}$$

Substituting (5.4.24) into (5.4.23), (5.4.20) and using the fact that $n_1 + n_2 = 1$ we

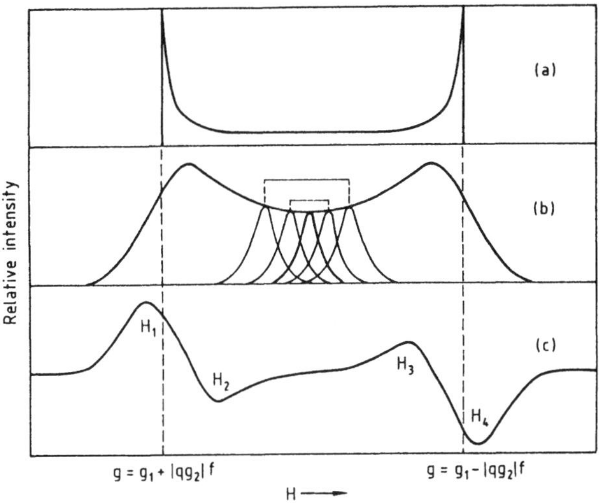

Fig. 5.25a–c. EPR line shape for the ground ^2E state taking into account the random strain averaged over all orientations: (**a**) under the assumption that the individual transitions contributing to the envelope correspond to a δ function; (**b**) the individual transitions are Gaussian bands with a halfwidth equal to 0.2Ω. (**c**) The first derivative of the function given in (**b**). (After [5.242])

finally obtain [5.241, 242]

$$F = \frac{1}{\pi} \frac{\theta[(qg_2\beta\mathscr{H}f)^2 - (h\nu - g_1\beta\mathscr{H})^2]}{\sqrt{(qg_2\beta\mathscr{H}f)^2 - (h\nu - g_1\beta\mathscr{H})^2}} \; . \tag{5.4.25}$$

The EPR line determined by the expression (5.4.25) is illustrated in Fig. 5.25a.

In order to consider the hyperfine interaction the following terms from (5.4.2a) and (5.4.2b) have to be added to the Hamiltonian:

$$A_1 \boldsymbol{IS}\hat{\sigma}_0 + \tfrac{1}{2}qA_2[(\boldsymbol{IS} - 3I_zS_z)\hat{\sigma}_z + \sqrt{3}(I_xS_x - I_yS_y)\hat{\sigma}_x] \; . \tag{5.4.26}$$

The diagonalization of the Hamiltonian including these terms can be performed in the same way as above by means of the perturbation theory using the fact that qg_2/g_1, qA_2/A_1, and $A_1/g_1\beta\mathscr{H}$ are small. As a result, we obtain resonances at the frequencies

$$h\nu_{1,2}(m_I) = (g_1\beta\mathscr{H} + A_1 m_I) \pm \tfrac{1}{2}q(g_2\beta\mathscr{H} + A_2 m_I)f\cos(\varphi - \alpha) \tag{5.4.27}$$

and instead of one line (5.4.25) we obtain the superposition of $2I + 1$ lines,

$$F(m_I) = \frac{1}{\pi} \frac{\theta[(qg_2\beta\mathscr{H} + gA_2m_I)^2 - (h\nu - g_1\beta\mathscr{H} - A_1m_I)^2]}{\sqrt{(qg_2\beta\mathscr{H} + qA_2m_I)^2 - (h\nu - g_1\beta\mathscr{H} - A_1m_I)^2}} \; . \tag{5.4.28}$$

Here m_I is the quantum number of the operator I_ζ, the projection of the nuclear spin on the axis of quantization.

5.4.3 Motional Narrowing of the EPR Broadened Band

The result (5.4.28) is obtained without consideration of the dissipative subsystem, that is, without considering the possible relaxation processes determined by the interaction of the Jahn-Teller system with the crystal or other media, for instance with the solvent. Due to the interaction with the thermal reservoir there is a nonzero probability of transitions between the spin-vibronic states with a simultaneous irradiation or absorption of phonons. In this case, the occupation numbers n_j begin to depend on time.

Strictly speaking the function $n_j(t)$ can be obtained only as a result of the solution of the Liouville equation for the density matrix. However, qualitatively $n_j(t)$ can be obtained using simple considerations of particle balance [5.243].

$$\dot{n}_j(t) = \sum_{k \neq j} W_{k \to j} n_k(t) - n_j(t) \sum_{k \neq j} W_{j \to k} \ , \tag{5.4.29}$$

where $W_{j \to k}$ is the probability per second of the transition $|j\rangle \to |k\rangle$.

The system of equations (5.4.29) was obtained under rather general assumptions and describes arbitrary relaxation processes. For Jahn-Teller systems the transitions without spin-flip, that is processes of orbital relaxation, are the most important. For the ^2E term under consideration these relaxation transitions are illustrated in Fig. 5.24d with wavy lines.

The kinetic equations (5.1.14) in this case have a simple form:

$$\dot{n}_1 = -R_1 n_1 + R_2 n_2 \ , \qquad \dot{n}_2 = R_1 n_1 - R_2 n_2 \ , \tag{5.4.30}$$

where $R_1 = W_{1 \to 2}$ and $R_2 \to W_{2 \to 1}$.

As the initial condition for the function $n_j(t)$, obviously the values (5.4.21) can be chosen. Strictly speaking, the quantum transitions between the states (5.4.11) change their equilibrium populations somewhat but these changes are negligible. Summing (5.4.30) we obtain $\dot{n}_1 + \dot{n}_2 = 0$, that is, $n_1(t) + n_2(t) = \text{const}$. It is obvious that one has to take const $= 1$ since from (5.4.21) it follows that $n_1(0) + n_2(0) = 1$.

In their physical sense the occupation numbers $n_j(t)$ are equal to the diagonal matrix elements of the reduced density matrix of the dynamic subsystem, $n_j(t) = \langle j | \hat{\varrho}_e(t) | j \rangle$. Therefore we can rewrite (5.4.29) in the different form

$$\dot{\varrho}_e(t) = \frac{1}{i\hbar} [\hat{H}_e, \hat{\varrho}_e(t)] + R\{\hat{\varrho}_e(t)\} \ , \tag{5.4.31}$$

where H_e is the Hamiltonian of the dynamic (electronic) subsystem, and R is a linear operator determined by the equation

$$\langle i | R\{\hat{\varrho}_e(t)\} | j \rangle = \sum_{k,n} R_{ij,kn} \langle k | \hat{\varrho}_e(t) | n \rangle \ . \tag{5.4.32}$$

Indeed, if one assumes that

$$R_{11,11} = -R_1 , \quad R_{11,22} = R_2 ,$$
$$R_{22,11} = R_1 , \quad R_{22,22} = -R_2 , \tag{5.4.33}$$

and all the other matrix elements $R_{ij,kn}$ are zero, then the system of equations (5.4.30) is obtained for the diagonal elements of the reduced density matrix $\hat{\varrho}_e(t)$ from (5.4.31) while for the nondiagonal matrix elements the usual Liouville equation remains.

The equation of motion (5.4.31) for the reduced density matrix is called the Bloch-Redfield equation [5.243], while the matrix composed from the numbers $R_{ij,kn}$ is called the relaxation matrix. As follows from the above discussion, the relaxation matrix takes into account the quantum transitions between the states of the dynamic subsystem induced by weak interaction with the thermal reservoir. In the absence of the latter all the $R_{ij,kn} = 0$, and (5.4.31) is reduced to the usual Liouville equation for the density matrix of the electrons.

The Bloch-Redfield equation can also be obtained directly from first principles if one considers the interaction with the thermal reservoir by means of perturbation theory [5.243]. In doing so an assumption about the essential continuity of the energy spectrum of the reservoir has to be introduced. In other words the density of states of the reservoir has to be a rather smooth function not containing narrow resonances. In the opposite case the possible occurrence of quantum beats between the states of the discrete energy spectrum of the dynamic subsystem and the reservoir not described by the Bloch-Redfield equation has to be considered. If the density of the thermostat states is a smooth function and the conditions for the occurrence of beats are not fulfilled then the local excitation of the reservoir dissipates rapidly enough (disperses). This is equivalent to short correlation times for excitation of the thermal reservoir.

Short correlation times are realized in a wide class of the so-called dilute magnetic systems. Jahn-Teller paramagnetic impurities in nonmagnetic crystals with relatively broad phonon bands correspond to these systems.

Considering the interaction of the radio wave with the electrons spins the operator (5.4.18) has to be added to the Hamiltonian H_e in (5.4.31), resulting in

$$\dot{\varrho}_e(t) = [H_e + H_{\text{int}}(t), \varrho_e(t)] + R\{\varrho_e(t)\} \ . \tag{5.4.34}$$

Following the theory of the linear response of the system to external perturbation [5.244] we look for the solution of (5.4.34) in the form $\varrho_e(t) = \varrho_0 + \varrho_1(t)$, where ϱ_0 is the solution of (5.4.34) in the absence of the external perturbation. It can be shown that ϱ_0 can be taken in the form

$$\varrho_0 = \frac{1}{Z_e} \exp\left(-\frac{H_e}{kT}\right) , \quad Z_e = \mathrm{Tr}\left\{\exp\left(-\frac{H_e}{kT}\right)\right\} , \tag{5.4.35}$$

that is, the usual density matrix in the case of thermodynamical equilibrium, and $R\{\varrho_0\} = 0$ [5.243]. Considering H_{int} as a small perturbation and keeping in

(5.4.34) the small terms not higher than the first order we can rewrite (5.4.34) in the form

$$\dot{\varrho}_1 - \frac{1}{i\hbar}[H_e, \varrho_1] - R\{\varrho_1\} = \frac{1}{i\hbar}[H_{\text{int}}, \varrho_0] \ . \tag{5.4.36}$$

If one writes this operator equation in a matrix form for the unknown functions $\langle j|\varrho_1(t)|j\rangle$ a system of linear inhomogeneous differential equations of the first order can be obtained. Taking into account that the time dependence of the inhomogeneous part has a simple exponential form (5.4.18) we look for a solution in the form

$$\varrho_1(t) = \mathrm{r}e^{2\pi i\nu t} + \mathrm{r}^\dagger e^{-2\pi i\nu t} \ . \tag{5.4.37}$$

Substituting (5.4.37) into (5.4.36) and equating the coefficients of equal exponents we obtain

$$2\pi i\nu\mathrm{r} - \frac{1}{i\hbar}[H_e, \mathrm{r}] - R\{\mathrm{r}\} = \beta\mathscr{H}_+^{(0)}[S_-, \varrho_0] \ . \tag{5.4.38}$$

If one passes to the matrix form of this operator equation one obtains a system of s^2 linear inhomogeneous algebraic equations of the first order in terms of s^2 unknowns $\langle i|\mathrm{r}|j\rangle$. Here s is the number of the states of the electronic subsystem taken into account in the calculations. In the above example of the ^2E terms, $s = 4$. The solution of the algebraic system of equations can be written in the matrix form

$$\hat{\mathrm{r}} = \beta\mathscr{H}_+^{(0)}(2\pi i\nu - i\hat{\omega} - \hat{\boldsymbol{R}})^{-1}[\hat{S}_-, \hat{\varrho}_0] \ , \tag{5.4.39}$$

where $\hat{\omega}$ and $\hat{\boldsymbol{R}}$ are $s^2 \times s^2$ matrices

$$\omega_{ij,kn} = \omega_{ij}\delta_{ik}\delta_{jn} = \frac{1}{\hbar}(E_j - E_i)\delta_{ik}\delta_{jn} \ .$$

The elements of the matrix $\hat{\boldsymbol{R}}$ are determined in (5.4.33) where it is assumed that the relaxation mixes only orbital states with the same spin quantum numbers. In other words it is assumed that the matrix $\hat{\boldsymbol{R}}$ is diagonal with respect to the spin quantum numbers[5].

The energy A which is absorbed over one period of the magnetic field $\mathscr{H}_\perp(t)$ of the radio wave can be found using the general formula of electrodynamics [5.245],

$$A = 2\beta \oint \mathscr{H}_\perp(t) d\langle \boldsymbol{S}_\perp(t)\rangle \ , \tag{5.4.40}$$

[5] This assumption is not obligatory. Considering the matrix elements to be nondiagonal in spin quantum numbers is equivalent to including the spin-flip relaxation transitions, i.e., the spin-lattice relaxation processes.

hence the intensity of the energy absorption at the frequency ν of the magnetic field is $K(\nu) = \nu A$. Using the above density matrix $\hat{\varrho}_e(t) = \hat{\varrho}_0 + \hat{\varrho}_1(t)$ where $\hat{\varrho}_1(t)$ has the form (5.4.37) for the calculation of $\langle \hat{S}_\perp(t) \rangle$, and bearing in mind that $\mathrm{Tr}\{\hat{\varrho}_0 \hat{S}_\perp\} = 0$ we obtain from (5.4.40) the relation

$$K(\nu) = -8\pi\beta\nu^2 \,\mathrm{Im}\{\mathscr{H}_-^{(0)} \mathrm{Tr}\{\hat{\mathrm{r}}\hat{S}_+\}\} . \qquad (5.4.41)$$

Substituting $\hat{\mathrm{r}}$ from (5.4.39) and assuming for simplicity that $R_1 \approx R_2 \approx R$ and $n_1(0) = n_2(0) = 1/2$ we obtain for the ^2E term considered above [5.246],

$$K(E) \sim \frac{1}{\pi} \frac{2R\gamma^2 h^2}{(E^2 - \gamma^2)^2 + 4E^2R^2h^2} , \qquad (5.4.42)$$

where

$$E = h\nu - g_1\beta\mathscr{H} , \qquad \gamma = \tfrac{1}{2}|h\nu_1 - h\nu_2| = \tfrac{1}{2}qg_2\beta\mathscr{H}f\,|\cos(\varphi - \alpha)| . \qquad (5.4.43)$$

The relaxation parameter R increases with temperature (see below). At not very high temperatures when $R \ll 2\gamma/h$ the expression (5.4.42) describes two Lorentz-shape lines centered at frequencies $h\nu = g_1\beta\mathscr{H} \pm \gamma$:

$$K(E) \sim \frac{h}{2\pi}\left(\frac{Rh}{(E-\gamma)^2 + R^2h^2} + \frac{Rh}{(E+\gamma)^2 + R^2h^2}\right) . \qquad (5.4.44)$$

At higher temperatures when $R \gg 2\gamma/h$ (5.4.42) describes one Lorentz-shape line centered at $h\nu = g_1\beta\mathscr{H}$:

$$K(E) \sim \frac{h}{\pi} \frac{hR_{\mathrm{eff}}}{E^2 + h^2R_{\mathrm{eff}}^2} , \qquad R_{\mathrm{eff}} = \frac{\gamma^2}{2Rh^2} . \qquad (5.4.45)$$

Thus as a result of relaxation transitions without spin-flip processes the EPR line which had a well-resolved doublet structure at low temperatures is narrowed and transforms into one intense narrow peak when the temperature rises. This phenomenon has a simple explanation. At low temperatures the relaxation transitions take place relatively seldom and the act of measurement is quick enough to "see" the paramagnetic center in one of the two Kramers doublets. The resulting EPR picture has two lines at the frequencies $h\nu_\pm = g_1\beta\mathscr{H} \pm \gamma$. At high temperatures the relaxation transitions take place so fast that the EPR is observed at an averaged frequency $\frac{1}{2}(h\nu_+ + h\nu_-) = g_1\beta\mathscr{H}$.

The *phenomenon of temperature narrowing of the* EPR *line* is in principle analogous to the known phenomenon of exchange narrowing in concentrated magnetic systems and to motional narrowing in liquids [5.247]. Similar temperature effects are observed in the hyperfine structure of the Mössbauer line.

The parameter γ in (5.4.41–45) as well as the $K(\mathscr{H})$ line shape is dependent on the orientation φ of the random strain. In order to evaluate the expected line shape, the function $K(\mathscr{H})$ has to be averaged over φ. The characteristic features of the averaged line shape are clear without performing the integration. At low

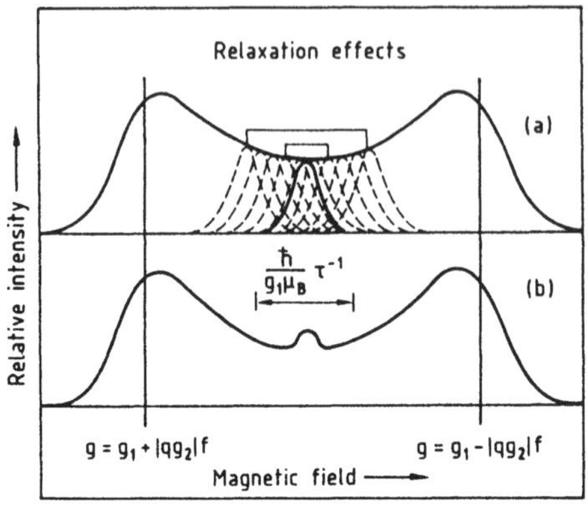

Fig. 5.26a, b. Influence of relaxation transitions without spin reversal on the EPR line shape—temperature narrowing. (**a**) The line shape at low temperatures when the number of pairs of transitions for which the condition for narrowing is fulfilled is small. (**b**) At higher temperatures an isotropic line occurs in the center of the band on account of the intensity of the wings. (After [5.242])

temperatures when for most angles φ the inequality $Rh \ll 2\gamma$ holds the EPR line has the shape given in Fig. 5.26a similar to that without considering relaxation (Fig. 5.25). The smoothed wings of the line are explained by the small broadening of the order of R of each of the lines (5.4.44) which form the inhomogeneous broadening of the EPR line. The center of the band at the frequency $h\nu = g_1 \beta \mathscr{H}$ has an isotropic angular dependence whereas the angular dependence of the wing lines is determined by the factor f from (5.4.9).

There are always regions of φ values for which $\varphi - \alpha \approx \pm\pi/2$ and the inequality $hR \ll 2\gamma$ is violated (5.4.43). However, at low temperatures R is relatively small and the contribution of these lines to the band in Fig. 5.26a is negligible.

With a rise in temperature the number of pairs of lines for which the criterion of narrowing, $hR \gg 2\gamma$, is fulfilled increases and, as a result, these lines sum their contributions, to form one isotropic line at the frequency $h\nu = g_1 \beta \mathscr{H}$ (Fig. 5.26b). Thus the increases in temperature results in the formation of an isotropic line in the central part of the band while the intensity of the wings decreases. In other words the intensity of the wings is transferred to the isotropic line, the integral intensity of the band remaining constant.

The calculation of R can be carried out by standard methods, e.g., by the method of Green's functions [5.246, 248, 249]. Here it is important to separate the dynamic subsystem from the remaining dissipative one weakly interacting with it. The physical meaning of R_1 and R_2 is that they are simple probabilities of radiationless transitions $|1\rangle \to |2\rangle$ and $|2\rangle \to |1\rangle$ between the states of the dynamic subsystem.

In the limiting case of weak vibronic coupling the spin-orbital states of the electrons in a rigid lattice can be chosen as the dynamic subsystem while the free lattice vibrations may serve as a thermal reservoir. The vibronic interaction in this case is reduced to the interaction of the dynamic subsystem with the reservoir. Using the usual perturbation theory expressions for the probabilities of the radiationless transitions $|1\rangle \to |2\rangle$ and $|2\rangle \to |1\rangle$ we obtain to first order [5.246]

$$R_1 = 2\pi q^2 V_E^2 \Delta^{-1}[\bar{n}(\Delta/\hbar) + 1]\varrho_E^{(0)}(\Delta/\hbar) \; ,$$
$$R_2 = 2\pi q^2 V_E^2 \Delta^{-1}\bar{n}(\Delta/\hbar)\varrho_E^{(0)}(\Delta/\hbar) \; , \tag{5.4.46}$$

where q is the vibronic reduction factor in the multimode $E \otimes e$ system (Sects. 4.6 and 4.7) and $\varrho_E^{(0)}(\omega)$ is the reference density of the E vibrations of the lattice,

$$\varrho_E^{(0)}(\omega) = \sum_{\kappa} a_\kappa^2(E\gamma)\delta(\omega - \omega_\kappa) \; , \tag{5.4.47}$$

$\bar{n}(\omega) = [\exp(\hbar\omega/kT) - 1]^{-1}$ are the Planck occupation numbers, V_E is the constant of linear vibronic coupling and $\Delta = 2qeP_E$ is the energy gap between the Kramers doublets (5.4.11) (small corrections of the order of g_2 are neglected).

In the case of relatively high temperatures, $kT \gg \Delta$, we have $R_1 \approx R_2 \sim T$. This is the normal result for the probability of direct one-phonon processes.

At higher temperatures the two-phonon Raman processes occurring in second-order perturbation theory become important. In this case [5.246]

$$R_1 \approx R_2 \approx \frac{4\pi q^4}{\hbar^2} V_E^4 \sum_{\kappa,\lambda} \bar{n}(\omega_\kappa)[\bar{n}(\omega_\lambda) + 1]\frac{a_\kappa^2(E\gamma)a_\lambda^2(E\gamma)}{\omega_\kappa^2\omega_\lambda^2}\delta\left(\frac{\Delta}{\hbar} + \omega_\kappa - \omega_\lambda\right) . \tag{5.4.48}$$

In the limit of very high temperatures $kT \gg \hbar\omega_{max}$, where ω_{max} is the largest frequency of lattice vibration, for two-phonon processes we have $R_{1,2} \sim T^2$. However, at such high temperatures the third, fourth and even higher order processes generally become important.

Exceptions are systems with extremely weak vibronic coupling for which the higher-order processes ($n > 2$) are quenched by the small factor V_E^{2n}. If the vibronic coupling is not weak enough the temperature dependence of the $R_{1,2}$ magnitude in the region of high temperatures differs from T^2 due to the higher-order processes.

In the region of intermediate temperatures, $\Delta \lesssim kT \lesssim \hbar\omega_{max}$, the temperature dependence of $R_{1,2}$ can be very complicated. For estimates, the Debye model of phonon dispersion is usually used, and in this model the integrals over κ and λ in (5.4.48) can be reduced to elementary functions. As a result, the temperature dependences T^3, T^5 and T^7 can be obtained (an appropriate bibliography can be found in [5.229]). However, the probability of Raman processes as distinct from the direct ones is determined not only by the long wavelength Debye region of the phonon band but also by the high-frequency regions, the features of the latter being significantly different from the Debye region. A rigorous treatment requires calculations of the integrals over κ and λ in (5.4.48) with a realistic phonon density, and this results in a temperature dependence determined by the structure of the phonon band.

In the case of strong linear vibronic coupling the energy spectrum of the impurity-phonon system is a superposition of the discrete spectrum of the free rotations along the trough of the lowest sheet of the adiabatic potential and the continuous spectrum of the radial vibrations (Sect. 4.6.2). Obviously the latter have to be chosen as a thermal reservoir while the spin-orbital vibronic states of the ground rotational level serve as a dynamic subsystem. The Hamiltonians of the subsystems are given in (4.6.24) in which only the strongest terms in the expansion with respect to the small parameter V_E^{-1} are kept. The neglected term linear in V_E^{-1} has the form [5.248]

$$\hat{H}_1 = \frac{\hbar}{V_E}\overline{(\omega^{-2})}_E \sum_{n \neq 1} (\omega^{-2})_{1n} P_{n\varepsilon} \hat{\sigma}_x \,, \quad \text{where} \tag{5.4.49}$$

$$(\omega^{-2})_{1n} = \sum_{\kappa} \omega_\kappa^{-2} a_\kappa(1\gamma) a_\kappa(n\gamma) \,, \tag{5.4.50}$$

and $\hat{\sigma}_x$ is the Pauli matrix determined in the basis of the rotational states $|\pm 1/2\rangle$ of the ground vibronic doublet. Considering the operator (5.4.49) as a small interaction between the dynamic subsystem and the reservoir we obtain for the probabilities of the direct processes a result similar to that in (5.4.46) [5.248]:

$$\begin{aligned} R_1 &= \hbar^2 V_E^{-2} \overline{(\omega^{-2})}_E^{-4} [\bar{n}(\Delta/\hbar) + 1] \varrho(\Delta/\hbar) \,, \\ R_2 &= \hbar^2 V_E^{-2} \overline{(\omega^{-2})}_E^{-4} \bar{n}(\Delta/\hbar) \varrho(\Delta/\hbar) \,, \end{aligned} \tag{5.4.51}$$

where $\varrho(\omega)$ is the density of the lattice states modified by the Jahn-Teller effect (4.6.29–31). In the high-temperature limit as in the case of weak coupling we obtain $R_{1,2} \sim T$. Although qualitatively this result is quite normal the quantitative changes are considerable owing to the significant modification of the dynamics of the lattice by the vibronic interaction in the vicinity of the paramagnetic center, which has a great effect on the relaxation process.

5.5 Electron Paramagnetic Resonance. The Static Jahn-Teller Effect

All the results of the previous section are related to situations where the excited vibronic states of the system are not populated, and where their mixing with the ground multiplet can be neglected.

On the other hand, Jahn-Teller systems possess hindered internal motions, which in the case of strong vibronic coupling transform into tunneling between equivalent minima of the adiabatic potential energy surface (Sect. 4.3). The tunneling gap between the ground and the first excited vibronic level decreases rapidly as the constant of vibronic coupling increases.

Hence the results of Sect. 5.4 are valid only when the vibronic coupling is relatively weak and the tunneling splitting is large compared with kT or the characteristic splittings caused by random strain and an external magnetic field.

5.5.1 Nature of the Isotropic EPR Spectrum of Cubic Paramagnetic Systems. Three-State Model

Consider first the consequences of increasing the temperature. In most cases important for experiments the first excited state, separated from the ground state by a tunneling gap, is a vibronic singlet (Sect. 4.3). For instance, for the case considered above of a cubic ^2E term the lowest excited state is $^2A_1$ or $^2A_2$.[6] The signs of the linear and quadratic vibronic constants determine which of these two terms is lower.

The spin Hamiltonian of an isolated vibronic singlet has an isotropic form:

$$H_A = g_1 \beta \mathscr{H} \mathbf{S} + A_1 \mathbf{I}\mathbf{S} \tag{5.5.1}$$

and allows $(2I + 1)$ transitions at the frequency of the radio-wavelength quantum. When the hyperfine interaction is neglected, all these transitions contribute to one isotropic line at the frequency $h\nu = g_1 \beta \mathscr{H}$. The intensity of this line is proportional to the Boltzmann population of the excited vibronic singlet, and therefore its ratio to the integral intensity of the inhomogeneously broadened EPR spectrum of the ground vibronic ^2E term is proportional to $\exp(-3\Gamma/kT)$, where 3Γ is the magnitude of the tunneling splitting.

Thus, there are two possible mechanisms for the occurrence of the isotropic EPR spectra with increasing temperature. The first is that of *relaxation narrowing* considered in Sect. 5.4, and the second is the *temperature population of the lowest vibronic singlet*. Of course, there are cases where both mechanisms are operative.

In general, the question of the origin of the isotropic EPR spectra is rather difficult and in every concrete case it requires a special investigation. Here the following consideration can be useful. The intensity of the isotropic line arising as a result of the relaxation narrowing of the partial contributions to the inhomogeneously broadened band is strongly dependent on the number of centers for which the condition for narrowing ($hR \gg 2\gamma$) is fulfilled. The width of the band is determined by the orientation of the vector $\mathscr{H}$, and if $\mathscr{H}$ is parallel to the cubic axis [111], it tends to zero. Thus when the direction of the vector $\mathscr{H}$ approaches the direction of the cubic axis [111] the number of centers for which the conditions for narrowing are satisfied increases and the intensity of the isotropic line increases. If the isotropic line is due to the temperature population of the tunneling vibronic singlet, its intensity is independent of the orientation of the magnetic field. However, the last statement is valid only in the case where anisotropic perturbations (random strain, Zeeman interaction) do not mix the ground states of the vibronic doublet with those of the vibronic singlet. As shown by direct calculation [5.242], if one considers such a mixing which is strongly dependent on the orientation of the magnetic field, the intensity of the isotropic spectrum caused only by the temperature population of the vibronic singlet also

[6] The more complicated picture of tunneling levels arising due to the tunneling between the six equivalent orthorhombic minima of the adiabatic potential of a T term (Sect. 4.3) is not considered here.

increases when the orientation of the vector $\mathscr{H}$ approaches the direction of the cubic axis [111]. Since such a mixing can never be excluded unambiguously, the above criterion can serve as a necessary but not sufficient condition for the relaxation origin of the isotropic spectrum.

Another criterion often used for choosing the possible origin of the isotropic spectrum is based on the results of (5.4.46) and (5.4.51). If one assumes that in the Debye region of the dispersion law, in which the frequency $\Delta/\hbar$ normally falls, we have for the density of states $\varrho(\omega) \sim \omega^4$, and at not very low temperatures $\bar{n}(\omega) + 1 \approx \bar{n}(\omega) \sim \omega^{-1}$; then $R_1 \approx R_2 \sim \Delta^2$. Thus at equal temperatures for equal paramagnetic samples the intensity of the isotropic spectrum in the case when it is caused by relaxations is larger, for larger Δ, the mean value of the energy gap between the Kramers doublets (5.4.11) arising due to random strain. The last value can be changed by growing crystals in vapors of light elements or under strong irradiation.

Consider now the consequences of mixing the states of the ground vibronic doublet with those of the excited singlet by low-symmetry perturbations. The perturbation operator can be written in the most general form as the tensor convolution (5.4.1) where, unlike in Sect. 5.4, the operators $G_{\Gamma\gamma}$ represent not only the irreducible tensors of the magnetic interaction (5.4.2), but also, for example, the appropriate components of the deformation tensor, cf. (5.4.10).

Let us pass now to the matrix representation of the electronic perturbation (5.4.1) in the basis of the exact vibronic states $\Psi_{E\theta}$, $\Psi_{E\varepsilon}$ and Ψ_A of the ground E term and the excited singlet A_1 (or A_2) which have similar energies. In accordance with the selection rules the contribution to the perturbation matrix is nonzero only for operators which transform as the irreducible representations A_1, A_2 and E:

$$\begin{array}{cccc} & |A_1\rangle & |E\theta\rangle & |E\varepsilon\rangle \\ \hat{H} = & \left(\begin{array}{c} 3\Gamma + G_{A_1} \\ \mathrm{r}G_{E\theta} \\ \mathrm{r}G_{E\varepsilon} \end{array}\right. & \begin{array}{c} \mathrm{r}G_{E\theta} \\ G_{A_1} - qG_{E\theta} \\ qG_{E\varepsilon} + \mathrm{i}pG_{A_2} \end{array} & \left.\begin{array}{c} \mathrm{r}G_{E\varepsilon} \\ qG_{E\varepsilon} - \mathrm{i}pG_{A_2} \\ G_{A_1} + qG_{E\theta} \end{array}\right) \end{array} . \tag{5.5.2}$$

Here 3Γ is the magnitude of the tunneling splitting and q, p and r are the appropriate vibronic reduction factors (Sect. 4.7). The matrix (5.5.2) is obtained under the assumption that the term nearest to the ground E term is the vibronic singlet A_1. If the first excited term is not A_1 but A_2, the perturbation matrix has the same form with the difference that the matrix elements $\langle \Psi_{A_2} | \hat{H} | \Psi_{E\varepsilon} \rangle = \langle \Psi_{E\varepsilon} | \hat{H} | \Psi_{A_2} \rangle$ are of opposite sign.

As follows from (5.5.2), the mixing of the states of the ground E term with that of the excited singlet can be neglected only in the case where the tunneling splitting 3Γ is much larger than the splitting caused by the operators $\mathrm{r}G_{E\gamma}$. In particular, if the main contribution to $G_{\Gamma\gamma}$ is due to random strain or the anisotropic Zeeman interaction, the condition for applicability of the results obtained in Sect. 5.4 is $\bar{\Delta}, g_2\beta\mathscr{H} \ll 3\Gamma$, where $\bar{\Delta} = 2qP_E\bar{e}$ is the mean value of the splitting of the E term by random strain.

Consider now the opposite case $3\Gamma \ll \bar{\Delta}, g_2\beta\mathscr{H}$. Since 3Γ decreases exponentially as V_E^2 increases (see Sect. 4.3.3), this case occurs for relatively strong vibronic coupling. Then $q \approx 1/2$, $p \approx 0$, $r \approx -q\sqrt{2}$ (Sect. 4.7). In the case under consideration the result obtained for $3\Gamma = 0$ can serve as a good initial approximation. Therefore assuming in (5.5.2) that $3\Gamma = 0$ and diagonalizing the matrix obtained, we find its eigenfunctions and eigenvalues as follows:

$$E_1 = G_{A_1} + \frac{1}{2}G_{E\theta} - \frac{\sqrt{3}}{2}G_{E\varepsilon}\,, \qquad \Psi_1 = \frac{1}{\sqrt{6}}(\sqrt{2}\Psi_{A_1} - \Psi_{E\theta} + \sqrt{3}\Psi_{E\varepsilon})\,,$$

$$E_2 = G_{A_1} + \frac{1}{2}G_{E\theta} + \frac{\sqrt{3}}{2}G_{E\varepsilon}\,, \qquad \Psi_2 = \frac{1}{\sqrt{6}}(\sqrt{2}\Psi_{A_1} - \Psi_{E\theta} - \sqrt{3}\Psi_{E\varepsilon})\,,$$

$$E_3 = G_{A_1} - G_{E\theta}\,, \qquad \Psi_3 = \frac{1}{\sqrt{3}}(\Psi_{A_2} + \sqrt{2}\Psi_{E\theta})\,. \qquad (5.5.3)$$

Thus the tetragonal perturbation $G_{E\gamma}$ removes the threefold degeneracy of the orbital states and results in three Kramers doublets with the energies (5.5.3). If one includes the magnetic interactions among the E-type anisotropic perturbations, then substituting the expressions (5.4.2) into (5.5.3) we obtain the spin Hamiltonian for each of the Kramers doublets:

$$H_1 = g_\parallel\beta\mathscr{H}_xS_x + g_\perp\beta(\mathscr{H}_yS_y + \mathscr{H}_zS_z) + A_\parallel S_xI_x + A_\perp(S_yI_y + S_zI_z)\,,$$

$$H_2 = g_\parallel\beta\mathscr{H}_yS_y + g_\perp\beta(\mathscr{H}_xS_x + \mathscr{H}_zS_z) + A_\parallel S_yI_y + A_\perp(S_xI_x + S_zI_z)\,, \qquad (5.5.4)$$

$$H_3 = g_\parallel\beta\mathscr{H}_zS_z + g_\perp\beta(\mathscr{H}_yS_y + \mathscr{H}_xS_x) + A_\parallel S_zI_z + A_\perp(S_xI_x + S_yI_y)\,, \quad \text{where}$$

$$g_\parallel = g_1 - g_2\,, \qquad g_\perp = g_1 + \tfrac{1}{2}g_2\,,$$

$$A_\parallel = A_1 - A_2\,, \qquad A_\perp = A_1 + \tfrac{1}{2}A_2\,. \qquad (5.5.5)$$

Analogous results can be obtained in the case when the lowest vibronic singlet is the A_2 term with the difference that g_2 and A_2 in (5.5.5) are of opposite sign.

The spin Hamiltonian (5.5.4) has axial symmetry with its axis directed along one of the three cubic axes of fourth order. Accordingly, in this case the EPR spectrum is a superposition of three axial spectra of tetragonally distorted systems.

The physical sense of this result can be understood if one goes back from the states (5.5.3) to the states

$$\Psi_{A_1} = \frac{1}{\sqrt{3}}(\Psi_1 + \Psi_2 + \Psi_3)\,, \qquad \Psi_{E\theta} = \frac{1}{\sqrt{6}}(2\Psi_1 - \Psi_2 - \Psi_3)\,,$$

$$\Psi_{E\varepsilon} = \frac{1}{\sqrt{2}}(\Psi_2 - \Psi_3)\,. \qquad (5.5.6)$$

From the comparison of (5.5.6) with (4.3.25) it follows that the states Ψ_1, Ψ_2, Ψ_3 correspond to the adiabatic states (4.3.20) in the minima of the adiabatic potential. The difference is due to the fact that the basis (5.5.3) unlike that of (4.3.20) is orthogonal. However, if one takes into account the fact that the assumption $3\Gamma = 0$ is equivalent to neglecting the overlap of the vibrational states in the minima, then this difference disappears.

Thus the diagonality of the electronic operator (5.5.2) in the basis (5.5.3) at $3\Gamma = 0$ is a consequence of neglecting the overlap of the vibrational states in different minima. The tetragonal electronic perturbation $G_{E\gamma}$ changes the surface of the adiabatic potential for the E term, lowering some of the minima and raising the other ones. The limitation of the basis to the three lowest vibronic states implies that the perturbation $G_{E\gamma}$ is relatively small, causing relatively small changes in the adiabatic potential and preserving the three-well nature of its surface. The equivalence of the three minima of the adiabatic potential is violated, the resonance between the vibrational states in the minima no longer takes place, and the adiabatic localized states in the minima become adapted zeroth-order states for the perturbation (5.5.2). Thus the energy (5.5.3) is simply the energy of the ground vibrational state in the minima, the depth of the latter being modified by the tetragonal perturbation $G_{E\gamma}$. Instead of the states Ψ_A, $\Psi_{E\theta}$ and $\Psi_{E\varepsilon}$ delocalized over different minima, the electronic perturbation $G_{E\gamma}$ forms the states Ψ_1, Ψ_2 and Ψ_3 localized in the minima. In other words, the perturbation $G_{E\gamma}$ "locks" the system in the minimum. This explains the occurrence of three tetragonal EPR spectra.

The situation when the anisotropic perturbation locks the Jahn-Teller system in one of the minima of the adiabatic potential is called the *static Jahn-Teller effect* to distinguish it from the *dynamic Jahn-Teller effect* for which the system is delocalized over all the minima of the potential surface. As follows from the results obtained above, the EPR spectra of Jahn-Teller systems in these two limiting cases differ drastically.

It is interesting to follow the changes in the parameters of the Jahn-Teller system leading to the transfer from the dynamic Jahn-Teller effect to the static one. The nature of the changes in the EPR spectrum is dependent on the kind of anisotropic perturbation, the Zeeman interaction or random strain, which locks the system in the minima.

Consider first the Zeeman interaction with the external magnetic field. In this case G_{A_1}, $G_{E\theta}$ and $G_{E\varepsilon}$ in (5.5.2) are the spin operators (5.4.2). Even if the hyperfine interaction is excluded from our consideration, the dimensionality of the matrix (5.5.2) is multiplied by the dimensionality of the spin space $(2S + 1)$. In particular, for the ^2E term the 6×6 matrix has to be diagonalized. However, this problem can be simplified if, as in Sect. 5.4, one uses perturbation theory with the small parameter g_2/g_1.

Passing to a new coordinate system with the unit vectors $\boldsymbol{\xi}$, $\boldsymbol{\eta}$, $\boldsymbol{\zeta}$ oriented in such a way that the ζ axis is the axis of quantization determined by the vector $\mathscr{H}$, we diagonalize the Hamiltonian of the zeroth approximation, $H_0 = g_1 \beta \mathscr{H} \boldsymbol{S}$.

The perturbation matrix in the space of the two reference electronic orbital states has the same form as in (5.4.7), but without the factor q. Passing to the matrix representation of the perturbation in the basis of the exact vibronic states Ψ_{A_1}, $\Psi_{E\theta}$ and $\Psi_{E\varepsilon}$ for each of the two values of the projection of the spin $S_\xi = \pm\frac{1}{2}$ we obtain a matrix of the type (5.5.2),

$$\frac{g_2\beta\mathcal{H}}{4}\begin{pmatrix} \dfrac{12\Gamma}{g_2\beta\mathcal{H}} & \pm \mathrm{r}f_\theta & \pm\sqrt{3}\mathrm{r}f_\varepsilon \\ \pm\mathrm{r}f_\theta & \mp q f_\theta & \pm q f_\varepsilon \\ \pm\sqrt{3}\mathrm{r}f_\varepsilon & \pm\sqrt{3}q f_\varepsilon & \pm q f_\theta \end{pmatrix}, \tag{5.5.7}$$

where the energy zero is the isotropic value $\pm g_1\beta\mathcal{H}/2$, $f_\theta = 3\zeta_z^2 - 1$, $f_\varepsilon = \zeta_x^2 - \zeta_y^2$. If the magnetic field is directed along the fourth-order axis (say, the z axis), $\zeta_x = \zeta_y = 0$, $\zeta_z = 1$ and the matrix (5.5.7) takes a simpler form; the values of the energy of the six levels are given by the expressions [5.250]

$$\begin{aligned} E_1^{(\pm)} &= \pm\tfrac{1}{2}\beta\mathcal{H}(g_1 + q g_2)\ . \\ E_2^{(\pm)} &= \tfrac{1}{2}(3\Gamma \pm g_1\beta\mathcal{H} \mp \tfrac{1}{2}q g_2\beta\mathcal{H}) \\ &\quad - \sqrt{\tfrac{1}{4}(3\Gamma \pm \tfrac{1}{2}q g_2\beta\mathcal{H})^2 + \tfrac{1}{4}\mathrm{r}^2 g_2^2\beta^2\mathcal{H}^2}\ , \\ E_3^{(\pm)} &= \tfrac{1}{2}(3\Gamma \pm g_1\beta\mathcal{H} \mp \tfrac{1}{2}q g_2\beta\mathcal{H}) \\ &\quad + \sqrt{\tfrac{1}{4}(3\Gamma \pm \tfrac{1}{2}q g_2\beta\mathcal{H})^2 + \tfrac{1}{4}\mathrm{r}^2 g_2^2\beta^2\mathcal{H}^2}\ , \end{aligned} \tag{5.5.8}$$

Evaluating the wave functions of these states and the probabilities of spin transitions between them, one can obtain the expected EPR spectrum in the case under consideration [5.251]. The analysis of this spectrum is conveniently carried out by dividing the region of the magnetic field variation into three parts (Fig. 5.27): I, the low-frequency region for which $g_2\beta\mathcal{H} \ll 3\Gamma$; II, the intermediate region $g_2\beta\mathcal{H} \sim 3\Gamma$; and III, the high-frequency region for which $g_2\beta\mathcal{H} \gg 3\Gamma$. In

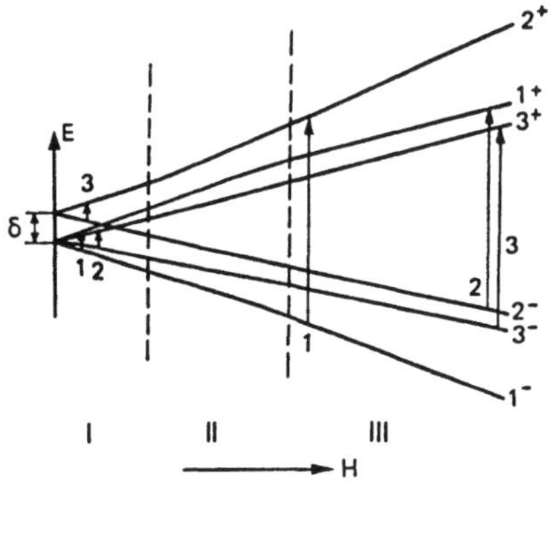

Fig. 5.27. Energy levels of the ground state $E \otimes e$ system in magnetic fields $\mathcal{H} \| z$ taking into account the tunneling splitting. The magnetic dipole allowed transitions in regions **I** and **III** are shown by arrows

the low-frequency region the three transitions shown in Fig. 5.27 by arrows are allowed with the usual intensities. Their frequencies are approximately given by the expressions (5.4.9) (transitions 1 and 2 in Fig. 5.27) and $g_1\beta\mathscr{H}$ (transition 3 in Fig. 5.27).

In the other limiting case $g_2\beta\mathscr{H} \gg 3\Gamma$, i.e., in the high frequency region, another three transitions are allowed with the g factors (5.5.5).

In the intermediate region the intensity of the low frequency spectrum is increasing and that of the high frequency one is decreasing as the intensity of the magnetic field decreases so for some $\mathscr{H}$ values both types of lines can be observed.

For directions of $\mathscr{H}$ not coinciding with the tetragonal axes the levels (5.5.8) approach each other so that the differences in the g factors decrease, and for the field directions along the trigonal axes all the g factor values for both low frequency and high frequency fields coincide and are equal to g_1.

5.5.2 Effect of Random Strain. Angular Dependence of Spectra

Consider now the changes in the EPR spectrum related to the influence of random strain, assuming for simplicity that the mixing of the vibronic singlet with the ground doublet by the magnetic field is negligibly small. Assume that the magnetic field is oriented in the plane [110]. If 3Γ is still large enough so that the random strain can be taken into account by perturbation theory, then in the initial approximation we obtain the results of Sect. 5.4; in particular, at low temperatures we obtain the EPR line (5.4.25).

Only those paramagnetic centers for which the anisotropic part of the g factor (5.4.17) is maximal, that is $\varphi - \alpha = 0$ and $\varphi - \alpha = \pi$, contribute to the most intense wings of the band (5.4.25). If one takes into account that for the chosen orientation of the magnetic field $\zeta_x^2 = \zeta_y^2$, then in accordance with (5.4.16) we have $\alpha = 0$ and the wings of the band are formed by the contributions of the centers with $\varphi = 0$ and $\varphi = \pi$. As follows from (5.4.12), these are the centers for which $e_\varepsilon = 0$, i.e., the random strain is oriented along the z axis of the cubic system. The matrix (5.5.2) in this case has the simple form

$$\begin{pmatrix} 3\Gamma & \mathrm{r}P_E e_\theta & 0 \\ \mathrm{r}P_E e_\theta & -qP_E e_\theta & 0 \\ 0 & 0 & qP_E e_\theta \end{pmatrix} . \tag{5.5.9}$$

In the first order of the perturbation theory with respect to $P_E e_\theta$ we obtain for the wave functions $\Psi'_{E\theta}$ and $\Psi'_{E\varepsilon}$ of the ground vibronic doublet [5.163]

$$\Psi'_{E\theta} \approx \Psi_{E\theta} - \frac{\mathrm{r}P_E e_\theta}{3\Gamma}\Psi_{A_1} , \qquad \Psi'_{E\varepsilon} \approx \Psi_{E\varepsilon} . \tag{5.5.10}$$

The spin Hamiltonians for the two corresponding Kramers doublets can be obtained by averaging the Zeeman Hamiltonian (5.4.1) over the states $\Psi'_{E\theta}$

and $\Psi'_{E\varepsilon}$:

$$H_\theta = g_1\beta\mathscr{H}\boldsymbol{S} - \frac{1}{2}\left(q + \frac{2r^2 P_E e_\theta}{3\Gamma}\right) g_2\beta(3\mathscr{H}_z S_z - \mathscr{H}\boldsymbol{S}) \ ,$$

$$H_\varepsilon = g_1\beta\mathscr{H}\boldsymbol{S} - \tfrac{1}{2} q g_2 \beta(\mathscr{H}\boldsymbol{S} - 3\mathscr{H}_z S_z) \ .$$

Using, as in Sect. 5.4, perturbation theory with respect to g_2/g_1 we find the corresponding g factors

$$g_\theta = g_1 - \frac{1}{2}\left(q + \frac{2r^2 P_E e_\theta}{3\Gamma}\right) g_2(3\zeta_z^2 - 1) \ ,$$
$$g_\varepsilon = g_1 + \tfrac{1}{2} q g_2 (3\zeta_z^2 - 1) \ . \tag{5.5.11}$$

The relative position of the two levels $\Psi'_{E\theta}$ and $\Psi'_{E\varepsilon}$ depends on the sign of e_θ. It is important here that the mixing with the vibronic singlet, independent of the sign of e_θ, shifts only one of the two EPR peaks which form the inhomogeneous broadened line. For instance, if $e_\theta > 0$, the ground Kramers doublet is $\Psi'_{E\theta}$, and if $g_2 > 0$ and $3\zeta_z^2 - 1 > 0$, only the EPR peak that is positioned on the right-hand side along the field, i.e., which corresponds to larger $\mathscr{H}$, is shifted. Since in this case q and $2r^2 P_E e_\theta/3\Gamma$ have the same sign, this shift takes place in the direction of larger field values, i.e., the anisotropy of the right-hand peak increases. For $e_\theta < 0$ the Kramers doublet $\Psi'_{E\theta}$ is excited but the EPR peak corresponding to it is also positioned on the right-hand side (at $g_2 > 0$, $3\zeta_z^2 > 1$). Since in this case the q and $2r^2 P_E e_\theta/3\Gamma$ values are of different signs, the shifts of the $\Psi'_{E\theta}$ peak is directed to the left, i.e., its anisotropy decreases.

Now we have to take into account the fact that the random strain is distributed not only in direction, but also in magnitude, $\varDelta = 2qP_E\sqrt{e_\theta^2 + e_\varepsilon^2}$. The appropriate distribution may have, say, Gaussian form [5.242, 252]:

$$P(\varDelta) \sim \exp[-(\varDelta - \bar{\varDelta})^2/2\sigma] \ , \tag{5.5.12}$$

where $\bar{\varDelta}$ is the mean value of $\varDelta$ and σ is an appropriate halfwidth of the distribution.

Considering this circumstance, the inhomogeneously broadened EPR line of the type (5.4.25) must also be averaged with respect to $\varDelta$, determining the position of the right-hand wing (at $g_2 > 0$, $3\zeta_z^2 > 1$) of the line. The result of the averaging is obvious: the appropriate wing of the line becomes smeared and lowered in intensity compared with the other wing.

If the lowest vibronic singlet is A_2 then its mixing with the ground vibronic doublet influences the position of the EPR line corresponding to the state $\Psi'_{E\varepsilon}$. The conclusions obtained by analogous considerations are given in Table 5.1 where the influence of the mixing with the excited vibronic singlet at $\mathscr{H} \parallel [110]$, $3\zeta_z^2 > 1$ is also shown. It is assumed that the magnetic field increases to the right. If $3\zeta_z^2 < 1$ all the data in the table change to the opposite sign [5.253].

Table 5.1 Shift of one of the two peaks (high and low field) in the vibronic EPR spectrum of a ^2E term under the influence of random-strain mixing with the excited ^2A level. $\mathscr{H} \| [110]$, $\zeta_x = \zeta_y$, $3\zeta_z^2 > 1$. In the case of $3\zeta_z^2 < 1$ all the results change to the opposite sign. (After [5.253])

Sign of g_2	Coupled singlet	Shifted peak	
		Position	Direction
Ground Kramers doublet			
+	A_1	High field	To high field
+	A_2	Low field	To low field
−	A_1	Low field	To low field
−	A_2	High field	To high field
First excited Kramers doublet			
+	A_1	High field	To low field
+	A_2	Low field	To high field
−	A_1	Low field	To high field
−	A_2	High field	To low field

As follows from Table 5.1, the mixing with the vibronic singlet always leads to a shift in the same peak for both the excited and the ground Kramers doublet, but they are in opposite directions. This somewhat masks the effect of the peak shifting since the resulting contribution of the ground and excited Kramers doublet is observed in the experiments simultaneously. However, if the temperature is low enough and only the ground Kramers doublet is populated, then observing the direction of the shift of the EPR spectrum and knowing the sign of g_2 enables one to determine the symmetry of the lowest vibronic singlet using the results of Table 5.1.

For large values of $rP_E e_\theta/3\Gamma$ the formulas of perturbation theory (5.5.10), (5.5.11) are not valid and the matrix (5.5.9) has to be diagonalized directly. The result of the corresponding averaging of the EPR spectrum over the orientations and the magnitude of the random strain in the case when the temperature is low enough and only the ground Kramers doublet is populated are illustrated in Fig. 5.28. As can be seen from this figure, if $\bar{\Delta}/3\Gamma \lesssim 0.01$, the EPR spectrum almost coincides with the results of (5.4.25) given in Fig. 5.26. However, a small broadening of the left extremum at $\mathscr{H}_2 = h\nu/(g_1 + qg_2)$ is already manifested. On increasing $\bar{\Delta}/3\Gamma$ this broadening increases very rapidly accompanied by a shift towards the smaller field. For large $\bar{\Delta}/3\Gamma$ values the velocity of the shift of the peak to the left decreases, and at $\bar{\Delta}/3\Gamma \approx 5$ it reaches its extreme left position at $\mathscr{H}_1 = h\nu/(g_1 + 2qg_2)$. Here the static Jahn-Teller effect is realized and the $g_\|$ value from (5.5.5) corresponds to the field $\mathscr{H}_1$. The intensity ratio 1 : 2 for the left and right peaks at $\mathscr{H} \| [001]$ also corresponds to the case of the static Jahn-Teller effect.

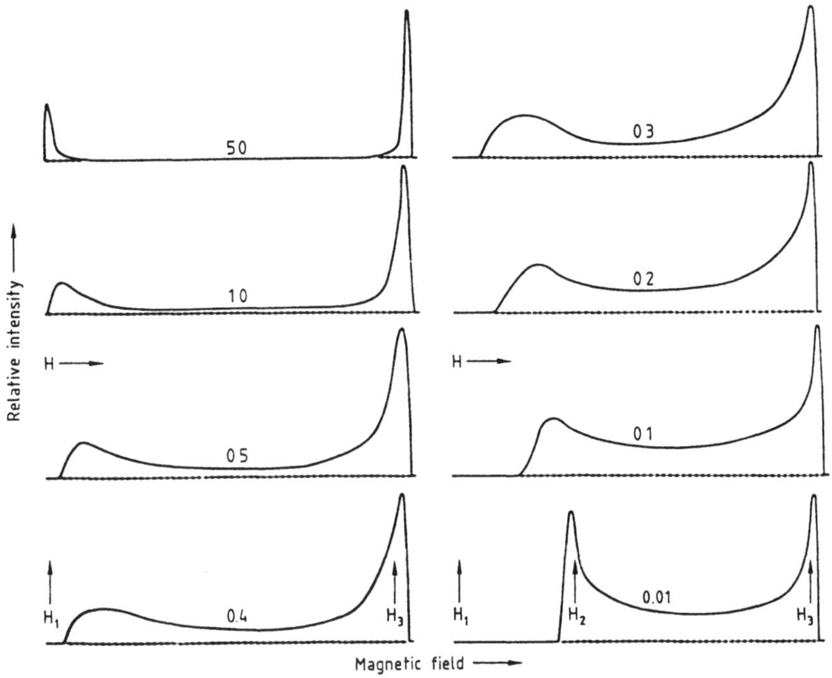

Fig. 5.28. EPR line shape calculated for a ^2E term in $\mathscr{H} \| [001]$ at low temperatures and different $\bar{\Delta}/3\Gamma$ values (indicated on the curves). The lowest vibronic singlet is assumed to be A_2, $g_1 = 2.0$, $qg_2 = 0.05$, $q = 1/2$, $r = -\sqrt{2q}$, $\beta\mathscr{H}_1 = \hbar\Omega/(g_1 + 2qg_2)$, $\beta\mathscr{H}_2 = \hbar\Omega/(g_1 + qg_2)$, $\beta\mathscr{H}_3 = \hbar\Omega/(g_1 - qg_2)$ (after [5.242])

On increasing the temperature, the first excited Kramers doublet becomes populated and the resulting EPR line, as mentioned above, is the result of the superposition of the contributions of two Kramers doublets, the ground and the excited ones (Fig. 5.29).

For lower 3Γ, one more Kramers doublet is populated, originating from the vibronic singlet. The EPR line shape in this case becomes even more complicated (Fig. 5.30).

It is interesting to investigate the angular dependence of the spectrum by changing the direction of the magnetic field in the plane [110]. By such rotations one can observe the spectral characteristics of EPR for $\mathscr{H} \|$ [001], [111], [110], i.e., along all the cubic symmetry axes. Figure 5.31 shows the angular dependence of the spectrum obtained as above by numerical diagonalization of the matrix (5.5.2) and a subsequent averaging of the spectrum over the orientations and the magnitude of random strain. It is assumed that the temperature is low enough so that the contribution to the EPR spectrum comes from the lowest Kramers doublet only. In addition, as earlier, it is assumed that the hyperfine structure is absent.

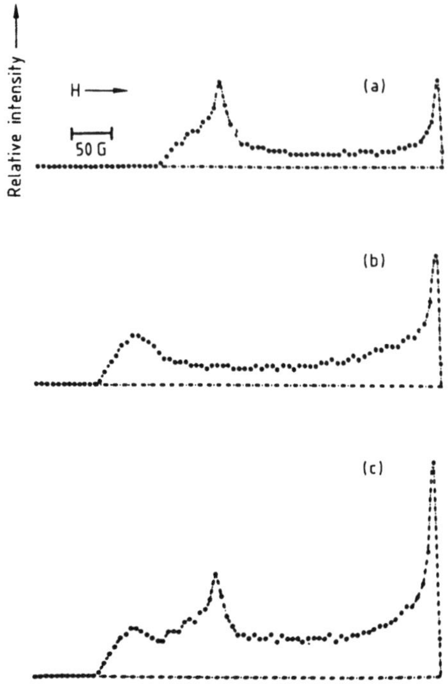

Fig. 5.29. Change in EPR line shape as a result of populating of the first excited Kramers doublet in the $E \otimes e$ system with random strain. (**a**) The line shape of the excited doublet. (**b**) The line shape of the ground doublet. (**c**) The total line ($T = 1.5$ K, $\bar{\Delta}/3\Gamma = 0.13$, $\Gamma = 1.0\ \mathrm{cm}^{-1}$, $qg_2 = 0.1$ and other data as in Fig. 5.28). (After [5.242])

Fig. 5.30. EPR line shape taking into account the contribution of all three Kramers doublets at $T = 1.3$ K and different values of tunneling splitting 3Γ (other data as in Fig. 5.29). (After [5.242])

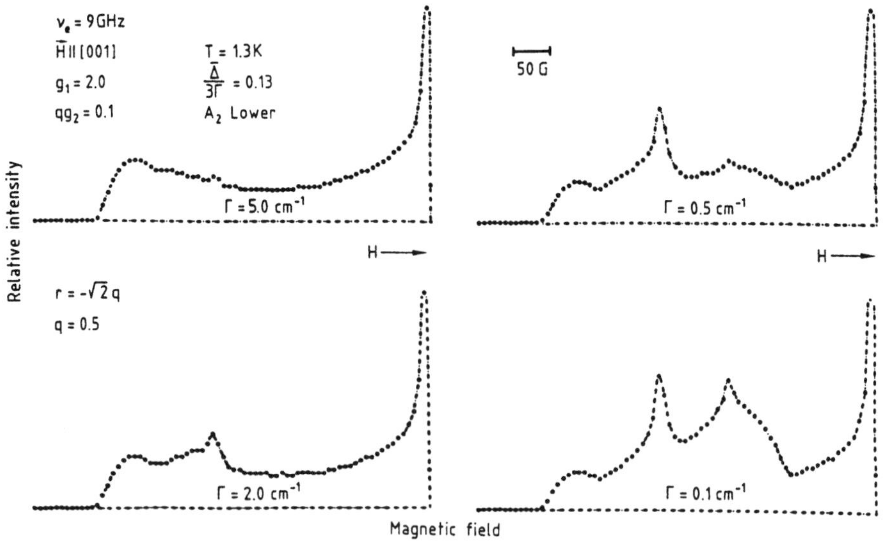

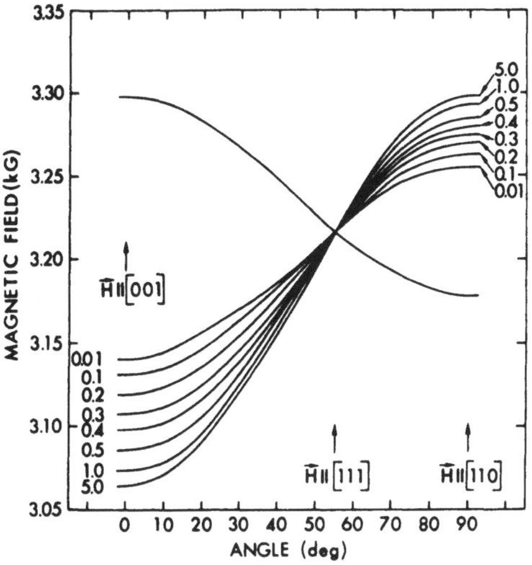

Fig. 5.31. Angular dependence of the low temperature EPR spectrum of a Jahn-Teller system in the ^2E state found by rotating the magnetic field $\mathscr{H}$ in the plane $[1\bar{1}0]$. The absolute values of $\mathscr{H}$ in kG are plotted on the ordinate. The lowest vibronic singlet is A_2. In the case of A_1 the picture has to be reflected about the $g - g_1 = 0$ axis. $\bar{\Delta}/3\Gamma$ for the different curves are shown on the left and right sides. (After [5.242])

If $\bar{\Delta}/3\Gamma = 0.01$, the angular dependence of the spectrum is symmetric about the magnetic field to which both peaks converge at $\mathscr{H} \parallel [111]$; this angular dependence is described exactly by the formula (5.4.17) for $\cos(\varphi - \alpha) = 1$. On increasing $\bar{\Delta}/3\Gamma$ the anisotropy of the "mixed" side of the EPR spectrum increases in accordance with the results of Table 5.1 up to the angular dependence which is characteristic for the three axial EPR spectra of the types [100], [010] and [001] obtained for the static Jahn-Teller effect [5.242]. This result is already valid at $\bar{\Delta}/3\Gamma \gtrsim 5$.

By means of the diagram in Fig. 5.31 one can determine the ratio $\bar{\Delta}/3\Gamma$ in the experimentally observed EPR spectrum. First, the experimental value for the resonance field for the unshifted side of the EPR spectrum fragment for different angles has to be matched with the single line in Fig. 5.31 (the upper line at $3\zeta_z^2 > 1$, $g_2 > 0$, the vibronic singlet A_2). This allows us to determine g_1, qg_2, A_1 and qA_2. Then by comparison of the angular dependence of the "mixed" wing of the EPR line with one of the lines in Fig. 5.31 we find the ratio $\bar{\Delta}/3\Gamma$ (following the numbers given at the left and right of Fig. 5.31).

It is interesting to consider also the angular dependence of the EPR spectrum which emerges by rotation of the magnetic field in the plane [001]. Figure 5.32 shows the angular dependence of the low temperature EPR spectrum obtained by numerical calculations for the same values $\bar{\Delta}/3\Gamma$ as used above when the values of resonance field for Fig. 5.31 were obtained. The angles 0° and 90° in Fig. 5.32 correspond to the directions $\mathscr{H} \parallel [100]$ and $\mathscr{H} \parallel [010]$ while the angle 45° corresponds to the direction $\mathscr{H} \parallel [110]$. If $\bar{\Delta}/3\Gamma = 0.01$, the angular dependence obtained is characteristic of the dynamic Jahn-Teller effect, i.e., by changing the orientation of the magnetic field the peaks of the EPR line are shifted in accordance with (5.4.17) if one inserts $\cos(\varphi - \alpha) = 1$. If $\bar{\Delta}/3\Gamma = 0.2$, a third line occurs, shown in Fig. 5.32 by a dashed line. For higher $\bar{\Delta}/3\Gamma$ the intensity of this

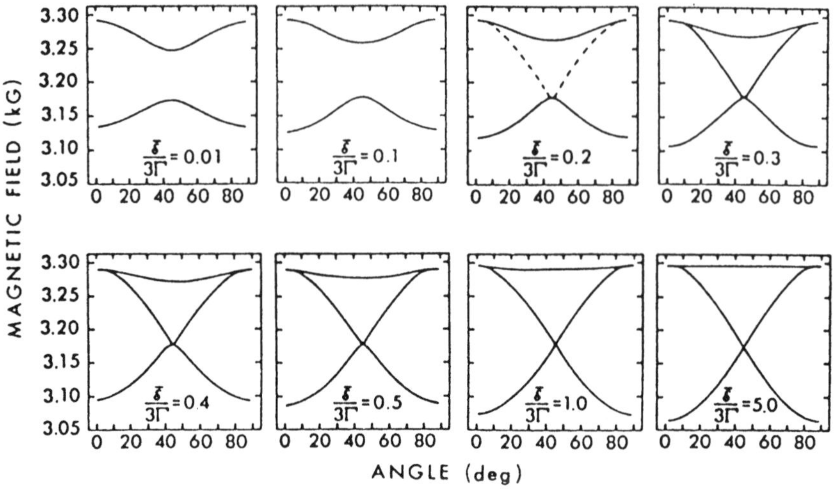

Fig. 5.32. Angular dependence of the low temperature EPR spectrum of a system in the ^2E state found by rotating the magnetic field $\mathscr{H}$ in the plane [001] ($\bar{\delta} = \bar{\Delta}$ other data as in Fig. 5.28–31). The angles 0, 45° and 90° correspond to the field directions $\mathscr{H}$ along [100], [110] and [010], respectively. At $(\bar{\Delta}/3\Gamma) \gtrsim 0.2$ a third line arises in the spectrum. At $\bar{\Delta}/3\Gamma \sim 5.0$ one of the lines becomes isotropic. (After [5.242])

line increases up to the values corresponding to the three axial EPR spectra of the static Jahn-Teller effect ($\bar{\Delta}/3\Gamma = 5$).

Although the angular dependence shown in Fig. 5.32 corresponds to the main and most acute resonances, the full EPR line in the case of an intermediate Jahn-Teller effect has an additional structure, additional lines with intensities depending on $\bar{\Delta}/3\Gamma$ (Fig. 5.33). In two extreme situations, $\bar{\Delta}/3\Gamma = 0.0033$ and $\bar{\Delta}/3\Gamma = 5$, the EPR line has a quite understandable shape corresponding to the cases of the dynamic and static Jahn-Teller effects, respectively, while in intermediate cases it is very complicated.

The EPR line obtained at $\mathscr{H} \parallel [111]$ in the case of the intermediate Jahn-Teller effect also has an abnormally complicated structure. In the two limiting cases of dynamic and static Jahn-Teller effects it transforms into one line of a Lorentz shape with the g factor [5.242]

$$g_{\text{dyn}} = [g_1^2 + \tfrac{1}{2}(qg_2)^2]^{1/2} , \qquad g_{\text{stat}} = [g_1^2 + 2(qg_2)^2]^{1/2} \tag{5.5.13}$$

In spite of this complication the above procedure of comparing the results of theory with the experimental data gives quite satisfactory agreement between the theory and the experiment in a series of cases. In Fig. 5.34 the experimental EPR spectrum of a Jahn-Teller Cu^{2+} : CaO impurity is illustrated [5.253]. At the bottom of this figure the comparison with the EPR spectrum obtained by numerical calculations (in the framework of the approximations described above) is given.

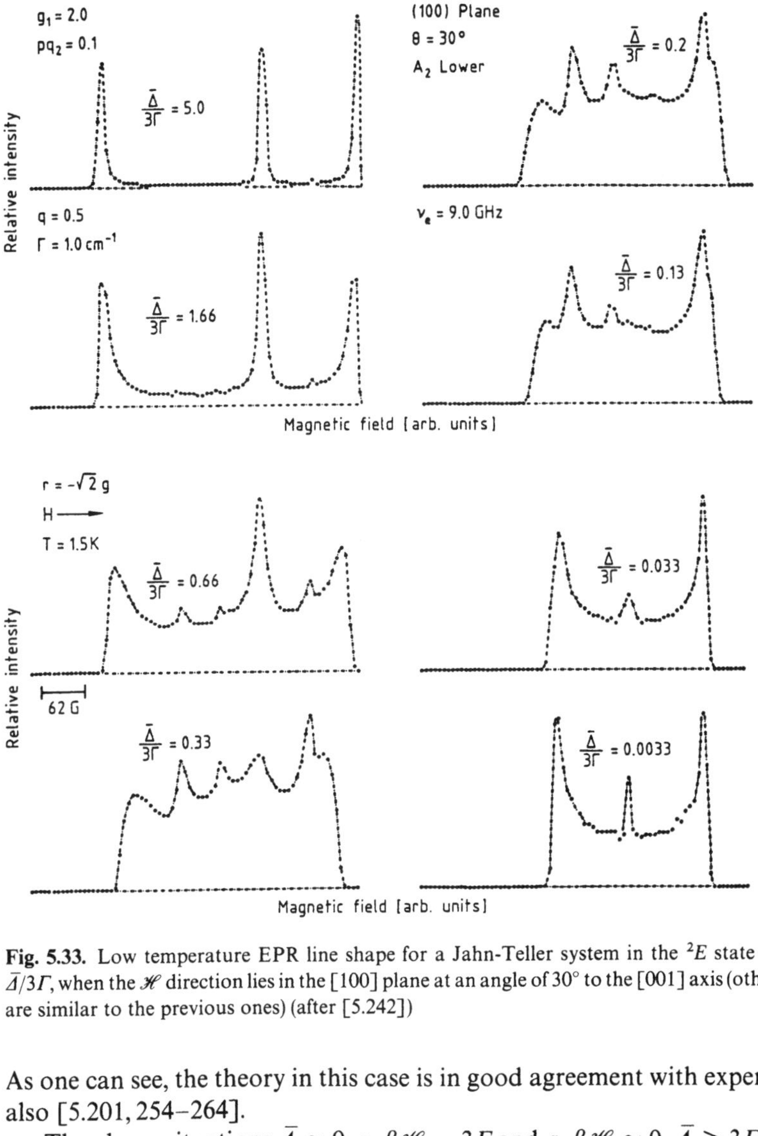

Fig. 5.33. Low temperature EPR line shape for a Jahn-Teller system in the ^2E state and different $\bar{\Delta}/3\Gamma$, when the $\mathscr{H}$ direction lies in the [100] plane at an angle of 30° to the [001] axis (other conditions are similar to the previous ones) (after [5.242])

As one can see, the theory in this case is in good agreement with experiment. See also [5.201, 254–264].

The above situations $\bar{\Delta} \approx 0$, $g_2\beta\mathscr{H} \sim 3\Gamma$ and $g_2\beta\mathscr{H} \approx 0$, $\bar{\Delta} \gtrsim 3\Gamma$ are exceptions rather than the rule. As mentioned above, $\bar{\Delta}$ depends on the method of sample preparation and $g_2\beta\mathscr{H}$ depends on the frequency of the electromagnetic field in the EPR apparatus. Therefore, depending on the properties of the crystalline sample and on the EPR equipment used, the relationship between $\bar{\Delta}$, $g_2\beta\mathscr{H}$ and 3Γ can be different even for the same impurity center in chemically similar crystals. Investigating the same sample by different EPR installations

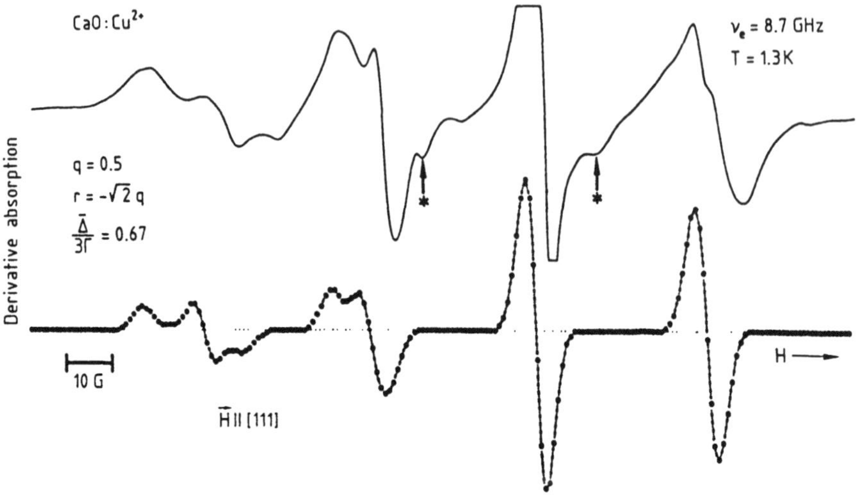

Fig. 5.34. Experimentally obtained (upper) and calculated (lower) EPR spectra of the Jahn-Teller ^2E $CaO:Cu^{2+}$ system at $T = 1.3$ K ($\nu = 8.7$ GHz). The vibronic parameters obtained by comparison of the experimental and theoretical angular dependence of the spectrum in the [110] plane (Fig. 5.31) are given on the left. The arrows indicate the forbidden transitions induced by quadrupole interactions (After [5.242])

Table 5.2 Vibronic parameters of some impurity $E \otimes e$ Jahn-Teller systems obtained from EPR measurements (after [5.265])

System	Lowest vibronic singlet	3Γ [cm^{-1}]	$\bar{\Delta}$ [cm^{-1}]	$\bar{\Delta}/3\Gamma$	g_1	qg_2	q
$SrO:Ag^{2+}$	A_2	—	—	—	2.049	0.032	0.682
$CaO:Ag^{2+}$	A_2	3.9	4.7	1.2	2.076	0.045	0.610
$MgO:Ag^{2+}$	A_2	4.8	0.62	0.13	2.099	0.056	0.577
$CaO:Cu^{2+}$	A_1	≈4.0	2.7	0.67	2.221	0.122	0.557
$MgO:Cu^{2+}$	A_2	—	—	—	—	—	—

with significantly different microwave frequencies, one can see the properties of the same sample in different situations [5.251]. For instance, if a paramagnetic system is the subject of an EPR investigation at a low frequency then $|g_2\beta\mathscr{H}| \ll \bar{\Delta}$, 3Γ. As mentioned above, in this case one can derive the ratio $\bar{\Delta}/3\Gamma$ from the experimental data. If the same sample is placed into another spectrometer with a considerably higher microwave frequency, then the situation of the high frequency spectrum occurs, which allows us to determine the ratio $g_2\beta\mathscr{H}/3\Gamma$. Combining this value with the one above we can determine the $\bar{\Delta}$ and 3Γ values separately. The data of Table 5.2 were obtained in this way [5.265].

All the foregoing concerns the low temperature EPR spectrum when the effects of relaxation can be neglected. In this case the processes leading to temperature averaging of the spectra are important and therefore the temperature has to be so small that the inequality $R \ll 2\pi\Delta\nu$ will hold, where R is the probability of radiationless transitions without spin flip, and $\Delta\nu$ is the difference between the EPR frequencies for different Kramers doublets. For axially symmetric spectra described by the spin Hamiltonians (5.5.4) in the case when the magnetic field is oriented along the [100] axis the frequency of the high temperature spectrum averaged by relaxation transition is

$$h\bar{\nu}(m_I) = (\tfrac{1}{3}g_\parallel + \tfrac{2}{3}g_\perp)\beta\mathscr{H} + (\tfrac{1}{3}A_\parallel + \tfrac{2}{3}A_\perp)m_I = g_1\beta\mathscr{H} + A_1 m_I \; . \qquad (5.5.14)$$

Thus the increase in the temperature in the cases of both dynamic and static Jahn-Teller effects results in a completely isotropic spectrum with the g factor equal to g_1. In the case when the magnetic field is directed along [111] all three distorted configurations are equivalent and $\Delta\nu = 0$. The temperature averaging of the spectrum in this case will take place at lower temperatures than in the case of $\mathscr{H} \parallel [100]$.

Analytical expressions for the shape of the EPR line in the case of fast reorientations of the Jahn-Teller distortions were obtained in [5.266].

Another important reason for the temperature dependence of the EPR spectrum is the change in the tunneling splitting 3Γ with the temperature. The physical nature of this phenomenon is connected with the temperature dependence of the elastic properties of the medium in which the Jahn-Teller cluster performs its pulse motions. For the paramagnetic impurity in the crystal this is due, in particular, to the lattice anharmonicity. Considering the temperature dependence of the frequencies of the normal vibrations, the height and width of the potential barriers dividing the minima of the adiabatic potential decrease with temperature (the medium softens), and this leads to the increase in 3Γ. Note that the tunneling splitting depends exponentially on the magnitude of the potential barriers (Sect. 4.3) and therefore is very sensitive even to small changes in these barriers.

Assuming a strong increase of 3Γ with temperature we come to the following picture of the temperature dependence of the EPR spectrum. At low temperatures the 3Γ magnitude is small and the spectrum characteristic of the static Jahn-Teller effect is observed. On increasing the temperature, 3Γ increases so that at some temperature the ratio $\bar{\Delta}/3\Gamma$ (or even $|g_2\beta\mathscr{H}|/3\Gamma$) becomes of the order of unity and the EPR spectrum corresponds to the intermediate Jahn-Teller effect. Finally, for high enough temperatures the region is reached where $\bar{\Delta}/3\Gamma \ll 1$ (or $|g_2\beta\mathscr{H}|/3\Gamma \ll 1$) and the EPR spectrum manifests the dynamic Jahn-Teller effect. Note that along with the described mechanisms of the temperature dependence of the EPR spectrum the relaxation processes leading to the averaging of the spectrum can also take place. It is obvious that the temperature dependence of 3Γ is the main reason for temperature dependence of the line shape provided the probability of relaxation transitions without spin-flip is small. Just this situation is presumably realized in the EPR spectra of some coordination systems (for more details see [5.267] and [Ref. 5.268; Table VI.4].

Passing to spin-flip relaxation processes we note that the paramagnetic anisotropy of the states in different minima (5.5.3) described by the Hamiltonian (5.5.4) has different main axes. As a result the state with the spin "up" in one minimum is not orthogonal with respect to the spin variables of the spin "down" state in the other minima. Therefore the rate of transitions with spin-flip and a simultaneous jump to another nuclear configuration may be considerably higher than that of spin-lattice relaxation of ions without the Jahn-Teller effect. A detailed discussion of processes of spin-lattice relaxation characteristic of Jahn-Teller ions is given in the review article [5.229].

5.5.3 Magnetic Resonance Spectra for Jahn-Teller Orbital Triplet Centers

The magnetic properties of Jahn-Teller triplets ^{2S+1}T are different from the properties of cubic ^{2S+1}E terms mainly because of the fact that the operator of the orbital momentum of the electrons in the basis of the orbital states of the T term has nonzero matrix elements. This circumstance results in two important consequences. First, the spin-orbit interaction is essential for the ^{2S+1}T term. As a result of the spin-orbit splitting considered in the first-order perturbation theory, a new system of electronic terms arises with energy gaps comparable in magnitude and even larger than the energy of the Jahn-Teller-active vibrational quanta. Strictly speaking, the pseudo-Jahn-Teller effect situation occurs, and the magnetic properties of the system at low temperatures are determined by the properties of the ground vibronic state. This case will not be considered here in more detailed; instead, we consider the situation for which the spin-orbit splitting is less than $\hbar\omega_\Gamma$, so that it can be taken into account by perturbation theory using the idea of vibronic reduction factors.

The second consequence of the orbital motion of the electrons in the T term states is due to the significant contribution of the orbital momentum to the Zeeman energy. In the case of relatively weak vibronic coupling, and considering only the lowest vibronic term ^{2S+1}T (see the discussion of the transformation properties of the ground state in Sect. 4.7), the Hamiltonian of the system can be written as

$$H = K(T_1)g_L\beta\mathscr{H}\boldsymbol{L} + g_1\beta\mathscr{H}\boldsymbol{S} + \lambda K(T_1)\boldsymbol{LS} \ , \qquad (5.5.15)$$

where g_L is the orbital g factor and $K(T_1)$ is the vibronic reduction factor of the operator of the orbital momentum L transforming as the irreducible representation T_1 (Sect. 4.7). In general, the effective Hamiltonian of the T term of a cubic system may also contain terms of the type $\mu(\boldsymbol{LS})^2$ and $\varrho(L_x^2S_x^2 + L_y^2S_y^2 + L_z^2S_z^2)$ arising due to the mixing of the excited electronic terms by the spin-orbit interaction in second-order perturbation theory. However, terms of a similar type occur due to the more important admixture of the excited vibronic states of the same T term, cf. (4.7.38). In first-order perturbation theory we can limit ourselves to the Hamiltonian (5.5.15). If the Zeeman splitting is much smaller than the spin-orbit one, the g factors are determined by the Landé formula with the modified value $\tilde{g}_L = g_L K(T_1)$,

$$g(J) = \frac{1}{2J(J+1)}\{\tilde{g}_L[J(J+1) + 2 - S(S+1]$$

$$+ g_S[J(J+1) + S(S+1) - 2]\} , \qquad (5.5.16)$$

where J is the quantum number of the total momentum, acquiring all the values from $|S-1|$ to $S+1$, and $\Delta J = 1$.

Since the vibronic reduction factor decreases rapidly as the constant of linear vibronic coupling increases, the spin-orbit interaction $\lambda K(T_1)\boldsymbol{LS}$ in the case of strong vibronic coupling may be small and the perturbation theory used when the result (5.5.16) was obtained is no longer applicable. Therefore the g factors obtained from (5.5.16) are valid for the weak and intermediate vibronic couplings only.

The importance of the vibronic reduction can be demonstrated by the example of the $^2T_2$ term of the electronic configuration $3d^1$ in an octahedral crystalline field [5.269]. Assuming in (5.5.15) that $g_1 \approx 2$, $g_L = -1$ [Ref. 5.232; Chap. 14, Sect. 2], we have for the operator of the Zeeman energy

$$H_Z = \beta[2\boldsymbol{S} - K(T_1)\boldsymbol{L}]\mathcal{H} . \qquad (5.5.17)$$

The spin-vibronic states $^2T_2 = \Gamma_6 + \Gamma_8$ composed of the wave functions of the $^2T_2$ term can be obtained by means of Clebsch-Gordan coefficients

$$|\Gamma\gamma\rangle = \sum_{\gamma_1\gamma_2} \langle\Gamma_6\gamma_1 T_2\gamma_2|\Gamma\gamma\rangle \Psi_{T_2\gamma_2}|\Gamma_6\gamma_1\rangle , \qquad (5.5.18)$$

where $\Psi_{T_2\gamma_2}$ are the vibronic states of the ground T_2 term and $|\Gamma_6\gamma_1\rangle$ are the spin functions transforming as the irreducible representation $D^{(1/2)} = \Gamma_6$. Thus, for example, from (5.5.18) we have

$$\left|\Gamma_8, -\frac{1}{2}\right\rangle = \frac{1}{\sqrt{2}}(\mathrm{i}\Psi_{T_2\xi} - \Psi_{T_2\eta})\left|\Gamma_6, -\frac{1}{2}\right\rangle .$$

Including the effective spin $S = 3/2$ for the spin-vibronic term Γ_8, the Hamiltonian (5.5.17) can be rewritten in the form

$$H_Z = \frac{4\beta}{9}[K(T_1) - 1][5\mathcal{H}\boldsymbol{S} - 2(\mathcal{H}_x S_x^3 + \mathcal{H}_y S_y^3 + \mathcal{H}_z S_z^3)] . \qquad (5.5.19)$$

As can be seen from this expression, in the absence of the Jahn-Teller effect, when

$K(T_1) = 1$ the splitting linear in the magnetic field equals zero. The splitting described by the Hamiltonian (5.5.19) is isotropic and the energy levels are equally spaced with the g factor

$$g = \tfrac{2}{3}[1 - K(T_1)] \ . \tag{5.5.20}$$

In the absence of vibronic coupling, when $K(T_1) = 1$ the matrix elements of the operator (5.5.17) calculated by the spin-vibronic states of the term Γ_8 are zero. This can be easily checked when one takes into account that the components of the orbital momentum $\boldsymbol{L}$ in the basis of the states $\psi_{T_2\gamma}(\mathrm{r})$ are represented by the matrices (3.3.3). Therefore in first-order perturbation theory the cubic quadruplet $\Gamma_8(^2T_2)$ is a nonmagnetic one, i.e., it does not possess the linear Zeeman effect because of the exact compensation of the orbital and spin contributions to the Zeeman energy [Ref. 5.232; Chap. 7, Sect. 8]. Considering the vibronic interaction, the orbital contribution is quenched by the vibronic reduction factor $K(T_1)$ and the compensation mentioned above is violated [5.269].

In the case of strong vibronic coupling the second-order spin-orbital interaction which takes into account the admixture of the excited vibronic states has to be added to the Hamiltonian (5.5.15). In the particular case of the $T \otimes e$ system the approximate terms of the Hamiltonian have the form (4.7.38). The same type of expressions can also be obtained in the more general case including the vibronic coupling to the E and T_2 vibrations, if in the sum (4.7.37) one takes into account only the upper sheets of the adiabatic potential in the configuration Q_0 of the minimum under consideration [see the discussion after (4.7.39)].

The orbital contribution to the Zeeman energy in the case of strong vibronic coupling is also reduced, and hence the second-order terms, multiplying the operators of spin-orbital and Zeeman interactions, become important. In the particular case of the $T \otimes e$ problem we obtain by means of analogous considerations

$$H_Z = -\frac{2\lambda g_L \beta}{3E_{\mathrm{JT}}(E)}(L_x^2 S_x \mathscr{H}_x + L_y^2 S_y \mathscr{H}_y + L_z^2 S_z \mathscr{H}_z) \ . \tag{5.5.21}$$

A similar result (with an accuracy up to a constant coefficient) can also be obtained in the general case of the $T \otimes (e + t_2)$ problem.

The total Hamiltonian of the T term including the second-order corrections (4.7.38) and (5.5.21) in addition to the first-order terms (5.5.15), results in different EPR spectra depending on the relations between the vibronic coupling constants with E and T_2 vibrations.

When the coupling to E vibrations is dominant the ground vibronic triplet consists of the ground oscillator wave functions in tetragonal minima. Neglecting the overlap of the states of different minima one obtains, as a result of the averaging of the total Hamiltonian over the states of each minimum, three axially symmetric spin-Hamiltonians, cf. (5.5.4), with axes along the cubic fourth-order axes:

$$H_x = D[S_x^2 - S(S+1)] + g_\parallel \beta \mathscr{H}_x S_x + g_\perp \beta(\mathscr{H}_y S_y + \mathscr{H}_z S_z) ,$$
$$H_y = D[S_y^2 - S(S+1)] + g_\parallel \beta \mathscr{H}_y S_y + g_\perp \beta(\mathscr{H}_x S_x + \mathscr{H}_z S_z) , \quad (5.5.22)$$
$$H_z = D[S_z^2 - S(S+1)] + g_\parallel \beta \mathscr{H}_z S_z + g_\perp \beta(\mathscr{H}_x S_x + \mathscr{H}_y S_y) ,$$ where

$$D = \frac{\lambda^2}{3E_{JT}(E)} , \qquad g_\parallel = g_S , \qquad g_\perp = g_S - \frac{2\lambda g_L}{3E_{JT}(E)} . \quad (5.5.23)$$

If the coupling to T_2 vibrations is dominant the situation is in large measure similar to the static Jahn-Teller effect for the E term. As in the case of the E term, the tunneling between the trigonal minima of the adiabatic potential results in the excited vibronic singlet being separated from the ground triplet by a small energy gap of 4Γ (tunneling splitting). But unlike in the E term case, the orbital interaction in the second order [the first order is reduced by the vibronic reduction factor $K(T_1)$] results in a splitting comparable in magnitude to that in the tunneling case. Hence a complicated picture of spin-tunneling levels occurs with a very complicated EPR spectrum [5.270, 271]. However, if the anisotropic perturbations (the Zeeman interaction or random strain) result in a considerably larger splitting, then, as in the case of the E term, the system is locked in the trigonal minima. The resulting EPR spectrum in this case is simply a superposition of four axially symmetric spectra of trigonal distorted centers. The principal axes of the g tensor coincide with the four threefold axes of the cubic system.

If the vibronic interaction with E and T_2 vibrations is comparable in magnitude, the Jahn-Teller effect for the T term, as shown in Sect. 3.3, may result in different types of absolute minima of the adiabatic potential which are lower in symmetry than the tetragonal or the trigonal ones. In particular, over a wide range of vibronic constants, including quadratic vibronic coupling, the absolute minima of the adiabatic potential may be orthorhombic. If the anisotropic perturbations are strong enough to lock the system in the orthorhombic minima, the EPR spectrum will be a superposition of six rhombic spectra of orthorhombic distorted centers. Such spectra testifying to the orthorhombic distortion were observed in different systems with T terms, e.g., in Ni^- : Ge, Pd^- : Si, Pt^- : Si [5.272].

The magnetic properties of a cubic ^{2S+1}T term are in large measure similar to the properties of a trigonal ^{2S+1}E term where first-order spin-orbit splitting is also possible. The EPR spectra of trigonal Jahn-Teller systems with the E term were investigated in [5.231, 273].

Let us summarize briefly the results obtained. The characteristic features of the EPR spectrum of a Jahn-Teller paramagnetic center strongly depend on the relation between the tunneling splitting magnitude δ and the parameters of the anisotropic perturbation $g_2\beta\mathscr{H}$ and $\bar{\Delta}$. If the latter are relatively small so that $|g_2|\beta\mathscr{H}/\delta < 0.1$ and $\bar{\Delta}/\delta < 0.1$, then at low temperatures the EPR lines characteristic of the dynamic Jahn-Teller effect (Sect. 5.4) are observed. If the anisotropic perturbations are relatively large, so that $|g_2|\beta\mathscr{H}/\delta > 5$ or $\bar{\Delta}/\delta > 5$, the system is locked in the minima of the adiabatic potential and the static Jahn-Teller effect occurs for which the EPR spectrum is a superposition of several (according to the number of equivalent minima) axial spectra with the axes directed along the axes of symmetry of the center.

The intermediate case $0.1 < |g_2|\beta\mathscr{H}/\delta < 5$, $0.1 < \bar{\Delta}/\delta < 5$ is manifest in the following properties of the low-temperature EPR spectrum: (1) selective broadening and shifting of one of the resonances of the EPR line, (2) angular dependence of the line differing from that of the limiting cases of the static and dynamic Jahn-Teller effects, and (3) an unusually complicated structure of the EPR line for arbitrary orientations of the magnetic field, in particular for $\mathscr{H} \parallel [111]$.

An increase in temperature results in the formation of an isotropic line. Simultaneously the intensity of the anisotropic components of the spectrum decreases to zero and the spectrum becomes averaged. The time of spin-lattice relaxation for Jahn-Teller paramagnetic centres is anomalously small. A review of the experiments on the magnetic resonance of Jahn-Teller systems with E and T_2 terms is given in [5.229–231]; see also [5.110, 274–277].

The use of magnetic resonance methods in the investigation of systems with the Jahn-Teller effect is not limited to the EPR method. Among the works on the application of the acoustic paramagnetic resonance to the problems under consideration one can note the review article [5.278], as well as [5.279–283] in which some aspects of the acoustic paramagnetic resonance theory of Jahn-Teller centers are discussed (see also [5.240]).

The application of the Mössbauer effect to the investigation of Jahn-Teller systems is constrained by the fact that not all the ions are subject to nuclear gamma resonance. The first experimental work in this area in which the quadrupole splitting of the Mössbauer line was explained by the Jahn-Teller effect appeared in the middle of the 1960s [5.284, 285]. The lowering of the quadrupole splitting with temperature was interpreted as being due to the tunneling between different nuclear configurations corresponding to the equivalent minima of the adiabatic potential which give different contributions to the gradient of the electric field on the nuclei. For quantitative descriptions of the temperature dependence in the nuclear gamma resonance spectra the Kubo-Anderson stochastic model is usually used, which is a phenomenological generalization of the Bloch-Redfield (Sect. 5.4.3) model. A presentation of this approach applied to the Mössbauer spectra of Jahn-Teller systems can be found in [5.286, 287]. This model, however, is criticized in [5.288] where an alternative approach is suggested based on the dominant role of random strain.

In a series of theoretical papers [5.289–294] the manifestation of the vibronic interactions in the nuclear gamma resonance spectra of systems with E and T terms of different degeneracy is investigated. In [5.294] the influence of relaxation transitions between vibronic states is taken into account in the generalized Kubo-Anderson model. In some studies the possibility of observing the vibronic reduction of the gradient of the electric field [5.294] and spin-orbit interaction [5.295] in the Mössbauer spectra is suggested.

6. Cooperative Phenomena. Structural Phase Transitions

In this chapter the ordering of Jahn-Teller and pseudo-Jahn-Teller local distortions in crystals is considered. In the case of Jahn-Teller distortions the cooperative Jahn-Teller effect considered in Sect. 6.1 takes place. The ordering of pseudo-Jahn-Teller distortions as well as Jahn-Teller distortions in some noncentrosymmetrical crystals may lead to spontaneous polarization of the lattice and to ferroelectricity (Sect. 6.2). In Sect. 6.3 the vibronic model of structural phase transformations in condensed media is discussed.

6.1 Ordering of Distortions in Crystals. Cooperative Jahn-Teller Effect

One of the conclusions obtained in Chaps. 3 and 4 is that the Jahn-Teller theorem should not be understood literally as a reason for a static distortion of the system which lowers the symmetry and reduces the problem to a nondegenerate one. The minimum of the adiabatic potential arising due to the Jahn-Teller effect does indeed correspond to a distorted nuclear configuration, however, there are several rather than just one such minima and they transform into one another under the operations of the reference high symmetry group of the system. Due to the equivalence of the minima the system is localized with equal probability in each of them, so that the averaged values of the low-syummetry atomic displacements are zero.

In the case of a polycenter Jahn-Teller effect the system acquires quite new features. The simplest case, the bioctahedron, was considered in detail in Sect. 3.6.1. In the crystal the translational displacements of the octahedra owing to the lattice restrictions are not possible, and the right-hand side of (3.6.4) can be taken to be zero. As a result the condition $Q_\theta(1) = -Q_\theta(2)$ holds (Sect. 3.6.1). If the intercenter interaction (or the second-order vibronic interaction, or some other low-symmetry interaction discussed in Sect. 3.6.1) is taken into account then the four minima occur at the bottom of the trough of the adiabatic potential of the bioctahedron corresponding to two possible distortions of the equilateral squares of the octahedra into rhomboids (Fig. 6.1). In the case when the intercenter interaction prefers similar orientations of the distortions (the constant of interaction being negative) the energies of the configurations $(++)$ and $(--)$

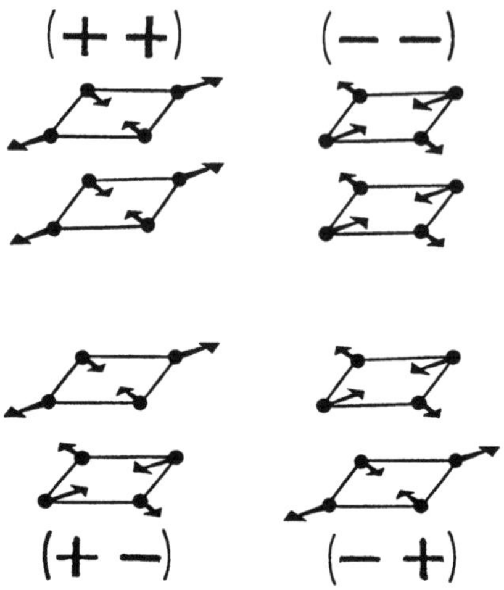

Fig. 6.1. Four types of possible packings of the distortions of two interacting Jahn-Teller centers, for each of which the $E \otimes b_1$ JTE (symmetry D_{4h}) is realized

are lower than that of the configurations $(+-)$ and $(-+)$. Thus here the Jahn-Teller distorted configurations of the centers are no longer equivalent since their energies depend on the distortions of the neighboring centers. Of course, in this case too, the initial symmetry is preserved: the configuration $(++)$ transforms into the isoenergetic one $(--)$ by rotations around the fourfold axis of symmetry of the system. However, now for such transitions a simultaneous reorientation of the distortions of both centers is needed [6.1].

The limiting polycenter situation occurs in crystals in which every elementary cell contains one or several Jahn-Teller centers. In other words, in this case one deals with crystals containing a sublattice of ions with a degenerate (or quasi-degenerate) electronic ground term which is not a Kramers doublet. Such crystals are called *Jahn-Teller crystals*. This name indicates that the Jahn-Teller effect is responsible for a series of characteristic and rather peculiar features of these crystals.

As in the above two-center situation, the energy of the crystal depends on the distortions of separate Jahn-Teller centers. For instance, if there is a Jahn-Teller effect on each center of the type $E \otimes b_1$ and the constant of intercenter interaction has an appropriate sign, the energy of the configurations of "ferro" type $(+ + + \cdots)$, $(- - - \cdots)$ is lower than the energy of other possible packings. In spite of the fact that there are several isoenergetic configurations, in this case it is clear that the transitions between them are not realized in crystals practically (quite analogously to the case of an isotropic magnetic sample where the flip of the macroscopic magnetic moment does not take place). Unlike in the case of the one- (two-, three-, four-)center problem, the symmetry of the ground state of the Jahn-Teller crystal is lower than the symmetry of the Hamiltonian (the effect of "broken symmetry") [6.1, 2].

Thus in crystals containing sublattices of Jahn-Teller ions, at low temperatures an ordering of the local distortions should take place. An ordering of the "ferro" type (called *ferrodistortive*) may lead to a macroscopic deformation of the crystal as a whole. Depending on the nature of the correlation between the local distortions, orderings of more complex types are possible, e.g., *antiferrodistortive*, *helicoidal*, etc., (Sects. 6.1.3 and 6.1.4). When the temperature rises and the thermal fluctuations destroy the correlation between the local distortions, the latter become disordered and the macrodeformation, if present in the low-temperature phase, disappears; the crystal undergoes a structural phase transition. This phenomenon is called the cooperative Jahn-Teller effect.

Structural phase transitions in Jahn-Teller crystals are at present subject to quite a number of investigations, both theoretical and experimental. This is due to the fact that in the 1970s quite unambiguous experimental confirmations of the Jahn-Teller nature of structural phase transitions in a large number of crystals were obtained, mainly in rare-earth compounds with the zircon structure, e.g., $TbVO_4$ (33 K), $DyVO_4$ (14 K), $TmVO_4$ (2.1 K), $TbAsO_4$ (27.7 K), $DyAsO_4$ (11.2 K), $TmAsO_4$ (6.1 K), $TbPO_4$ (3.5 K) (the temperature of the structural phase transition is shown in parentheses), and in spinels $NiCr_2O_4$ (274 K), $FeCr_2O_4$ (135 K), $CuCr_2O_4$ (860 K), Mn_3O_4 (1170 K), $CuFe_2O_4$ (360 K), Fe_2TiO_4 (142 K), as well as in some other crystals.

The recent status of investigations into the cooperative Jahn-Teller effect is presented in [6.3, 4]. The results of spectroscopic investigations of Jahn-Teller crystals are given in [6.1]. Anomalous elastic properties are discussed in detail in [6.5].

6.1.1 The Hamiltonian of the Jahn-Teller Crystal. Correlation of Local Distortions with Electronic States of Jahn-Teller Ions

The Hamiltonian of the Jahn-Teller crystal can be written as [cf. (3.6.8)]

$$\hat{H} = \sum_{\boldsymbol{n}} \hat{H}_{\mathrm{JT}}(\boldsymbol{n}) + \tfrac{1}{2} \sum_{\substack{\boldsymbol{n}\ \boldsymbol{m} \\ (\boldsymbol{n} \neq \boldsymbol{m})}} \hat{Q}^{+}(\boldsymbol{n}) \hat{K}(\boldsymbol{n}-\boldsymbol{m}) \hat{Q}(\boldsymbol{m}) \ , \tag{6.1.1}$$

where $\boldsymbol{n} = (n_1, n_2, n_3)$ and $\boldsymbol{m} = (m_1, m_2, m_3)$ label the elementary cells of the crystal, $H_{\mathrm{JT}}(\boldsymbol{n})$ is the vibronic Hamiltonian of the elementary cell, its electronic matrices being determined in the space of localized site states, while the last term in (6.1.1) describes the generally accepted bilinear interaction between the atomic displacements of different elementary cells in lattice dynamics. Here, as in Sect. 3.6.1, vector notation is used: $\hat{Q}(\boldsymbol{n})$ is a column vector having $3s$ components of $Q_{\Gamma\gamma}(\boldsymbol{n})$, where s is the number of atoms in the elementary cell. Accordingly, $\hat{K}(\boldsymbol{n}-\boldsymbol{m})$ is a $3s \times 3s$ matrix with $K_{\Gamma_1\gamma_1,\Gamma_2\gamma_2}(\boldsymbol{n}-\boldsymbol{m})$ elements. The terms with $\boldsymbol{n} = \boldsymbol{m}$ describe the elastic properties of isolated elementary cells, and therefore they are separated from the double sum in (6.1.1) and included in $H_{\mathrm{JT}}(\boldsymbol{n})$.

For simplicity, it is assumed that there are no atoms on the border between the elementary cells (that is true, e.g., for molecular crystals). The complication of the theory which arises due to the nonorthogonality of some of the $Q_{\Gamma\gamma}(\boldsymbol{n})$ for neighboring values of $\boldsymbol{n}$ in cases when there are border atoms belonging simultaneously to two elementary cells is considered in [6.6] (see also Sect. 3.6.1).

Another approximation used in (6.6.1) is based on the vibronic coupling of the electrons of each elementary cell to the atomic displacements of the nearest-neighbor coordination spheres of the same elementary cell only. Since, as mentioned above, the Jahn-Teller ions in a crystal with a cooperative Jahn-Teller effect are usually transition metals or rare-earth elements having small radii, this assumption is, as a rule, justified.

Finally it is assumed here that the intersite resonance interaction of localized electronic states forming the energy bands is negligibly small. In most Jahn-Teller crystals the width of the electronic bands formed by atomic d and f orbitals is much smaller than the characteristic phonon quantum, and this justifies the above approximation. Additional effects related to the finite electronic bandwidth are considered in [6.7].

Note that the Hamiltonian (6.1.1) also includes the cases when there are several identical Jahn-Teller ions in the elementary cell. The resonance interaction between the electronic states of different ions of the same cell results in a small splitting of the electronic terms (the Davydov splitting); strictly speaking, here the pseudo-Jahn-Teller effect, and not the Jahn-Teller effect, takes place.

Using the Fourier transformation

$$\hat{Q}(\boldsymbol{n}) = \frac{1}{\sqrt{N}} \sum_{\boldsymbol{q}} \hat{Q}(\boldsymbol{q}) \mathrm{e}^{\mathrm{i}\boldsymbol{q}\cdot \boldsymbol{R}(\boldsymbol{n})} , \qquad \hat{Q}(\boldsymbol{q}) = \frac{1}{\sqrt{N}} \sum_{\boldsymbol{q}} \hat{Q}(\boldsymbol{n}) \mathrm{e}^{-\mathrm{i}\boldsymbol{q}\cdot \boldsymbol{R}(\boldsymbol{n})} , \tag{6.1.2}$$

where N is the number of elementary cells in the crystal and $\boldsymbol{R}(\boldsymbol{n})$ is the lattice vector, the Hamiltonian (6.1.1) can be transformed to the form

$$\hat{H} = \tfrac{1}{2} \sum_{\boldsymbol{q}\nu} [P_\nu(\boldsymbol{q}) P_\nu(-\boldsymbol{q}) + \omega_\nu^2(\boldsymbol{q}) Q_\nu(\boldsymbol{q}) Q_\nu(-\boldsymbol{q})] + \sum_{\boldsymbol{n}} \hat{V}(\boldsymbol{n}) , \tag{6.1.3}$$

where $\hat{V}(\boldsymbol{n})$ is the operator of the vibronic interaction in the $\boldsymbol{n}$th elementary cell, and the normal coordinates $Q_\nu(\boldsymbol{q})$ are related to the components $Q_{\Gamma\gamma}(\boldsymbol{q})$ of the vector $\hat{Q}(\boldsymbol{q})$ by the orthogonal transformation

$$Q_{\Gamma\gamma}(\boldsymbol{q}) = \sum_{\nu=1}^{3s} \mathcal{O}_{\Gamma\gamma,\nu} Q_\nu(\boldsymbol{q}) , \tag{6.1.4}$$

which diagonalizes the dynamic matrix

$$\hat{\boldsymbol{K}}(\boldsymbol{q}) = \sum_{\boldsymbol{n}} \hat{\boldsymbol{K}}(\boldsymbol{n}) \mathrm{e}^{\mathrm{i}\boldsymbol{q}\cdot \boldsymbol{R}(\boldsymbol{n})} . \tag{6.1.5}$$

The index ν labels the branches of vibrations, $\omega_\nu^2(\boldsymbol{q})$ are the eigenvalues of the dynamic matrix, and $P_\nu(\boldsymbol{q})$ are momenta conjugate to the coordinates $Q_\nu(\boldsymbol{q})$. Substituting (6.1.4) into (6.1.2) we obtain

$$Q_{\Gamma\gamma}(\boldsymbol{n}) = \sum_{\boldsymbol{q},\nu} a_{\boldsymbol{q},\nu}(\boldsymbol{n}|\Gamma\gamma) Q_\nu(\boldsymbol{q}) , \tag{6.1.6}$$

where $a_{\boldsymbol{q},\nu}(\boldsymbol{n}|\Gamma\gamma) = \mathcal{O}_{\Gamma\gamma,\nu}(\boldsymbol{q}) \exp[\mathrm{i}\boldsymbol{q}\boldsymbol{R}(\boldsymbol{n})]/\sqrt{N}$ are the Van Vleck coefficients (Sect. 3.5).

The Hamiltonians (6.1.1) and (6.1.3) are equivalent and should in principle lead to the same result, provided the problem is solved exactly. Since, however, exact solution is impossible in real situations and approximations are necessarily used, the Hamiltonians (6.1.1) and (6.1.3) have different advantages and faults.

The Hamiltonian (6.1.1) containing the bilinear interaction of the displacements describes explicitly the correlation of local distortions in different elementary cells. Any simplification of this bilinear term results in an inaccurate description of the lattice dynamics (the phonon subsystem) determined by this bilinear interaction of the nuclear displacements.

On the other hand, the normal coordinates in which the Hamiltonian (6.1.3) is expressed describe harmonic normal modes of vibrations which are uncorrelated in the harmonic approximation. Nevertheless the phase transition takes place due to the correlation of the electronic states at different centers interacting between them via the phonon field. In some cases when the Jahn-Teller effect does not influence the vibrations and leads only to a shift of the equilibrium configuration of the vibrating atoms ($E \otimes b_1$ and $T \otimes e$ systems) the unitary shift transformation changing the zero for normal coordinates like (4.1.35) reveals the correlation between the electronic states, while the phonon subsystem is separated (Sect. 6.1.5). In this approach the lattice dynamics is considered exactly, and all the approximations concern the ordering in the electronic subsystem. Hence this second approach is more exact, although less general.

The existence of two approaches in the theory of the cooperative Jahn-Teller effect is due to the existence of two classes of Jahn-Teller polyatomic systems, with or without significant nonadiabaticity. The first group is inherent in the $E \otimes e$, $T \otimes t_2$, etc. systems, while the second one corresponds to the $E \otimes b_1$, $T \otimes e$ and similar systems mentioned above.

In some works using the Hamiltonian (6.1.1) as a starting point some additional approximations are introduced. The displacements $Q_{\Gamma\gamma}(\boldsymbol{n})$, inactive in the Jahn-Teller effect in the isolated elementary cell, are excluded (see, e.g., the review [6.4]). As in the one-center Jahn-Teller system, this greatly simplifies the solution. For ferrodistortive phase transitions in which the primary role is played by normal modes with $\boldsymbol{q} = 0$ such an approximation is justified and does not influence the results. For antiferrodistortive and more complicated types of phase transitions where the modes with $\boldsymbol{q} \neq 0$ are important the neglected displacements are very important, since the corresponding terms in (6.1.1) contribute strongly to $\omega_\nu^2(\boldsymbol{q})$, $\boldsymbol{q} \neq 0$. In this sense the results obtained by means of (6.1.3), where the exact dependence $\omega_\nu(\boldsymbol{q})$ is used, are more accurate. In the case of antiferrodistortive ordering some of the degrees of freedom may nevertheless be excluded from the consideration, if one passes to the enlarged elementary cell of double dimensions. The same approach can be used in the case of ordering with tripling, quadrupling, etc., of the lattice period, although the simplification gained in this way ceases to compensate the increasing complexity of the cell. Finally, an ordering is possible with a transition to the so-called incommensurate phase for which the crystal lattice period is not a simple multiple of the period of the high-temperature lattice. In these cases all the degrees of freedom of the elementary cell should be taken into account, and the Hamiltonian (6.1.1) loses its advantages over that of (6.1.3).

6.1.2 Mean Field Approximation

The main idea usually used for description of phase transitions in solid states is that each elementary cell in the low-temperature phase is actually affected by a low-symmetry field due to all the other elementary cells. This field lowers the symmetry of the cell under consideration. As a result the latter becomes a source of the low-symmetry field contributing to the resulting low-symmetry crystal field. Due to the thermal motion of the cell atoms (mainly the one nearest to that under consideration) the low-symmetry crystal field fluctuates. The approximation which neglects these fluctuations and takes into account only the averaged self-consistent crystal field is called the *mean field approximation* (MFA).

As mentioned above (Chap. 3), in the absence of vibronic interactions the electronic density in degenerate electronic states rotates around the Jahn-Teller center and can freely orient along the field. These properties of degenerate electronic states can be described by means of the operators of energy spin (pseudospin). The low-symmetry distribution of the electronic density creates an anisotropic electric field. For instance, the *p* orbitals have a dipole moment, the *d* orbitals have a quadrupole moment, etc. The appropriate dipole-dipole, quadrupole-quadrupole and similar interactions of the electronic states of different centers may lead to an ordering in the electronic subsystem. The electric fields of separate cells in this case are summed to an averaged electric field. A similar ordering in the electronic subsystem results from the non-Coulomb (exchange, superexchange and so on) interaction of the electronic states. In the absence of vibronic coupling this ordering will not influence the nuclear behavior, that is, it will not result in a structural phase transition. As already mentioned in this section, we neglect the intersite interaction of the electronic states and hence we exclude the possibility of an ordering in the electronic subsystem without the participation of the vibrational degrees of freedom of the lattice. Such effects were considered in a number of works, e.g., in [6.8, 9].

In principle, the ordering of local distortions may take place without the participation of the electronic subsystem. The mechanism of such phase transitions is assumed to be due to the proper anharmonicity of the lattice [6.10] (see, however, Sect. 6.2). The mean field determining these phase transitions is the field of deformations. In some cases, when the lowering of the symmetry of the cell is accompanied by the formation of an electric (magnetic) dipole moment, the main contribution to the mean field results from the electric (magnetic) field of the cell. The Hamiltonian (6.1.1), as well as that of (6.1.3), does not take into account the proper lattice anharmonicity, and hence this mechanism of structural ordering is also excluded from consideration.

The phase transitions discussed below are due to the simultaneous action of the Jahn-Teller distorting forces within each elementary cell and the elastic intercell interaction. The former result in vibronic anharmonicity of the potential energy of the nuclei (Chap. 3), while the latter leads to an indirect interaction of the electronic states of different sites.

In the high-temperature phase when the mean field is absent the mean values of atomic displacements, as mentioned above, are zero, $\langle \hat{Q}(\boldsymbol{n})\rangle = 0$. In the low-temperature phase, $\langle \hat{Q}(\boldsymbol{n})\rangle \neq 0$, and therefore it is worthwhile to make a transformation of the variables $\hat{Q}(\boldsymbol{n}) = \langle \hat{Q}(\boldsymbol{n})\rangle + \Delta\hat{Q}(\boldsymbol{n})$. The time dependence of the new dynamic variables $\Delta\hat{Q}(\boldsymbol{n})$ determines the fluctuations of the crystal field, and therefore in the framework of the MFA they are considered to be small. As a result of this substitution the operator of the intercell interaction in (6.1.1) becomes

$$-\tfrac{1}{2}\sum_{\substack{n\; m\\(n\neq m)}} \langle \hat{Q}^\dagger(\boldsymbol{n})\rangle \hat{K}(\boldsymbol{n}-\boldsymbol{m})\langle \hat{Q}(\boldsymbol{m})\rangle + \sum_{n} \hat{f}(\boldsymbol{n})\hat{Q}(\boldsymbol{n})$$

$$+\tfrac{1}{2}\sum_{\substack{n\; m\\(n\neq m)}} \Delta\hat{Q}^\dagger(\boldsymbol{n})\hat{K}(\boldsymbol{n}-\boldsymbol{m})\Delta\hat{Q}(\boldsymbol{m}) \;, \tag{6.1.7}$$

where

$$\hat{f}(\boldsymbol{n}) = \sum_{\substack{m\\(m\neq n)}} \hat{K}(\boldsymbol{n}-\boldsymbol{m})\langle \hat{Q}(\boldsymbol{m})\rangle \;. \tag{6.1.8}$$

The first term in (6.1.7) is a constant (it does not depend on the dynamic variables) and can be excluded from the Hamiltonian. The last term can be neglected in the framework of the MFA. As a result the Hamiltonian of the crystal is reduced to a sum of commutative terms each of which describes an isolated elementary cell in the external mean field:

$$\hat{H}_{\mathrm{MFA}} = \sum_{n} \hat{H}(\boldsymbol{n}) \;, \qquad \hat{H}(\boldsymbol{n}) = \hat{H}_{\mathrm{JT}}(\boldsymbol{n}) + \sum_{\Gamma\gamma} f_{\Gamma\gamma}(\boldsymbol{n}) Q_{\Gamma\gamma}(\boldsymbol{n}) \;. \tag{6.1.9}$$

At this stage, provided the quadratic vibronic interaction mixing the Jahn-Teller vibrations with the other vibrations is not considered, the non-Jahn-Teller modes in the Hamiltonian $\hat{H}(\boldsymbol{n})$ are separated, resulting in harmonic vibrations near the equilibrium positions $Q^{(0)}_{\Gamma\gamma}(\boldsymbol{n}) = -f_{\Gamma\gamma}(\boldsymbol{n})/\omega^2_\Gamma$. The density matrix of the crystal described by the Hamiltonian $\hat{H}_{\mathrm{MFA}}$ from (6.1.9) is factored into a product of density matrices of separate cells and the latter is factored into a product of density matrices of separated degrees of freedom, and therefore for non-Jahn-Teller vibrations $\langle Q_{\Gamma\gamma}(\boldsymbol{n})\rangle = -f_{\Gamma\gamma}(\boldsymbol{n})/\omega^2_\Gamma$. In the absence of vibronic coupling the same result is obtained for the Jahn-Teller modes. By substituting $f_{\Gamma\gamma}(\boldsymbol{n})$ from (6.1.8) one obtains a system of transcendental equations in $\langle Q_{\Gamma\gamma}(\boldsymbol{n})\rangle$, which has at arbitrary temperatures only one solution: $\langle Q_{\Gamma\gamma}(\boldsymbol{n})\rangle = 0$. The nontrivial solution corresponding to the low-symmetry phase arises when the vibronic coupling in the elementary cells is considered, resulting in a molecular Jahn-Teller system described by the Hamiltonian $\hat{H}(\boldsymbol{n})$ from (6.1.9). See also [6.11, 12].

6.1.3 A Simple Example: The $E \otimes b_1$ Jahn-Teller Crystal

As an example consider first a crystal with a simple Jahn-Teller $E \otimes b_1$ system in each elementary cell (Sect. 3.1). The vibronic Hamiltonian of the separated Jahn-Teller mode Q_1 in this case has the form [cf. (3.1.41)]

$$\hat{H}(\boldsymbol{n}) = \tfrac{1}{2}(P_1^2 + \omega_1^2 Q_1^2) + V_1 Q_1 \hat{\sigma}_z + f_1(\boldsymbol{n}) Q_1 \ , \tag{6.1.10}$$

where $\omega_1^2 = K_{B_1B_1}(0)$, see (6.1.5), $f_1(\boldsymbol{n}) = f_{B_1}(\boldsymbol{n})$. The corresponding wave functions and energies are

$$\begin{aligned} \Psi_{1,2}^{(p)} &= \psi_{1,2}(\mathrm{r}) \Phi_p(Q_1 - Q_1^{(1,2)}) \ , \\ E_p^{(1,2)} &= \hbar\omega_1 \left(p + \frac{1}{2}\right) - \frac{1}{2\omega_1^2}(f_1 \pm V_1)^2 \ , \end{aligned} \tag{6.1.11}$$

where $Q_1^{(1,2)} = -(f_1 \pm V_1)/\omega_1^2$ and Φ_p are the wave functions of harmonic oscillators. From (6.1.11) we have

$$\begin{aligned} \langle Q_1(\boldsymbol{n})\rangle &= \frac{1}{Z_{\boldsymbol{n}}} \sum_{\alpha=1,2} \sum_{p=0}^{\infty} \exp\left(-\frac{E_p^{(\alpha)}}{kT}\right) \langle \Psi_\alpha^{(p)} | Q_1(\boldsymbol{n}) | \Psi_\alpha^{(p)} \rangle \\ &= -\omega_1^{-2}\left[f_1(\boldsymbol{n}) + V_1 \tanh\left(\frac{V_1 f_1(\boldsymbol{n})}{\omega_1^2 kT}\right)\right] , \end{aligned} \tag{6.1.12}$$

where $Z_{\boldsymbol{n}}$ is the statistical sum of the vibronic states of the nth elementary cell. Substituting $f_1(\boldsymbol{n})$ from (6.1.8) we obtain a system of equations in $\langle Q_{\Gamma\gamma}(\boldsymbol{n})\rangle$ determining the self-consistent mean field:

$$\begin{aligned} \sum_{\boldsymbol{m}} \hat{K}(\boldsymbol{n}-\boldsymbol{m})\langle \hat{Q}(\boldsymbol{m})\rangle = \hat{I}(B_1) V_1 \tanh\Big[& V_1 \omega_1^{-2} \sum_{\boldsymbol{m}} \hat{I}^{+}(B_1) \\ & \times \hat{K}(\boldsymbol{n}-\boldsymbol{m})\langle \hat{Q}(\boldsymbol{m})\rangle (\delta_{\boldsymbol{mn}} - 1)/kT \Big] , \end{aligned} \tag{6.1.13}$$

where $\hat{I}(B_1)$ is the column vector with components $I_{\Gamma\gamma}(B_1) = \delta_{\Gamma B_1}$.

The same system of equations can be obtained by direct minimization of the free energy of the crystal with respect to $\langle Q_{\Gamma\gamma}(\boldsymbol{n})\rangle$,

$$\mathscr{F} = \tfrac{1}{2} \sum_{\boldsymbol{n},\boldsymbol{m}} \langle \hat{Q}^{\dagger}(\boldsymbol{n})\rangle \hat{K}(\boldsymbol{n}-\boldsymbol{m}) \langle \hat{Q}(\boldsymbol{m})\rangle (\delta_{\boldsymbol{nm}} - 1) - kT \sum_{\boldsymbol{n}} \ln Z_{\boldsymbol{n}} \ , \tag{6.1.14}$$

where the first term of (6.1.7) is also taken into account.

In the case of ferrodistortive ordering when $\langle Q(\boldsymbol{n})\rangle = \langle \hat{Q}\rangle$, we obtain from (6.1.8, 11)

$$\begin{aligned} \mathscr{F} = & -\tfrac{1}{2} N \langle \hat{Q}^{\dagger}\rangle \sum_{\boldsymbol{n}\neq 0} \hat{K}(\boldsymbol{n}) \langle \hat{Q}\rangle \\ & - kTN \ln\left\{2\cosh\left[V_1 \omega_1^{-2} \hat{I}^{\dagger}(B_1) \sum_{\boldsymbol{n}\neq 0} \hat{K}(\boldsymbol{n})\langle \hat{Q}\rangle / kT\right]\right\} . \end{aligned} \tag{6.1.15}$$

Here the translational invariance of the crystal is used, due to which $\sum_{n} \hat{K}(n-m)(\delta_{nm}-1)$ is independent of the index m, and N is the number of elementary cells in the crystal.

Being the invariant of the point group of the crystal, the matrix $\sum_{n \neq 0} \hat{K}(n)$ decomposes into blocks which can be classified by the irreducible representations $\sum_{n \neq 0} K_{\Gamma_1\gamma_1\Gamma_2\gamma_2}(n) \sim \delta_{\Gamma_1\Gamma_2}\delta_{\gamma_1\gamma_2}$. This allows us to simplify the argument of the hyperbolic cosine in (6.1.15):

$$\hat{\Gamma}^{\dagger}(B_1) \sum_{n \neq 0} \hat{K}(n)\langle\hat{Q}\rangle = \bar{Q} \sum_{n \neq 0} K(n) , \tag{6.1.16}$$

where for the sake of brevity the following notation is introduced: $\langle Q_1 \rangle = \bar{Q}$, $K_{B_1B_1}(n) = K(n)$.

The first term in (6.1.15) is a quadratic function of $\langle Q_{\Gamma\gamma} \rangle$ while the logarithmic term at large $\bar{Q}$ behaves as a linear function. Therefore, for the free energy (6.1.15) to have a minimum, it is necessary that the matrix $\sum_{n \neq 0} \hat{K}(n)$ be positively defined and that, in particular,

$$\sum_{n \neq 0} K(n) > 0 . \tag{6.1.17}$$

If this condition is not fulfilled, the ferrodistortive ordering cannot lead to a thermodynamic equilibrium, and hence in such a crystal the occurrence of a more complicated low-symmetry structure has to be expected.

Equating the derivatives of $\mathscr{F}$ with respect to $\langle Q_{\Gamma\gamma} \rangle$ we obtain that in thermodynamic equilibrium $\langle Q_{\Gamma\gamma} \rangle = 0$ for all $\Gamma \neq B_1$, while $\langle Q_1 \rangle = \bar{Q}$ can be obtained from the transcendental equation

$$\bar{Q} \sum_{n} K(n) = V_1 \tanh\left[V_1 \omega_1^{-2} \sum_{n \neq 0} K(n)/kT \right] . \tag{6.1.18}$$

The ferrodistortive ordering is realized in the subspace of the normal coordinates $\hat{Q}(q)$ with $q = 0$, and therefore the order parameter belongs to the irreducible representation B_1 of the point group of the crystal. The symmetrized cube of the representation B_1, $[B_1^3]$, does not contain the totally symmetric representation A_1, and therefore in accordance with the Landau theory [Ref. 6.13; Sect. 145] in the case under consideration a phase transition of the second order can take place. This means that near the point of the phase transition, $\bar{Q}$ is small, and this allows us to keep only the linear term in the expansion of the right-hand side of (6.1.18) in a series in $\bar{Q}$. From this we find the temperature of the phase transition,

$$kT_0 = 2\lambda E_{\mathrm{JT}} , \qquad \lambda = \sum_{n \neq 0} K(n) \Big/ \sum_{n} K(n) , \tag{6.1.19}$$

where $E_{\mathrm{JT}} = V_1^2/2\omega_1^2$. The value of λ is obviously determined by the damping of the correlations between the atomic displacements with the increase of the distance between the crystal cells.

In the low-symmetry mean field the electronic degeneracy of the elementary cells is removed and this can be interpreted as the orientation of the pseudospin along the field. As follows from (6.1.11), $\langle\hat{\sigma}_z(\boldsymbol{n})\rangle$ is determined by the expression

$$\langle\hat{\sigma}_z(\boldsymbol{n})\rangle = V_1^{-1}[f_1(\boldsymbol{n}) + \langle Q_1(\boldsymbol{n})\rangle\omega_1^2] \ . \tag{6.1.20}$$

In the case of ferrodistortive ordering, considering (6.1.18) we have

$$\langle\hat{\sigma}_z\rangle = \bar{Q}V_1^{-1}\sum_{\boldsymbol{n}} K(\boldsymbol{n}) = \tanh(\langle\hat{\sigma}_z\rangle T_0/T) \ . \tag{6.1.21}$$

Keeping the cubic term in the expansion of the hyperbolic tangent in a series in $\langle\hat{\sigma}_z\rangle$, one can find the temperature dependence of $\langle\hat{\sigma}_z\rangle$ in the neighborhood of the temperature T_0, $\langle\hat{\sigma}_z\rangle \approx \sqrt{3(T_0 - T)/T_0}$ at $T \leq T_0$. The temperature dependence of $\langle\hat{\sigma}_z\rangle$ obtained by a numerical solution of the transcendental equation (6.1.21) is given in Fig. 6.2.

The quantity $\langle\hat{\sigma}_z\rangle$ determines the splitting

$$\Delta E = 2f_1 V_1/\omega_1^2 = 4\langle\hat{\sigma}_z\rangle\lambda E_{\mathrm{JT}} \tag{6.1.22}$$

of all the vibronic states (6.1.11). This splitting influences various properties of crystals with the cooperative Jahn-Teller effect and can be measured by different methods. A typical example of the cooperative Jahn-Teller effect of the $E \otimes b_1$ type under consideration with ferrodistortive ordering is the phase transition in the $TmVO_4$ crystal. It was investigated by optical absorption experiments for the transition to the lowest singlet level of the excited electronic term ${}^1G_4$ ($\hbar\Omega = 20\,940$ cm^{-1}). As can be seen from Fig. 6.2 where the results of this experiment are given, the mean field approximation gives a quite adequate description of the phase transition in $TmVO_4$.

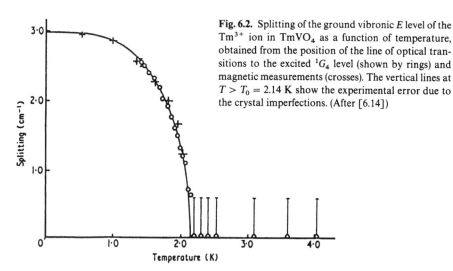

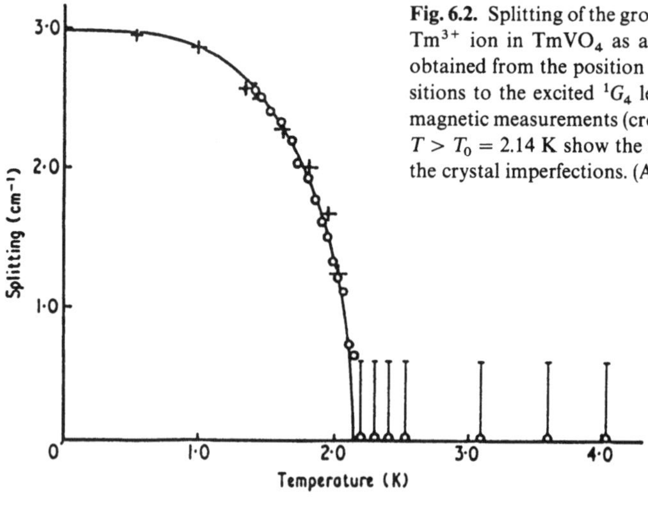

Fig. 6.2. Splitting of the ground vibronic E level of the Tm^{3+} ion in $TmVO_4$ as a function of temperature, obtained from the position of the line of optical transitions to the excited ${}^1G_4$ level (shown by rings) and magnetic measurements (crosses). The vertical lines at $T > T_0 = 2.14$ K show the experimental error due to the crystal imperfections. (After [6.14])

The structure of the low-symmetry crystalline phase is to a large degree dependent on the function $\hat{K}(\boldsymbol{n})$, $\boldsymbol{n} \neq 0$, which determines the nature of the weakening of the correlations of the local distortions with increasing distance between the centers. For instance, a helicoidal ordering is possible for which [6.15]

$$\langle \hat{Q}(\boldsymbol{n}) \rangle = \langle \hat{Q} \rangle \cos[\boldsymbol{q}_0 \boldsymbol{R}(\boldsymbol{n}) + \alpha] \ . \tag{6.1.23}$$

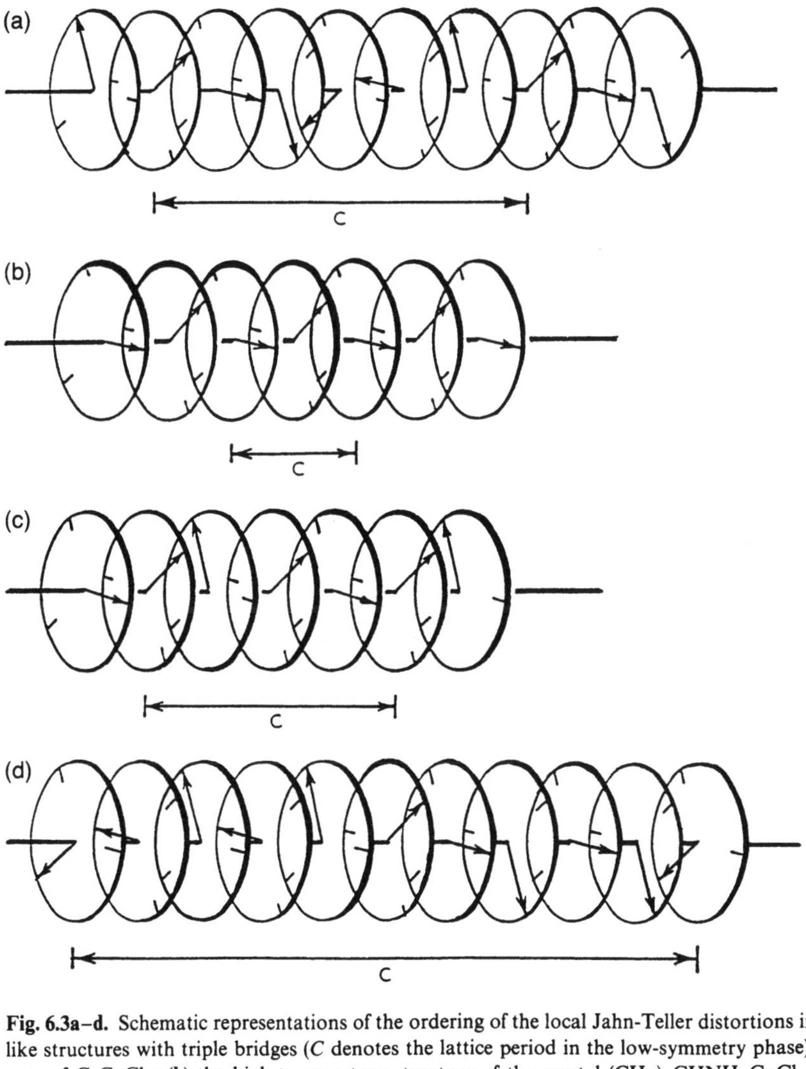

Fig. 6.3a–d. Schematic representations of the ordering of the local Jahn-Teller distortions in chain-like structures with triple bridges (C denotes the lattice period in the low-symmetry phase): (**a**) the case of $CsCuCl_3$; (**b**) the high-temperature structure of the crystal $(CH_3)_2CHNH_3CuCl_3$; (**c**) the low-temperature phase of the crystal $RbCuCl_3$; (**d**) the crystal $(CH_3)_4NCuCl_3$. (After [6.16])

Substituting (6.1.23) into (6.1.14) and minimizing the free energy with respect to the parameters $\langle Q_{\Gamma\gamma}\rangle$, $\boldsymbol{q}_0$ and α one can, in particular, come to a structure with a period which is incommensurate with the period of the high-temperature lattice. The periodic distribution of the distortions in the crystal described by the expression (6.1.23), as follows from (6.1.20), induces a periodic distribution of the electronic density described by the wave $\langle\hat{\sigma}_z(\boldsymbol{n})\rangle$ with the same period.

A similar ordering of local distortions has been found in chain-like structures of the type $A\mathrm{CuCl}_3$, where A is a monovalent cation [6.16–20]. In these compounds the octahedral polyhedra are interlinked by triple bridges, as in the bioctahedron with a common face (Fig. 3.14c), and form chains in which the Cu^{2+} ions lie on the common threefold axis. The Jahn-Teller effect in the ground ^2E_g term results in the elongation of two of the six bonds Cu–Cl corresponding to one of the three wells at the bottom of the circular trough of the adiabatic potential (Fig. 3.3b). The structure of the compounds under consideration is equivalent to replacement of the minimum configurations by saddle-point ones, and vice versa, when moving along the chain. Therefore the diagram of equipotential cross sections of the lowest surface of the adiabatic potentials for each octahedron can be obtained from that of the previous one by a 60° rotation about the threefold axis (Fig. 6.3). The system can thus be described by a one-dimensional Ising model for coupled spins $S = 1/2$ [6.21].

The structure of the crystal lattice arising due to the ordering of the local distortions can be obtained by a periodic repetition (along the chain) of the possible variants of localization of the distortions of the octahedron in one of the three wells of the adiabatic potential. Four examples of such ordering are illustrated in Fig. 6.3.

The type of structural ordering is determined by the free energy value in the state of thermodynamic equilibrium at a given temperature, the phase with the lowest free energy being preferred.

6.1.4 Cooperative Jahn-Teller Effect in Cases of Weak, Strong and Intermediate Vibronic Coupling

Going back to the general Hamiltonian (6.1.9) we note that the vibronic properties of the one-center system are determined by the dimensionless parameter $E_{\mathrm{JT}}/\hbar\omega$. Depending on the value of this parameter the cases of weak, strong and intermediate coupling occur (Chap. 4). In an external field the situation is, strictly speaking, determined by two more parameters, $E_f/\hbar\omega$ and E_f/E_{JT}, where $E_f = f_{\Gamma\gamma}^2/2\omega_\Gamma^2$ is the energy of the stabilization of the low-symmetry configuration of the system in the external field in the absence of vibronic coupling. However, one has to keep in mind that the $f_{\Gamma\gamma}$ value is related to the correlation of local distortions, so that in the case under consideration only one parameter λ from (6.1.19) is added. In fact the largest value of $\bar{Q}$ achieved at $T = 0$ K can be found from (6.1.21), if one assumes that $\langle\hat{\sigma}_z\rangle = 1$. Substituting this value in (6.1.8) we find that $E_f \lesssim \lambda E_{\mathrm{JT}}$. This limits the number of possible situations.

a) *Weak vibronic coupling, $E_{\rm JT} \ll \hbar\omega$, moderate correlations, $\lambda \sim 1$.* Considering the aforesaid, $E_f \ll \hbar\omega$, and $kT_0 \ll \hbar\omega$. The latter condition allows us to limit the treatment to the ground vibronic term only and to represent the operators in the Hamiltonian by matrices determined in the basis of the states of the ground vibronic term. In accordance with the theory of vibronic amplification of vibrational operators (Sect. 4.7) the coordinates $Q_{\Gamma\gamma}(\boldsymbol{n})$ are represented by the matrices

$$\hat{Q}_{\Gamma\gamma}(\boldsymbol{n}) = Q_\Gamma \hat{C}_{\Gamma\gamma}(\boldsymbol{n}) \ , \qquad Q_\Gamma = -\frac{V_\Gamma}{\omega_\Gamma^2} K(\Gamma) \ . \tag{6.1.24}$$

Here $K(\Gamma)$ is the vibronic reduction factor. Considering that $\hat{H}_{\rm JT}(\boldsymbol{n})$ is given by the unit matrix $\hat{C}_A(\boldsymbol{n})$ determining only the energy zero, the Hamiltonian of the Jahn-Teller crystal (6.1.1) in this representation has an Ising form:

$$\hat{H} = -\tfrac{1}{2} \sum_{\boldsymbol{n},\boldsymbol{m}} \sum_{\Gamma_1\gamma_1} \sum_{\Gamma_2\gamma_2} Q_{\Gamma_1} Q_{\Gamma_2} K_{\Gamma_1\gamma_1\Gamma_2\gamma_2}(\boldsymbol{n}-\boldsymbol{m}) \hat{C}_{\Gamma_1\gamma_1}(\boldsymbol{n}) \hat{C}_{\Gamma_2\gamma_2}(\boldsymbol{m}) \ . \tag{6.1.25}$$

If there is a low-symmetry crystal field in the crystal which removes the degeneracy of the ground vibronic term, the following perturbation has to be added to the Hamiltonian (6.1.25):

$$\hat{W} = \sum_{\boldsymbol{n}} \sum_{\Gamma\gamma} \Delta_{\Gamma\gamma} \hat{C}_{\Gamma\gamma}(\boldsymbol{n}) K(\Gamma) \ , \tag{6.1.26}$$

where $\Delta_{\Gamma\gamma}$ are constants determining the magnitude and the orientation of the crystal field.

The phase transition in the system described by the Hamiltonian $\hat{H} + \hat{W}$, where $\hat{H}$ and $\hat{W}$ have the matrix forms (6.1.25), (6.1.26), can be considered in an ordinary way, say, in the framework of the MFA. The matrices $\hat{C}_{\Gamma\gamma}(\boldsymbol{n})$ can in principle be expressed by the pseudospin matrices $\hat{S}_x(\boldsymbol{n})$, $\hat{S}_y(\boldsymbol{n})$ and $\hat{S}_z(\boldsymbol{n})$ of appropriate dimensionality. In this sense the Hamiltonian (6.1.25), (6.1.26) is mathematically equivalent to the Hamiltonian of some spin lattice with an anisotropic exchange interaction. An exhaustive investigation of phase transitions in spin systems, as well as possible spin structures, is given in [6.22, 23] and this eliminates the need to present this topic in more detail.

b) *Intermediate vibronic coupling, $E_{\rm JT} \sim \hbar\omega$, weak correlations, $\lambda \ll 1$.* Since the intersite correlations of local distortions are conditioned by the elastic interaction of the lattice atoms, that is, by the same mechanism which leads to phonon dispersion, the condition of weak correlations can be fulfilled for crystals in which the width of the phonon bands formed by Jahn-Teller active vibrations is relatively small. In this case, as in the previous one, $kT \ll \hbar\omega$ and across the whole temperature range in which the existence of the low-symmetry phase is possible, only the ground vibronic terms are populated. Passing to the matrix representation of the operators $Q_{\Gamma\gamma}(\boldsymbol{n})$ in the basis of these states we obtain again the Hamiltonian (6.1.25, 26).

c) *Strong vibronic coupling, $E_{\rm JT} \gg \hbar\omega$, weak correlations, $\lambda \ll 1$.* The vibronic spectrum (in the absence of the mean field) is characterized by a small tunneling splitting, $\delta \ll \hbar\omega$ (Sect. 4.3) and because of weak correlations $\delta \lesssim kT_0 \ll \hbar\omega$. In this case the temperature population of the tunneling states near the ground vibronic term has to be considered, while the population of all the other vibronic states can be neglected. The latter allows us to limit the basis to the lowest tunneling states (their number, s, is equal to the number of equivalent minima of the adiabatic potential), and to exchange all the operators $Q_{\Gamma\gamma}(\boldsymbol{n})$ for $s \times s$ matrices in this basis.

For instance, for a crystal with a pseudo-Jahn-Teller effect on each site described by the Hamiltonian (3.4.1) in the basis of the states localized in the minima we obtain (to within overlap integrals that are unimportant here)

$$\hat{Q}(\boldsymbol{n}) = Q_0 \hat{\sigma}_z(\boldsymbol{n}) \ , \tag{6.1.27}$$

where $Q_0^2 = (V/\omega^2)^2 - (\Delta/V)^2$, see the passage after (3.4.3). The site Hamiltonian in this basis has the form

$$\hat{H}_{\mathrm{JT}}(\boldsymbol{n}) = \Gamma \hat{\sigma}_x(\boldsymbol{n}) \ , \tag{6.1.28}$$

where Γ is the amplitude of the probability of tunneling between the minima determining the magnitude of the tunneling splitting. Substituting (6.1.27, 28) into (6.1.1) and considering the Jahn-Teller vibrations only we obtain

$$\hat{H} = \Gamma \sum_{\boldsymbol{n}} \hat{\sigma}_x(\boldsymbol{n}) + \tfrac{1}{2} \sum_{\substack{\boldsymbol{n},\boldsymbol{m} \\ (\boldsymbol{n} \neq \boldsymbol{m})}} Q_0^2 K(\boldsymbol{n}-\boldsymbol{m}) \hat{\sigma}_z(\boldsymbol{n}) \hat{\sigma}_z(\boldsymbol{m}) \ . \tag{6.1.29}$$

The Hamiltonian (6.1.29) corresponds to an Ising system in a transverse field. The phase transition in such a system, its thermodynamic and magnetic properties, as well as the spectrum of elementary excitations, have been well studied [6.24].

For a crystal with a quadratic $E \otimes e$ Jahn-Teller effect in the basis of states localized in the minima (4.3.20) in the same approximation we have

$$\hat{Q}_{E\gamma}(\boldsymbol{n}) = Q_E \hat{C}_{E\gamma}(\boldsymbol{n}) \ , \qquad Q_E = \frac{|V_E|}{\omega_E^2 - 2|W_E|} \ , \tag{6.1.30}$$

where the 3×3 matrices $\hat{C}_{\Gamma\gamma}$ have the same form as in (3.3.2). The site Hamiltonian in this basis is

$$\hat{H}_{\mathrm{JT}} = -\Gamma(\hat{C}_{T_2\xi} + \hat{C}_{T_2\eta} + \hat{C}_{T_2\zeta}) \ . \tag{6.1.31}$$

Substituting (6.1.30, 31) into (6.1.1) and considering the Jahn-Teller modes only we obtain a Hamiltonian of the same type as in (6.1.29), but with 3×3 matrices $\hat{C}_{\Gamma\gamma}(\boldsymbol{n})$ instead of $\hat{\sigma}_x(\boldsymbol{n})$ and $\hat{\sigma}_z(\boldsymbol{n})$. If we make the substitution

$$\hat{C}_{E\theta} = \frac{1}{2}(2\hat{S}_z^2 - \hat{S}_x^2 - \hat{S}_y^2) \ , \qquad \hat{C}_{E\varepsilon} = \frac{\sqrt{3}}{2}(\hat{S}_x^2 - \hat{S}_y^2) \ ,$$

$$\hat{C}_{T_2\xi} = \hat{S}_y\hat{S}_z + \hat{S}_z\hat{S}_y \ , \qquad \hat{C}_{T_2\eta} = \hat{S}_x\hat{S}_z + \hat{S}_z\hat{S}_x \ , \qquad \hat{C}_{T_2\zeta} = \hat{S}_x\hat{S}_y + \hat{S}_y\hat{S}_x \ ,$$

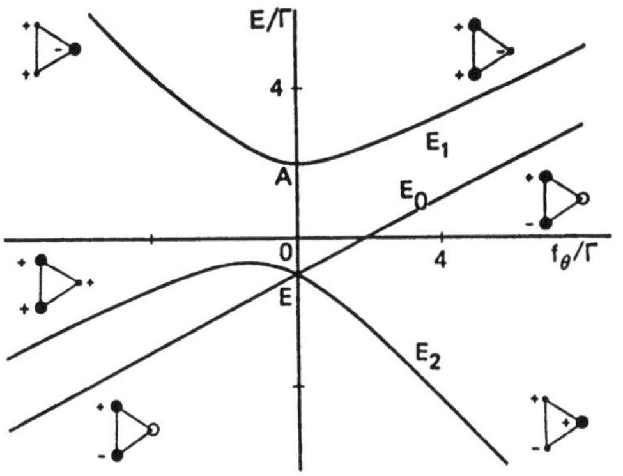

Fig. 6.4. Tunneling energy levels of the Jahn-Teller $E \otimes e$ system in the tetragonal field f_θ. The dimensions of the dots at the corners of the triangle indicate the relative probability of the corresponding distortion (zero probability is shown by rings), while "+" and "−" indicate the phase of the wave function. (After [6.25])

where $\hat{S}_i$ are 3×3 matrices (3.3.3), it acquires the form of the Hamiltonian of a spin system with an anisotropic biquadratic exchange interaction.

The Hamiltonian of the elementary cell for the $E \otimes e$ system under consideration in the framework of the MFA has the form

$$\hat{H}(\boldsymbol{n}) = -\Gamma(\hat{C}_\xi + \hat{C}_\eta + \hat{C}_\zeta) + Q_E(f_\theta \hat{C}_\theta + f_\varepsilon \hat{C}_\varepsilon) \ . \tag{6.1.32}$$

The eigenvalues and eigenfunctions of this Hamiltonian in the particular case of the tetragonal mean field ($f_\varepsilon = 0$) are illustrated in Fig. 6.4. It is seen that for positive f_θ, when the minimum at the Q_θ axis is the deepest one, the system in the ground state is locked in this minimum. In this case $\langle Q_\theta \rangle \approx Q_E$, $\langle Q_\varepsilon \rangle = 0$. If $f_\theta < 0$ and the two other minima are deeper, the system is locked in each of them with equal probability and $\langle Q_\theta \rangle \approx -Q_E/2$, $\langle Q_\varepsilon \rangle = 0$. These values corresponding to the ground state determine the low symmetry of the crystal at $T = 0$ K. Populating the three vibronic states according to Boltzmann, one can obtain equations of the type (6.1.13) determining the self-consistent values of the order parameters $\langle Q_\theta(\boldsymbol{n}) \rangle$ and $\langle Q_\varepsilon(\boldsymbol{n}) \rangle$.

The simplest result is obtained in case of ferrodistortive tetragonal order at $\Gamma = 0$, i.e., in the limiting case of very strong vibronic coupling. The only order parameter in this case $\langle Q_\theta \rangle = Q_E \langle \hat{C}_\theta \rangle$ is determined from the equation [6.26]:

$$\frac{ukT}{\lambda E_{\mathrm{JT}}} = \frac{\mathrm{e}^u - \mathrm{e}^{-2u}}{2\mathrm{e}^u + \mathrm{e}^{-2u}} \ , \tag{6.1.33}$$

where $u = \lambda \langle \hat{C}_\theta \rangle E_{\mathrm{JT}}/kT$ and $\lambda = \sum_{\boldsymbol{n} \neq 1} K(\boldsymbol{n})/(\omega_E^2 - 2|W_E|)$ is the parameter of ferrodistortive correlations analogous to that introduced in (6.1.19), and E_{JT} is determined by (3.1.21).

The order parameters $\langle Q_\theta \rangle$ and $\langle Q_\varepsilon \rangle$ in the case of ferrodistortive ordering belong to the irreducible representation E of the point group of the crystal (point $q = 0$ of the Brillouin zone). The symmetrized cube of this representation contains the totally symmetric representation A_1, and therefore in accordance with the Landau theory the corresponding phase transition should be of the first order. The temperature of the phase transition is determined from the condition of equal values of free energy in the ordered and disordered phases,

$$\mathscr{F}(\langle Q_\theta \rangle) = \mathscr{F}(0) \ . \tag{6.1.34}$$

Substituting (6.1.14) and considering that all $\langle \hat{Q}(\boldsymbol{n}) \rangle = \langle \hat{Q} \rangle$ we obtain [6.26]

$$\frac{u^2 kT}{2\lambda E_{\mathrm{JT}}} = \ln\left(\frac{1}{3}(\mathrm{e}^u + 2\mathrm{e}^{-u/2})\right) \ . \tag{6.1.35}$$

The simultaneous solution of (6.1.33) and (6.1.35) gives the temperature of the phase transition and the value of the order parameter [6.26],

$$kT_0 = 3\lambda E_{\mathrm{JT}}/8\ln 2 \ , \qquad \langle Q_\theta \rangle = \tfrac{1}{2} Q_E \ . \tag{6.1.36}$$

The antiferrodistortive ordering in the cooperative $E \otimes e$ Jahn-Teller effect within the three-state model under consideration was investigated in [6.25]. Depending on the nature of the antiferrodistortive correlations the existence of two low-symmetry phases is possible in this case. In one of them the tetragonal distortions of the sublattices are the same, however, they are oriented along different cubic axes in such a way that the resulting structure is similar to that of spin polyaxial ordering (Fig. 6.5a). In the other phase the distortions in the sublattices are different due to the different sign of f_θ in (6.1.32). The positive f_θ for one sublattice stabilizes the usual tetragonal distortion $\langle Q_\theta \rangle \approx Q_E, \langle Q_\varepsilon \rangle = 0$. The negative f_θ value for the other sublattice stabilizes two other minima leaving them equivalent (see the left-hand side of Fig. 6.4). The resulting structure can be called ferridistortive, since it is similar to ferrimagnetic spin ordering (Fig. 6.5b).

Note that the above classification is related to cubic crystals, whose investigation is reported in [6.25]. The same situation in crystals with other orientations of

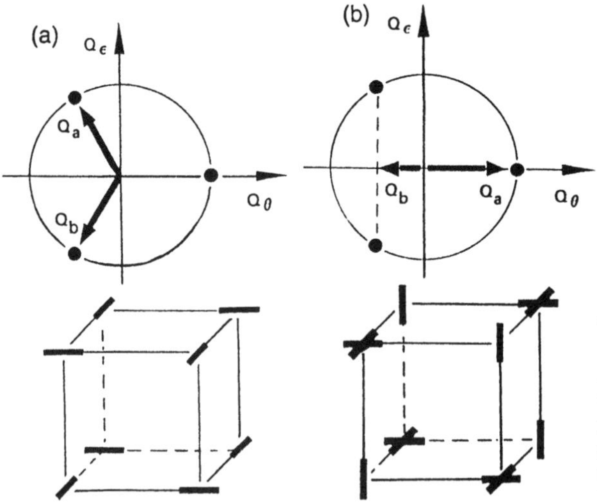

Fig. 6.5a, b. Two types of antiferrodistortive ordering of the Jahn-Teller distortions due to the cooperative Jahn-Teller effect in the case of an $E \otimes e$ system for each center: (**a**) spin polyaxial, (**b**) ferridistortive. Corresponding ordering in a perovskite-type structure is illustrated schematically. (After [6.25, 27])

symmetry axes corresponds to other structures. For instance, the case illustrated in Fig. 6.5b in a trigonal crystal corresponds to a noncollinear weak ferrodistortive ordering.

As in the cases discussed above, the resulting structure is determined by the competition of many $K(\boldsymbol{n} - \boldsymbol{m})$ values, and therefore it may have a more complicated nature than pure ferro- or antiferrodistortive ordering. For instance, the inclusion of the interaction of Jahn-Teller centers with the second neighbors [performed in [6.27] in the framework of the three-state model without taking account of tunneling in order to explain the structural phase transitions in the $K_2PbCu(NO_2)_6$ crystal] results in seven different phases, one of which is helical and two are sinusoidal with incommensurate periods. The stability and the order of occurrence of these phases as the temperature increases are determined by the relationship between the intra- and intercell elastic constants. In the experiments on these crystals three phases were observed: the high-temperature cubic phase at 281 K, the intermediate helical umbrella type phase at $273 \text{ K} < T < 281 \text{ K}$ and the low-temperature polyaxial phase at $T < 273$ K [6.28].

d) *Strong correlations,* $\lambda \gg 1$, *intermediate vibronic coupling,* $E_{\mathrm{JT}} \sim \hbar\omega$. If the order parameter is significantly different from zero across the whole temperature range of the low-symmetry phase, including the temperature of the phase transition (as holds for first-order phase transitions), the splitting of the Jahn-Teller term by the mean field E_f is much larger than $E_{\mathrm{JT}} \sim \hbar\omega$. In the solution of the vibronic problem corresponding to the Hamiltonian (6.1.9), the strongest perturbation, that is the low-symmetry mean field, has to be taken into account first. The electronic degeneracy is removed and we come to the problem of a weak pseudo-Jahn-Teller effect (Sect. 3.4, Fig. 3.8). The vibronic interaction results in a small anharmonicity of the potential surfaces, which is the reason for the structural phase transition. Phase transitions which are due to the anharmonicity of potential surfaces are considered in detail in the theory of ferroelectricity [6.10] and will not be discussed here. The analytical solution of the vibronic problem in a strong molecular field for Jahn-Teller $E \otimes e$ systems is given in [6.26]. In this work the temperature of the ferrodistortive phase transition is also determined.

e) *General case of arbitrary vibronic coupling and arbitrary correlations.*

This can be investigated numerically [6.26, 29]. The vibronic levels of the elementary cell with the $E \otimes e$ Jahn-Teller effect in a tetragonal mean field for intermediate values of the constant of the quadratic vibronic interaction can be obtained by computer [6.26]. The numerical results of the diagonalization of the vibronic Hamiltonian (6.1.9) can be used for determining the temperature averages $\langle Q_{\Gamma\gamma}(\boldsymbol{n})\rangle$. Substituting the latter into (6.1.14) and minimizing the free energy together with the condition $\mathscr{F}(\langle\hat{Q}(\boldsymbol{n})\rangle) = \mathscr{F}(0)$ we obtain the temperature of the phase transition and the magnitude of the equilibrium values of the order parameters $\langle Q_{\Gamma\gamma}(\boldsymbol{n})\rangle$. In [6.26] the ferrodistortive ordering in spinels was investigated in this way, and in [6.29] a similar investigation of antiferrodistortive ordering in perovskites is given.

6.1.5 Jahn-Teller Crystals with Insignificant Nonadiabaticity. Shift Transformation

In some cases when the vibronic Hamiltonian does not contain noncommutative electronic matrices $\hat{C}_{\Gamma\gamma}(\boldsymbol{n})$ the "exchange" interaction between the pseudospins can be obtained without the additional assumptions discussed in the above subsections a)–e) (Sect. 6.1.4), namely, by means of a unitary shift transformation [6.30]:

$$\hat{S} = \exp\left(\frac{i}{\hbar}\sum_{\boldsymbol{q},\nu} \hat{\mathscr{V}}_\nu(-\boldsymbol{q})P_\nu(\boldsymbol{q})\right) , \tag{6.1.37}$$

where (6.1.6)

$$\hat{\mathscr{V}}_\nu(\boldsymbol{q}) = \sum_{\boldsymbol{n}}\sum_{\Gamma\gamma} V_\Gamma a_{\boldsymbol{q},\nu}(\boldsymbol{n}|\Gamma\gamma)\omega_\nu^{-2}(\boldsymbol{q})\hat{C}_{\Gamma\gamma}(\boldsymbol{n}) . \tag{6.1.38}$$

The transformation (6.1.37) results in the substitution

$$Q_\nu(\boldsymbol{q}) \to \hat{S}^\dagger Q_\nu(\boldsymbol{q})\hat{S} = Q_\nu(\boldsymbol{q}) - \hat{\mathscr{V}}_\nu(\boldsymbol{q}) , \tag{6.1.39}$$

leading to the form

$$\hat{H} = \hat{H}_0 - \tfrac{1}{2}\sum_{\boldsymbol{n}}\sum_{\substack{\boldsymbol{m}\\(\boldsymbol{n}\neq\boldsymbol{m})}} \hat{\boldsymbol{C}}^\dagger(\boldsymbol{n})\hat{J}(\boldsymbol{n}-\boldsymbol{m})\hat{\boldsymbol{C}}(\boldsymbol{m}) - E_{\mathrm{JT}} \tag{6.1.40}$$

for the Hamiltonian (6.1.3). Here $\hat{\boldsymbol{C}}(\boldsymbol{n})$ is the column vector with components $\hat{C}_{\Gamma\gamma}(\boldsymbol{n})$, H_0 is the Hamiltonian of free phonons of the form (6.1.3) without the last terms. The elements of the matrix $\hat{J}(\boldsymbol{n}-\boldsymbol{m})$ are given by the expression

$$J_{\Gamma_1\gamma_1\Gamma_2\gamma_2}(\boldsymbol{n}-\boldsymbol{m}) = \sum_{\boldsymbol{q},\nu} V_{\Gamma_1}V_{\Gamma_2}\omega_\nu^{-2}(\boldsymbol{q})a_{\boldsymbol{q},\nu}(\boldsymbol{n}|\Gamma_1\gamma_1)a_{-\boldsymbol{q},\nu}(\boldsymbol{m}|\Gamma_2\gamma_2) , \tag{6.1.41}$$

and E_{JT} is the total Jahn-Teller stabilization energy for all the elementary cells,

$$E_{\mathrm{JT}} = \tfrac{1}{2}\sum_{\Gamma\gamma}\sum_{\boldsymbol{q},\nu}\sum_{\boldsymbol{n}} V_\Gamma^2\omega_\nu^{-2}(q)|a_{\boldsymbol{q},\nu}(\boldsymbol{n}|\Gamma\gamma)|^2\hat{C}_{\Gamma\gamma}^2(\boldsymbol{n}) . \tag{6.1.42}$$

The phonon variables in the Hamiltonian (6.1.40) can be separated, and we come to the "exchange" Hamiltonian of the same type as (6.1.25).

The separation of variables at this stage can be realized only because the Hamiltonian does not contain noncommutative matrices $\hat{C}_{\Gamma\gamma}(\boldsymbol{n})$. In some cases, however, when such matrices are present in the Hamiltonian the separation of the variables can be performed approximately using the fact that the coefficients of some of the noncommutative matrices are small. For instance, the cooperative $E\otimes b_1$ pseudo-Jahn-Teller effect corresponds to the Hamiltonian $H_{\mathrm{JT}}(\boldsymbol{n})$ of the form (3.4.1). The unitary transformation (6.1.37) in this case transforms (6.1.1) to the form [6.30]

$$\hat{H} = \hat{H}_0 - \frac{1}{2} \sum_{\substack{n \\ (n \neq m)}} \sum_{m} J(n-m)\hat{\sigma}_x(n)\hat{\sigma}_x(m) + \Delta \sum_{n} [\hat{\sigma}_z(n) \cos g_n - \hat{\sigma}_y(n) \sin g_n] \ , \tag{6.1.43}$$

where

$$g_n = \frac{2V}{\hbar} \sum_{q,\nu} a_{q,\nu}(n|B_1)\omega_\nu^{-2}(q)P_\nu(-q) \ . \tag{6.1.44}$$

If $\Delta = 0$, the phonon variables may be separated. If $\Delta \ll E_{JT}$, that is, the strong Jahn-Teller effect occurs (Fig. 3.8b), the variables can be separated approximately by replacing $\cos g_n$ and $\sin g_n$ in (6.1.43) by their temperature average over the states of the phonon Hamiltonian H_0, $\langle \cos g_n \rangle = \gamma_n$ and $\langle \sin g_n \rangle = 0$. The quantity $2\gamma_n \Delta$ is a statistically averaged splitting of all the vibronic states of the $E \otimes b_1$ system by the weak perturbation $\Delta\hat{\sigma}_z$. At $T = 0$ K it is equal to $2\Delta K(B_2)$, the splitting of the ground vibronic level reduced by the vibronic reduction factor $K(B_2)$ (Sect. 4.7). The quantity γ_n can therefore be interpreted as a generalization of the vibronic reduction factor to the case $T \neq 0$ K.

The Hamiltonian of the separated electronic subsystem is equivalent to the Hamiltonian (6.1.29) with $\Gamma = \gamma\Delta$, $Q_0^2 K(n) = -J(n)$. A simple rotation of the coordinate axes results in the substitution $\hat{\sigma}_x \rightleftarrows \hat{\sigma}_z$. Thus the system is reduced to the well-studied Ising model in a transverse field [6.24].

The theory of the cooperative Jahn-Teller effect starting with the Hamiltonian (6.1.3) and based on the unitary shift transformation (6.1.37) is presented in [6.3], where quite a number of applications to real systems in rare-earth compounds with a zircon structure and comparisons with experimental data are also given. The comparison with experiment and good agreement with the theory obtained in these cases are very important for the development of the area as a whole, since it demonstrates convincingly the usefulness of the vibronic approach under consideration to the problem of structural phase transitions.

6.1.6 Dynamics of Elementary Excitations

In the mean field approximation the crystal has a discrete energy spectrum, since the total energy is the sum of the energies of the elementary cells each of which has a discrete spectrum. Physically this means that the elementary excitations are localized in the elementary cells of the lattice and cannot migrate through the crystal. In order to take account of migration of the elementary excitations one has to go out of the framework of the MFA, that is to take into account the intercell correlation of the fluctuations described by the last term in (6.1.7).

The dispersion of elementary excitations is described by the poles of the retarded Green's function $D_{nm} = \langle\langle \Delta\hat{Q}^\dagger(n)|\Delta\hat{Q}(m)\rangle\rangle = \langle\langle \hat{Q}(n)|\hat{Q}(m)\rangle\rangle$. However, $D_{nm}(t)$ is the susceptibility describing the linear response of the system to an external perturbation [Ref. 6.13; Sect. 126],

$$\hat{V}(t) = \sum_{n} \hat{Q}^{\dagger}(\boldsymbol{n})\hat{\varphi}_{\boldsymbol{n}}(t) \ , \tag{6.1.45}$$

where $\hat{\varphi}_n(t)$ is the appropriate force with a given time dependence. The perturbation (6.1.45) causes an increase in all the $\langle \hat{Q}(\boldsymbol{n}) \rangle$ values by $\hat{\kappa}_n(t)$. As a result all the f values in (6.1.8) change:

$$\hat{F}(\boldsymbol{n}) = \hat{f}(\boldsymbol{n}) + \Delta\hat{f}(\boldsymbol{n}, t) \ , \qquad \text{where} \tag{6.1.46}$$

$$\Delta\hat{f}(\boldsymbol{n}, t) = \sum_{m} \hat{K}(\boldsymbol{n} - \boldsymbol{m})(1 - \delta_{\boldsymbol{n},\boldsymbol{m}})\hat{\kappa}_{\boldsymbol{m}}(t) \tag{6.1.47}$$

is an additional mean field created in the crystal by the perturbation (6.1.45). The effective field acting at each elementary cell is the sum of the external and induced internal fields,

$$\hat{\varphi}_{\boldsymbol{n}}^{(\mathrm{eff})}(t) = \hat{\varphi}_{\boldsymbol{n}}(t) + \Delta\hat{f}(\boldsymbol{n}, t) \ . \tag{6.1.48}$$

According to the theory of linear response [Ref. 6.13; Sect. 123], $\hat{\kappa}_n(t)$ is linearly related to $\varphi_{\boldsymbol{n}}^{(\mathrm{eff})}$,

$$\hat{\kappa}_{\boldsymbol{n}}(t) = \int_{0}^{\infty} \hat{D}_0(\tau)\hat{\varphi}_{\boldsymbol{n}}^{(\mathrm{eff})}(t - \tau)\, d\tau \ , \tag{6.1.49}$$

where $\hat{D}_0(\tau)$ is the retarded Green's function of the elementary cell in the mean field described by the vibronic Hamiltonian $\hat{H}_{\mathrm{JT}}(\boldsymbol{n})$ from (6.1.9). Here, for simplicity, it is assumed that ferrodistortive ordering takes place so that the $\hat{H}_{\mathrm{JT}}(\boldsymbol{n})$ and hence the $\hat{D}_0$ do not depend on $\boldsymbol{n}$. Using the Fourier representation (5.1.5) and (6.1.2) one can exclude $\Delta\hat{f}$ and $\hat{\varphi}^{(\mathrm{eff})}$ from the system of coupled equations (6.1.47–49). As a result we obtain

$$\hat{\kappa}_{\boldsymbol{q}}(\omega) = 2\pi\hat{D}_0(\omega)[\hat{\varphi}_{\boldsymbol{q}}(\omega) + \hat{K}(\boldsymbol{q})\hat{\kappa}_{\boldsymbol{q}}(\omega)] \ , \tag{6.1.50}$$

where

$$\hat{K}(\boldsymbol{q}) = \sum_{n} \hat{K}(\boldsymbol{n})(1 - \delta_{\boldsymbol{n},0})\mathrm{e}^{-\mathrm{i}\boldsymbol{q}\cdot\boldsymbol{R}(\boldsymbol{n})} \ . \tag{6.1.51}$$

From (6.1.50) it follows that

$$\hat{\kappa}_{\boldsymbol{q}}(\omega) = [\hat{1} - 2\pi\hat{D}_0(\omega)\hat{K}(\boldsymbol{q})]^{-1}2\pi\hat{D}_0(\omega)\hat{\varphi}_{\boldsymbol{q}}(\omega) \ . \tag{6.1.52}$$

Since the retarded Green's function is the same as the generalized susceptibility of the system, it follows that the expression for the Fourier transform of the Green's function $\hat{D}_{\boldsymbol{nm}}(t)$ is [6.4]

$$\hat{D}_{\boldsymbol{q}}(\omega) = [\hat{1} - 2\pi\hat{D}_0(\omega)\hat{K}(\boldsymbol{q})]^{-1}2\pi\hat{D}_0(\omega) \ . \tag{6.1.53}$$

The matrix equation $2\pi\hat{D}_0(\omega)\hat{K}(\boldsymbol{q}) = \hat{I}$ determining the poles of the Green's function (6.1.53) and hence the spectrum of elementary excitations can be reduced to the following system of equations:

$$2\pi \sum_{\bar{\Gamma}\bar{\gamma}} \langle\langle Q_{\Gamma_1\gamma_1} | Q_{\bar{\Gamma}\bar{\gamma}} \rangle\rangle_\omega K_{\bar{\Gamma}\bar{\gamma},\Gamma_2\gamma_2}(\boldsymbol{q}) = \delta_{\Gamma_1\Gamma_2}\delta_{\gamma_1\gamma_2} , \tag{6.1.54}$$

or in the representation in which the matrix $\hat{K}(\boldsymbol{q})$ is diagonal,

$$2\pi \sum_{\Gamma_1\gamma_2} \sum_{\Gamma_2\gamma_2} \langle\langle Q_{\Gamma_1\gamma_1} | Q_{\Gamma_2\gamma_2} \rangle\rangle_\omega a_{-\boldsymbol{q}\nu}(\boldsymbol{n}|\Gamma_1\gamma_1) a_{\boldsymbol{q}\mu}(n|\Gamma_2\gamma_2)[\omega_\nu^2(\boldsymbol{q}) - \omega_\nu^2(0)]$$

$$= \delta_{\nu\mu}/N . \tag{6.1.55}$$

In particular, in the absence of the Jahn-Teller effect when $\langle\langle Q_{\Gamma_1\gamma_1} | Q_{\Gamma_2\gamma_2} \rangle\rangle_\omega = \delta_{\Gamma_1\Gamma_2}\delta_{\gamma_1\gamma_2}/2\pi(\omega^2 - \omega_{\Gamma_1}^2)$, the initial phonon dispersion $\omega^2 = \omega_\nu^2(\boldsymbol{q})$ can be obtained from (6.1.55). When including the Jahn-Teller effect, equations (6.1.55) result in a complicated continuous spectrum, the elementary excitations having a hybrid vibronic nature.

The last circumstance can be illustrated by the example of a simple Jahn-Teller situation, $E \otimes b_1$. Equations (6.1.55) in this case can be reduced to one transcendental equation,

$$2\pi\langle\langle Q_1 | Q_1 \rangle\rangle_\omega = [\omega_\nu^2(\boldsymbol{q}) - \omega_\nu^2(0)]^{-1} . \tag{6.1.56}$$

The graphical solution of this equation is given in Fig. 6.6 from which it is seen that each vibronic level of the elementary cell is transformed into an energy band. Thus considering the last term in (6.1.7) the vibronic excitation of the elementary cell can migrate over the crystal. The corresponding wave is described by a linear combination of one-cell excitations with the phases $\exp[\mathrm{i}\boldsymbol{q} \cdot \boldsymbol{R}(\boldsymbol{n})]$.

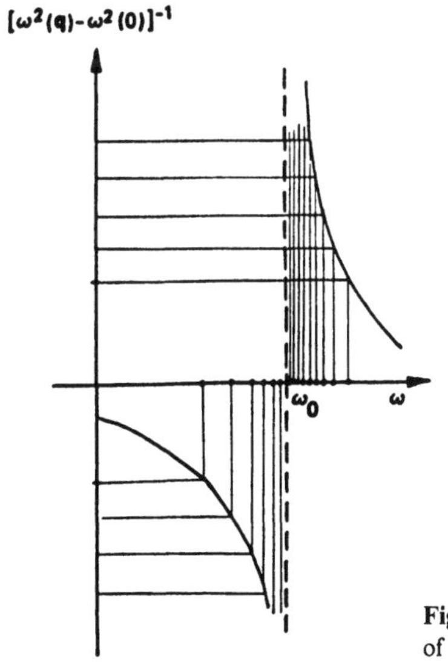

Fig. 6.6. Graphical solution of (6.1.56) in the vicinity of one of the poles ω_0 of the cell Green's function $\langle\langle Q_1 | Q_1 \rangle\rangle$

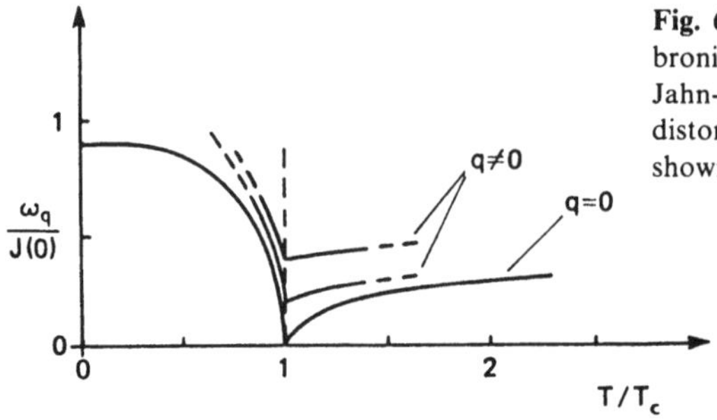

Fig. 6.7. Temperature dependence of the vibronic modes in a crystal with a cooperative Jahn-Teller effect ($E \otimes b_1$ system) and ferrodistortive ordering. The soft mode frequency is shown by a full line. (After [6.31])

The mean field in the crystal changes with temperature. Accordingly, the vibronic spectrum of each elementary cell described by the Hamiltonian $\hat{H}_{\mathrm{JT}}(\boldsymbol{n})$ from (6.1.9) becomes temperature dependent. The temperature dependence of the poles of the Green's function $\langle\langle Q_{\Gamma_1\gamma_1}|Q_{\Gamma_2\gamma_2}\rangle\rangle_\omega$ results in a temperature dependence of the spectrum of collective vibronic modes determined by the system of equations (6.1.55). The reduction of the frequency ω_0 (Fig. 6.6) results in a shift of the whole vibronic band towards low frequencies. The phase transition takes place at a temperature when the energy gap separating the bottom of the band from zero vanishes. Physically this means that the crystal loses its elasticity with respect to deformations described by the appropriate mode. In other words, the crystal becomes soft in this direction. The presence of such a soft mode in the spectrum of elementary excitations is a characteristic feature of crystals with structural phase transitions (Fig. 6.7).

The type of phase transition and the resulting structure of the crystal are characterized by the wave number $\boldsymbol{q}_0$ which determines the soft mode. As can be seen from Fig. 6.6, the problem of determining $\boldsymbol{q}_0$ is reduced to the determination of the absolute minimum of the expressions $\omega_\nu^2(\boldsymbol{q}) - \omega_\nu^2(0)$. If the minimum is achieved at $\boldsymbol{q}_0 = 0$, a ferrodistortive phase transition occurs. For other $\boldsymbol{q}_0$ values more complicated transitions to commensurate or incommensurate phases are possible. The last circumstance has to be taken into account in (6.1.49), of course.

In accordance with the Landau theory, the type of phase transition is determined by the irreducible representation of the group of the wave vector $\boldsymbol{q}_0$ of the soft mode. For instance, the ferrodistortive phase transition ($\boldsymbol{q}_0 = 0$) of second order is possible only in a Jahn-Teller crystal of $E \otimes b_1$ type. From both E- and T_2-type vibrations an invariant of the third order can be constructed, and therefore the cooperative $E \otimes e$ or $T \otimes (e + t_2)$ Jahn-Teller effect results in a first-order ferrodistortive phase transition. In the case of antiferrodistortive ordering the third-order invariants are absent and for all types of the cooperative Jahn-Teller effect a second-order phase transition is possible.

The external forces $\varphi_{\boldsymbol{n}}(t)$ in the approach presented above are taken into account only through the additional mean field induced in the crystal by the perturbation (6.1.45), and therefore the dispersion law of vibronic modes determined by (6.1.55) is not exact. This is seen even from the fact that the elementary excitations obtained have no relaxation broadening. Meanwhile the Jahn-Teller effect results in a strong anharmonicity of the potential surfaces, and this causes a limiting lifetime for the elementary excitations.

It can be shown that the approximate approach leading to (6.1.55) is equivalent to the *random phase approximation*. The majority of investigations of the vibronic dispersion in crystals with the cooperative Jahn-Teller effect were performed in the framework of this approximation in its different modifications [6.3]. The comparison of the results obtained with the experimental data on Raman and Mandelstam-Brillouin scattering for Jahn-Teller crystals shows that the random phase approximation describes the vibronic dispersion quite satisfactorily [6.1].

Attempts based on the use of the Green's functions methods to go beyond this approximation are described in [6.32–34], where the relaxation widths for vibronic excitations were obtained. A semiphenomenological approach to the problem is suggested in [6.35].

Interesting results of investigations [6.15, 36, 37] on the dispersion of vibronic modes and structural low-temperature phases including the incommensurate ones have been obtained based on the direct use of the variational principle for the free energy [Ref. 6.38; Sect. 20].

6.1.7 Interaction with Macroscopic Deformations. Ferroelastic Phase Transitions

The structural phase transition, as mentioned above, can be accompanied by a macroscopic deformation of the crystal. The contribution of homogeneous deformations to the total energy is described by the operator

$$\hat{H}_{\text{str}} = \tfrac{1}{2}N\Omega\sum_{\Gamma\gamma} c_\Gamma\varepsilon_{\Gamma\gamma}^2 - \sum_{\Gamma\gamma} h_\Gamma\sqrt{c_\Gamma\Omega}\sum_{\boldsymbol{n}}\varepsilon_{\Gamma\gamma}Q_{\Gamma\gamma}(\boldsymbol{n}) - \sum_{\Gamma\gamma} g_\Gamma\sqrt{c_\Gamma\Omega}\sum_{\boldsymbol{n}}\varepsilon_{\Gamma\gamma}\hat{C}_{\Gamma\gamma}(\boldsymbol{n}) \ . \tag{6.1.57}$$

Here the first term corresponds to the energy of static elastic deformations, c_Γ being the modulus of elasticity of the crystal with respect to the symmetrized deformations $\varepsilon_{\Gamma\gamma}$, and Ω is the volume of the elementary cell.

The second term in (6.1.57) describes the direct interaction of the deformation with the lattice vibrations. The nature of this interaction can be understood if one takes into account that the homogeneous deformation results in a shift of the equilibrium positions of the atoms by $\sim\varepsilon_{ik}$. As a result of the substitution $Q_{\Gamma\gamma}(\boldsymbol{n}) \to Q_{\Gamma\gamma}(\boldsymbol{n}) + \text{const}\cdot\varepsilon_{\Gamma\gamma}$ these terms of (6.1.57) can be separated from the last term of (6.1.1). The factor $(c_\Gamma\Omega)^{1/2}$ is separated for convenience. From this discussion it is clear that $h_\Gamma \sim \omega_\Gamma^2$, where $\omega_\Gamma = \omega_\nu(\boldsymbol{q})$ at $\boldsymbol{q}\to 0$ (at the point $\boldsymbol{q} = 0$ the symmetry group of the wave vector coincides with that of the point group of the crystal, and therefore $\omega_\nu(0)$ may be classified by the irreducible representations of this symmetry group).

The last term in (6.1.57) describes the electron-deformation interaction. It emerges from the operator of the vibronic coupling as a result of the substitution

$Q_{\Gamma\gamma}(\boldsymbol{n}) \to Q_{\Gamma\gamma}(\boldsymbol{n}) + \text{const}\cdot\varepsilon_{\Gamma\gamma}$. It is clear that $g_\Gamma \sim V_\Gamma$, so in the absence of the vibronic coupling this kind of interaction with deformations is absent.

Note that the two types of influence of deformations are difficult to separate, since they lead to the same physical effects. Indeed, as was shown in Sect. 4.7, the operator linear in $Q_{\Gamma\gamma}$ acts in the same way as the electronic operator of the same symmetry. Therefore without loss of generality one can consider the influence of only one kind of deformation. In the approach using the site Hamiltonian (6.1.1) it is more convenient to use the interaction $\varepsilon_{\Gamma\gamma}Q_{\Gamma\gamma}$. The substitution

$$\varepsilon_{\Gamma\gamma} = e_{\Gamma\gamma} + h_\Gamma N^{-1}(c_\Gamma\Omega)^{-1/2}\sum_{\boldsymbol{n}} Q_{\Gamma\gamma}(\boldsymbol{n})$$

leads to the form ($g_\Gamma = 0$)

$$H_{\text{str}} = \frac{1}{2}N\Omega\sum_{\Gamma\gamma} c_\Gamma e_{\Gamma\gamma}^2 - \frac{1}{2N}\sum_{\boldsymbol{n},\boldsymbol{m}} \hat{Q}^\dagger(\boldsymbol{n})\hat{h}^2\hat{Q}(\boldsymbol{m}) \ , \tag{6.1.58}$$

for (6.1.57), where the matrix $\hat{h}$ with the elements $h_{\Gamma_1\gamma_1\Gamma_2\gamma_2} = h_{\Gamma_1}\delta_{\Gamma_1\Gamma_2}\delta_{\gamma_1\gamma_2}$ is introduced. By comparing (6.1.58) with (6.1.1) we find that the deformation interaction provides an additional contribution to the correlation of the local distortions. The $\hat{K}(\boldsymbol{n}-\boldsymbol{m})$ are modified,

$$\hat{K}(\boldsymbol{n}-\boldsymbol{m}) \to \hat{K}(\boldsymbol{n}-\boldsymbol{m}) - \frac{1}{N}\hat{h}^2 \ .$$

Accordingly, the mean field from (6.1.8) is also modified,

$$\hat{f}(\boldsymbol{n}) \to \hat{f}(\boldsymbol{n}) - \frac{1}{N}\hat{h}^2\sum_{\boldsymbol{m}\neq\boldsymbol{n}} \langle\hat{Q}(\boldsymbol{m})\rangle \ . \tag{6.1.59}$$

From (6.1.59), the following deformation contribution is added to the free energy (6.1.14):

$$\mathscr{F}_{\text{str.}} = \tfrac{1}{2}N\Omega\sum_{\Gamma\gamma} c_\Gamma\varepsilon_{\Gamma\gamma}^2 - \sum_{\Gamma\gamma} h_\Gamma\sqrt{c_\Gamma\Omega}\,\varepsilon_{\Gamma\gamma}\sum_{\boldsymbol{n}}\langle Q_{\Gamma\gamma}(\boldsymbol{n})\rangle \ . \tag{6.1.60}$$

Minimizing this expression with respect to $\varepsilon_{\Gamma\gamma}$ we find the equilibrium magnitude of deformation,

$$\bar{\varepsilon}_{\Gamma\gamma} = h_\Gamma N^{-1}(\Omega c_\Gamma)^{-1/2}\sum_{\boldsymbol{n}}\langle Q_{\Gamma\gamma}(\boldsymbol{n})\rangle \ . \tag{6.1.61}$$

From this it follows that homogeneous deformation can arise only in the case of ferrodistortive ordering (otherwise $\sum_{\boldsymbol{n}}\langle Q_{\Gamma\gamma}(\boldsymbol{n})\rangle = 0$). Structural phase transitions which are accompanied by a spontaneous homogeneous deformation of the whole crystal are called *ferroelastic transitions*, and crystals in which such transitions are possible are called *ferroelastics*.

From (6.1.61) it also follows that the temperature dependence of

$$\bar{\varepsilon}_{\Gamma\gamma} = h_\Gamma (c_\Gamma \Omega)^{-1/2} \langle Q_{\Gamma\gamma} \rangle \tag{6.1.62}$$

is completely determined by the temperature dependence of the order parameter $\langle Q_{\Gamma\gamma} \rangle$. In the high-temperature phase $\langle Q_{\Gamma\gamma} \rangle = 0$ and the macroscopic deformation is absent.

If the vibronic Hamiltonian does not contain noncommutative matrices [or their commutator has a small coefficient, as in (6.1.43)] and the approach based on the unitary shift transformation (6.1.37) is used (Sect. 6.1.5), the electron-deformation interaction of the type $\varepsilon_{\Gamma\gamma}\hat{C}_{\Gamma\gamma}$ is more conveniently considered. The substitution

$$\varepsilon_{\Gamma\gamma} = e_{\Gamma\gamma} + g_\Gamma \sqrt{c_\Gamma \Omega} N^{-1} \sum_{\boldsymbol{n}} \hat{C}_{\Gamma\gamma}(\boldsymbol{n})$$

changes (6.1.57) to the form ($h_\Gamma = 0$)

$$\hat{H}_{\rm str} = \frac{1}{2} N\Omega \sum_{\Gamma\gamma} c_\Gamma e^2_{\Gamma\gamma} - \frac{1}{2N} \sum_{\boldsymbol{n}} \sum_{\substack{\boldsymbol{m} \\ (\boldsymbol{n} \neq \boldsymbol{m})}} \hat{C}^\dagger(\boldsymbol{n}) \hat{g}^2 \hat{C}(\boldsymbol{m}) \ , \tag{6.1.63}$$

where the matrix $\hat{g}$ with the elements $g_{\Gamma_1\gamma_1\Gamma_2\gamma_2} = g_{\Gamma_1}\delta_{\Gamma_1\Gamma_2}\delta_{\gamma_1\gamma_2}$ is introduced. The terms with $\boldsymbol{n} = \boldsymbol{m}$ are excluded from (6.1.63), since they result in an additive constant in the energy, changing its zero only. By comparing (6.1.63) with (6.1.40) we see that the deformation interaction gives an additional contribution to the correlations. The expression (6.1.41) is modified:

$$\hat{J}(\boldsymbol{n} - \boldsymbol{m}) \to \hat{J}(\boldsymbol{n} - \boldsymbol{m}) + \frac{1}{N} \hat{g}^2 \ .$$

The equilibrium deformation in the ferroelastic transition can be obtained, as above, by minimizing the appropriate contribution to the free energy with respect to $\varepsilon_{\Gamma\gamma}$ resulting in

$$\bar{\varepsilon}_{\Gamma\gamma} = g_\Gamma (c_\Gamma \Omega)^{-1/2} \langle \hat{C}_{\Gamma\gamma} \rangle \ . \tag{6.1.64}$$

The magnitude $\bar{\varepsilon}_{\Gamma\gamma}$ thus changes with temperature in the same way as the order parameter $\langle \hat{C}_{\Gamma\gamma} \rangle$, and at $T > T_0$ it is therefore equal to zero.

The elastic properties of Jahn-Teller crystals are described by the elasticity moduli. In accordance with the definition they are the reciprocals of the principal components of the elasticity tensor,

$$T_{\Gamma\gamma\Gamma'\gamma'} = \left. \frac{\partial \bar{\varepsilon}_{\Gamma\gamma}}{\partial P_{\Gamma'\gamma'}} \right|_{\text{all } P_{\Gamma\gamma}=0} , \tag{6.1.65}$$

where $P_{\Gamma\gamma}$ are symmetrized combinations of the components of the stress tensor. For instance, for a Jahn-Teller $E \otimes b_2$ crystal, considering the energy of the external stress $-P_z\varepsilon_z$ and homogeneous deformations (6.1.57), the Hamiltonian

(6.1.40) acquires the form

$$\hat{H} = -\frac{1}{2}\sum_{\substack{n\ m\\(n\neq m)}}\sum [J(n-m) + N^{-1}g_z^2]\hat{\sigma}_z(n)\hat{\sigma}_z(m) + \frac{g_z P_z}{N\sqrt{c_z\Omega}}\sum_n \hat{\sigma}_z(n)\ . \tag{6.1.66}$$

The ferroelastic phase transition in this case is described by the order parameter $\langle\hat{\sigma}_z\rangle$, which in the framework of the MFA is determined by the transcendental equation

$$\langle\hat{\sigma}_z\rangle = \tanh\left[\left(\frac{1}{kT}\langle\hat{\sigma}_z\rangle(J + g_z^2) - \frac{g_z P_z}{N\sqrt{c_z\Omega}}\right)\right], \tag{6.1.67}$$

where $J = \sum_{n\neq 0} J(n)$. Using (6.1.64, 67) one can easily obtain the modulus of elasticity of the crystal per unit volume,

$$\tilde{c}_z = \left[N\Omega\left(\frac{\partial\varepsilon_z}{\partial P_z}\right)_{P_z=0}\right]^{-1} = c_z\frac{(T_0 - T) - T_0\bar{\sigma}_z^2}{(T_0 - T) - T_0\bar{\sigma}_z^2 - g_z^2(1-\bar{\sigma}_z^2)k^{-1}}\ , \tag{6.1.68}$$

where $\bar{\sigma}_z$ is the order parameter $\langle\hat{\sigma}_z\rangle$ in the absence of external stress, determined by (6.1.67) at $P_z = 0$, and T_0 is the temperature of the phase transition, $kT_0 = J + g_z^2$. At $T \geq T_0$ when $\bar{\sigma}_z = 0$ we have

$$\tilde{c}_z = c_z\frac{k(T - T_0)}{g_z^2 + k(T - T_0)}\ . \tag{6.1.69}$$

If $T \lesssim T_0$ and $\bar{\sigma}_z^2 \approx 3(T_0 - T)/T_0$, it follows from (6.1.68) that

$$\tilde{c}_z = 2c_z k(T_0 - T)/g_z^2\ . \tag{6.1.70}$$

Thus at the point of the ferroelastic phase transition, one of the elastic moduli of the Jahn-Teller crystal equals zero. The crystal loses its elasticity with respect to deformations of the corresponding symmetry, that is, it becomes soft. In this sense the corresponding deformational degree of freedom plays the role of a soft mode.

The correlation of local distortions through deformations is especially important in cases when the alternative mechanism of Jahn-Teller ordering through the phonon field is relatively weak. As follows from (6.1.41), the latter is possible in crystals with weak vibronic interactions. Just this situation is realized, for example, in some rare-earth compounds with zircon structure where the radius of the f orbitals responsible for the vibronic coupling is relatively small. For instance, in the $TmVO_4$ crystal where the above $E\otimes b_2$ situation exists, $kT_0 = J + g_z^2 \approx 1.5\ \mathrm{cm}^{-1}$. The contributions of J and g_z^2 can be separated by means of piezospectroscopic measurements [6.39]. In the case of $TmVO_4$ the analysis gives $J = -0.75\ \mathrm{cm}^{-1}$, $g_z^2 = 2.25\ \mathrm{cm}^{-1}$.

The sign of J can be understood, if one writes J in the form [6.1]

$$J = V_z^2 \sum_\nu \left(N \lim_{q \to 0} \frac{|a_{q\nu}(z)|^2}{\omega_\nu^2(q)} - \sum_q \frac{|a_{q\nu}(z)|^2}{\omega_\nu^2(q)} \right) .$$

From this it follows that the contribution of the acoustic phonons to J is negative. Indeed, such phonons at $q = 0$ correspond to lattice translations, and the electrons do not interact with them (in a significant way), therefore $\lim_{q \to 0} |a_{q\nu}(z)|^2/\omega_\nu^2(q) = 0$ and $J_{\text{acoust}} < 0$. Thus the positive contribution to J comes only from the optical phonons. If the interaction with them is absent, then it follows from the Debye model that $J = -g_z^2/3$. This condition is satisfied by the values of J and g_z^2 found above, and this shows that in the case of $TmVO_4$ the interaction of acoustic phonons with electrons is dominant. This condition also testifies that the main contribution to the correlation comes from the interaction with the deformations having an infinite radius of correlations. All this allows the successful explanation of the experimental data in the framework of the MFA.

The predominant role of the interactions with the deformations is used in [6.40–42] for a simple model treatment of the cooperative Jahn-Teller effect in the case of ferrodistortive ordering. In these works only the coupling of the elementary cells with the deformations is considered, and the phonon mechanism is excluded. Using symmetry arguments this model provides a good thermodynamic description of the crystal properties by a minimal number of phenomenological parameters.

In some cases the interaction of electrons with the crystal deformation in Jahn-Teller ferroelastics is strong enough to give rise to an interesting possibility of influencing the deformations via electrons, and, vice versa, special types of unaxial stress may significantly change the electronic properties of the crystal. These characteristic properties of Jahn-Teller crystals are expected mostly at temperatures near T_c [6.1, 3, 5]. The effect goes beyond the critical phenomena when the value of the appropriate electronic perturbation is of the same order of magnitude as the energy gap in the electron spectrum caused by the mean field, i.e., of the order of kT_c. Such a situation may occur if the corresponding electronic susceptibility is large enough and if the value of kT_c is relatively small. This is the case in some rare-earth compounds with zircon structure, where the g factor is of the order of 10 and T_c is less than 10 K. The resulting possibility of an anomalously large magnetostriction effect in some rare-earth orthovanadates is discussed in [6.43–48]. An interesting case of strong anomalous magnetostriction is possible when there is an antiferrodistortive ordering in the low-temperature phase. In this case the external magnetic (or electric) field can reverse one of the sublattices, resulting in a ferrodistortive ordering accompanied by a macroscopic deformation of the crystal. As shown by estimates, the magnitude of the magnetic (or electric) field needed for this striction effect in some zircons is relatively small and quite accessible under laboratory conditions [6.45, 49, 50].

6.2 Pseudo-Jahn-Teller Mechanism of Spontaneous Polarization of Crystals. The Vibronic Theory of Ferroelectricity

Crystals are called ferroelectrics if in the absence of external electric fields in a certain interval of temperatures and stresses a spontaneous polarization occurs, the direction of which can be changed by means of an electric field and in some cases by means of external stress.

The first theory with a fairly full description of ferroelectric phase transitions was the phenomenological theory of Ginsburg and Devonshire [6.51–53]. Considering the spontaneous polarization P_s as an order parameter, this theory postulates the expansibility of the thermodynamic potential near the point of the phase transition in a power series in P_s and investigates the temperature dependence of several first terms (in accordance with the ideas of the Landau theory). Although the origin of temperature dependence of the expansion coefficients is not explained, this theory remains up to now a convenient scheme for the description of the experimental facts.

The next stage in the development of the theory of ferroelectricity is related to the dynamic theory of Ginsburg, Anderson and Cochran (GAC) [6.54–58]. This theory relates the structural phase transitions to the loss of stability of the crystal with respect to some normal mode. For a ferroelectric phase transition, according to this theory, it is necessary that the lattice loses its stability with respect to the boundary phonons (vibrations with the wave vector $\boldsymbol{q} = 0$) of the transverse optical branch. The GAC theory postulates that the symmetric configuration of the nuclei at $T = 0$ K is unstable (i.e., the force constant is negative). The anharmonicity of even orders and an appropriate choice of the signs of the constants provide the stabilization of the symmetric configuration in the region of high temperatures. By lowering the temperature the stabilizing contribution of the anharmonicity falls. At a certain temperature this contribution becomes equal to the destabilizing elastic forces and the phase transition takes place. Thus the GAC theory is based on the assumption that in ferroelectric crystals an optical vibration exists that is strongly dependent on temperature, its frequency going to zero at the point of the phase transition. This vibration is called a "soft" mode. The consequent experimental study of the phonon spectra of ferroelectric crystals confirmed the existence of a soft mode in most cases.

The early experimental investigations of semiconductive properties of ferroelectric crystals performed in the 1960s showed that the electronic subsystem strongly influences the ferroelectric phase transitions, and vice versa, the phase transition in the phonon subsystem is accompanied by a modification of the electronic spectrum. These facts suggest the important role of the electronic subsystem and electron-phonon interactions in the origin of the lattice instability of ferroelectrics.

The ferroelectric properties described above coincide with the general properties of structural phase transitions conditioned by the cooperative Jahn-Teller effect (Sect. 6.1) with one special feature: the distortion of the lattice resulting in the loss of the central symmetry leading to spontaneous polarization of the crystals (the case of crystals with a noncentral symmetric paraphase is considered below) corresponds to odd boundary optical vibrations. On the other hand, Jahn-Teller-active boundary vibrations belonging to the irreducible representations of the point group of the crystal are even (Sect. 2.3). Only terms with opposite parity can be mixed by odd displacements in the vibronic interaction (Sect. 3.4). Therefore the cause of ferroelectric distortion of the lattice has to be sought in the vibronic mixing of different electronic band states. In other words, the ferrolectric phase transition is determined by the ordering of local distortions of dipole-unstable elementary cells due to the pseudo-Jahn-Teller effect (Sect. 3.4). Following the terminology taken in the previous section, this phenomenon can be called the *cooperative pseudo-Jahn-Teller effect.*

The first work in which the pseudo-Jahn-Teller mechanism of spontaneous polarization and ferroelectric phase transitions was investigated was published in 1966 [6.59]. Some particular statements of this kind were known from earlier work [6.60, 61]. Subsequently these ideas were developed in the *vibronic theory of ferroelectricity* (see the review articles [6.62–67]).

It will be shown below that both postulates of the theory of the soft mode, the instability of the paraphase at $T = 0$ K and the presence of a significant stabilizing anharmonicity of the lattice at higher temperatures, are direct consequences of the pseudo-Jahn-Teller effect. As will be shown later, the vibronic theory relates the parameters of the soft mode theory with the details of the electronic structure of the atoms forming the crystal.

6.2.1 The Simplest Case: Two-Band Model

Consider first the simplest case, the cooperative pseudo-Jahn-Teller effect for two nondegenerate terms of different parity mixed by one odd vibration (in each elementary cell). The Hamiltonian of the crystal can be written in the form

$$\hat{H} = \hat{H}_e + H_{ph} + \hat{H}_{int} , \tag{6.2.1}$$

where H_e is the Hamiltonian of the electronic subsystem,

$$\hat{H}_e = \Delta \sum_{\boldsymbol{n}} \hat{\sigma}_z(\boldsymbol{n}) = \sum_{\alpha, \boldsymbol{n}} \varepsilon_\alpha a^\dagger_{\alpha \boldsymbol{n}} a_{\alpha \boldsymbol{n}} . \tag{6.2.2}$$

Here the method of secondary quantization is used: $a^\dagger_{\alpha \boldsymbol{n}}$ and $a_{\alpha \boldsymbol{n}}$ are operators of creation and annihilation of the site electronic states with the energies ε_α, $\alpha = 1$, 2; $\varepsilon_2 - \varepsilon_1 = 2\Delta$.

If one takes into account the resonant intercell interaction of the electronic site states which broadens the electronic terms into bands, one has to pass to the Fourier representation:

$$a_{\alpha k}^{\dagger} = \frac{1}{\sqrt{N}} \sum_{n} a_{\alpha n} e^{ik \cdot R(n)} , \qquad a_{\alpha k} = \frac{1}{\sqrt{N}} \sum_{n} a_{\alpha n} e^{-ik \cdot R(n)} . \tag{6.2.3}$$

The electronic Hamiltonian in this representation is diagonal,

$$\hat{H}_e = \sum_{\alpha, k} \varepsilon_\alpha(k) a_{\alpha k}^{\dagger} a_{\alpha k} . \tag{6.2.4}$$

The Hamiltonian of the phonon subsystem H_{ph} is described by the usual expression

$$H_{ph} = \sum_{q, \nu} \hbar\omega_\nu(q) b_{q\nu}^{\dagger} b_{q\nu} , \tag{6.2.5}$$

where $\omega_\nu(q)$ are the initial (primary) frequencies and $b_{q\nu}^{\dagger}$ and $b_{q\nu}$ are the operators of the phonon field by which the normal coordinates of the lattice can be expressed:

$$Q_\nu(q) = \sqrt{\frac{\hbar}{2\omega_\nu(q)}} (b_{q\nu} + b_{-q\nu}^{\dagger}) . \tag{6.2.6}$$

The third term in (6.2.1) describes the interband electron-phonon interaction,

$$\begin{aligned} \hat{H}_{int} &= V \sum_{n} Q(n) \hat{\sigma}_x(n) = V \sum_{k, q, \nu} \sqrt{\frac{\hbar}{2N\omega_\nu(q)}} \\ &\times (b_{q\nu} + b_{-q\nu}^{\dagger})(a_{1,k}^{\dagger} a_{2,k-q} + a_{2,k}^{\dagger} a_{1,k-q}) . \end{aligned} \tag{6.2.7}$$

Here the expressions (6.2.2), (6.2.3) and (6.2.6) are used.

Note that the vibronic interaction (6.2.7) does not have the most general form. The operator (6.2.7) includes only the localized vibronic interaction of the site electronic states with one mode of vibrations of nearest-neighbor atoms, i.e., in its own elementary cell. In the general case H_{int} has the form

$$\begin{aligned} \hat{H}_{int} &= \sum_{k, q, \nu} \Gamma_{q\nu}(k) \sqrt{\frac{\hbar\omega_\nu(q)}{2N}} (b_{q\nu} + b_{-q\nu}^{\dagger}) \\ &\times (a_{1,k}^{\dagger} a_{2,q-k} + a_{2,k}^{\dagger} a_{1,q-k}) . \end{aligned} \tag{6.2.8}$$

with an arbitrary dispersion of the vibronic constant $\Gamma_{q\nu}(k)$. The approximation of the localized vibronic coupling corresponds to $\Gamma_{q\nu}(k) = V/\omega_\nu(q)$. For simplicity, henceforth we neglect the dependence of the vibronic constant on the wave vector k of the electrons, $\Gamma_{q\nu}(k) = \Gamma_{q\nu}$. In addition we consider the vibronic coupling to only one transverse optical branch, and thus the index ν in (6.2.5, 8) is omitted.

The phonon Green's function obeys the Dyson equation $\hat{D} = \hat{D}_0 + \hat{D}_0 \hat{\Pi} \hat{D}$, so that for the renormalized frequencies we have

$$\tilde{\omega}_{\boldsymbol{q}}^2 = \omega_{\boldsymbol{q}}^2[1 + \Pi_{\boldsymbol{q}}(\tilde{\omega}_{\boldsymbol{q}})] \ , \tag{6.2.9}$$

where $\hat{\Pi} = \hat{\Pi}_{12} + \hat{\Pi}_{21}$ is the polarization operator which can be calculated using, say, the diagram expansion in perturbation theory [6.68]:

$$\Pi_{12}(\boldsymbol{q}) =$$

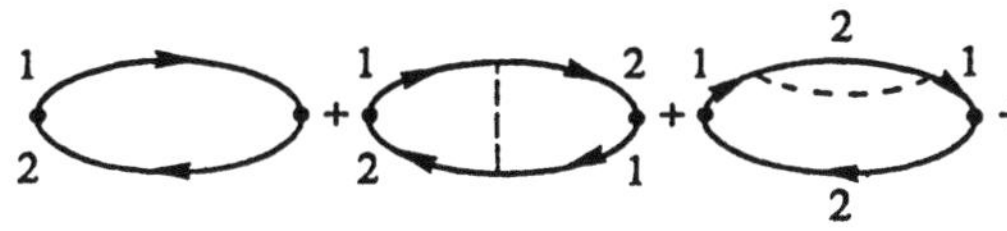

$+ \ldots$ (6.2.10)

Here the full and dashed lines correspond to free-electron and phonon Green's functions, respectively. The indices of the full lines correspond to the numbers of the electronic bands.

The first diagram in (6.2.10) corresponds to the expression

$$\Pi_{\boldsymbol{q}}^{(2)}(\omega) = -2\Gamma_{\boldsymbol{q}}^2\Omega \int \frac{d\boldsymbol{k}}{(2\pi)^3} \frac{[\varepsilon_1(\boldsymbol{k}) - \varepsilon_2(\boldsymbol{k}-\boldsymbol{q})][n_2(\boldsymbol{k}-\boldsymbol{q}) - n_1(\boldsymbol{k})]}{[\varepsilon_1(\boldsymbol{k}) - \varepsilon_2(\boldsymbol{k}-\boldsymbol{q})]^2 + \omega^2} \ , \tag{6.2.11}$$

where Ω is the volume of the elementary cell, $n_\alpha(\boldsymbol{k})$ is the Fermi distribution in the αth electronic band, $\varepsilon_\alpha(\boldsymbol{k})$ is taken as zero at the middle of the forbidden band and the chemical potential μ equals zero. From (6.2.11) it follows that the second-order polarization operator is negative. If at $T = 0$ K for some $\boldsymbol{q}$ values the following condition is fulfilled:

$$1 + \Pi_{\boldsymbol{q}}^{(2)} < 0 \ . \tag{6.2.12}$$

Then the symmetrical configuration of the lattice at low temperatures is unstable with respect to the nuclear displacements corresponding to these $\boldsymbol{q}$. The magnitude of the vector $\boldsymbol{q}$ which labels the unstable mode determines the structure of the low-symmetry phase. The ferroelectric ordering corresponds to the instability of the crystal with respect to the boundary optical phonons ($\boldsymbol{q} = 0$). From now on we consider only this case.

Near the temperature of the phase transition $\tilde{\omega}_{\boldsymbol{q}} \approx 0$ and the expression for $\Pi_{\boldsymbol{q}}^{(2)}(\tilde{\omega}) \approx \Pi_{\boldsymbol{q}}^{(2)}(0)$ acquires a simpler form,

$$\Pi_{\boldsymbol{q}}^{(2)} = -2\Gamma_{\boldsymbol{q}}^2\Omega \int \frac{d\boldsymbol{k}}{(2\pi)^3} \frac{n_2(\boldsymbol{k}-\boldsymbol{q}) - n_1(\boldsymbol{k})}{\varepsilon_1(\boldsymbol{k}) - \varepsilon_2(\boldsymbol{k}-\boldsymbol{q})} \ . \tag{6.2.13}$$

The contribution to the polarization operator from the second and third diagrams in (6.2.10) is proportional to the fourth order of the vibronic constant and is given by the expression

$$\Pi_{\boldsymbol{q}}^{(4)} = \frac{2\Gamma_{\boldsymbol{q}}^2}{N^2} \sum_{\boldsymbol{k},\boldsymbol{q}_1} \Gamma_{\boldsymbol{q}_1}^2 \omega_{\boldsymbol{q}_1} \coth\left(\frac{\hbar\omega_{\boldsymbol{q}_1}}{2kT}\right) \{[\varepsilon_1(\boldsymbol{k}+\boldsymbol{q}_1) - \varepsilon_2(\boldsymbol{k})]$$
$$\times [\varepsilon_1(\boldsymbol{k}-\boldsymbol{q}) - \varepsilon_2(\boldsymbol{k})][\varepsilon_1(\boldsymbol{k}-\boldsymbol{q}) - \varepsilon_2(\boldsymbol{k}-\boldsymbol{q}-\boldsymbol{q}_1)]\}^{-1} \ . \tag{6.2.14}$$

Thus the softening of the phonon frequency is determined by the contribution to the polarization operator of the second order in $\Gamma_{\boldsymbol{q}}$, that is by processes determining the formation of electron-hole pairs. On the other hand, the contribution proportional to $\Gamma_{\boldsymbol{q}}^4$ describing the interband scattering of electron-hole pairs stabilizes the lattice ($\Pi_{\boldsymbol{q}}^{(4)} > 0$), this stabilizing contribution increasing with temperature.

The physical sense of this result can be easily understood if one goes back to the particular case of localized vibronic coupling ($\Gamma_{\boldsymbol{q}} = V/\omega_{\boldsymbol{q}}$) and passes to the limit of the absence of dispersion of electrons and phonons. For the sake of illustration let us assume that $\varepsilon_\alpha(\boldsymbol{k}) = \varepsilon_\alpha$, $\varepsilon_2 - \varepsilon_1 = 2\Delta$, $\omega_{\boldsymbol{q}} = \omega_0$. Considering that in this case $\Pi^{(2)} = -V^2 \tanh(\Delta/2kT)/\omega_0^2\Delta$, at $T = 0$ K we have (Sect. 3.4.1)

$$\tilde{\omega}_0^2 = \omega_0^2 - \frac{V^2}{\Delta} . \tag{6.2.15}$$

The absence of dispersion means that the elementary cells of the crystal can be considered as independent molecules with the pseudo-Jahn-Teller effect in each of them.[1] The curvature of the lowest sheet of the adiabatic potential in the vicinity of the high-symmetry configuration ($Q = 0$) can be found, say, by means of the Öpik and Pryce procedure (Sect. 3.1) using perturbation theory with VQ as a small parameter. The result of such an approach (Sect. 3.4.1) obviously agrees with (6.2.15). This explains the validity of perturbation theory with respect to $\Gamma_{\boldsymbol{q}}$ in the situation of structural instability corresponding to a strong pseudo-Jahn-Teller effect. In fact, however, in the investigation of the curvature the small parameter of the theory is not $\Gamma_{\boldsymbol{q}}$, but $\Gamma_{\boldsymbol{q}} Q(\boldsymbol{q})$, where the magnitude $Q(\boldsymbol{q})$ can be chosen as small as needed.

In the framework of the same approximation of localized vibronic coupling and the absence of dispersion the expression (6.2.14) acquires the form

$$\Pi^{(4)} = \frac{3\hbar\omega_0}{8\Delta} [\Pi^{(2)}]^2 \coth\left(\frac{\hbar\omega_0}{2kT}\right) . \tag{6.2.16}$$

The same temperature dependence of the stabilizing contribution arises from the normal lattice anharmonicity [6.10]. This means that (6.2.16) can be interpreted as the contribution of the vibronic anharmonicity. Indeed, the initial Hamiltonian (6.2.1), (6.2.4), (6.2.5), (6.2.8) does not contain anharmonic terms, but includes the vibronic mixing of the electronic bands. The consideration of vibronic interactions in second-order perturbation theory is equivalent to considering additional harmonic terms $\sim \Gamma_{\boldsymbol{q}}^2 Q^2(\boldsymbol{q})$ to the electronic energy. The fourth order of the perturbation theory giving the $\Pi^{(4)}$ term is equivalent to considering the additional term $\Gamma_{\boldsymbol{q}}^4 Q^4(\boldsymbol{q})$. In other words, the vibronic coupling results in significant anharmonicity of the lowest sheet of the adiabatic potential

[1] Strictly speaking, in the system of independent elementary cells a phase transition is impossible. Therefore by the absence of dispersion we mean that the bandwidth is negligibly small compared with 2Δ and $\hbar\omega_0$ but large enough for the existence of cooperative phenomena.

which is expressed by the results (6.2.14, 16). The vibronic anharmonicity of the adiabatic potential follows also from (3.4.2). This can be checked easily by expanding (3.4.2) in a power series in Q. The stabilizing nature of the vibronic anharmonicity is seen from Fig. 3.8: the branches of the lowest adiabatic sheet are turned up.

Thus the vibronic coupling leads to a new stabilizing mechanism, the vibronic anharmonicity, which is different from the usual lattice anharmonicity.

However, the results obtained above contain one more possibility of stabilization of the paraphase with temperature. This mechanism is related to the temperature reoccupation of the bands, which is in principle impossible in the one-band theories with lattice anharmonicity [6.10]. As seen from Fig. 3.8b, only the electrons which occupy the lowest sheet of the adiabatic potential (the lowest band) destabilize the paraphase. The electrons which occupy the upper sheet (the upper band), however, stabilize the undistorted configuration of the lattice. This explains the temperature dependence of $\Pi^{(2)}$ in (6.2.13) which is included in the difference of the Fermi occupation numbers of the two bands. Considering this temperature dependence, the ferroelectric phase transition can take place even without the stabilizing contribution of $\Pi^{(4)}$ at a temperature determined by the equation $1 + \Pi^{(2)} = 0$.

In a simple model of localized vibronic coupling and the absence of dispersion when the crystal can be considered as a set of noninteracting molecules with adiabatic potentials (3.1.11), the free energy of one elementary cell has the form

$$\mathscr{F} = -kT\ln\left\{2\left[1 + \cosh\left(\frac{\sqrt{\Delta^2 + V^2Q^2}}{kT}\right)\right]\right\} + \frac{1}{2}\omega_0^2 Q^2 \, . \tag{6.2.17}$$

The force constant $\tilde{\omega}_0^2 = (\partial^2\mathscr{F}/\partial Q^2)_{Q=0}$ in this case is

$$\tilde{\omega}_0^2 = \omega_0^2\left[1 - \frac{V^2}{\omega_0^2\Delta}\tanh\left(\frac{\Delta}{2kT}\right)\right] , \tag{6.2.18}$$

whence we obtain the same result for $\Pi^{(2)}$ as above, $\Pi^{(2)} = -V^2\tanh(\Delta/2kT)/\omega_0^2\Delta$.

The equilibrium values of the normal coordinate $\langle Q\rangle$ are found from the condition $\partial\mathscr{F}/\partial Q = 0$. This results in the transcendental equation

$$\langle Q\rangle^2 = \frac{V^2}{\omega_0^4}\tanh^2\left(\frac{\sqrt{\Delta^2 + V^2\langle Q\rangle^2}}{2kT}\right) - \frac{\Delta^2}{V^2} \, . \tag{6.2.19}$$

The temperature dependence of $\langle Q\rangle$ resulting from (6.2.19) has the form characteristic of a phase transition of the second order (Fig. 6.2). The Curie temperature can be obtained from the condition $\langle Q\rangle = 0$. Then from (6.2.19) we obtain $kT_c = \Delta[\operatorname{arctanh}(\omega_0^2\Delta/V^2)]^{-1/2}$. Substituting this value of kT into (6.2.18) results in $\tilde{\omega} = 0$ at the point of the phase transition, that is, the mode under consideration is indeed a soft mode.

Since this mechanism is based on a thermal redistribution of the band occupation numbers, it leads to $kT_c = \Delta$, as was also obtained above. In cases of practical interest $kT_c \sim 10^{-2}$ eV, and the redistribution of the band occupation is important only for narrow-gap ferroelectric semiconductors.

In the case of ferroelectrics with large forbidden bands (say, of the $BaTiO_3$ type where $2\Delta \approx 3$ eV) the temperature redistribution of the population of the electronic bands is inefficient. Here the phase transition is due to an harmonicity, but this anharmonicity has a vibronic origin.

Assuming in (6.2.13) that $n_2 = 0$, $n_1 = 1$ we find that the condition of dipole instability (6.2.12) is reduced to the inequality

$$2\Gamma_0^2\Omega \int \frac{d\boldsymbol{k}}{(2\pi)^3}[\varepsilon_2(\boldsymbol{k}) - \varepsilon_1(\boldsymbol{k})]^{-1} \gg 1 \ . \tag{6.2.20}$$

This inequality can be called the vibronic criterion of dipole instability of the lattice. This criterion was formulated for the first time in [6.59]. The inequality (6.2.20) provides the limits of the parameter values of crystals for which ferroelectricity can occur. This criterion is obeyed better for larger vibronic constants and smaller primary frequency and energy gap between the bands. In fact for a simple model of localized vibronic coupling and in the absence of dispersion (6.2.20) is reduced to the inequality (3.4.3) which contains three parameters: V, Δ and ω_0.

The quantity $\Pi^{(2)}$ can be calculated for a concrete type of electronic spectrum. For instance, if $\varepsilon_2(\boldsymbol{k}) = -\varepsilon_1(\boldsymbol{k}) = \Delta + \varepsilon(\boldsymbol{k})$, where $\varepsilon(\boldsymbol{k}) = -2J[\cos(ak_x) + \cos(ak_y) + \cos(ak_z)]$, then considering the bands to be narrow enough ($J \ll \Delta$) at $T = 0$ K we find

$$\tilde{\omega}^2(\boldsymbol{q}) = \omega^2(\boldsymbol{q})\left[1 - \frac{2\Gamma_{\boldsymbol{q}}^2}{\Delta}\left(1 + \frac{6J}{\Delta}\right) + \frac{\pi^2\Gamma_{\boldsymbol{q}}^2(aq)^2}{9\Delta}\left(\frac{6J}{\Delta}\right)^2\right] . \tag{6.2.21}$$

From this expression two consequences emerge. First, the broadening of the electronic term into a band results in a larger destabilization of the lattice with respect to the nuclear displacements with $\boldsymbol{q} = 0$. Second, for vibrations with $\boldsymbol{q} \neq 0$ the criterion of instability $\tilde{\omega}^2(\boldsymbol{q}) < 0$ is less well satisfied. Both these consequences are related to the modification of the effective energy gap between the bands. Indeed, the increase in J results in a decrease in the energy gap between the bands. On the other hand, the phonons with $\boldsymbol{q} \neq 0$ mix the band states with larger energy gaps having a lower destabilizing contribution.

Equations (6.2.9–21) describe the softening of the ferroelectric vibration in the paraphase on lowering the temperature to T_c. The situation at $T < T_c$, when in accordance with (6.2.18) $\tilde{\omega}_0^2 < 0$, has to be investigated by other methods taking into account the occurrence of the anomalous average $\langle b_0^+ + b_0\rangle$ and interband electronic Green's functions $\langle\langle a_{1\boldsymbol{k}}^+ | a_{2\boldsymbol{k}}\rangle\rangle$. In [6.69] it was shown that the stabilization of the lattice in the ferroelectric phase is also due to the vibronic mixing of electronic bands and that the temperature T_c is proportional to the constant B determining the phonon dispersion. In accordance with this result $T_c \to 0$ in the absence of phonon dispersion, and this corresponds to the physical arguments given in the footnote on page 348.

In the general case T_c is obviously dependent not only on the phonon dispersion but also on the dispersion (dependence on momentum) of the vibronic constants; for some crystals this dependence is rather significant.

In the papers [6.70, 71] considering the same two-band vibronic model of ferroelectrics it was shown that going beyond the mean field approximation and considering fluctuations of the order parameter $\langle b_0^+ + b_0 \rangle$ results in significant changes of the picture of the phase transition. The Curie temperature becomes lower than in MFA and under certain conditions the phase transition may become a first-order transition.

A comparison of the relative contribution of the proper anharmonicity of the lattice with that of the vibronic anharmonicity was carried out in [6.71], and it was concluded that in the case of narrow-gap ferroelectrics the latter is dominant.

6.2.2 Adiabatic Approach to Spontaneous Polarization

The configuration instability of a polyatomic system with a pseudo-Jahn-Teller effect, as mentioned above, can be elucidated without solving the dynamic equation of the atoms. For this purpose it is enough to investigate the adiabatic potentials considering their vibronic mixing. In a crystal containing an infinite number of atoms and, consequently, an infinite number of vibrational degrees of freedom, such an approach is generally speaking unacceptable, since in this case it is impossible to solve the electronic equation for arbitrary atomic positions. In the ferroelectric phase transition, however, the equilibrium positions of only the boundary optical vibrations with $\boldsymbol{q} = 0$ change. The number of such coordinates in the crystal is $3s - 3$ where s is the number of atoms in the elementary cell. The investigation of the dependence of the electronic bands on such a number of normal coordinates is a solvable problem. In addition, this approach allows us to relate the main parameters of the vibronic Hamiltonian to the electronic structure of the atoms forming the crystal.

By way of example consider the influence of the vibronic mixing on the properties of a crystal which in the paraphase has the centrosymmetrical structure of rock salt or perovskite [6.72]. As will be shown below, this case is a crystal analog of the pseudo-Jahn-Teller effect in an octahedral molecule (Sect. 3.4).

In the absence of vibronic mixing the dependence of the one-electron energy on nuclear displacements is described by the adiabatic potentials

$$\varepsilon_\alpha(\boldsymbol{k}, Q) = \varepsilon_\alpha(\boldsymbol{k}) + \tfrac{1}{2} \sum_{\boldsymbol{q}, \nu} \omega_\nu^2(\boldsymbol{q}) Q_\nu(\boldsymbol{q}) Q_\nu(-\boldsymbol{q}) \; , \qquad (6.2.22)$$

where, as above, $\varepsilon_\alpha(\boldsymbol{k})$ is the energy of one-electron states in the frozen lattice corresponding to the configuration of the paraphase.

The operator of localized linear vibronic coupling can be written in the form

$$\sum_{n}\sum_{\Gamma\gamma} Q_{\Gamma\gamma}(n) V_{\Gamma\gamma}(\mathbf{r} - R(n)) \ . \tag{6.2.23}$$

Substituting $Q_{\Gamma\gamma}(n)$ from (6.1.2) and considering that for a ferroelectric phase transition only the boundary optical displacements ($q = 0$) are important, we neglect the vibronic coupling with other modes. Then expressions (6.2.23) may be replaced by

$$\frac{1}{\sqrt{N}} \sum_{\Gamma\gamma} Q_{\Gamma\gamma}(0) \sum_{n} V_{\Gamma\gamma}(\mathbf{r} - R(n)) \ . \tag{6.2.24}$$

It is important that in this approximation the operator of the vibronic coupling mixes the states with the same k only. The eigenvalues $\varepsilon_\alpha(k, Q)$ of the appropriate matrices of the potential energy describe the vibronic modification of the electronic bands, while the minima of the adiabatic potentials $\varepsilon_\alpha(k, Q)$ in the space of the nuclear displacements $Q_{\Gamma\gamma}(0)$ determine the desired equilibrium coordinates of the lattice atoms.

Let us start with the consideration of crystals of the GeTe type with a rock-salt-type lattice and two atoms in the elementary cell. The ground and the first excited states of the elementary cell which form the valence and conduction bands are constructed mainly from the six p orbitals of the two atoms of the cell. The matrix of the operator (6.2.24) can be constructed by strong coupling Bloch functions,

$$\begin{aligned} P_{1\alpha}(k) &= \frac{1}{N} \sum_{n} P_{1\alpha}(\mathbf{r} - R(n)) \exp[ik \cdot R(n)] \ , \\ P_{2\alpha}(k) &= \frac{1}{N} \sum_{n} P_{2\alpha}(\mathbf{r} - A - R(n)) \exp[ik \cdot R(n)] \ , \end{aligned} \tag{6.2.25}$$

where $\alpha = x, y, z$, the indices 1 and 2 distinguish the two atoms of the cell, and $A = (a, a, a)$ is the distance between them. As already mentioned, the operator (6.2.24) is diagonal with respect to k and decomposes into N sixth-order matrices for each k. The matrix elements describing the interaction with the dipole T_{1u} vibrations, $Q_x(0) = Q_x$, $Q_y(0) = Q_y$ and $Q_z(0) = Q_z$, which are responsible for the displacements of the sublattices along the axes x, y and z, are given in [6.72]. In the limit of only σ bonds we obtain

$$\begin{aligned} \varepsilon_x^{(\pm)} = \frac{1}{2}[1 - S^2\cos^2(k_x a)]^{-1} \Bigg(&\Delta - 2hS\cos(k_x a) \pm \Bigg\{ [\Delta - 2hS\cos(k_x a)]^2 \\ &+ 4[1 - S^2\cos^2(k_x a)]\left[h^2\cos^2(k_x a) + \frac{V^2}{N} Q_x^2 \sin^2(k_x a)\right]\Bigg\}^{1/2}\Bigg) \end{aligned} \tag{6.2.26}$$

and in addition similar expressions for $\varepsilon_y^{(\pm)}$ and $\varepsilon_z^{(\pm)}$ which differ by the replacement of k_x, Q_x by k_y, Q_y and k_z, Q_z, respectively. Here h is the parameter determining the width of the electronic bands, V is the reduced matrix element of the one-electron operator $V_{T_1\gamma}(\mathbf{r} - \boldsymbol{R}(\boldsymbol{n}))$ calculated by the p functions of strong coupling, i.e., it is the constant of vibronic interaction, Δ is the energy gap between the p states of Ge and Te atoms, and S is the overlap integral for the p functions in σ bonding. In crystals of the type under consideration (GeTe, etc.) there are six outer p electrons in the elementary cell, and therefore in the ground state the lowest $3N$ one-electron states corresponding to the levels $\varepsilon_\alpha^{(-)}(\boldsymbol{k})$ are occupied. The potential surface corresponding to this electronic configuration is [6.72]

$$\varepsilon(Q_x, Q_y, Q_z) = \frac{\omega_0^2}{2}(Q_x^2 + Q_y^2 + Q_z^2) + \sum_{\alpha,\boldsymbol{k}} \varepsilon_\alpha^{(-)}(\boldsymbol{k}, Q) \; . \tag{6.2.27}$$

Subsituting (6.2.26) into (6.2.27) and keeping only the terms no higher than second order with respect to S, and remembering that $h \sim S$, we obtain the following condition for instability of the high-symmetry configuration $Q_x = Q_y = Q_z = 0$:

$$\frac{V^2}{2\Delta}\left(1 - \frac{h(h - S\Delta)}{2\Delta^2}\right) > \omega_0^2 \; . \tag{6.2.28}$$

If this inequality is fulfilled, the surface (6.2.27) has the same extrema as the adiabatic potential of the cluster TiO_6^{8-} (Sect. 3.4). The criterion (6.2.27) is analogous in a physical sense to the condition (3.4.3). The role of the effective energy gap between the mixing bands is played here by the following quantity averaged over the bands:

$$\Delta_{\text{eff}} = \Delta\left(1 - \frac{h(h - S\Delta)}{2\Delta^2}\right)^{-1} \; . \tag{6.2.29}$$

One can easily see that since the sum of the roots of the electronic equation is independent of the nuclear displacements, the interactions between the fully occupied bands and within each fully occupied band do not contribute to the instability of the symmetric configuration of the lattice. Hence the instability is due to the interaction of fully occupied bands with empty ones. From the criterion (6.2.28) it follows that the effect is stronger, the closer in energy the appropriate states are (the lower the value of Δ is). Therefore the main contribution to the instability comes from the interaction of the nearest neighboring bands. This means that for an estimate of the effect one can limit the consideration to mixing of the valence band with the conduction band.

6.2.3 Perovskites

Let us pass now to crystals with the perovskite structure, to which the most intensely studied crystal, $BaTiO_3$, belongs. The conduction band and valence band are formed here by five atomic $3d$ functions of the metal and nine $2p$ functions of three atoms of oxygen in each elementary cell. The appropriate Bloch functions have a form analogous to (6.25). Let us take into account in the vibronic interaction operator the terms linear in Q_x, Q_y and Q_z describing, as above, the coupling to the T_{1u} vibrations which shift the sublattice of the metal along the x, y and z axes. The matrix of the vibronic coupling decomposes into N 14th-order matrices corresponding to N possible values of $\boldsymbol{k}$. If the condition of the type (6.2.28) is satisfied, the potential surface has the same extrema points as in the case discussed above (Sect. 3.4). These extrema points correspond to the paraphase ($Q_x = Q_y = Q_z = 0$) and such states of the crystal for which the polarization has a rhombohedral, orthorhombic or tetragonal direction [6.59, 72].

Indeed, since only the trigonal extrema points are minima of the adiabatic potential, the totally ordered phase of the crystal corresponds to the displacements of all the active ions along the equivalent trigonal directions [111] resulting in the low-temperature rhombohedral structure. On increasing the temperature, the lowest barrier with a saddle-point on the second-order crystallographic axes [110] is overcome first, the system averaging over two nearest-neighbor minima, [111] and [11$\bar{1}$], and becoming macroscopically equivalent to the orthorhombic phase. In the latter the crystal is thus disordered along one of its axes, remaining ordered along the other two. By a further increase in temperature, the higher barrier with the tetragonal saddle-point [100] on its top is reached, the crystal transforming into the tetragonal phase with two directions of disorder. Finally, at high enough temperatures, the maximum of the adiabatic potential at $Q_x = Q_y = Q_z = 0$ is overcome, the system is averaged over all eight trigonal minima, and the crystal reaches the completely disordered paraphase.

This picture of phase transitions in $BaTiO_3$ (as well as in other ferroelectric crystals) which emerges from the vibronic theory [6.59] shows that these phase transitions are order-disorder transitions. It contradicts the wide-spread idea that the phase transitions in such crystals are displacive, for which all the phases are completely ordered. Note that the vibronic theory also allows transitions which are macroscopically manifest as displacive. This can take place when the energy barrier between the minima of the adiabatic potential is very low, and here the distinction between the two types of phase transitions, displacive and order-disorder, becomes conventional. However, in this case the possibility of more than one phase transition in the same crystal is rather doubtful.

The vibronic picture of phase transitions in such crystals as $BaTiO_3$, $KNBO_3$, etc. was confirmed later by direct experiments on diffuse x-ray scattering [6.73] and Raman scattering of light [6.74]. Recently, additional confirmation of the above picture was obtained in a series of fine experiments on EPR spectra of paramagnetic probing ions inserted into the ferroelectric matrices [6.75]. In the case of $BaTiO_3$ the probing ion Mn^{4+} was choosen to follow almost exactly the Ti^{4+}. It was shown [6.75] that in the completely ordered low temperature rhombohedral phase the Ti^{4+} ion occupies the position in one of the minima along the trigonal axes [111], as predicted by the theory, and on passing to the orthorhombic phase the EPR signal disappears due to the fast reorientations of the ion between the two nearest-neighbour minima, the frequency of these reorientations being estimated as lying between $10^9 s^{-1}$ and $10^{10} s^{-1}$. In the isostructural crystal $SrTiO_3$, where there are no such phase transitions, the EPR signal is seen well up to high temperatures. In the works [6.73, 75] it is assumed that in the high temperature disordered phases there still remain correlated short chains containing 10 or more elementary cells depending on the crystal imperfections.

The presence of disordered phases in ferroelectrics of the type of $BaTiO_3$ is of primary interest, since it elucidates the role of long-range correlations in ferroelectric (and other structural) phase transitions. Indeed, if the instability of the high symmetry configuration of the lattice is due to long-range forces, that is to the dominance of the energy of ordering over the energy of local distortion, than disordered phases are impossible. The experimental observation of disordered phases confirms, in principle, the vibronic origin of instability for which long-range forces are not needed. On the other hand, as mentioned above, the vibronic theory also includes the cases when the disordered phases are not observed explicitly. Indeed, when the pseudo-Jahn-Teller effect is small enough and the energy barriers between the minima of the adiabatic potential are small, the averaging over the minima is achieved not by disorder but by vibrations of large amplitudes.

The physical meaning of the criterion of vibronic instability (6.2.28) can be elucidated in a similar way to the origin of the pseudo-Jahn-Teller effect (Sect. 3.4): the instability arises due to the gain of energy in the formation of additional covalent bonds and the polarization of the atoms (ions) of the environment by distortions. To illustrate this statement for $BaTiO_3$ consider again the TiO_6^{8-} cluster (Sect. 3.4). When the Ti^{4+} ion is in the center of the octahedron the overlap of the orbitals d_{xy}, d_{xz}, d_{yz} transforming after the representation T_{2g} with the orbital of the T_{1u} type formed by the $2p$ functions of the oxygen atoms is zero by symmetry due to parity. The displacement of this ion from the inversion center removes this symmetry, resulting in a non-zero overlap and the formation of an additional π bond between the Ti ion and the oxygen atoms. Therefore the meaning of the criterion (6.2.28) is that the instability under consideration arises if and when the formation of the new covalent bonds is dominant in energy over the deterioration of the previous bonds. Note that this statement has nothing to do with the compensation of local repulsive forces by long-range (ordering) attractive ones; the local repulsive forces in the vibronic theory are compensated by the local formation of new covalent bonds (or nearest-neighbor polarization, see below).

As mentioned above, the vibronic theory also considers the effects of polarization of the electronic cloud by distortion, represented by the terms mixing atomic states with their own excited states of different parity. The estimate in Sect. 3.4 for the NH_3 molecule obtained by nonempirical calculations show that the contribution of the polarization effect to the curvature of the adiabatic potential is $-6\ \mathrm{Nm}^{-1}$ whereas the contribution of new covalent bond formation is $-62\ \mathrm{Nm}^{-1}$. A similar ratio for these contributions has been found for the TiO_6^{8-} cluster in the $BaTiO_3$ crystal [6.64, 65]: the polarization contribution to the instability at the point $Q_x = Q_y = Q_z = 0$ is an order of magnitude smaller than the contribution of new covalent bonds between the Ti and oxygen atoms.

These estimates explain the failure of the theory of Slater [6.76] who tried to explain the origin of instability by polarization effects only. In order to get the appropriate instability Slater had to assume unusually high polarizabilities of the oxygen ion ("polarization catastrophe"). From the point of view of the vibronic theory the reason for this negative result is quite clear: in the Slater theory the most significant effect of additional covalent bond formation was not taken into account, so unusual polarizabilities were required. A similar argument is applicable to the theory of *Bilz* [6.77, 78] in which the main role in the instability is attributed to the nonlinear polarizability of oxygen octahedrons.

Emphasizing the predominantly covalent (local) origin of instability of the high symmetry configuration of the crystal, or its elements, we note that the phase transition itself cannot take place without long-range interactions. The temperature of the phase transition depends on both the parameters of vibronic instability V, Δ_{eff}, ω_0^2 and the parameters of the interaction of dynamically unstable elements.

The investigation of the criteria for the occurrence of the extrema points allows us to relate the presence of different types of polarized phases to the details of the electronic structure of the perovskites. The changes in the electronic structures calculated for a series of similar crytals permit the determination of the ferroelectric properties related to these changes. For instance, the increase of Δ_{eff} leads to a deterioration of the probability of occurrence of spontaneously polarized states. The analysis shows that the increase of Δ_{eff} is due to the increase in the overlap of d and p orbitals of the neighboring metal ions and oxygen. For perovskites of titanium, zirconium and hafnium this overlap increases because of the change of the electronic state of the metal ion, while for the barium, strontium and calcium titanate it increases due to the decrease of the metal-oxygen distance. Hence in these series the ferroelectric properties should deteriorate [6.59]. This conclusion is in good agreement with the experimental data.

In the same way that the ferroelectric structures result from the displacement of the equilibrium position of the optical vibration with $\boldsymbol{q} = 0$, the antiferroelectric ordering is due to the displacement with $\boldsymbol{q} = (\pi/2a, \pi/2a, \pi/2a)$. The analysis of the situation in this case is somewhat more complicated than in the case of ferroelectric ordering due to the doubling of the lattice period. In [6.79] such an analysis was carried out for crystals like GeTe considering the orthogonalized π bonding. The electronic energy as a function of the appropriate nuclear vibration has the form (the indices f and a distinguish ferro and antiferro situations)

$$\varepsilon_{\mathrm{f}}^{(\pm)}(k_\alpha, Q_{\mathrm{f}\alpha}) = \pm\left(\Delta^2 + h^2\cos^2(k_\alpha a) + \frac{V^2}{N} Q_{\mathrm{f}\alpha}^2 \sin^2(k_\alpha a)\right)^{1/2},$$

$$\varepsilon_{1\alpha a}^{(\pm)} = -\varepsilon_{2\alpha a}^{(\pm)} = \left[\Delta^2 + \frac{h^2}{2} + \frac{V^2}{N} Q_{a\alpha}^2 \sin^2(k_\alpha a)\right.$$
$$\left. \pm \left(\frac{V^2}{N} Q_{a\alpha}^2 h^2 \sin^2(k_\alpha a)[1 - \sin(2k_\alpha a)] + \frac{1}{4} h^4 \cos^2(2k_\alpha a)\right)^{1/2}\right]^{1/2} , \tag{6.2.30}$$

where $\alpha = x, y, z$. The criteria for instability determined from the condition $(\partial^2 E/\partial Q_\alpha^2) < 0$ have the form

$$\omega_f^2 < \sum_k \frac{V^2}{N} \sin^2(k_\alpha a)[\Delta^2 + h^2 \cos^2(k_\alpha a)]^{1/2} ,$$
$$\omega_a^2 < \frac{1}{2}\frac{V^2}{N} \sum_k \sin^2(k_\alpha a)\left[\frac{1}{A_\alpha} + \frac{1}{B_\alpha} + \left(\frac{1}{A_\alpha} - \frac{1}{B_\alpha}\right)\frac{\cos(k_\alpha a) - \sin(k_\alpha a)}{\cos(k_\alpha a) + \sin(k_\alpha a)}\right] , \tag{6.2.31}$$

where

$$A_\alpha = [\Delta^2 + h^2 \cos^2(k_\alpha a)]^{1/2} , \qquad B_\alpha = [\Delta^2 + h^2 \sin^2(k_\alpha a)]^{1/2} .$$

It can easily be seen that at $h = 0$, that is, in the case of extremely narrow electronic bands, the right-hand sides of the inequalities (6.2.31) are the same. This means that the negative vibronic contributions to ferroelectric and antiferroelectric frequency are the same. The preferability of one or another ordering in this case is determined by the nature of the dispersion of the optical vibrations of the lattice. The vibration with the lower frequency is less stable. The analysis for the case $h \neq 0$ shows that the destabilizing contribution is larger for ferroelectric distortions. In other words, if $\omega_a = \omega_f$, the ferroelectric ordering is preferred. Antiferroelectric ordering can be realized only if $\omega_a < \omega_f$, and this inequality should be strong enough to compensate the difference in the vibronic contribution to the frequency.

The adiabatic electronic bands (6.2.26), (6.2.30) obtained above correspond quite well to the real electronic bands to the extent that the nonadiabatic effects can be neglected. In a good approximation to these expressions it is enough to substitute $Q_{\Gamma\gamma}$ by the equilibrium temperature-dependent values $\langle Q_{\Gamma\gamma}\rangle$. The same expressions of the type

$$\langle \varepsilon_\pm(\boldsymbol{k})\rangle = \tfrac{1}{2}[\varepsilon_1(\boldsymbol{k}) + \varepsilon_2(\boldsymbol{k})] \pm \sqrt{\tfrac{1}{4}[\varepsilon_1(\boldsymbol{k}) - \varepsilon_2(\boldsymbol{k})]^2 + \Gamma_0^2\langle Q\rangle^2} \tag{6.2.32}$$

can be obtained also in a more rigorous self-consistent treatment [6.80]. The temperature dependence of the order parameter $\langle Q\rangle$ determines the characteristic temperature dependence of the band structure and, in particular, the effective mass of the charge carriers [6.81]. The change of temperature may lead to such a great modification of the electronic bands that the number of extrema and their symmetry will change. In its turn this influences the Van Hove singularities in the density of states. The latter should be manifest in the interband absorption which can serve as an exact method of investigation of the dispersion in vibronic problems. The anomalies of the temperature and frequency dependence of the coefficient of interband light absorption are investigated in [6.82, 83]. In [6.84] spontaneous birefringence in a vibronic ferroelectric was investigated. The contribution to the optical dielectric susceptibility, renormalized by the vibronic interaction of the active bands, was calculated. Using the tetragonal phase of $BaTiO_3$, it was shown that the main aspects of spontaneous birefrigence can be thought of as natural consequences of the vibronic theory. In particular, the temperature dependence of the difference in the refractive indices is determined by the square of the order parameter in accordance with the well-known phenomenological theory often confirmed in the experiments.

We emphasize that all these important conclusions are essentially based on the vibronic mixing of the electronic bands and cannot be obtained from the one-band model with lattice anharmonicity. Therefore they can serve as a base for experimental confirmation of the pseudo–Jahn-Teller origin of ferroelectricity.

Another example of this kind is the influence of the magnetic field on the ferroelectric phase transition. In the model of lattice anharmonicity, the magnetic field, which directly influences only the electronic subsystem, should not influence the phase transition. In other words, if as a result of the influence of the magnetic field upon the crystal, the ferroelectric characteristics change, then this testifies to the electron-vibrational nature of the phase transition.

The influence of the magnetic field $\mathscr{H}$ on the soft mode frequency in ferroelectrics was investigated in [6.85, 86] by the Green's functions method. Since the quantizing magnetic field contracts the band states into Landau levels the influence of the magnetic field on the spontaneous polarization and Curie temperature is essentially determined by the dispersion of the vibronic constant $\Gamma_{\boldsymbol{q}}(\boldsymbol{k})$ If the vibronic constant does not increase with $|\boldsymbol{k}|$, then the contraction of the band states leads to an effective increase in the energy gap between the bands, resulting in a decrease of the softening vibronic contribution to the soft mode frequency. The resulting increase in the frequency (and hence in the Curie temperature T_c) has an oscillating behavior because of the ejection of the Landau levels from the band. A strong effect should be expected for large-gap vibronic ferroelectrics in which the phase transition is determined by the vibronic anharmonicity. The magnetic field corrections to the component $\Pi^{(4)}$ of the polarization operator can be neglected. The corrections to the component $\Pi^{(2)}$ are also small but near the point of the phase transition when $1 + \Pi^{(2)} \approx 0$ the influence of the magnetic field is significant. Accordingly, the corrections to T_c are also significant. Estimates made for an activated $SrTiO_3$ crystal result in a significant correction of T_c even for magnetic fields of the order of 10^5 Oe. Such a change of the ferroelectric properties under the influence of an external magnetic field is confirmed experimentally [6.87, 88].

The vibronic theory was also confirmed in a set of experimental observations including the changes of the band structure of ferroelectrics at the phase transition as well as in photo-induced effects in ferroelectrics, in particular, in the shift of the Curie temperature under irradiation [6.81, 89, 90].

Some new results and experimental evidence within the vibronic model of ferroelectricity have been obtained recently, including the interband photogalvanic effect possesing nonclassical properties, polar thermo-emf and thermal conductivity, the magnetization by illumination, and different effects produced by intensive electromagnetic irradiation [6.91–93] (see also [6.66, 67, 94]).

The list of results of the vibronic theory of ferroelectricity discussed here is not exhaustive. For instance, we did not consider the impurity properties of the vibronic ferroelectrics [6.95], and, in particular, the statement confirmed experimentally [6.96] that the Jahn-Teller impurities promote the antiferroelectric ordering [6.97]. Note also the attempt in [6.98] to describe the domain structure of the crystal in terms of tunneling between the equivalent minima.

In this section we have considered only centrosymmetric crystals. This allowed us to exclude the even vibrations and the related cooperative Jahn-Teller effect from the discussion. A treatment is given in [6.99, 100] where the ferroelectricity in noncentrosymmetric crystals is explained by the cooperative Jahn-Teller effect.

Another large field of applications of the vibronic theory is provided by crystal chemistry. Some results of this field are reviewed in [6.101]. More detailed discussion can be found in [6.102, 103].

6.3 The Vibronic Theory of Structural Transformations in Condensed Media

A widely held belief about the origin of structural phase transitions in crystals mentioned in the previous section is that the instability of the crystal lattice with respect to any crystalline mode is due to the compensation of the repulsive forces preventing local distortions by attractive long range ones favoring the simultaneous occurrence and ordering of these distortions, provided the gain of energy due to the ordering of the latter is larger than the loss of energy in the local distortions. However, in the two types of structural phase transitions considered above—general Jahn-Teller (Sect. 6.1) and ferroelectric (Sect. 6.2)—the lattice instability occurs as a result of only short range forces in each elementary cell. In these cases the gain of energy by distortion is achieved due to the vibronic interactions resulting in the formation of additional chemical bonds that are forbidden in the high symmetry configuration by symmetry restrictions. Certainly, for the phase transition to occur, the cooperative interactions of the distortions are very important, determining (together with the magnitude of the local distortion) the transition temperature, but the cooperative properties of the crystal are not required for an explanation of the origin of the instability itself at low temperatures, and it will be shown in this section that without vibronic interactions these cooperative effects do not result in instability.

The temperature-dependent frequency of the soft mode is determined by the second derivative of the free energy $\mathscr{F}$ of the system on the corresponding normal coordinate (Sect. 6.2.1). Since $\mathscr{F} = \varepsilon_0 - TS$, where ε_0 is the adiabatic energy of the lattice and S is the entropy, it is clear that at $T = 0$ the negative curvature of the free energy in the direction of the soft mode is determined by the negative curvature of the adiabatic potential. This fact, together with the pseudo-Jahn-Teller origin of the negative curvature of the adiabatic potential in polyatomic systems proved above (Sect. 3.4.3), raises the possibility that any[2] structural phase transition may have a vibronic origin [6.104, 105]. The Jahn-Teller origin of structural phase transitions in crystals containing a sublattice of ions with degenerate electronic terms is obvious (Sect. 6.1). In the absence of degeneracy the problem becomes less trivial. The ordering of local distortions of dipolarly unstable elementary cells was considered in Sect. 6.2. In Sect. 6.3.1 a general proof of the vibronic origin of instabilities initating structural phase transitions in dielectric crystals is given. The vibronic mechanism of instability is considered in detail, and it is shown that the main contribution to the effect of instability is due to the formation of new covalent bonds by local distortions. In Sect. 6.3.2 the vibronic origin of Peierls transitions in metals is considered. A general (unique) model of all the structural transformations in condensed media controlled by the electronic structure through the vibronic coupling, is given in Sect. 6.3.3.

6.3.1 Vibronic Origin of Instability of the High Symmetry Configuration of Dielectric Crystals

The general Hamiltonian of the crystal is determined by (2.1.1). Its Coloumb interaction term can be written as

$$V(\mathrm{r}, Q) = W(Q) + \sum_{\alpha=1}^{s} Z_\alpha V_\alpha(\mathrm{r}, Q) , \tag{6.3.1}$$

where $W(Q)$ is the operator of internuclear repulsion,

$$W = \sum_{\boldsymbol{n},\boldsymbol{m}}{}' \sum_{\alpha,\beta=1}^{s}{}' \frac{e^2 Z_\alpha Z_\beta}{|\boldsymbol{R}_\alpha(\boldsymbol{n}) - \boldsymbol{R}_\beta(\boldsymbol{m})|} , \tag{6.3.2}$$

the indices $\boldsymbol{n} = (n_x, n_y, n_z)$ and $\boldsymbol{m} = (m_x, m_y, m_z)$ number the elementary cells of the crystal, each containing s different atoms, α and β label the different atoms in the same cell, Z_α is the nuclear charge (in atomic units), $\boldsymbol{R}_\alpha(\boldsymbol{n})$ is the position vector of the αth nucleus in the $\boldsymbol{n}$th cell, and $V_\alpha(\mathrm{r}, Q)$ in (6.3.1) describes the Coloumb interactions of the electrons with the nuclei of the αth sublattice:

[2] We do not consider here possible cases where there are two or more minima of the free energy of the system arising due to the entropy term only; these possibilities need a special treatment.

$$V_\alpha(\mathrm{r}, Q) = -e^2 \sum_i \sum_{\boldsymbol{n}} |\mathbf{r}_i - \boldsymbol{R}_\alpha(\boldsymbol{n})|^{-1} \ , \tag{6.3.3}$$

i labels the electrons, and $\mathbf{r}_i$ is the position vector of the ith electron.

Consider first the instability of a cubic dielectric crystal with respect to the limiting (boundary) nuclear displacements describing the shift of the sublattices corresponding to the change of the crystal structure in the phase transition under consideration [6.64, 65]. The sublattice coordinates can be taken as

$$\boldsymbol{R}_\alpha = \frac{1}{\sqrt{N}} \sum_{\boldsymbol{n}} \boldsymbol{R}_\alpha(\boldsymbol{n}) \ . \tag{6.3.4}$$

Then the normal coordinates of the lattice formed by these $3s$ limiting coordinates are

$$Q_\alpha^{(j)} = \sum_{\beta=1}^{s} C_{\alpha\beta} X_\beta^{(j)} \ , \tag{6.3.5}$$

where $X_\beta^{(j)}$ is the jth component of the vector $\boldsymbol{R}_\beta$, $j = x, y, z$; $C_{\alpha\beta}$ are the elements of the matrix of the orthogonal transformation that performs the transition to the normal coordinates $Q_\alpha^{(j)}$. Three of these coordinates describe the displacement of the crystal as a whole, the remaining $3s - 3$ give the relative shifts of the sublattices.

In the coordinates (6.3.5) the nonvibronic contribution (3.4.21) to the force constant $K_{j\alpha}$ corresponding to the shift $Q_\alpha^{(j)}$ is

$$K_{j\alpha}^{(0)} = \left\langle 0 \left| \left(\frac{\partial^2 V(\mathrm{r}, Q)}{\partial Q_\alpha^{(j)2}} \right)_0 \right| 0 \right\rangle \ , \tag{6.3.6}$$

where, allowing for (6.3.1),

$$\frac{\partial^2 V}{\partial Q_\alpha^{(j)2}} = \sum_{\beta=1}^{s} C_{\alpha\beta}^2 Z_\beta \frac{\partial^2 V_\beta}{\partial X_\beta^{(j)2}} + \frac{\partial^2 W}{\partial Q_\alpha^{(j)2}} \ . \tag{6.3.7}$$

Note that for any polyatomic system

$$\frac{\partial^2 V_\beta}{\partial X_\beta^{(j)2}} = \frac{e^2}{N} \sum_{i,\boldsymbol{n}} \left(\frac{4}{3} \pi \delta(\mathbf{r}_i - \boldsymbol{R}_\beta(\boldsymbol{n})) - \frac{3(x_i^{(j)} - X_\beta^{(j)})^2 - (\mathbf{r}_i - \boldsymbol{R}_\beta(\boldsymbol{n}))^2}{|\mathbf{r}_i - \boldsymbol{R}_\beta(\boldsymbol{n})|^5} \right) \ ,$$

where $x_i^{(j)}$ means the jth component of the vector $\mathbf{r}_i$. It follows that

$$\left\langle 0 \left| \left(\frac{\partial^2 V_\beta}{\partial X_\beta^{(j)2}} \right)_0 \right| 0 \right\rangle = \frac{4\pi e}{3} \varrho_\beta + e Q_{jj}^{(\beta)} \ , \tag{6.3.8}$$

where ϱ_β is the density of electronic charge at the nuclei of the β sublattice, and $Q_{jj}^{(\beta)}$ is the component of the gradient of the electric field produced by the electrons at the nuclei of the β sublattice. Since $Q_{xx}^{(\beta)} + Q_{yy}^{(\beta)} + Q_{zz}^{(\beta)} = 0$,

$$\langle 0|(\nabla_\beta^2 V_\beta)_0|0\rangle = 4\pi e\varrho_\beta \geq 0 \ , \tag{6.3.9}$$

where $\nabla_\beta = \partial/\partial R_\beta$. For a cubic crystal all three terms of the Laplacian in (6.3.9) are equal to each other, and hence

$$\left\langle 0\left|\left(\frac{\partial^2 V_\beta}{\partial X_\beta^{(j)2}}\right)_0\right|0\right\rangle = \frac{4}{3}\pi e\varrho_\beta \geq 0 \ . \tag{6.3.10}$$

In the expression (6.3.2) all the denominators are nonzero, and therefore one can easily prove that

$$\sum_{j=x,y,z}\left(\frac{\partial^2 W}{\partial X_\alpha^{(j)}(\boldsymbol{n})\partial X_\beta^{(j)}(\boldsymbol{m})}\right)_0 = 0 \ ,$$

where $X_\alpha^{(j)}(\boldsymbol{n})$ are the components of the vector $\boldsymbol{R}_\alpha(\boldsymbol{n})$, $j = x, y, z$. Hence

$$\sum_{j=x,y,z}\left(\frac{\partial^2 W}{\partial Q_\alpha^{(j)2}}\right)_0 = 0 \ . \tag{6.3.11}$$

In a cubic lattice the directions x, y, z are equivalent, and therefore the three terms in the sum (6.3.11) are equal, and hence each of them is zero:

$$\left(\frac{\partial^2 W}{\partial Q_\alpha^{(j)2}}\right)_0 = 0 \ , \qquad j = x, y, z \ . \tag{6.3.12}$$

From (6.3.6), (6.3.7), (6.3.10) and (6.3.12) we obtain the final inequality [6.64, 65]

$$K_{j\alpha}^{(0)} \geq 0 \ , \qquad j = x, y, z \ , \tag{6.3.13}$$

for all the $\alpha = 1, 2, \ldots, s$.

Based on this inequality, the following conclusion (similar to that given in Sect. 3.4.3) can be formulated: if the curvature of the adiabatic potential in the high symmetry configuration of the crystal lattice is negative with respect to any normal coordinate, describing a relative shift of the sublattices, then the only source of this instability is the vibronic contribution to the curvature (3.4.20), i.e., the pseudo-Jahn-Teller effect [6.64, 65, 104, 105]. In other words, the only possible source of instability of the high symmetry configuration of the crystal lattice resulting in structural phase transitions is the vibronic (pseudo-Jahn-Teller) mixing of the ground electronic state with the excited ones by appropriate nuclear displacements. Since the proof, given above, does not depend on the number of sublattices s, then by considering an extended elementary cell it can be applied to displacements describing transitions to superstructures.

For a linear chain which can be considered as a limiting case of an extremely anisotropic crystal, an analogous proof of the above statement is possible [6.64, 65]. Consider, for simplicity a diatomic chain, for which the coordinates (6.3.5) can be reduced to six degrees of freedom:

$$Q = \frac{1}{\sqrt{2}}(\boldsymbol{R}_1 - \boldsymbol{R}_2) , \qquad \boldsymbol{q} = \frac{1}{\sqrt{2}}(\boldsymbol{R}_1 + \boldsymbol{R}_2) ,$$

where the coordinates $\boldsymbol{R}_1$ and $\boldsymbol{R}_2$ are determined by (6.3.4). The vector $\boldsymbol{Q}$ describes the relative shift of the two sublattices, whereas $\boldsymbol{q}$ corresponds to a shift of the coordinate origin. Assuming that in the high symmetry configuration the chain is along the z axis, we obtain the nonvibronic contribution (6.3.6) to the force constant of the displacement Q_j ($j = x, y, z$),

$$K_j^{(0)} = \frac{1}{2}\left\langle 0 \left| Z_1 \left(\frac{\partial^2 V_1}{\partial X_1^{(j)2}}\right)_0 + Z_2 \left(\frac{\partial^2 V_2}{\partial X_2^{(j)2}}\right)_0 \right| 0 \right\rangle + \lambda_j \frac{Z_1 Z_2}{N} \sum_{n,m} R_{nm}^{(0)^{-3}} , \tag{6.3.14}$$

where $\lambda_x = \lambda_y = -2$, $\lambda_z = 4$, $R_{nm}^{(0)} = |\boldsymbol{R}_1^{(0)}(\boldsymbol{n}) - \boldsymbol{R}_2^{(0)}(\boldsymbol{m})|$ is the distance between the nth atom of the first sublattice and the mth atom of the second one in the high symmetry configuration, and V_1 and V_2, $X_1^{(j)}$ and $X_2^{(j)}$ are determined by (6.3.3) and (6.3.5), respectively. Let us show first that

$$\left\langle 0 \left| Z_1 \left(\frac{\partial^2 V_1}{\partial X_1^{(j)2}}\right)_0 + Z_2 \left(\frac{\partial^2 V_2}{\partial X_2^{(j)2}}\right)_0 \right| 0 \right\rangle \geq 0 . \tag{6.3.15}$$

Following [6.106], we use the general expressions for adiabatic potentials of polyatomic systems derived from the formula $\varepsilon_0 = \langle \Psi_0 | H | \Psi_0 \rangle$. From the Helmann-Feynman theorem [6.107, 108] we have

$$\frac{\partial \varepsilon_0}{\partial X} = \left\langle \Psi_0 \left| \frac{\partial H}{\partial X} \right| \Psi_0 \right\rangle ,$$

$$\frac{\partial^2 \varepsilon_0}{\partial X^2} = \left\langle \Psi_0 \left| \frac{\partial^2 H}{\partial X^2} \right| \Psi_0 \right\rangle + \left\langle \frac{\partial \Psi_0}{\partial X} \left| \frac{\partial H}{\partial X} \right| \Psi_0 \right\rangle + \left\langle \Psi_0 \left| \frac{\partial H}{\partial X} \right| \frac{\partial \Psi_0}{\partial X} \right\rangle .$$

The last two terms here represent the second-order perturbation theory correction to the adiabatic energy of the electrons, provided the operator $(\partial H/\partial X)X$ (the first-order term in the expansion of the Hamiltonian H with respect to X) is used as a perturbation. As corrections to the ground state these terms are negative, and therefore

$$\left\langle \Psi_0 \left| \frac{\partial^2 H}{\partial X^2} \right| \Psi \right\rangle \geq \frac{\partial^2 \varepsilon_0}{\partial X^2} . \tag{6.3.16}$$

Using this inequality when differentiating with respect to q_j (the components of $\boldsymbol{q}$), and taking into account that ε_0 is independent of q_j, we come to the inequality (6.3.15). Since $\lambda_z = 4$ and the last term in (6.3.14) for $j = z$ is positive, we obtain the inequality sought for: $K_z^{(0)} \geq 0$.

A similar inequality for $K_x^{(0)}$ and $K_y^{(0)}$ cannot be obtained directly, since $\lambda_x = \lambda_y = -2$ and the last term in (6.3.14) for $j = x$, y is negative. However, in these cases, too, assuming, for instance, that the energy of the chain depends on the absolute values of the interatomic distances R_{nm} only, and using the inequality (6.3.16), we have

$$\left\langle 0 \left| \left(\frac{\partial^2 H}{\partial X_1^2}\right)_0 \right| 0 \right\rangle \geq \sum_{n,m} \left[\left(\frac{\partial \varepsilon_0}{\partial R_{nm}}\right)_0 \left(\frac{\partial^2 R_{nm}}{\partial X_1^2}\right)_0 + \left(\frac{\partial^2 \varepsilon_0}{\partial R_{nm}^2}\right)_0 \left(\frac{\partial R_{nm}}{\partial X_1}\right)_0^2 \right].$$

In the high symmetry configuration $(\partial\varepsilon_0/\partial R_{nm})_0 = 0$ due to the condition of equilibrium, while $(\partial R_{nm}/\partial X_1)_0 = 0$ due to the fact that for the chain along the z axis $X_1^{(0)}(n) = X_2^{(0)}(n) = 0$. Hence $K_x^{(0)} = K_y^{(0)} \geq 0$.

For anisotropic crystals the negative term in (6.3.14) is multiplied by the anisotropy factor $\gamma \sim 10^{-1}$–10^{-2}. In the general case this term is of the order of $\gamma Z_\alpha Z_\beta R_0^{-3}$, where R_0 is the shortest distance between the different kinds of atoms in the high symmetry configuration of the lattice. The positive term in (6.3.14) is of the order of $Z_\alpha \varrho_\alpha$, see (6.3.7, 6.3.10), and for $Z_\alpha > 1$ it is much larger than the negative terms. Indeed, the charge density of s electrons at the nuclei $\varrho_\alpha \sim Z_\alpha^3$, and therefore the positive terms in (6.3.14) are of the order of Z_α^4, their ratio to the negative term being $\sim Z^2 R_0^3 \gamma^{-1}$, i.e., much greater than unity. This estimate is also valid when the inner electrons are included in the atomic cores; in this case Z_α is the effective nuclear charge. In the case of a purely ionic lattice, the reasons given above are invalid, since the condition for equilibrium of the high symmetry configuration is not satisfied.

In [6.104, 105] a less rigorous but more general proof that the nonvibronic contribution (6.3.6) to the force constant of the relative shift of the sublattices is positive is given. This proof is valid for any crystal lattice of a dielectric crystal; it is based on the representation of the electronic wave function of the ground state in the form of a Slater determinant in the MO LCAO approximation, $|0\rangle = N^{-1/2} \det|\psi_1 \psi_2 \ldots \psi_N|$, where $\psi_i = \sum_\mu C_{i\mu}\varphi_\mu$, $C_{i\mu}$ are the LCAO coefficients, and φ_μ is the appropriate atomic function (in the tight binding approximation). Calculating the diagonal matrix element (6.3.6) with this function, and making a reasonable approximation, one can see that $K_{j\alpha}^{(0)}$ consists of two contributions: positive terms of the order of $\sum_\alpha q_\alpha \zeta_\alpha^3 Z_\alpha$, which are due to the interaction of the displacing nuclei with their own electronic shells, and negative terms of the order of $\sum_{\alpha,\beta} Z_\alpha Z_\beta R_{\alpha\beta}^{-3}$ due to the nuclear repulsions. Here ζ_α is the parameter of the Slater 1s orbital of the α atom (or any other s orbital of the first ns shell not included in the core when the inner and valence electrons are separated), q_α is the electronic population of the 1s orbital, and $R_{\alpha\beta} = |R_\alpha^{(0)}(n) - R_\beta^{(0)}(n)|$. Since $\zeta_\alpha \sim Z_\alpha$ (this is true also when Z_α is the effective charge of the core), $R_{\alpha\beta} \approx 2$ a.u., $q_\alpha \approx 2$, the positive term $\sim \sum_\alpha Z_\alpha^4$ is much larger than the negative one $\sim \sum_{\alpha,\beta} Z_\alpha Z_\beta$, provided $Z_\alpha > 1$. Thus $K_{j\alpha}^{(0)}$ is also positive in the general case of arbitrary structure and symmetry, and this proves the pseudo-Jahn-Teller origin of the instability of any crystal lattice, provided it exists in the high symmetry configuration.

The case $Z_\alpha = 1$ when the displacing nuclei are hydrogen nuclei (in all other cases, including that of the core-valence electron separation approximation,

$Z_\alpha > 1$) requires additional consideration. For this case some nonempirical calculations of $K_{j\alpha}^{(0)}$ were carried out together with an evaluation of the vibronic contribution to the curvature [6.109] for a series of simple molecules (H_3, CH_4, NH_3, etc.) in high symmetry configurations that are unstable with respect to symmetrized nuclear displacements including displacements of the hydrogen atoms. In every case $K_{j\alpha}^{(0)}$ calculated by (3.4.21) is positive.

Thus the dynamic instability of the crystal lattice with respect to any crystalline mode can occur only when there is vibronic mixing of the ground electronic state of the lattice with the excited states under the nuclear displacements of the mode under consideration, provided the contribution of this vibronic mixing is larger than the nonvibronic terms of opposite sign. This means that the instability of the equilibrium configuration of the lattice triggering structural phase transitions in dielectric crystals is always of pseudo-Jahn-Teller origin, and in the framework of this approach, based on the Born-Oppenheimer approximation of the Schrödinger equation *no alternative explanation of the origin of lattice instability is possible.*

6.3.2 The Peierls-Fröhlich Theorem. The Band Jahn-Teller Effect

The discussion in the previous section concerns dielectric (and semiconductor) systems in which the nondegenerate ground state is formed by a completely occupied valence band. In the case of Jahn-Teller crystals with orbitally degenerate centers, considered in Sect. 6.1, a simplifying assumption was made, neglecting the intercenter resonance interactions which broaden the electronic terms into bands. This assumption allows us to reduce the problem to the consideration of vibronic interactions in a separate elementary cell, and to include the translational symmetry of the crystal at a later stage.

However, in the general case the translational symmetry has to be taken into account from the very beginning. The appropriate group-theoretical analysis of the matrix elements of the operator of vibronic coupling for space groups and the consequent conclusion about the possible cooperative Jahn-Teller effect in crystals with degenerate electronic bands was performed in [6.110–112]. The mixing of electronic states with the same wave vector is produced by the limiting vibrations ($\boldsymbol{q} = 0$) (corresponding to a relative shift of the crystal sublattices) which in its final stage can be reduced to the local Jahn-Teller effect in a separate elementary cell [6.112]. In general, however, such a localization of the vibronic interaction is a consequence of the presence of localized chemical bonds, which are characteristic of dielectric crystals. Later in this section we consider another case, namely, a crystal with a relatively broad band, occupied only partly and thus having electrical conductivity. The delocalized nature of the metallic bonds determines the behavior (in the vibronic mixing) of phonons with nonzero wave vectors. As shown below, the vibronic interaction in these cases, too, plays the determinant role in the structural instability of the lattice.

Consider first the simplest example—a linear monatomic chain with one electron in the valence s orbital of each atom. The Hamiltonian of the system can be written in the form (6.2.1). Its phonon part has the same form (6.2.5) as in Sect. 6.2.1, with the distinction that there is only one acoustic branch of vibrations and the wave vector q is a scalar (wavenumber), $-\pi/a \le q \le \pi/a$ (a is the lattice constant). In the case where only the nearest-neighbor atoms interact,

$$\omega(q) = 2\sqrt{\frac{A}{M}}\sin(|q|a/2) \ , \tag{6.3.17}$$

where A is the constant of elastic coupling between the neighbors, and M is the nuclear mass.

The electronic Hamiltonian has the form (6.2.4), which is also analogous to that given in Sect. 6.2.1, with the distinction that here α labels two spin states of the electron, the one-electron energy levels do not depend on α, and the wave vector k is a scalar (wavenumber), $-\pi/a \le k \le \pi/a$. If the energy zero is taken as the electronic energy level of the atom, and the resonance interaction of only the nearest-neighbor atoms are taken into account, then

$$\varepsilon(k) = -2h_0\cos(ka) \ , \tag{6.3.18}$$

where h_0 is the appropriate resonance integral. Note that all the energy levels (6.3.18) are doubly degenerate with respect to the direction of the electron momentum, $\varepsilon(k) = \varepsilon(-k)$.

The operator of electron-phonon coupling differs somewhat from (6.2.8). First, the phonons cannot excite the electron into a state with the opposite spin. Second, as in Sect. 6.2.1, we neglect the dependence of the vibronic constant on the wave vector of the electrons. With these restrictions

$$H_{\text{int}} = \sum_{q,k}\sum_{\alpha} \Gamma(q)\sqrt{\frac{\hbar\omega(q)}{2N}}(b_q + b_{-q}^{\dagger})(a_{\alpha,k}^{\dagger}a_{\alpha,k-q} + a_{\alpha,k+q}^{\dagger}a_{\alpha,k}) \ . \tag{6.3.19}$$

Since the number of states in the band is $2N$, while the number of electrons is N, the band is half filled and the Fermi surface is at the middle of the band. *Peierls* [6.113] and *Fröhlich* [6.114] (see also [6.115]) noted that in this case the crystal is unstable with respect to spontaneous deformation of the lattice towards a doubling of its period (Fig. 6.8a). Under this antiferrodistortive distortion the conduction band splits into two subbands. The lowest subband is fully occupied and goes down in energy, whereas the upper one is empty and goes up in energy (the total energy thus being lowered), and this explains the origin of the distortion under consideration.

The doubling of the lattice period—the alternation of the bond lengths (Fig. 6.8)—can be described by the nuclear displacements $Q_n = (-1)^n Q = Q\exp(\pm i\pi n)$. Substituting this expression into (6.1.2) and using the identity

$$\frac{1}{N}\sum_n e^{ian(k-q)} = \delta_{k,q} \ ,$$

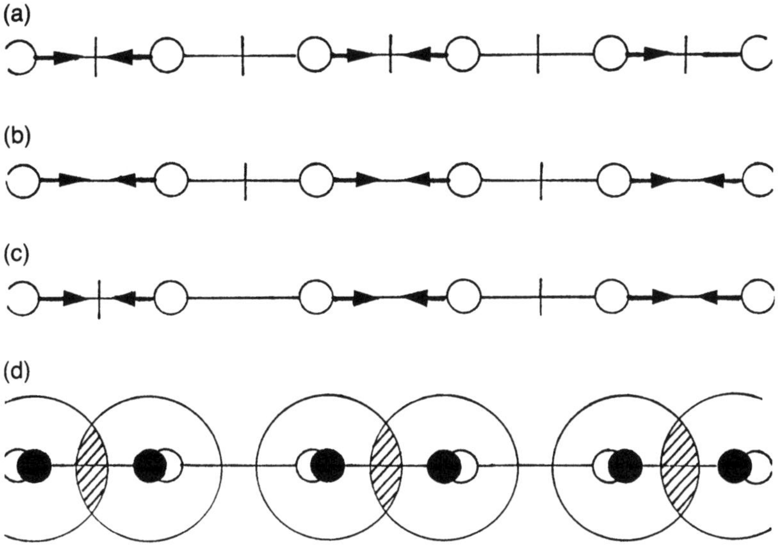

Fig. 6.8a–d. The Peierls transition in a one-dimensional lattice with a half filled band. (**a**) Doubling of the lattice period—antiferrodistortive ordering. The vertical lines separate the elementary cells. (**b**) Ferrodistortive isostructural distortions in a lattice with a doubled period. (**c**) Odd antiferro-distortive distortions in a lattice with a tripled period. (**d**) Formation of additional covalent bonds in the Peierls doubling of the lattice period. The rings denote the atomic s orbitals. The areas of increased overlap of nearest neighbor atomic orbitals due to the distortion are shown by hatching

we have

$$Q(q) = \sqrt{N} Q \delta_{q, \pm q_0} , \tag{6.3.20}$$

where $q_0 = 2k_F = \pi/a$; k_F is the Fermi momentum of the electrons. Then we write the operator (6.3.19) in the normal coordinates (6.2.6) and substitute (6.3.20), which means that we take into account only one mode doubling the lattice period, and this yields

$$H_{\text{int}} = VQ \sum_{k,\alpha} (a^\dagger_{\alpha,k} a_{\alpha,k-q_0} + a^\dagger_{\alpha,k} a_{\alpha,k+q_0} + a^\dagger_{\alpha,k+q_0} a_{\alpha,k} + a^\dagger_{\alpha,k-q_0} a_{\alpha,k}) , \tag{6.3.21}$$

where $V = \Gamma(q_0)\omega(q_0)$ is the local (elementary cell) vibronic constant for the active mode Q. The operator (6.3.21) mixes the state with a given value of the wavenumber k with another electronic state from the first Brillouin zone with wave-number $k + \pi/a$ or $k - \pi/a$. The appropriate electronic matrix

$$H_e + H_{\text{int}} = \begin{pmatrix} \varepsilon(k) & VQ \\ VQ & \varepsilon(k + 2k_F) \end{pmatrix} \tag{6.3.22}$$

has a characteristic form inherent to systems with the pseudo-Jahn-Teller effect

(Sect. 3.4). Thus the electron-phonon interaction in a linear chain with a half-filled electronic band, taking account of one active vibrational mode with wavenumber $\pm 2k_F$, is reduced to the pseudo-Jahn-Teller effect for different pairs of one-electron states mixed by one nondegenerate vibration. The energy gap between these states is $4h_0|\cos(ka)|$. Note that there are always two mixing states with an energy gap of zero, and for them the mixing is equivalent to the Jahn-Teller effect. These are the states with opposite momenta at the Fermi surface, $k = k_F$ and $k = -k_F$. There are also quite a number of pairs of mixing states with a very small energy gap between them (the states near the Fermi surface).

The eigenvalues of the matrix (6.3.22), the one-electron contributions to the adiabatic potential, are [cf. (6.2.32]

$$\varepsilon_{\pm}(k) = \tfrac{1}{2}[\varepsilon(k) + \varepsilon(k + 2k_F)] \pm \sqrt{\tfrac{1}{4}[\varepsilon(k) - \varepsilon(k + 2k_E)]^2 + V^2Q^2} \ . \quad (6.3.23)$$

Considering the symmetry of the electronic band with respect to the Fermi surface ε_F,

$$\varepsilon(k + 2k_F) - \varepsilon_F = -[\varepsilon(k) - \varepsilon_F] \ ,$$

the expression (6.3.23) can be reduced to a simpler one:

$$\varepsilon_{\pm}(k) = \varepsilon_F \pm \sqrt{[\varepsilon_F - \varepsilon(k)]^2 + V^2Q^2} \ . \quad (6.3.24)$$

It follows, from comparison with (3.4.2), that the depth of the occupied energy level under the Fermi surface is half the energy gap between the mixing states in the pseudo-Jahn-Teller effect.

Neglecting the Coloumb and exchange corrections, the adiabatic energy of the ground state of the crystal $\varepsilon_0(Q)$ is the sum of the Hartree-Fock one-electron energies of the occupied states and the elastic energy of the active mode:

$$\varepsilon_0(Q) = \frac{N}{2}\omega_0^2 Q^2 + 2 \sum_{k=-k_F}^{k_F} \varepsilon_{-}(k) \ ,$$

where $\omega_0 = \sqrt{2}\omega(2k_F)$. Substituting here the expression for $\varepsilon_{-}(k)$ from (6.3.24) and introducing the density of one-electron states in the band

$$\varrho_{el}^{(0)}(\varepsilon) = \frac{1}{N}\sum_k \delta(\varepsilon - \varepsilon(k)) \ , \quad (6.3.25)$$

we obtain

$$\varepsilon_0(Q) = N\varepsilon_F + \frac{N}{2}\omega_0^2 Q^2 - 2N \int_{\varepsilon_{min}}^{\varepsilon_F} \varrho_{el}^{(0)}(\varepsilon)\sqrt{(\varepsilon_F - \varepsilon)^2 + V^2Q^2}\, d\varepsilon \ , \quad (6.3.26)$$

where ε_{min} is the energy at the bottom of the band. The expression (6.3.26) for $\varepsilon_0(Q)$ is a rather complicated function with a nonanalytical behavior at $Q = 0$

arising due to the superposition of the Jahn-Teller and pseudo-Jahn-Teller effects at this point (the energy of the states at the Fermi surface has a discontinuity in the derivative at $Q = 0$).

Taking the energy zero at the electronic energy level in the atoms, and considering the resonance interaction of the nearest-neighbor atoms only, we have $\varepsilon_F = 0$, and hence $\varepsilon(k)$ can be substituted from (6.3.18) into (6.3.25). Passing from summation over k to the appropriate integration we obtain

$$\varrho_{el}^{(0)}(\varepsilon) = \frac{\theta(4h_0^2 - \varepsilon^2)}{\pi\sqrt{4h_0^2 - \varepsilon^2}} , \tag{6.3.27}$$

where $\theta(z)$ is a single step function. Substituting the density of states (6.3.27) into (6.3.26), assuming in accordance with (6.3.18) that $\varepsilon_{min} = -2h_0$, and performing the integration we come to the following expression for the adiabatic potential:

$$\varepsilon_0(Q) = \frac{N}{2}\omega_0^2 Q^2 - \frac{N}{\pi}\sqrt{4h_0^2 + V^2Q^2}\, E\left(\frac{2h_0}{\sqrt{4h_0^2 + V^2Q^2}}\right) , \tag{6.3.28}$$

where $E(z)$ is the full second-order elliptic integral,

$$E(z) = \int_0^{\pi/2} \sqrt{1 - z^2 \sin^2\varphi}\, d\varphi .$$

Using the expansion of elliptic integrals near the point $z = 1$,

$$E(z) \approx 1 + \frac{1}{2}\left(\ln\frac{4}{z'} - \frac{1}{2}\right) z'^2 ,$$

where $z' = \sqrt{1 - z^2}$, we obtain for the function $\varepsilon_0(Q)$ near the point $Q = 0$ (corresponding to the high symmetry nuclear configuration)

$$\varepsilon_0(Q) = E_0 + \frac{N}{2}\left(\omega_0^2 - \frac{V^2}{\Delta_{eff}}\right) Q^2 + \frac{2Nh_0}{\pi}\left(\frac{VQ}{2h_0}\right)^2 \ln\left|\frac{VQ}{2h_0}\right| . \tag{6.3.29}$$

Here $E_0 = -4Nh_0/\pi$ is the electronic energy in the undistorted lattice, and $\Delta_{eff} \approx 1.6656\, h_0$.

The analysis of the expression (6.3.29) shows that for arbitrary values of the vibronic constant the branches of the curve $\varepsilon_0(Q)$ in the close vicinity of the point $Q = 0$ are turned down, while the derivative of $\varepsilon_0(Q)$ at this point is zero. At $V^2 \geq \omega^2 \Delta_{eff}$ an additional instability occurs determined by the first terms of (6.3.29). Nevertheless, at large values of $|Q|$ the logarithmic term in (6.3.29) becomes dominant, and the wings of the curve $\varepsilon_0(Q)$ turn up. Being thus similar in shape to the lowest sheet of the adiabatic potential for systems with the pseudo-Jahn-Teller effect (Fig. 3.8b), the curve $\varepsilon_0(Q)$ obtained here is free from the restriction of the type (3.4.3) required for an instability of the high symmetry con-

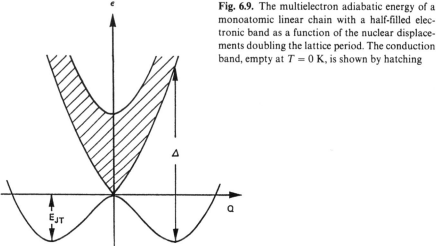

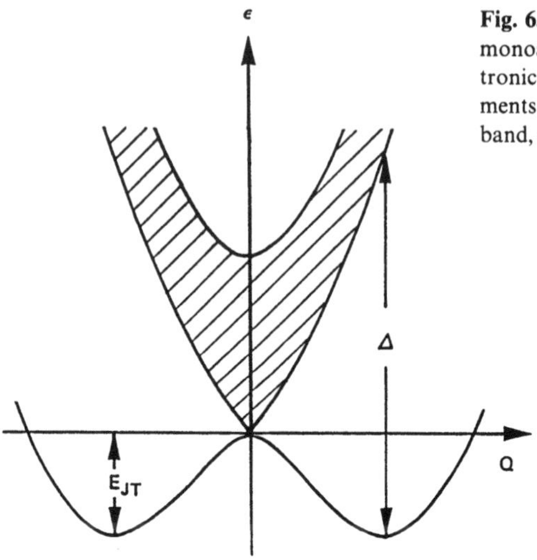

Fig. 6.9. The multielectron adiabatic energy of a monoatomic linear chain with a half-filled electronic band as a function of the nuclear displacements doubling the lattice period. The conduction band, empty at $T = 0$ K, is shown by hatching

figuration in the usual molecular case. This result (the instability at $Q = 0$ for arbitrary values of the vibronic constant and other lattice parameters) is a characteristic feature of the Jahn-Teller effect, and arises here due to the significant contribution of the mixing of states with energies at or near the Fermi surface.

The shape of the curve $\varepsilon_0(Q)$ corresponding with the definition (6.3.26) is shown schematically in Fig. 6.9. The adiabatic potential of the ground state has a symmetrical double-well shape characteristic of Jahn-Teller systems. As in the usual Jahn-Teller cases, the distortion of the lattice at the minima $\pm Q_0$ and the Jahn-Teller stabilization energy E_{JT} increase with the vibronic constant V. However, in the case under consideration this dependence is more complex. As a new feature, the dependence of the steepness of the potential barrier between the minima on the density of states near the Fermi surface has to be noted. Indeed, as follows from (6.3.26), the most important contribution to the instability at $Q = 0$ determined by the degenerate or quasi-degenerate states with $\varepsilon \approx \varepsilon_{\text{F}}$, depends on $\varrho_{\text{el}}^{(0)}(\varepsilon_{\text{F}})$. Since, as mentioned at the beginning of Sect. 6.3, the free energy of the crystal at $T = 0$ K coincides with $\varepsilon_0(Q)$, the double-well adiabatic potential means that the low temperature phase corresponds to a distorted lattice with a doubled period (Fig. 6.8). Thus a linear monatomic lattice with a half-filled electronic band with respect to distortions of wavenumber $q_0 = 2k_{\text{F}}$ when vibronic interactions are taken into account is unstable. This statement is known as the *Peierls-Fröhlich theorem*, while the corresponding structural phase transition[3] is the *Peierls transition*.

[3] We neglect here the role of fluctuations which are known to quench phase transitions in one-dimensional systems. However, real systems are never ideally one-dimensional. A weak interaction between the chains or any spatial complication of the chain itself usually removes the fluctuation restriction.

Note that the vibronic constant V, as well as the resonance integral h_0, is proportional to the overlap integral S of the atomic orbitals of the nearest-neighbor lattice sites, $V \sim S$, $h_0 \sim S$. If $S \to 0$, then the electronic band becomes infinitely narrow, and the vibronic constant tends to zero. In other words, in the case of infinitely narrow bands the Peierls instability is absent. Conversely, the broader the partly occupied electronic band, the greater the vibronic constant, and hence the greater the Jahn-Teller stabilization energy in the minima configurations. Therefore the phenomenon under consideration is often called *the band Jahn-Teller effect.*

Similar to the cases discussed above, the gain of energy in the Peierls transition is due to the formation of new covalent bonds by distortion (by the atoms pairing in the linear chain, Fig. 6.8d). This results from the one-electron Hamiltonian. The diagonalization of the matrix (6.3.21) leads to the formation of bonding and antibonding orbitals $\psi_{\pm}(k) = (\psi_k \pm \psi_{k-q_0})/\sqrt{2}$. Transformed back to localized orbitals by (6.2.3) they result in bonding "molecular" orbitals of pairs of atoms, shown in Fig. 6.8d. Note, that here, as above, while the cooperative properties of the chain are very important for the phase transition, the instability itself of the high symmetry configuration is determined not by cooperative, but by short range local forces of chemical bonding.

The change of the electronic structure of the crystal due to the Peierls transition can be interpreted as the occurrence of a standing longitudinal wave of electron density with wavenumber $q_0 = 2k_F$ in the low symmetry phase. Simultaneously, in the electronic spectrum a forbidden band of width $\Delta \gtrsim E_{JT}$ arises (Fig. 6.9). At $T = 0$ K, when only the ground state is occupied, the crystal becomes dielectric. Therefore the Peierls transition is also a metal-dielectric or a metal-semiconductor transition, depending on the size of the energy gap Δ. In the treatment of the Peierls transition with a doubling of the lattice period one can start with the extended elementary cell containing two atoms (Fig. 6.8b). In this case the transition is equivalent to a ferrodistortive ordering of local totally symmetric distortions of the cells, i.e., the transition is isostructural. However, the pseudo-Jahn-Teller origin of the Peierls instability is more obvious when one starts with the elementary cell containing three atoms (Fig. 6.8c). In each elementary cell of three atoms one can form three molecular orbitals, two even and one odd. They mix under the symmetrized odd displacements shown in Fig. 6.8c. As a result we have the pseudo-Jahn-Teller effect in a linear triatomic system of the type *BAB* for three states mixed by one odd vibration, quite similar to the two-level case considered at the beginning of Sect. 3.4.1. If the vibronic constant is large enough, the pseudo-Jahn-Teller effect is strong, and the high symmetry configuration in the elementary cell becomes unstable with respect to the odd distortion under consideration. By interaction through the phonon field the local distortions form an antiferrodistortive ordering (Fig. 6.8c). The active mode here is the odd optical vibration with wavenumber $q = 0$. In this sense the Peierls transition with a doubling of the lattice period is similar to the antiferroelectric ordering considered in Sect. 6.2.3.

The peculiarity of the band Jahn-Teller effect here consists in that the intercell interaction, which broadens the local energy levels into bands, is determined by the same parameter h_0 that determines the energy gap between the bonding and antibonding orbital of the elementary cell itself. Therefore the forbidden band Δ between the valence and conduction bands in the high symmetry configuration is "accidentally" equal to zero, and some of the mixing states are degenerate. This results in the combination mentioned above of the pseudo-Jahn-Teller effect with the Jahn-Teller one. The analogy between the Peierls transformation and antiferroelectric phase transitions was noted by *Anderson* and *Blount* [6.116].

The arguments resulting in the Peierls instability are applicable also to one-dimensional metals with the conduction band other than half occupied. It is clear that in this case a distortion of the lattice can be found for which the change of the lattice period results in a band splitting such that its occupied part goes down in energy, leading to an energy gain. For instance, if the electronic band is filled up to the states with the Fermi momentum $k_F = \pi/6a$, the active mode (in the sense discussed above) is a normal vibration having the wavenumber $q_0 = \pm 2k_F$ which triples the lattice period. For other electronic band occupancies one can arrive, in particular, at transformations into structures with periods that are incommensurate with the initial one.

Following Peierls [6.113], one can assume that a similar instability of vibronic origin is inherent to two- and three-dimensional metallic systems. The noncoincidence of the shape of the Wigner-Seitz cell and the Fermi surface in combination with a strong enough vibronic coupling and great density of states at the Fermi surface may be the reason for the rather complicated crystal structures often observed in metals of relatively simple composition (e.g. see [6.117]).

The details of what can actually happen in these cases are discussed in [6.118]. Usually the observed structure is a result of one of several ways of stabilizing the initial (reference) structure by its deformation. Some special cases of layered crystals of the type PbFCl (ZrSiS, BiOCl, Co_2Sb, Fe_2As) were investigated in [6.119–121]. Three possible ways to stabilize a square lattice suggested in [6.122] are presented in Fig. 6.10. A one-atomic cubic lattice was shown to have no less than 36 different means of stabilization by lattice distortion [6.123, 124]. Two of these, shown in Fig. 6.11, correspond to black phosphorus and arsenic structures. There are also other possibilities.

The deformation pulls down some of the states from the Fermi level region. But because the shape of the Wigner-Seitz cell does not coincide with the Fermi surface, as mentioned above, it may not be possible to remove all the states from the Fermi level region. Some of the electrons remain in the conductive band and some of the holes remain in the valence band. Therefore the material may still be a conductor.

The one-particle Hamiltonian (6.2.4) does not take into account the effects of electron correlation. The latter, by means of coupling the orbital and spin states of different lattice sites, may lead to an ordering in the electronic subsystem accompanied by a wave of charge density [6.125, 126]. The Mott transitions,

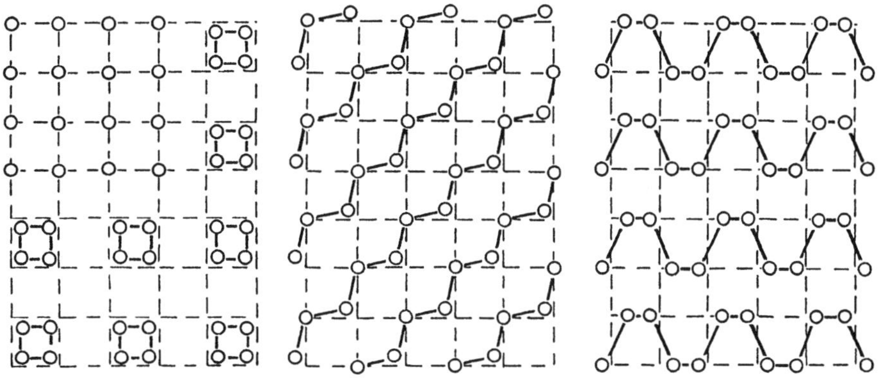

Fig. 6.10. Three examples of alternating bond lengths in a monoatomic square-planar lattice. (After [6.122])

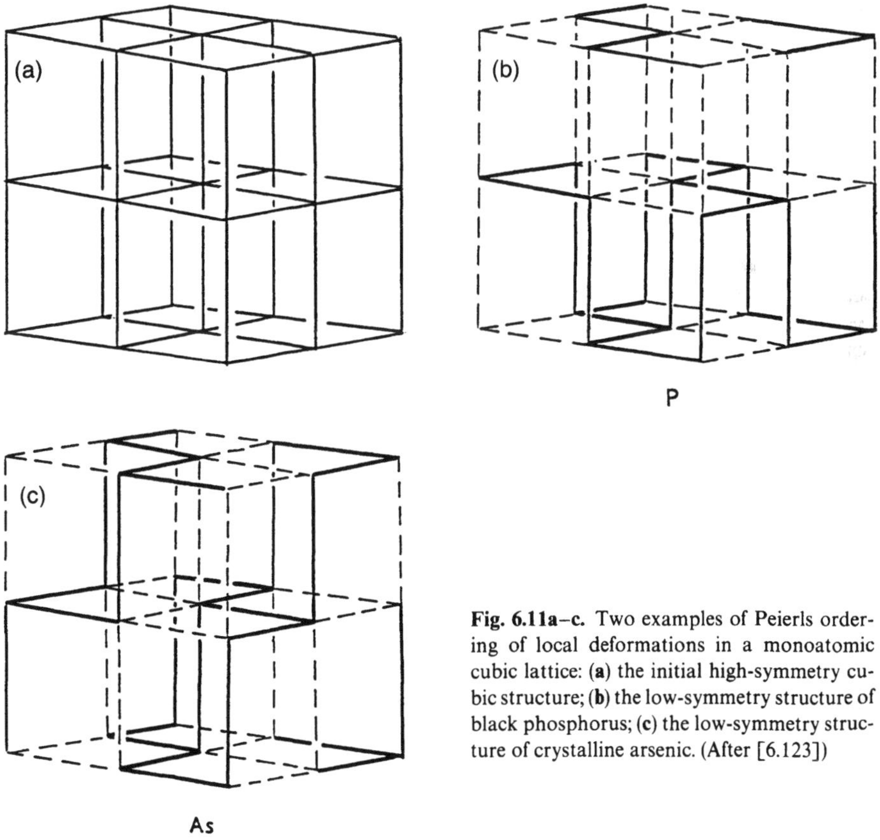

Fig. 6.11a–c. Two examples of Peierls ordering of local deformations in a monoatomic cubic lattice: (**a**) the initial high-symmetry cubic structure; (**b**) the low-symmetry structure of black phosphorus; (**c**) the low-symmetry structure of crystalline arsenic. (After [6.123])

as well as bond alternation in conjugated hydrocarbons are examples of this kind of phenomenon. Also related to these phenomena is the Bardeen-Cooper-Schrieffer (BCS) interaction resulting in Cooper pairing in superconductivity, as well as the antiferromagnetic instability giving rise to the giant spin wave in crystals [6.127].

In all these cases the phase transition in the electronic subsystem is accompanied by the formation of an energy gap in the electronic spectrum that may strongly influence the vibronic interactions. For instance, the Mott transition, trasnforming a metal into a dielectric, removes the degeneracy of the one-electron states at the Fermi surface, and hence the special combination of the Jahn-Teller and pseudo-Jahn-Teller effects mentioned above disappears. Here a rather complex problem of the joint consideration of both the vibronic interaction and electronic correlation effects arises. The use of the mean field approximation [6.128–131] for the solution of this problem shows that these two instabilities cannot coexist, and under certain conditions the Peierls transition may be quenched. However, the mean field approximation is too rough here. Variational calculations [6.132–134] testify that the Peierls instability and electronic correlations are not always competing effects, and sometimes the correlations may even enhance the transition to the Peierls state. This conclusion is also confirmed by perturbation theory with respect to the mean field [6.135] and to the constant of electron-electron interaction [6.136], as well as by Monte Carlo calculations [6.137], and exact computations for finite systems [6.138].

On the other hand, taking the electron-phonon interaction into account, for instance in second-order perturbation theory, results in an effective electron-electron interaction of the types mentioned above. Therefore the Peierls instability causes an instability in the electronic subsystem, too. For example, it was shown in [6.125] that the Peierls instability is always accompanied by BCS instability, i.e., by the occurrence of Cooper pairs; and vice versa, allowing for the electron-phonon interaction, the phase transition in the electronic subsystem is inevitably accompanied by a reconstruction of the nuclear configuration. In these circumstances the question whether the interelectronic interaction or the electron-phonon (vibronic) one is responsible for the phase transition, loses its meaning. But one has to note that without taking account of vibronic interaction, the phase transition in the nuclear subsystem (the structural phase transition) cannot take place.

In the last 20 years the investigation of one-dimensional structures has grown into a wide ranging branch of scientific activity, mostly due to the Little hypothesis about the possibility of high-temperature superconductivity in some one-dimensional system [6.139]. For a long time the efforts of researchers were directed to quasi-one-dimensional compounds like V_3Si having A-15 structure, as well as to organic compounds, like TTF-TCNQ[4], which demonstrate struc-

[4] TTF is tetrathiofulvalene, $[(C_3S_2)HCH_3]_2$, and TCNQ is tetracyanoquinondimethane, $(CN)_2C(C_6H_6)C(CN)_2$

tural phase transitions and anomalies in the temperature dependence of their electrical properties. Different models developed for the description of the structural transformations in these compounds (the model of Labbé, Friedel and Barišić [6.140–145], the model of Gor'kov [6.146–148], etc.) employ, in one form or another, the above ideas based on the vibronic nature of the Peierls instability. In particular, they link the spontaneous deformation of the lattice with the high density of states at the Fermi surface.

A striking example of a quasi-one-dimensional conducting system which undergoes a Peierls transition into a dielectric phase is the *trans*-modification of the polyacetylene polymer $(CH)_x$. There are two theoretical models describing the properties of this polymer well [6.149–151]. Both are based on the vibronic mechanism of the structural instability of the chain *trans*-$(CH)_x$. The reader interested in more details on the problems discussed in this section can consult the review articles [6.7, 152–158] and [Ref. 6.159; Sect. 4.4].

6.3.3 General Model of Structural Transformations in Condensed Matter

The discussion given above, in particular the proof of the vibronic origin of the instability of high symmetry configurations of any polyatomic system (see also Sect. 3.4.3), allows a more general view of all structural phase transformations in condensed matter [6.64, 65]. From this point of view the phase transitions gas → liquid (gas condensation), liquid → crystal (liquid crystallization) and structural phase transitions in crystals, which take place consecutively on decreasing the temperature, form a set of similar electronically controlled sharp symmetry breaks accompanied (and determined) by the formation of new chemical bonds (that are forbidden in the previous phase at higher symmetry). The effect of the formation of new bonds by distortion is described by the operator of vibronic interaction and therefore the micro-mechanism of all these symmetry breaks is the same: the Jahn-Teller or pseudo-Jahn-Teller effect.

Denote the high symmetry and low symmetry (at the Jahn-Teller minima) configurations by $Q = 0$ and $Q = Q_0$, respectively. Then the Jahn-Teller stabilization energy $E_{JT} = \varepsilon(0) - \varepsilon(Q_0)$ and, similarly, the Jahn-Teller stabilization entropy $\Delta S = S(0) - S(Q)$. Since the entropy of the high symmetry configuration is always larger than that of the low symmetry one, $\Delta S > 0$. Introducing

$$T_0 = \frac{E_{JT}}{\Delta S} \tag{6.3.30}$$

we find from the free energy difference $\mathscr{F}(0) - \mathscr{F}(Q) = (T_0 - T)\Delta S$ that at low temperatures, $T < T_0$, the low symmetry configuration at the minima of the Jahn-Teller (or pseudo-Jahn-Teller) adiabatic potential is more stable, whereas at $T > T_0$ the high symmetry one is favorable. Macroscopically this means that at $T = T_0$ a symmetry break, known as a structural phase transition, takes place (E_{JT} includes the interactions of the Jahn-Teller units).

Thus the Jahn-Teller (pseudo-Jahn-Teller) effect may serve as a trigger mechanism for symmetry breaking and structural phase transitions. Since it has been proved that the pseudo-Jahn-Teller effect is the only possible source of instability of the high symmetry configuration of any polyatomic system (the Jahn-Teller effect may be considered as a limiting case of the pseudo-Jahn-Teller effect at $\Delta = 0$), this mechanism of symmetry breaking is unique and serves as a base for all kinds of structural phase transitions. It means that they are controlled by the electronic states (ground and low-lying excited) through the vibronic coupling in the way predicted by the Jahn-Teller (pseudo-Jahn-Teller) effect. Visually, the phenomenon as a whole is due to the formation of new covalent bonds by distortion, whereas at higher temperature the entropy contribution favors the high symmetry configuration.

So far only structural phase transitions in crystals have been considered in detail. As shown earlier in this chapter, the micromechanism of these transitions is of Jahn-Teller (pseudo-Jahn-Teller) origin. Recently the liquid–solid and gas–liquid transitions were also treated in this way [6.64, 65, 160]. This became possible due to progress in understanding the nature of the liquid state. In accordance with [6.161, 162] the liquid state, as a rule, consists of mobile and relatively unstable symmetrical polyatomic aggregates that in the case of a monatomic liquid have icosahedral symmetry.

At sufficiently high temperatures the matter is in the atomic gas state with the highest symmetry. Consider 13 identical noninteracting atoms (a virtual icosahedral cluster with one atom in the center). The Longuet-Higgins symmetry group of such a system is $G = \pi(13) \times C_i \times R(3)$, where $R(3)$ is the group of three-dimensional rotations of a free atom, $\pi(13)$ is the group of permutations of 13 identical atoms, and C_i is a group of two elements, the identity and inversion. The group G possesses high-dimensional irreducible representations, in particular, 13-fold ones. To this representation appertain, for instance, the molecular orbitals combined from atomic s orbitals, provided the interatomic interactions can be neglected. Due to the vibronic interactions these highly degenerate states are unstable with respect to the nuclear displacements that transform the system into an icosahedron (or a system with lower symmetry) generating covalent bonds between the atoms. In this process the 13-fold degenerate energy levels mentioned above split, the bonding and antibonding orbitals going down and up, respectively, and the potential energy lowers, provided the number of electrons at the bonding orbitals is larger than at the antibonding ones. This process is quite similar to that taking place in the Jahn-Teller effect for chemically bonded systems with the distinction that in the latter case new covalent bonds are formed by distortion in addition to those already existing (and in a quantity sufficient to overcome the losses in energy due to the distortion) whereas in the formation of molecules (aggregates of atoms) chemical bonds are formed by displacements between noninteracting atoms (or simpler molecules). It is clear that when the cooperative effects of interactions between the icosahedral aggregates (which depend on the density of the matter) and the contribution of the entropy factor are included, the symmetry break takes place at a certain temperature as a sudden process of condensation of the gas into the liquid state. If the interaction between the icosahedral clusters, or aggregates of lower symmetry, is not strong enough, then the transformation of the gas of atoms into a gas of molecules takes place first, and then at the next symmetry break at lower temperatures (controlled by the appropriate electronic states), the transition to the liquid phase follows. If on the other hand the interaction between the icosahedra is very strong then the transition of the gas of atoms directly into the solid state may occur.

Consider now the liquid–solid transition [6.160]. The local nearest-neighbor ordering in the liquid state preserves high symmetry local formations (clusters), say, of icosahedral symmetry, which due to the Jahn-Teller or pseudo-Jahn-Teller effect may be unstable with respect to low symmetry nuclear displacements. In this case the adiabatic potential of the molecular aggregates possesses several equivalent minima. Localization of nuclear dynamics in one of these minima corresponds to one of the equivalent low symmetry distortions of the icosahedron accompanied by the formation of new chemical bonds that are forbidden in the high symmetry nuclear configuration. At high temperatures, $T > T_0$, when the population of excited vibronic states is large the dynamics of the system may be described by localized vibrations in the minima accompanied by tunneling between the neighboring equivalent minima and over-barrier reflections. The distorted nuclear configuration at the minimum periodically changes its orientation in space in accordance with the symmetries of the other equivalent minima, and the averaged nuclear configuration remains at high symmetry, say, icosahedral. Note that it is impossible to compose a crystal from icosahedral elementary cells, therefore at high enough temperature the high symmetry configuration is preferable due to the entropy contribution to the free energy and the system remains in the amorphous liquid state.

On cooling the system the lower symmetry configuration becomes lower in free energy. The interaction between the local aggregates in the liquid state result in a sudden symmetry break at $T = T_0$ describing the liquid $\rightarrow$ solid phase transition. The details of this process depend on the relation between the tunneling splitting energy gap in the icosahedron, δ, and the interaction energy between the icosahedra in the liquid state, E_i. If $\delta < E_i$ then by cooling the system the tunneling transitions between different equivalent minima of the adiabatic potential of the icosahedron can be quenched, and the latter become frozen in one of the minima before the interaction between the icosahedra can order them into a crystal lattice. In this case we come to a phase of disordered frozen Jahn-Teller local distortions, which means the amorphous state of a Jahn-Teller glass. In the opposite case of $\delta > E_i$ the ordering of the local Jahn-Teller distortions produces a molecular field which locks every molecular aggregate, say, in the same Jahn-Teller minimum (ferrodistortive ordering), and here a direct transition from the liquid state to the crystal one takes place.

In the case when the icosahedron is rather stable (no Jahn-Teller effect or weak pseudo-Jahn-Teller effect) only the transition to the glass state may take place.

Using the simple formula (6.3.30) we can obtain some further simple relations. Indeed, in the case when the transition leads to the formation of the amorphous Jahn-Teller glass phase,

$$T_G \cong E_i/k \;, \tag{6.3.31}$$

whereas in the case of crystallization (melting)

$$T_M \cong E_{JT}/\Delta S \;, \tag{6.3.32}$$

where ΔS is the entropy change on crystallization of the liquid.

In [6.160] some estimates of E_{JT}, E_i, and ΔS as functions of the masses of the atoms and the correlation radius of the chemical interaction of the atoms in the crystal state were obtained. The correlation radius was evaluated by means of electronic structure calculations in the Slater-Johnson X_α scattered wave approximation (SCF X_α-SW). As a result the melting and vitrification temperatures of a series of monatomic crystals (Ar, Al, Cd, Co, Pb, Fe, Ru, Si, W) were estimated with fairly good agreement with experimental data.[5]

Finally consider briefly some structural phase transformations which at first sight seem to have no relation to the Jahn-Teller effect. These are transitions in which the crystal changes not its symmetry, but the interatomic distances, i.e., isostructural transitions. Such a transition arises due to the mixing of electronic states (ground and excited) of the same symmetry by totally symmetric nuclear displacements. Since in this case the excited state, though having the same symmetry as the ground state, may have a different charge distribution, its admixture to the ground state can result in better covalent bonding and hence in a lowering of the potential energy, $\varepsilon_0(Q)$. For instance, the $\alpha \to \gamma$ isostructural transition in the Ce crystal [6.163] results in a change of the delocalization of the f electrons (with the appropriate consequences for the observable quantities). The totally symmetric nature of the active nuclear displacements should not delude the reader about the possibility of the pseudo-Jahn-Teller effect playing a part in this case. In Sect. 6.3.2 we have shown that passing to an extended elementary cell allows us to consider the isostructural ferrodistortive ordering (Fig. 6.8b) as an antiferrodistortive ordering of odd low symmetry distortions (Fig. 6.8c). Similarly, any isostructural transition, by passing to an extended elementary cell, can be considered as a cooperative pseudo-Jahn-Teller effect accompanied by an ordering of low-symmetry local distortions.

Special interest should be paid to the high temperature superconductivity revealed recently in Jahn-Teller crystals [6.164–167]. The relation between

[5] Not all the assumptions of [6.160] are clear, but the main idea of the vibronic origin of the liquid-solid phase transition seems to be quite reasonable.

superconductivity and the Jahn-Teller effect is determined by the nonadiabatic nature of the electron-phonon interaction needed for the Cooper pairing of the electrons [6.168–177]. This is also confirmed by the fact that the superconductivity transition in Jahn-Teller crystals is accompanied by a structural phase transition [6.164, 165] which, as stated above, must be of vibronic nature (see also the discussion at the end of Sect. 6.3.2).

To conclude this chapter, we emphasize once more that in all the structural phase transitions considered above, there is a common feature. The formation of additional covalent bonds lowers the local symmetry of the system and also the potential energy, provided the gain of energy in the new bond is larger than its loss in the distortion of the previous ones (if any). Just this effect is taken into account by the operator of vibronic interactions. In all cases the instability of the high symmetry phase is determined by the Jahn-Teller or pseudo-Jahn-Teller effect, i.e., by the special electronic structure of the high symmetry configuration, while the temperature of the phase transition depends also on the parameters of interaction of the vibronically unstable centers. The general approach to the problem suggested here gives a key to the understanding of these processes as a whole, allows us to formulate the criteria for different structural transformation in condensed matter, and opens a way to their more detailed investigation.

Appendix A. Expansion of the Full Vibrational Representation in Terms of Irreducible Ones

As mentioned in Sect. 2.2.1, the general principles of construction of symmetrized nuclear displacements and the expansion of the full vibrational representation in terms of irreducible ones are given in manuals on group theory and on molecular vibrations (see, for instance, [A.1]). However, in these books the atomic structure and symmetry of the molecule is assumed to be known a priori; since for each symmetry group there may be an infinite number of different molecular structures, the expansion of the full vibrational representation into irreducible ones has to be carried out for each molecule individually.

Jahn and *Teller* [A.2] have solved this problem in a general way by reduction of the infinite number of different molecular structures to a finite number of cases. In Table A.1 the results [A.2] are presented for all possible molecular structures. To explain how it was obtained let us first give some definitions.

Identical atoms in a molecule are equivalent if under symmetry operators of the corresponding point group G_0 they transform into each other. The set of equivalent atoms forms a transitive set. The number of transitive sets can be larger than the number of kinds of atoms, i.e., there may be several transitive sets formed by atoms of the same kind. For example, in the diborane molecule B_2H_6 having D_{2h} symmetry there are two transitive sets of hydrogen atoms. The first one contains four atoms that occupy symmetric positions on the plane of symmetry σ_h, while the second set contains the remaining two hydrogen atoms on the perpendicular plane σ_v. By definition, there are no symmetry operations in the group G_0 which transfer the atoms of one transitive set into positions of the atoms of other transitive sets.

If an atom is placed at an arbitrary point of space, and then by symmetry operations of the group G_0 is transferred to new positions, then all these new positions, if occupied by identical atoms, form a molecule with G_0 symmetry. The resulting molecule depends on the initial position of the atom from which the procedure started. If the starting point does not lie on a symmetry axis or plane of the group G_0, then the symmetry transformation results in a new point which belongs to the same transitive set, the number of equivalent positions (atoms) being equal to the order of the group G_0. But if the initial point is on one of the symmetry elements (axes and planes) or one of their intersections, then some of the symmetry operations do not create new points, and the resulting molecule of G_0 symmetry has a smaller number of atoms.

For each point group G_0 there are a finite number of transitive sets, which differ, in principle, in the number of equivalent atoms. An example is given in Fig. 3.5, where three types of transitive sets of C_{3v} symmetry are shown (Sect. 3.2.2). Any molecular structure can be represented by several transitive sets chosen from those possible for its symmetry group G_0. Therefore the problem under consideration will be solved if the expansion of the full vibrational representation into irreducible ones for each transitive set of the given symmetry group is obtained, and this has been done by *Jahn* and *Teller* [A.2].

The first coloumn of Table A.1 gives the notation for the group. Coloumns II and III contain the classifications in terms of irreducible representations of translational and rotational degrees of freedom, respectively. Column IV indicates the different transitive sets, column V gives the number of atoms in each transitive set, and coloumn VI indicates the symmetry elements (if any) on which the equivalent atoms lie. Finally, coloumn VII contains the expansion of the complete set of nuclear displacements, including translational, rotational and internal motions, into irreducible representations. Subtracting the representations of the translational and rotational degrees of freedom from those listed in coloumn VII, one obtains the irreducible representations of the full vibrational representation for the transitive set in question.

Table A.1 Reduction of the complete set of nuclear displacements into irreducible representations for all possible molecular structures

I	II	III	IV	V	VI	VII
$C_{\infty,v}$	$A_1 + E_1$	E_1	a	1	All	$A_1 + E_1$
$D_{\infty,h}$	$A_{1u} + E_{1u}$	E_{1g}	a	2	C_∞, σ_v	$A_{1g} + A_{1u} + E_{1g} + E_{1u}$
			b	1	All	$A_{1u} + E_{1u}$
C_{2p+1}	$A + E_1$	$A + E_1$	a	$2p + 1$	None	$3(A + E_1 + \cdots + E_p)$
			b	1	C_{2p+1}	$A + E_1$
C_{2p}	$A + E_1$	$A + E_1$	a	$2p$	None	$3(A + B + E_1 + \cdots + E_{p-1})$
			b	1	C_{2p}	$A + E_1$
D_{2p+1}	$A_2 + E_1$	$A_2 + E_1$	a	$2(2p + 1)$	None	$3(A_1 + A_2) + 6(E_1 + E_2 + \cdots + E_p)$
			b	$2p + 1$	C_2	$A_1 + 2A_2 + 3(E_1 + E_2 + \cdots + E_p)$
			c	2	C_{2p+1}	$A_1 + A_2 + 2E_1$
			d	1	All	$A_2 + E_1$
D_{2p}	$A_2 + E_1$	$A_2 + E_1$	a	$4p$	None	$3(A_1 + A_2 + B_1 + B_2)$ $+ 6(E_1 + E_2 + \cdots + E_{p-1})$
			b	$2p$	C_2	$A_1 + 2A_2 + B_1 + 2B_2$ $+ 3(E_1 + E_2 + E_3 + \cdots + E_{p-1})$
			c	$2p$	C_2	$A_1 + 2A_2 + 2B_1 + B_2$ $+ 3(E_1 + E_2 + \cdots + E_{p-1})$
			d	2	C_{2p}	$A_1 + A_2 + 2E_1$
			e	1	All	$A_2 + E_1$
$C_{2p+1,i} = S_{2(2p+1)}$	$A_u + E_{1u}$	$A_g + E_{1g}$	a	$2(2p - 1)$	None	$3(A_g + A_u + E_{1g} + E_{1u} + \cdots + E_{pg} + E_{pu})$
			b	2	C_{2p+1}	$A_g + A_u + E_{1g} + E_{1u}$
			c	1	All	$A_u + E_{1u}$

$C_{2p,i} = C_{2p,\mathrm{h}}$	$A_u + E_{1u}$	$A_g + E_{1g}$	a	$4p$	None	$3(A_g + A_u + B_g + B_u + E_{1g} + E_{1u} + \cdots + E_{p-1,g} + E_{p-1}$
			b	$2p$	σ_{h}	$2A_g + A_u + \begin{Bmatrix} B_g + 2B_u \text{ (odd } p) \\ 2B_g + B_u \text{ (even } p) \end{Bmatrix} + 2E_{1u} + E_{1g} + 2E_{2g} + E_{2u} + \cdots + \begin{Bmatrix} 2E_{p-1,g} + E_{p-1,u} \text{ (odd } p) \\ E_{p-1,g} + 2E_{p-1,u} \text{ (even } p) \end{Bmatrix}$
			c	2	C_{2p}	$A_g + A_u + E_{1g} + E_{1u}$
			d	1	All	$A_u + E_{1u}$
$D^i_{2p+1} = D^d_{2p+1}$	$A_{2u} + E_{1u}$	$A_{2g} + E_{1g}$	a	$4(2p + 1)$	None	$3(A_{1g} + A_{1u} + A_{2g} + A_{2u}) + 6(E_{1g} + E_{1u} + \cdots + E_{pg} + E_{pu})$
			b	$2(2p + 1)$	σ_d	$2A_{1g} + A_{1u} + A_{2g} + 2A_{2u} + 3(E_{1g} + E_{1u} + \cdots + E_{pg} + E_{pu})$
			c	$2(2p + 1)$	C_2	$A_{1g} + A_{1u} + 2A_{2g} + 2A_{2u} + 3(E_{1g} + E_{1u} + \cdots + E_{pg} + E_{pu})$
			d	2	C_{2p+1}, σ_d	$A_{1g} + A_{2u} + E_{1g} + E_{1u}$
			e	1	All	$A_{2u} + E_{1u}$
$D_{2p,i} = D_{2p,h}$	$A_{2u} + E_{1u}$	$A_{2g} + E_{1g}$	a	$8p$	None	$3(A_{1g} + A_{1u} + A_{2g} + A_{2u} + B_{1g} + B_{1u} + B_{2g} + B_{2u}) + 6(E_{1g} + E_{1u} + \cdots + E_{p-1,g} + E_{p-1,u})$
			b	$4p$	σ	$2A_{1g} + A_{1u} + A_{2g} + 2A_{2u} + \begin{Bmatrix} B_{1g} + 2B_{1u} + 2B_{2g} + B_{2u} \text{ (odd } p) \\ 2B_{1g} + B_{1u} + B_{2g} + 2B_{2u} \text{ (even } p) \end{Bmatrix} + 3(E_{1g} + E_{1u} + \cdots + E_{p-1,g} + E_{p-1,u})$

Table A.1 (*cont.*)

I	II	III	IV	V	VI	VII
			c	$4p$	σ'	$2A_{1g} + A_{1u} + A_{2g} + 2A_{2u} + \begin{Bmatrix} 2B_{1g} + B_{1u} + B_{2g} + 2B_{2u} \text{ (odd } p) \\ B_{1g} + 2B_{1u} + 2B_{2g} + B_{2u} \text{ (even } p) \end{Bmatrix} + 3(E_{1g} + E_{1u} + \cdots + E_{p-1,g} + E_{p-1,u})$
			d	$4p$	σ_h	$2A_{1g} + A_{1u} + 2A_{2g} + A_{2u} + \begin{Bmatrix} B_{1g} + 2B_{1u} + B_{2g} + 2B_{2u} \text{ (odd } p) \\ 2B_{1g} + B_{1u} + 2B_{2g} + B_{2u} \text{ (even } p) \end{Bmatrix} + 2E_{1g} + 4E_{1u} + 4E_{2g} + 2E_{2u} + \cdots + \begin{cases} 4E_{p-1,g} + 2E_{p-1,u} \text{ (odd } p) \\ 2E_{p-1,g} + 4E_{p-1,u} \text{ (even } p) \end{cases}$
			e	$2p$	σ_h, C_2, σ	$A_{1g} + A_{2g} + A_{2u} + \begin{Bmatrix} B_{1u} \text{ (odd } p) \\ B_{1g} \text{ (even } p) \end{Bmatrix} + B_{2g} + B_{2u} + E_{1g} + 2E_{1u} + 2E_{2g} + E_{2u} + \cdots + \begin{cases} 2E_{p-1,g} + E_{p-1,u} \text{ (odd } p) \\ E_{p-1,g} + 2E_{p-1,u} \text{ (even } p) \end{cases}$
			f	$2p$	σ_h, C_2', σ'	$A_{1g} + A_{2g} + A_{2u} + B_{1g} + B_{1u} + \begin{Bmatrix} B_{2u} \text{ (odd } p) \\ B_{2g} \text{ (even } p) \end{Bmatrix} + E_{1g} + 2E_{1u} + 2E_{2g} + E_{2u} + \cdots + \begin{cases} 2E_{p-1,g} + E_{p-1,u} \text{ (odd } p) \\ E_{p-1,g} + 2E_{p-1,u} \text{ (even } p) \end{cases}$

			g	2	C_{2p}, σ, σ'	$A_{1g} + A_{2u} + E_{1g} + E_{1u}$
			h	1	All	$A_{2u} + E_{1u}$
$C_{2p+1,\mathrm{h}} = S_{2p+1}$	$A'' + E'_1$	$A' + E''_1$	a	$2(2p + 1)$	None	$3(A' + A'' + E'_1 + E''_1 + \cdots + E'_p + E''_p)$
			b	$2p + 1$	σ_h	$2A' + A'' + 2E'_1 + E''_1 + 2E'_2 + E''_2 + \cdots + 2E'_p + E''_p$
			c	2	C_{2p+1}	$A' + A'' + E'_1 + E''_1$
			d	1	All	$A'' + E'_1$
$D_{2p+1,h} = D_{2p+1,\mathrm{v}}$	$A''_2 + E'_1$	$A'_2 + E''_1$	a	$4(2p + 1)$	None	$3(A'_1 + A''_1 + A'_2 + A''_2) + 6(E'_1 + E''_1 + \cdots + E'_p + E''_p)$
			b	$2(2p + 1)$	σ_v	$2A'_1 + A''_1 + A'_2 + 2A''_2 + 3(E'_1 + E''_1 + \cdots + E'_p + E''_p)$
			c	$2(2p + 1)$	σ_h	$2A'_1 + A''_1 + 2A'_2 + A''_2 + 4E'_1 + 2E''_1 + \cdots + 4E'_p + 2E''_p$
			d	$2p + 1$	$\sigma_\mathrm{h}, \sigma_\mathrm{v}, C_2$	$A'_1 + A'_2 + A''_2 + 2E'_1 + E''_1 + \cdots + 2E'_p + E''_p$
			e	2	C_{2p+1}	$A'_1 + A''_2 + E'_1 + E''_1$
			f	1	All	$A''_2 + E'_1$
$C_{2p+1,\mathrm{v}}$	$A_1 + E_1$	$A_2 + E_1$	a	$2(2p + 1)$	None	$3(A_1 + A_2) + 6(E_1 + \cdots + E_p)$
			b	$2p + 1$	σ	$2A_1 + A_2 + 3(E_1 + \cdots + E_p)$
			c	1	All	$A_1 + E_1$
$C_{2p,v}$	$A_1 + E_1$	$A_2 + E_1$	a	$4p$	None	$3(A_1 + A_2 + B_1 + B_2) + 6(E_1 + \cdots + E_{p-1})$
			b	$2p$	σ	$2A_1 + A_2 + 2B_1 + B_2 + 3(E_1 + \cdots + E_{p-1})$
			c	$2p$	σ'	$2A_1 + A_2 + B_1 + 2B_2 + 3(E_1 + \cdots + E_{p-1})$
			d	1	All	$A_1 + E_1$

Table A.1 (*cont.*)

I	II	III	IV	V	VI	VII
S_{4p}	$B + E_1$	$A + E_{2p-1}$	a	$4p$	None	$3(A + B + E_1 + \cdots + E_{2p-1})$
			b	2	C_{2p}	$A + B + E_1 + E_{2p-1}$
			c	1	All	$B + E_1$
$S_{4p,v} = D_{2p,d}$	$B_1 + E_1$	$A_2 + E_{2p-1}$	a	$8p$	None	$3(A_1 + A_2 + B_1 + B_2)$ $+ 6(E_1 + E_2 + \cdots + E_{2p-1})$
			b	$4p$	σ_v	$2A_1 + A_2 + 2B_1 + B_2$ $+ 3(E_1 + E_2 + \cdots + E_{2p-1})$
			c	$4p$	C_2	$A_1 + 2A_2 + 2B_1 + B_2$ $+ 3(E_1 + E_2 + \cdots + E_{2p-1})$
			d	2	C_{2p}	$A_1 + B_1 + E_1 + E_{2p-1}$
			e	1	All	$B_1 + E_1$
T	T	T	a	12	None	$3A + 3E + 9T$
			b	6	C_2	$A + E + 5T$
			c	4	C_3	$A + E + 3T$
			d	1	All	T
T_d	T_2	T_1	a	24	None	$3A_1 + 3A_2 + 6T + 9T_1 + 9T_2$
			b	12	σ	$2A_1 + A_2 + 3E + 4T_1 + 5T_2$
			c	6	C_2, σ	$A_1 + E + 2T_1 + 3T_2$
			d	4	C_3, σ	$A_1 + E + T_1 + 2T_2$
			e	1	All	T_2

T_h	T_u	T_g	a	24	None	$3A_g + 3A_u + 3E_g + 3E_u + 9T_g + 9T_u$
			b	12	σ	$2A_g + A_u + 2E_g + E_u + 4T_g + 5T_u$
			c	8	C_3	$A_g + A_u + E_g + E_u + 3T_g + 3T_u$
			d	6	C_2, σ	$A_g + E_g + 2T_g + 3T_u$
			e	1	All	T_u
O	T_1	T_1	a	24	None	$3A_1 + 3A_2 + 6E + 9T_1 + 9T_2$
			b	12	$6C_2$	$A_1 + 2A_2 + 3E + 5T_1 + 4T_2$
			c	8	$3C_2$	$A_1 + A_2 + 2E + 3T_1 + 3T_2$
			d	6	$3C_2, C_4$	$A_1 + E + 3T_1 + 2T_2$
			e	1	All	T_1
O_h	T_{1u}	T_{1g}	a	48	None	$3(A_{1g} + A_{1u} + A_{2g} + A_{2u}) + 6(E_g + E_u)$ $+ 9(T_{1g} + T_{1u} + T_{2g} + T_{2u})$
			b	24	σ_n	$2(A_{1g} + A_{2g}) + A_{1u} + A_{2u} + 4E_g + 2E_u$ $+ 4(T_{1g} + T_{2g}) + 5(T_{1u} + T_{2u})$
			c	24	σ_d	$2A_{1g} + A_{1u} + A_{2g} + 2A_{2u} + 3E_g + 3E_u$ $+ 4T_{1g} + 5T_{1u} + 5T_{2g} + 4T_{2u}$
			d	12	C_2, σ_h, σ_d	$A_{1g} + A_{2g} + A_{2u} + 2E_g + E_u + 2T_{1g}$ $+ 3T_{1u} + 2T_{2g} + 2T_{2u}$
			e	8	C_3, σ_d	$A_{1g} + A_{2u} + E_g + E_u + T_{1g} + 2T_{1u} + 2T_{2g}$ $+ T_{2u}$
			f	6	$C_2, C_4, \sigma_h, \sigma_d$	$A_{1g} + E_g + T_{1g} + 2T_{1u} + T_{2g} + T_{2u}$
			g	1	All	T_{1u}
I	T_1	T_1	a	60	None	$3A + 9(T_1 + T_2) + 12U + 15V$
			b	30	C_2	$A + 5(T_1 + T_2) + 6U + 7V$
			c	20	C_3	$A + 3(T_1 + T_2) + 4U + 5V$
			d	12	C_5	$A + 3T_1 + T_2 + 2U + 3V$
			e	1	All	T_1

Table A.1 (*cont.*)

I	II	III	IV	V	VI	VII
I_h	T_{1u}	T_{1g}	a	120	None	$3(A_g + A_u) + 9(T_{1g} + T_{1u} + T_{2g} + T_{2u}) + 12(U_g + U_u) + 15(V_g + V_u)$
			b	60	σ	$2A_g + A_u + 4T_{1g} + 5T_{1u} + 4T_{2g} + 5T_{2u} + 6(U_g + U_u) + 8V_g + 7V_u$
			c	30	C_2, σ	$A_g + 2T_{1g} + 3T_{1u} + 2T_{2g} + 3T_{2u} + 3(U_g + U_u) + 4V_g + 3V_u$
			d	20	C_3, σ	$A_g + T_{1g} + 2T_{1u} + T_{2g} + 2T_{2u} + 2(U_g + U_u) + 3V_g + 2V_u$
			e	12	C_5, σ	$A_g + T_{1g} + 2T_{1u} + T_{2u} + U_g + U_u + 2V_g + V_u$
			f	1	All	T_{1u}

Appendix B. Expansion of the Symmetric and Antisymmetric Products of Degenerate Representations

In order to answer the question whether a given low-symmetric vibration $Q_{\Gamma\gamma}$ is Jahn-Teller active in a given electronic state $\psi_{\bar{\Gamma}\bar{\gamma}}(\mathrm{r})$, it is necessary to know whether the symmetric product $[\bar{\Gamma}^2]$ contains Γ (Sect. 2.3). If $\bar{\Gamma}$ is a double group representation, then the antisymmetric product $\{\bar{\Gamma}^2\}$ has to be examined instead of the symmetric one. In the first coloumn of Table A.2 the group of symmetry is indicated, while in the second one the expansion of $[\bar{\Gamma}^2]$ or $\{\bar{\Gamma}^2\}$ (for double groups) is given.

Table A.2 Expansion of symmetric and antisymmetric products of irreducible representations

Group	$[\Gamma^2], \{\Gamma^2\}$
$C_{\infty,\mathrm{v}}$	$[E_k^2] = A_1 + E_{2k} \quad (k = 1, 2, \dots)$
$D_{\infty,\mathrm{h}}$	$[E_{kg}^2] = [E_{ku}^2] = A_{1g} + E_{2k,g} \quad (k = 1, 2, \dots)$
C_{2p+1}	$[E_k^2] = \begin{cases} A + E_{2k} & k \le p/2 \\ A + E_{2p+1-2k} & k > p/2 \end{cases} \quad (k = 1, 2, \dots, p)$
C_{2p}	$[E_k^2] = \begin{cases} A + E_{2k} & k < p/2 \\ A + 2B & k = p/2 \\ A + E_{2p-2k} & k > p/2 \end{cases} \quad (k = 1, 2, \dots, p-1)$
D_{2p+1} $C_{2p+1,\mathrm{v}}$	$[E_k^2] = \begin{cases} A_1 + E_{2k} & k \le p/2 \\ A_1 + E_{2p+1-2k} & k > p/2 \end{cases} \quad (k = 1, \dots, p)$
D_{2p} $C_{2p,\mathrm{v}}$	$[E_k^2] = \begin{cases} A_1 + E_{2k} & k < p/2 \\ A_1 + B_1 + B_2 & k = p/2 \\ A_1 + E_{2p-2k} & k > p/2 \end{cases} \quad (k = 1, \dots, p-1)$
$C_{2p+1,i}$	$[E_{kg}^2] = [E_{ku}^2] = \begin{cases} A_g + E_{2k,g} & k \le p/2 \\ A_g + E_{2p+1-2k,g} & k > p/2 \end{cases} \quad (k = 1, \dots, p)$
$C_{2p,i}$	$[E_{kg}^2] = [E_{ku}^2] = \begin{cases} A_g + E_{2k,g} & k < p/2 \\ A_g + 2B_g & k = p/2 \\ A_g + E_{2p-2k,g} & k > p/2 \end{cases} \quad (k = 1, \dots, p-1)$
$D_{2p+1,i}$	$[E_{kg}^2] = [E_{ku}^2] = \begin{cases} A_{1g} + E_{2k,g} & k \le p/2 \\ A_{1g} + E_{2p+1-2k,g} & k > p/2 \end{cases} \quad (k = 1, \dots, p)$
$D_{2p,i}$	$[E_{kg}^2] = [E_{ku}^2] = \begin{cases} A_{1g} + E_{2k,g} & k < p/2 \\ A_{1g} + B_{1g} + B_{2g} & k = p/2 \\ A_{1g} + E_{2p-2k,g} & k > p/2 \end{cases} \quad (k = 1, \dots, p-1)$

Table A.2 *(cont.)*

Group	$[\Gamma^2], \{\Gamma^2\}$
$C_{2p+1,\mathrm{h}}$ $D_{2p+1,\mathrm{h}}$	$[E_k'^2] = [E_k''^2] = \begin{cases} A' + E'_{2k} & k \le p/2 \\ A' + E'_{2p+1-2k} & k > p/2 \end{cases} \quad (k = 1, \ldots, p)$
S_{4p}	$[E_k^2] = \begin{cases} A + E_{2k} & k < p \\ A + 2B & k = p \\ A + E_{4p-2k} & k > p \end{cases} \quad (k = 1, \ldots, 2p - 1)$
$S_{4p,\mathrm{v}}$	$[E_k^2] = \begin{cases} A_1 + E_{2k} & k < p \\ A_1 + B_1 + B_2 & k = p \\ A_1 + E_{4p-2k} & k > p \end{cases} \quad (k = 1, \ldots, 2p - 1)$
T	$[E^2] = A + E$ $[T^2] = A + E + T$ $\{\Gamma_8^2\} = A + E + T$
$T_d = 0$	$[E^2] = A + E$ $[T_{1,2}^2] = A + E + T_2$ $\{\Gamma_8^2\} = A_1 + E + T_2$
T_{h}	$[E_g^2] = [E_u^2] = A_g + E_g$ $[T_g^2] = [T_u^2] = A_g + E_g + T_g$ $\{\Gamma_{8g}^2\} = \{\Gamma_{8u}^2\} = A_g + E_g + T_g$
O_{h}	$[E_g^2] = [E_u^2] = A_{1g} + E_g$ $[T_{1g}^2] = [T_{1u}^2] = [T_{2g}^2] = [T_{2u}^2] = A_{1g} + E_g + T_{2g}$ $\{\Gamma_{8g}^2\} = \{\Gamma_{8u}^2\} = A_{1g} + E_g + T_{2g}$
I	$[T_1^2] = [T_2^2] = A + V$ $[U^2] = A + U + V$ $[V^2] = A + U + 2V$ $\{\Gamma_8^2\} = A + V$ $\{\Gamma_9^2\} = A + U + V$
I_{h}	$[T_{1g}^2] = [T_{1u}^2] = [T_{2g}^2] = [T_{2u}^2] = A_g + V_g$ $[U_g^2] = [U_u^2] = A_g + U_g + V_g$ $[V_g^2] = [V_u^2] = A_g + U_g + 2V_g$ $\{\Gamma_{8g}^2\} = \{\Gamma_{8u}^2\} = A_g + V_g$ $\{\Gamma_{9g}^2\} = \{\Gamma_{9u}^2\} = A_g + U_g + V_g$

Appendix C. Proof of the Jahn-Teller Theorem[1]

Let us formulate the Jahn-Teller theorem as follows. For all possible molecules, except linear ones, and for all degenerate terms $\bar{\Gamma}$, except the twofold Kramers degeneracy, there are nuclear displacements $Q_{\Gamma\gamma}$ for which the matrix elements $\langle\psi_{\bar{\Gamma}\gamma}(\mathrm{r})|\, V_{\Gamma\gamma}(\mathrm{r})\,|\psi_{\bar{\Gamma}\gamma_2}(\mathrm{r})\rangle$ are nonzero. Here, as in Sect. 2.2, $V_{\Gamma\gamma}(\mathrm{r})$ is the operator of the linear vibronic coupling to $Q_{\Gamma\gamma}$ and the $\psi_{\bar{\Gamma}\gamma}(\mathrm{r})$ are electronic wave functions of the degenerate electronic term $\bar{\Gamma}$. For the matrix element $\langle\psi_{\bar{\Gamma}\gamma_1}|\, V_{\Gamma\gamma}\,|\psi_{\bar{\Gamma}\gamma}\rangle$ to be nonzero it is necessary that the expansion

$$[\bar{\Gamma}^2] = \sum_{i=1}^{k} n_i \Gamma_i \qquad \text{(A.1)}$$

contains one of the low-symmetry representations Γ of the full vibrational representation. In (A.1) all the irreducible representations of the group G_0 are labelled by i, Γ_1 is the totally symmetric representation, k is the number of different irreducible representations of the group G_0, equal to the number of classes, and n_i is the coefficient of Γ_i in $[\bar{\Gamma}^2]$:

$$n_i = \frac{1}{g_0} \sum_{G_0} [\chi^2(G)]\chi_i(G)\ . \qquad \text{(A.2)}$$

Here g_0 is the order of the group G_0, $[\chi^2(G)]$ is the character of the symmetrized (or antisymmetrized, for double group representations) square of the representation $\bar{\Gamma}$, and $\chi_i(G)$ is the character of the representation Γ_i. All the characters are real due to the fact that the representations of the group G_0 are real or can be made real (Sect. 2.3).

For polyatomic systems possessing orbital degeneracy the reference symmetry group G_0 has at least one n-fold rotational axis C_n or one axis of rotation-reflection S_n with $n \geq 3$. This means that among the transitive sets of atoms of the molecule under consideration there is at least one which contains s atoms that do not lie on the symmetry axis C_n or S_n, and $s \geq 3$. Let us assume that G_0 is a point group, and consider the displacement of one of the atoms of this transitive set along its position vector with respect to the center of the system. Under the symmetry operations of the group G_0 this displacement transforms into other equivalent displacements generating the basis of some, in general,

[1] The proof given in this Appendix follows mainly [A.3, 4].

reducible representation Γ_D. Since any operation from G_0 which changes the position of the given nucleus transforms its displacement into the displacement of another equivalent nucleus (from the remaining $s-1$ ones), and the displacements of different nuclei are linearly independent, the dimension of the representation Γ_D is s. In the case of three and more equivalent atoms the radial displacements $u_1, u_2, \ldots$ forming the basis of Γ_D cannot produce rotations of the system as a whole, but these displacements contain, in most cases, degrees of freedom which remove the degeneracy. Although the representation Γ_D contains also translational degrees of freedom (in addition to the vibrational ones) these, as shown below, are not significant.

The expansion for Γ_D is

$$\Gamma_D = \sum_{i=1}^{k} m_i \Gamma_i \ . \tag{A.3}$$

It follows from (A.3) that for any element G from G_0

$$\chi_D(G) = \sum_{i=1}^{k} m_i \chi_i(G) \ . \tag{A.4}$$

Multiplying (A.2) by m_i, summing over i, and using (A.4) we have

$$\sum_{i=1}^{k} n_i m_i = \frac{1}{g_0} \sum_{G_0} [\chi^2(G)] \chi_D(G) \ , \tag{A.5}$$

or, separating the term $n_1 m_1$, we obtain

$$\sum_{i=2}^{k} n_i m_i = \frac{1}{g_0} \sum_{G_0} [\chi^2(G)] \chi_D(G) - n_1 m_1 \ . \tag{A.6}$$

The numbers n_i and m_i, being coefficients of Γ_i from G_0 in the reducible representations $[\bar{\Gamma}^2]$ and Γ_D, respectively, are either positive integers or zero, and therefore the sum on the left-hand side of (A.6) is also either positive or zero. The latter case is possible if the low-symmetric representations of the expansion (A.1) are not present in the expansion (A.3).

The fact that this sum is positive means that there are radial displacements which remove the degeneracy in first-order perturbation theory. Since Γ_D belongs to the full vibrational representation, the Jahn-Teller theorem will be proved if one proves the inequality

$$\frac{1}{g_0} \sum_{G_0} [\chi^2(G)] \chi_D(G) > n_1 m_1 \ . \tag{A.7}$$

For all the point groups the symmetrized square of the irreducible representation $[\bar{\Gamma}^2]$ (or antisymmetrized representation for double group representations $\{\bar{\Gamma}^2\}$) contains the totally symmetric representation Γ_1 only once. This means that $n_1 = 1$, see (A.1, 2). Therefore

$$\frac{1}{g_0}\sum_{G_0}[\chi^2(G)]\chi_D(G) > m_1 \ . \tag{A.8}$$

It follows from (A.3) and from the fact that the Γ_i are real that

$$m_1 = \frac{1}{g_0}\sum_{G_0}\chi_D(G)\chi_1(G) \ ,$$

and, since $\chi_1(G) = 1$ for all the elements G of the group G_0,

$$m_1 = \frac{1}{g_0}\sum_{G_0}\chi_D(G) \ . \tag{A.9}$$

Consequently, the inequality (A.8) can be written as

$$\frac{1}{g_0}\sum_{G_0}[\chi^2(G)]\chi_D(G) > \frac{1}{g_0}\sum_{G_0}\chi_D(G) \ . \tag{A.10}$$

Consider the representation Γ_D. In general, operation G from G_0 transforms each of the s displacements u_i into a linear combination as follows:

$$Gu_i = \sum_{j=1}^{s} G_{ji}u_i \ .$$

The set of coefficients G_{ji} forms the matrix of the representation Γ_D. On the other hand, the symmetry operation over the radial displacements just interchanges their positions, and hence $G_{ji} = \delta(u_j, Gu_i)$, where $\delta(x, y)$ is the Kronecker symbol [$\delta(x, y) = 0$ for $x \neq y$, and $\delta(x, y) = 1$ for $x = y$]. Therefore

$$\chi_D(G) = \sum_{i=1}^{s} G_{ii} = \sum_{i=1}^{s} \delta(u_i, Gu_i) \ . \tag{A.11}$$

Consider now the right-hand side of (A.10). Using (A.11), it can be rewritten

$$m_1 = \frac{1}{g_0}\sum_{i=1}^{s}\sum_{G_0}\delta(u_i, Gu_i) \ . \tag{A.12}$$

Among the elements of the group G_0 there are operations which leave the ith atom unmoved. The set of these operations H_i is a subgroup of the group G_0, while all the s sets of H_i with different i values form s classes: H_1, $H_2 = G_2H_1$, $H_3 = G_3H_1, \ldots$, where $G_2, G_3, \ldots$ are the elements of the group which transform the position of atom 1 to that of $2, 3, \ldots, s$, respectively. The order h of each of the isomorphous subgroups H_i is g_0/s since s is the index of the subgroup H_i in the group G_0. Taking into account the isomorphism of the subgroups H_i one can rewrite (A.12) as

$$m_1 = \frac{s}{g_0}\sum_{H_a}\delta(u_a, Gu_a) = \frac{s}{g_0}h = 1 \ . \tag{A.13}$$

Here the index a corresponds to any of the equivalent atoms of the transitive set under consideration.

Insert (A.11) into the left-hand side of (A.10). Using an argument similar to that employed above we have

$$\frac{s}{g_0}\sum_{G_0}[\chi^2(G)] = \frac{1}{h}\sum_{H_a}[\chi^2(G)]\chi_1^{(a)}(G) \ . \tag{A.14}$$

Here the notation $\chi_i^{(a)}(G)$ has been introduced for the characters of the irreducible representations $\Gamma_i^{(a)}$ of the group H_a, $\Gamma_1^{(a)}$ being the totally symmetric representation. Taking account of (A.13) and (A.14), the inequality (A.10) can be rewritten

$$\frac{1}{h}\sum_{H_a}[\chi^2(G)]\chi_1^{(a)}(G) > 1 \ . \tag{A.15}$$

Thus, if the symmetrized square $[\bar{\Gamma}^2]$ contains more than one totally symmetric representation $\Gamma_1^{(a)}$ of the group H_a, then the left-hand side of (A.6) is positive and the Jahn-Teller theorem is proved. There are two possible cases:

A) The representation $\bar{\Gamma}$ of the group G_0 is reducible in the subgroup H_a. In this case the expansion

$$\bar{\Gamma} = \sum_{j=1}^{k_a} l_j \Gamma_j^{(a)} \ , \tag{A.16}$$

where k_a is the number of different irreducible representations of H_a, contains more than one term. It follows from the expansion (A.16) that

$$\bar{\Gamma} \times \bar{\Gamma} = \sum_{i=1}^{k_a}\sum_{j=1}^{k_a} l_i l_j \Gamma_i^{(a)} \times \Gamma_j^{(a)} \ . \tag{A.17}$$

Hence in $[\bar{\Gamma}^2]$ there are terms of the type

$$\sum_{i=1}^{k_a} l_i^2 [\Gamma_i^{(a)2}] \ . \tag{A.18}$$

Each symmetrized square $[\Gamma_i^{(a)2}]$ contains one totally symmetric representation $\Gamma_1^{(a)}$, and taking into account that in (A.16) [and hence in (A.18)] there is more than one term, we come to the conclusion that the symmetrized (antisymmetrized) square $[\bar{\Gamma}^2]$ contains more than one totally symmetric representation. Thus the inequality (A.15) and the Jahn-Teller theorem are proved for this case.

B) The representation $\bar{\Gamma}$ of the group G_0 is irreducible in the subgroup H_a. In this case the expansion (A.16) contains only one term, i.e., $\bar{\Gamma} = \Gamma_i^{(a)}$, and $[\bar{\Gamma}^2] = [\Gamma_i^{(a)2}]$ contains only one totally symmetric representation $\Gamma_1^{(a)}$. Consequently,

$$\frac{1}{h}\sum_{H_a}[\chi^2(G)]\chi_1^{(a)}(G) = 1 \ ,$$

and (A.15) is invalid. This means that within the space of radial displacements there are no totally symmetric ones which remove the degeneracy of $\bar{\Gamma}$ in first-order perturbation theory. In this case the remaining vibrational degrees of freedom have to be tried to see whether they contain displacements which remove the degeneracy in the first order. So far no general procedure for the solution of this problem has been worked out but, as shown below, there are only three cases belonging to the case *B*, and the problem for these cases can be solved by direct verification, as has been done by Jahn and Teller.

For situations belonging to case *B*, the group H_a is either an axial group (for linear molecules) or its subgroups C_n or C_{nv}, $n \geq 3$. All these groups possess no irreducible representations of more than two dimensions, and hence the case *B* contains no more than doubly degenerate terms. Several cases are possible here.

a) $\bar{\Gamma}$ is one of the two-valued representations of a double group. The case of a two-dimensional representation corresponds to the Kramers degeneracy that cannot be removed by nuclear displacements. It is one of the exceptions to the Jahn-Teller theorem mentioned above. The double-valued representations of higher dimensionality are not irreducible in the subgroup H_a, and therefore they belong to the case *A* considered above.

b) The molecule is linear. The representations *E* of the axial group are characterized by the quantum number *L* of the momentum describing the rotations around the axis of the molecule, and $L \geq 1$. In this case $[E_L^2] = A_1 + E_{2L}$, whereas all the non-totally symmetric nuclear displacements transform as E_L with $L = 1$. (See the end of Sect. 2.2.1.) Thus, linear molecules are the second exception to the Jahn-Teller theorem.

c) The subgroup H_a coincides with C_n and C_{nv}. Here, in addition to the limitation $n \geq 3$ mentioned above, there is another one: $n \leq 5$. This is due to the fact that there are no regular polyhedra with equivalent apexes (in the sense mentioned above) with one lying on the axis of symmetry C_n for $n > 5$. Hence there are three possible cases: $n = 3, 4, 5$. The fifth order axis is possible only for icosahedral groups *I* and I_h that have no other two-dimensional representations than Kramers double-valued ones. Therefore icosahedral groups and hence the case $n = 5$ can be excluded from consideration.

The cases $n = 3, 4$ are present only for cubic groups T, T_d, T_h, O, O_h. In all these groups the cases $n = 4$ (i.e., when the subgroup H_a coincides with C_4 and C_{4v}) belong to the case *A*, since their *E* representations are not irreducible—they decompose into two one-dimensional representations. Thus, only the cases $n = 3$ remain, i.e., the cases of an *E* term in a cubic molecule in which the equivalent atoms lie on the threefold axis of symmetry. This case is possible when four atoms of the transitive set occupy the apexes of a regular cube so as to form a regular tetrahedron. The direct verification of these two cases by means of the method of Jahn and Teller proves that there exist non-totally symmetric vibrational degrees of freedom which remove the electronic degeneracy of the E term in first-order perturbation theory.

To prove the theorem as a whole, it remains to show that the inequality (A.10) is fulfilled due to the contribution of vibrational degrees of freedom to the representation Γ_D. The rotational degrees of freedom, as mentioned above, are excluded. To exclude translational degrees of freedom one can use symmetry considerations as follows. Since all the positions of a polyatomic systems in space are equivalent, we can state that its Hamiltonian is invariant with respect to translation, rotation or inversion of the space (not to be confused with the point symmetry of the molecular framework). This means that the full symmetry group G_0 of the system contains the subgroup of continous three-dimensional orthogonal transformations including inversion. This statement becomes convincing when one passes to the center of mass coordinate system. If the additional operation of space inversion is taken into account, all the irreducible representations are also characterized by parity.

The translational displacement of the system is defined by a vector, i.e., all the translational degrees of freedom change sign on inversion of the space, and hence they belong to odd representations. The representation $[\bar{\Gamma}^2]$ is defined in the basis of products $\psi^*_{\bar{\Gamma}\gamma_1}(\mathrm{r})\psi_{\bar{\Gamma}\gamma_2}(\mathrm{r})$ which do not change sign on inversion, i.e., the expansion (A.1) contains only even representations. Thus the odd representations in the expansion (A.3) are not present in the left-hand side of (A.5) and hence the translational degrees of freedom do not contribute to the inequality (A.10). The Jahn-Teller theorem is proved.

References

Chapter 1

1.1 M. Born, J.R. Oppenheimer: Ann. Phys. (Leipzig) **84**, 457–484 (1927)
1.2 M. Born, K. Huang: *Dynamical Theory of Crystal Lattices* (Oxford University Press, New York 1954) Sect. 14, Appendices VII and VIII
1.3 E. Teller: "An Historical Note", in *The Jahn-Teller Effect in Molecules and Crystals*, ed. by R. Englman (Wiley, New York 1972)
1.4 J. von Neumann, E. Wigner: Phys. Z. **30**, 467–470 (1929)
1.5 L.D. Landau, E.M. Lifshitz: *Quantum Mechanics: Non-Relativistic Theory*, 3rd ed., Course of Theoretical Physics, Vol. 3 (Pergamon, Oxford 1977)
1.6 H.A. Jahn, E. Teller: Proc. R. Soc. London, Ser. A **161**, 220–235 (1937)
1.7 J.H. Van Vleck: J. Chem. Phys. **7**, 61–71, 72–84 (1939)
1.8 W. Low: *Paramagnetic Resonance in Solids* (Academic, New York 1960)
1.9 M.D. Sturge: "The Jahn-Teller Effect in Solids", in *Solid State Physics*, ed. by Seitz, D. Turnbull, H. Ehrenreich, Vol. 20 (Academic, New York 1967) pp. 91–211
1.10 A. Abragam, B. Bleaney: *Electron Paramagnetic Resonance of Transition Ions* (Clarendon, Oxford 1970) Chap. 21
1.11 I.B. Bersuker: *The Jahn-Teller Effect and Vibronic Interactions in Modern Chemistry* (Plenum, New York 1984)
1.12 F.S. Ham: "Jahn-Teller Effects in Electron Paramagnetic Resonance Spectra", in *Electron Paramagnetic Resonance*, ed. by S. Geschwind (Plenum, New York 1972) pp. 1–119
1.13 R. Englman: *The Jahn-Teller Effect in Molecules and Crystals* (Wiley, New York 1972)
1.14 G.A. Gehring, K.A. Gehring: Rep. Prog. Phys. **38**, 1–89 (1975)
1.15 I.B. Bersuker, V. Z. Polinger: Adv. Quantum Chem. **15**, 85–160 (1982)
1.16 I.B. Bersuker: Coord. Chem. Rev. **14**, 357–412 (1975)
1.17 Yu.E. Perlin, B.S. Tsukerblat: *The Effects of Electron-Vibrational Interactions in the Optical Spectra of Paramagnetic Impurity Ions* (Shtiintsa, Kishinev 1974) [in Russian]
1.18 C.A. Bates: Phys. Rep. **35**, 187–304 (1978)
1.19 I.B. Bersuker, B.G. Vekhter: Ferroelectrics **19**, 137–150 (1978)
1.20 Yu.E. Perlin, M. Wagner: *The Dynamical Jahn-Teller Effect in Localized Systems* (North-Holland, Amsterdam 1984)
1.21 J.S. Slonczewski: Phys. Rev. **131**, 1596–1610 (1963)
1.22 C. Abulaffio, J. Irvine: Phys. Lett. **B38**, 492–494 (1972)
1.23 B.R. Judd: Can. J. Phys. **52**, 999–1044 (1974)
1.24 B.S. Lee: J. Phys. A **9**, 573–580 (1976)
1.25 L. Allen, J.H. Eberly: *Optical Resonance and Two-Level Atoms* (Wiley, New York 1975)
1.26 B. Duwall, V. Celli: Phys. Rev. **181**, 276–286 (1969)
1.27 I.B. Bersuker (ed.): *The Jahn-Teller Effect. A Bibliographic Review* (IFI/Plenum, New York 1984)
1.28 I.B. Bersuker, I.Ya. Ogurtsov: Adv. Quantum Chem. **18**, 1–84 (1986)

Chapter 2

2.1 M. Born, J.R. Oppenheimer: Ann. Phys. (Leipzig) **84**, 457–484 (1927)
2.2 M. Born, K. Huang: *Dynamical Theory of Crystal Lattices* (Oxford University Press, New York 1954) Sect. 14, Appendices VII and VIII

2.3 A.A. Kiselev: Can. J. Phys. **56**, 615–647 (1978)
2.4 S. Benk, E. Sigmund: J. Phys. C **18**, 533–550 (1985)
2.5 E.E. Nikitin: Adv. Chem. Phys. **28**, 317–377 (1975)
2.6 A. Tohru, M. Kazuo: Photochem. Photobiol. **25**, 315–326 (1977)
2.7 J. von Neumann, E. Wigner: Phys. Z. **30**, 467–470 (1929)
2.8 L.D. Landau, E.M. Lifshitz: *Quantum Mechanics: Non-Relativistic Theory*, 3rd ed. (Pergamon, Oxford 1977)
2.9 H.A. Jahn, E. Teller: Proc. R. Soc. London, Ser. A **161**, 220–235 (1937)
2.10 C.M. Longuet-Higgins: Mol. Phys. **6**, 445–460 (1963)
2.11 E.B. Wilson, Jr., J.C. Decius, P.C. Cross: *Molecular Vibrations* (McGraw-Hill, New York 1955)
2.12 C.J. Bradley, A.P. Cracknell: *The Mathematical Theory of Symmetry in Solids: Representation Theory for Point Groups and Space Groups* (Clarendon, Oxford 1972)
2.13 J.S. Griffith: *The Irreducible Tensor Method for Molecular Symmetry Groups* (Prentice-Hall, Englewood Cliffs, NJ 1962)
2.14 H.A. Jahn: Proc. R. Soc. London, Ser. A **164**, 117–131 (1938)
2.15 E. Ruch, A. Schönhofer: Theor. Chim. Acta **3**, 291–304 (1965)
2.16 E.J. Blount: J. Math. Phys. (NY) **12**, 1890–1896 (1971)
2.17 I.V.V. Raghavacharyulu: "Simple Proof of the Jahn-Teller Theorem", in Proc. Nucl. Phys. and Solid State Phys. Symp., Bangelore 1973 (Dept. of Atomic Energy, New Delhi 1973) pp. 296–298
2.18 I.V.V. Raghavacharyulu: "The Jahn-Teller Theorem for Space Groups", in Proc. Nucl. Phys. and Solid State Phys. Symp., Bangelore 1973 (Dept. of Atomic Energy, New Delhi 1973) p. 299
2.19 I.V.V. Raghavacharyulu: J. Phys. C **6**, L455–L457 (1973)
2.20 J. Birman: Phys. Rev. **125**, 1959–1961 (1962); ibid. **127**, 1093–1106 (1962)
2.21 R. Berenson, J.N. Kotzev, D.B. Litvin: Phys. Rev. B **25**, 7523–7543 (1982)
2.22 S. Aronowitz: Phys. Rev. A **14**, 1319–1325 (1976)
2.23 W.L. Clinton, B. Rice: J. Chem. Phys. **30**, 542–546 (1959)
2.24 R. Renner: Z. Phys. **92**, 172–193 (1934)

Chapter 3

3.1 L.D. Landau, E.M. Lifshitz: *Quantum Mechanics: Non-Relativistic Theory*, 3rd ed., Course of Theoretical Physics, Vol. 3 (Pergamon, Oxford 1977)
3.2 A.D. Liehr: J. Phys. Chem. **67**, 389–471 (1963)
3.3 W.J.A. Maaskant: J. Phys. C **22**, 3987–3998 (1984)
3.4 R.N. Porter, R.M. Stevens, M. Karplus: J. Chem. Phys. **49**, 5163–5178 (1968)
3.5 U. Öpik, M.H.L. Pryce: Proc. R. Soc. London, Ser. A **238**, 425–447 (1957)
3.6 H. Hellmann: *Einführung in die Quantenchemie* (Franz Deuticke, Leipzig 1937)
3.7 R.P. Feynmann: Phys. Rev. **56**, 340–343 (1939)
3.8 M. Bacci: Phys. Rev. B **17**, 4495–4498 (1978)
3.9 B.R. Judd: Can. J. Phys. **52**, 999–1044 (1974)
3.10 B.R. Judd: "Group Theoretical Approach", in *The Dynamical Jahn-Teller Effect in Localized Systems*, ed. by Yu.E. Perlin, M. Wagner (North-Holland, Amsterdam 1984) pp. 87–116
3.11 R. Dagis, I.B. Levinson: "Group Theoretical Properties of the Adiabatic Potential in Molecules", in Opt. Spectrosk.: Sb. Statei, Vol. 3 (Nauka, Leningrad 1967) pp. 3–8 [Russian ed.]; A. Ceulemans: J. Chem. Phys. **87**, 5374–5385 (1987)
3.12 M. Hamermesh: *Group Theory and Its Application to Physical Problems* (Addison-Wesley, Reading, Mass. 1964)
3.13 R.P. Feynman, F.L. Vernon, P.W. Helwarth: J. Appl. Phys. **28**, 49–52 (1957)
3.14 W. Moffitt, W. Thorson: Phys. Rev. **108**, 1251–1255 (1957)
3.15 C.H. Leung, W.H. Kleiner: Phys. Rev. B. **10**, 4434–4436 (1974)
3.16 D.R. Pooler: J. Phys. A **11**, 1045–1055 (1978)
3.17 D.R. Pooler: J. Phys. C **13**, 1029–1042 (1980)

3.18 B.R. Judd: "The Jahn-Teller Effect for Degenerate Systems", in *Spectroscopie des elements lourds dans les solides*, Colloq. Int. CNRS **255**, 127–132 (1977)
3.19 M. Escribe, A.E. Hughes: J. Phys. C **4**, 2537–2549 (1971)
3.20 L.M. Kushkuley, Yu.E. Perlin, B.S. Tsukerblat, G.R. Engelgardt: Zh. Eksp. Teor. Fiz. **70**, 2226–2235 (1976) [English transl.: Sov. Phys.—JETP]
3.21 J. Duran: Semicond. Insulators **3**, 329–350 (1978)
3.22 M.C.M. O'Brien: Phys. Rev. **187**, 407–418 (1969)
3.23 M.C.M. O'Brien: J. Phys. C **4**, 2524–2536 (1971)
3.24 V.P. Khlopin, V.Z. Polinger, I.B. Bersuker: Theor. Chim. Acta **48**, 87–101 (1978)
3.25 B.R. Judd: J. Chem. Phys. **68**, 5643–5646 (1978)
3.26 I.B. Bersuker, V.Z. Polinger: Phys. Lett. A **44**, 495–496 (1973)
3.27 I.B. Bersuker, V.Z. Polinger: Zh Eksp. Teor. Fiz. **66**, 2078–2091 (1974) [English transl.: Sov. Phys.—JETP]
3.28 S. Muramatsu, T. Iida: J. Phys. Chem. Solids **31**, 2209–2216 (1970)
3.29 M. Bacci, M.P. Fontana, A. Ranfagni, G. Viliani: Phys. Rev. B **12**, 5907–5911 (1975)
3.30 A. Ranfagni, D. Mugnai, M. Bacci, M. Montagna, O. Pilla, J. Villiani: Phys. Rev. **20**, 5358–5365 (1979)
3.31 H.A. Jahn, E. Teller: Proc. R. Soc. London, Ser. A **161**, 220–235 (1937)
3.32 R.G. Pearson: *Symmetry Rules for Chemical Reaction: Orbital Topology and Elementary Processes* (Wiley-Interscience, New York 1976)
3.33 I.B. Bersuker, N.N. Gorinchoi, V.Z. Polinger: Theor. Chim. Acta **66**, 161–172 (1984)
3.34 A. Tohru, M. Kazuo: Photochem. Photobiol. **25**, 315–326 (1977)
3.35 M. Wagner: Phys. Status. Solidi B **115**, 427–435 (1983)
3.36 E. Teller: J. Phys. Chem. **41**, 109–116 (1937)
3.37 G. Herzberg, H.C. Longuet-Higgins: Discuss. Faraday Soc. **35**, 77–82 (1963)
3.38 G. Herzberg: *Molecular Spectra and Molecular Structure*, Vol. 3: Electronic Spectra and Electronic Structure of Polyatomic Molecules (Van Nostrand, New York 1966) p. 422
3.39 T. Carrington: Discuss. Faraday Soc. 53, 27–34 (1972); Acc. Chem. Res. **7**, 20–25 (1974)
3.40 E.R. Davidson: J. Am. Chem. Soc. **99**, 397–402 (1977)
3.41 H. Köppel, W. Domcke, L.S. Cederbaum: "Multimode Molecular Dynamics Beyond the Born-Oppenheimer Approximation", in Adv. Chem. Phys., Vol. 57 ed. by I. Prigogine, S.A. Rice (Wiley, New York 1984) pp. 59–246
3.42 L.S. Cederbaum, W. Domcke, H. Köppel, W. von Niessen: Chem. Phys. **26**, 169–177 (1977)
3.43 H. Köppel, W. Domcke, L.S. Cederbaum, W. von Niessen: J. Chem. Phys. **69**, 4252–4263 (1978)
3.44 C.F. Jackels, E.R. Davidson: J. Chem. Phys. **64**, 2908–2917 (1976)
3.45 H. Köppel, L.S. Cederbaum. W. Domcke, W. von Niessen: Chem. Phys. **37**, 303–317 (1979)
3.46 I.B. Bersuker: Phys. Lett. **20**, 589–590 (1966)
3.47 I.B. Bersuker, B.G. Vekhter, M.L. Rafalovich: "Dipole Moments of Symmetrical Molecular Systems", in Abstr. of Allunion Conf. on Dipole Moments and Molecular Structure, Rostov-na-Donu, 1967, p. 9 [Russian ed.]
3.48 I.B. Bersuker: Teor. Eksp. Khim. 5, 293–299 (1969) [Theor. Exp. Chem. (Engl. Transl.)]
3.49 N. Ohnishi: Ferroelectrics **46**, 167–173 (1983)
3.50 I.B. Bersuker, B.G. Vekhter: Fiz. Tverd. Tela **9**, 2652–2660 (1967) [English transl.: Sov. Phys.—Solid State]
3.51 I.B. Bersuker, B.G. Vekhter, G.S. Danilchuk, L.S. Kremenchugskiy, A.A. Muzalevskiy: Fiz. Tverd. Tela **11**, 2452–2458 (1969) [English transl.: Sov. Phys.—Solid State]
3.52 I.B. Bersuker: Nouv. J. Chim. **4**, 139–145 (1980)
3.53 I.B. Bersuker: Teor. Eksp. Khim. **16**, 291–299 (1980) [Theor. Exp. Chem. (Engl. Transl.)]
3.54 I.B. Bersuker: Fiz. Tverd. Tela **30**, 1738–1744 (1988)
3.55 G.G. Deleo, W.B. Fowler, G.D. Watkins: Phys. Rev. B **29**, 3193–3209 (1984)
3.56 I.B. Bersuker, S.S. Stavrov: Chem. Phys. **54**, 331–340 (1981)
3.57 T. Obert, I.B. Bersuker: Czech. J. Phys. B **33**, 568–572 (1983)

3.58 S.A. Borshch, I.Ya. Ogurtsov, I.B. Bersuker: Zh. Strukt. Khim. **23**, 7–12 (1982) [English Transl.: J. Struct. Chem. (USSR)]
3.59 P.N. D'yachkov, N.V. Kharchevnikova, A.A. Levin: Khim. Fiz. **9**, 1172–1176 (1983)
3.60 P.N. D'yachkov: Dokl. Akad. Nauk SSSR **258**, 655–659 (1981)
3.61 I.B. Bersuker: *The Jahn-Teller Effect and Vibronic Interactions in Modern Chemistry* (Plenum, New York 1984)
3.62 R.B. Woodward, R. Hoffmann: *The Conservation of Orbital Symmetry* (Academic, New York 1970)
3.63 T.F. George, J. Ross: J. Chem. Phys. **55**, 3851–3866 (1971)
3.64 N.D. Sokolov, V.D. Sutula: Teor. Eksp. Khim. **5**, 620–630 (1969) [Theor. Exp. Chem. (Engl. Transl.)]
3.65 F.D. Mango: Adv. Catal. **20**, 291–325 (1969)
3.66 R.P. Messmer, A.J. Bennett: Phys. Rev. B **6**, 633–638 (1972)
3.67 I.B. Bersuker: Kinet. Katal. **18**, 1268–1282 (1977) [English transl.: Kinet. Catal. (USSR)]
3.68 I.B. Bersuker: Teor. Eksp. Khim. **14**, 3–12 (1978) [Theor. Exp. Chem. (Engl. Transl.)]
3.69 I.B. Bersuker: "Modern Aspects of Structure/Reactivity Problems for Coordination Compounds", in *Coordination Chemistry*—20 (IUPAC), ed. by D. Banerjea (Pergamon, Oxford 1980) pp. 201–218
3.70 I.B. Bersuker: Chem. Phys. **31**, 85–93 (1978)
3.71 Y. Toyozawa, M. Inoue: J. Phys. Soc. Jpn. **20**, 1289–1290 (1965); ibid. **21**, 1663–1679 (1966)
3.72 J.H. Van Vleck: Phys. Rev. **57**, 426–427 (1940)
3.73 Yu.E. Perlin, B.S. Tsukerblat: *The Effects of Electron-Vibrational Interactions in the Optical Spectra of Paramagnetic Impurity Ions* (Shtiintsa, Kishinev 1974) [in Russian]
3.74 K.W.H. Stewens: J. Phys. C **2**, 1934–1946 (1969)
3.75 F.S. Ham, W.M. Schwarz, M.C.M. O'Brien: Phys. Rev. B **185**, 584–567 (1969)
3.76 B. Halperin, R. Englman: J. Phys. C **8**, 3975–3987 (1975)
3.77 J.R. Fletcher: J. Phys. C **5**, 852–862 (1972)
3.78 M.D. Glinchuk, M.F. Deigen, A. Karmazin: Fiz. Tverd. Tela **15**, 2048–2052 (1973) [English transl.: Sov. Phys.—Solid State]
3.79 G.I. Bersuker, V.Z. Polinger, V.P. Khlopin: Fiz. Tverd. Tela **24**, 2471–2476 (1982) [English transl.: Sov. Phys.—Solid State]
3.80 G.I. Bersuker, V.Z. Polinger: Phys. Status Solidi B **125**, 401–408 (1984)
3.81 G.I. Bersuker, V.Z. Polinger: Fiz. Tverd. Tela **26**, 2549–2551 (1984) [English transl.: Sov. Phys—Solid State]
3.82 G.I. Bersuker, V.Z. Polinger: Chem. Phys. **86**, 57–66 (1984)
3.83 B.S. Tsukerblat, M.I. Belinskiy, A.V. Ablov: Dokl. Akad. Nauk SSSR **201**, 1409–1413 (1971); ibid. **207**, 125–129 (1972) [English transl.: Sov. Phys.—Dokl.]
3.84 A.V. Vaisleib, Yu.B. Rosenfeld, B.S. Tsukerblat: Fiz. Tverd. Tela **18**, 1864–1873 (1976) [English transl.: Sov. Phys.—Solid State]
3.85 T. Murao: Phys. Lett. A **49**, 33–35 (1974)
3.86 V.Z. Polinger, L.F. Chibotaru, I.B. Bersuker: Phys. Status Solidi B **129**, 615–624 (1985)
3.87 B.S. Tsukerblat, M.I. Belinskiy: *Magnetochemitry and Radiospectroscopy of Exchange Clusters* (Shtiintsa, Kishinev 1983) [Russian ed.]
3.88 J. Catterick, P. Thornton: "Structure and Physical Properties of Polynuclear Carboxylates", in Adv. Inorg. Chem. Radiochem. **20**, 291–362 (1977)
3.89 L.L. Lohr: Proc. Natl. Acad. Sci. USA **59**, 720–725 (1968)
3.90 R. Englman, B. Halperin: Phys. Rev. B **2**, 75–94 (1970)
3.91 A. Raizman, J. Barak, R. Englman, J.T. Suss: Phys. Rev. B **24**, 6262–6273 (1981)
3.92 I.B. Bersuker, B.G. Vekhter, M.L. Rafalovich: Kristallografiya **18**, 11–18 (1973) [English transl.: Sov. Phys.—Crystallogr.]
3.93 P. Novak: J. Phys. Chem. Solids **30**, 2357–2364 (1969)
3.94 M.V. Eremin, Yu.V. Yablokov, T.A. Ivanova, R.M. Gumerov: Zh. Eksp. Teor. Fiz. **87**, 220–227 (1984) [English transl.: Sov. Phys.—JETP]

3.95 M. Kurzyński: J. Phys. C **8**, 2749–2759 (1975)
3.96 P. Novak, K.W.H. Stevens: J. Phys. C **3**, 1703–1710 (1970)
3.97 M.C.G. Passeggi, K.W.H. Stevens: J. Phys. C **6**, 98–108 (1973)
3.98 T. Fujiwara: J. Phys. Soc. Jpn. **34**, 36–43 (1973)
3.99 I.B. Bersuker, B.G. Vekhter, M.L. Rafalovich: Cryst. Lattice Defects **6**, 1–6 (1975)
3.100 V.Z. Polinger, L.F. Chibotaru, I.B. Bersuker: Teor. Eksp. Khim. **20**, 1–9 (1984) [Theor. Exp. Chem. (Engl. Transl.)]
3.101 V.Z. Polinger, L.F. Chibotaru, I.B. Bersuker: Mol. Phys. **52**, 1271–1289 (1984)
3.102 D.B. Brown (ed.): *Mixed-Valence Compounds*, NATO Adv. Study Inst. (Riedel, Dordrecht 1975)
3.103 W.E. Hartfield: "Magnetism of Mixed-Valence Compounds", in [142]
3.104 C. Zener: Phys. Rev. **82**, 403–405 (1951)
3.105 P.W. Anderson, H. Hasegava: Phys. Rev. **100**, 675–681 (1955)
3.106 P.W. Anderson: "Exchange in Insulators: Superexchange, Direct Exchange, and Double Exchange", in *Magnetism: A Treatise on Modern Theory and Materials*, ed. by G.T. Rado, H. Suhl, Vol. 1 (Academic, New York 1963) Chap. 2, pp. 25–83
3.107 P.-G. de Gennes: Phys. Rev. **118**, 141–154 (1960)
3.108 S.A. Borshch, I.N. Kotov, I.B. Bersuker: Chem. Phys. Lett. **111**, 264–270 (1984)
3.109 S.A. Borshch, I.N. Kotov, I.B. Bersuker: Chem. Phys. Lett. **89**, 381–384 (1982)
3.110 B.S. Tsukerblat, M.I. Belinskiy: Fiz. Tverd. Tela **25**, 3512–3514 (1983) [English transl.: Sov. Phys.—Solid State]
3.111 M.I. Belinskiy, B.S. Tsukerblat: Fiz. Tverd. Tela **26**, 758–764 (1984) [English transl.: Sov. Phys.—Solid State]
3.112 I.B. Bersuker: *Electronic Structure and Properties of Coordination Compounds: Introduction to the Theory* 2nd ed. (Khimiya, Leningrad 1976) [Russian ed.]
3.113 R.L. Flurry, Jr.: *Symmetry Groups. Theory and Chemical Applications* (Prentice-Hall, Englewood Cliffs, NJ 1980)
3.114 S.A. Borshch, I.N. Kotov, I.B. Bersuker: Khim. Fiz. **3**, 667–671 (1984)
3.115 S.A. Borshch, I.N. Kotov, I.B. Bersuker: Teor. Eksp. Khim. **20**, 675–681 (1984) [Theor. Exp. Chem. (Engl. transl.)]
3.116 S.A. Borshch: Dokl. Akad. Nauk SSSR **280**, 652–656 (1985) [English transl.: Sov. Phys.—Dokl.]
3.117 I.B. Bersuker: Coord. Chem. Rev. **14**, 357–412 (1975)
3.118 P.N. Schatz: "A Vibronic Coupling Model for Mixed-Valence Compounds and Its Application to Real Systems", in [Ref. 3.102; pp. 115–150]

Chapter 4

4.1 A.S. Davydov: *Quantum Mechanics* (Pergamon, Oxford 1976)
4.2 B.S. Tsukerblat, Yu.B. Rosenfeld, V.Z. Polinger, B.G. Vekhter: Zh. Eksp. Teor. Fiz. **68**, 1117–1126 (1975) [English transl.: Sov. Phys.—JETP]
4.3 W. Moffitt, W. Thorson: Phys. Rev. **108**, 1251–1255 (1957)
4.4 I.B. Bersuker, V.Z. Polinger: Phys. Status Solidi B **60**, 85–96 (1973)
4.5 W. Moffitt, W. Thorson: "Some Calculations Related to Jahn-Teller Effect", in *Calcul des fonctions d'onde moleculaire*, ed. by R.A. Daudel (CNRS, Paris 1958) pp. 141–148
4.6 B. Weinstock, G.L. Goodman: Adv. Chem. Phys. **9**, 169–316 (1965)
4.7 A. Witkowski: Rocz. Chem. **35**, 1409–1418 (1961)
4.8 R.L. Fulton, M. Gouterman: J. Chem. Phys. **35**, 1059–1072 (1961)
4.9 W. Moffitt, A.D. Liehr: Phys. Rev. **106**, 1195–1200 (1957)
4.10 J.A. Pople: Mol. Phys. **3**, 16–22 (1960)
4.11 H.C. Longuet-Higgins, U. Öpik, M.H.L. Pryce, R.A. Sack: Proc. R. Soc. London, Ser. A **244**, 1–16 (1958)
4.12 J.S. Slonczewski: Phys. Rev. **131**, 1596–1610 (1963)

4.13 I.B. Bersuker, B.G. Vekhter: Fiz. Tverd. Tela **9**, 2652–2660 (1967) [English transl.: Sov. Phys.—Solid State]
4.14 M.C.M. O'Brien: Phys. Rev. **187**, 407–418 (1969)
4.15 M.C.M. O'Brien: J. Phys. C **4**, 2524–2536 (1971)
4.16 B.R. Judd: J. Chem. Phys. **68**, 5643–5646 (1978)
4.17 L.D. Landau, E.M. Lifshitz: *Quantum Mechanics: Non-Relativistic Theory*, 3rd ed., Course of Theoretical Physics, Vol. 3 (Pergamon, Oxford 1977)
4.18 Yu.E. Perlin, M. Wagner: "Introduction", in *The Dynamical Jahn-Teller Effect in Localized Systems* ed. by Yu.E. Perlin, M. Wagner (North-Holland, Amsterdam 1984) pp. 1–20
4.19 A.I. Voronin, S.P. Karkach, V.I. Osherov, V.G. Ushakov: Zh. Eksp. Teor. Phys. **71**, 884–895 (1976) [English transl.: Sov. Phys.—JETP]
4.20 M.C.M. O'Brien: J. Phys. C **9**, 2375–2382 (1976)
4.21 M.C.M. O'Brien: Proc. R. Soc. London, Ser. A **281,** 323–339 (1964)
4.22 M. Abramowitz, I.A. Stegun (eds.): *Handbook of Mathematical Functions with Formulas, Graphs and Mathematical Tables* (Dover, Mineola, NY 1964)
4.23 G.M.S. Lister, M.C.M. O'Brien: J. Phys. C **17**, 3975–3986 (1984)
4.24 I.B. Bersuker: Opt. Spectrosk. **11**, 319–324 (1961) [English transl.: Opt. Spectrosc. (USSR)]
4.25 I.B. Bersuker: Zh. Eksp. Teor. Fiz. **43**, 1315–1322 (1962) [English transl.: Sov. Phys.—JETP]
4.26 I.B. Bersuker (ed.): *The Jahn-Teller Effect. A Bibliographic Review* (IFI/Plenum, New York 1984)
4.27 I.B. Bersuker: *Electronic Structure and Properties of Coordination Compounds: Introduction to the Theory* 2nd ed. (Khimiya, Leningrad 1976) [Russian ed.]
4.28 I.B. Bersuker, V.Z. Polinger: Adv. Quantum Chem. **15**, 85–160 (1982)
4.29 I.B. Bersuker: Coord. Chem. Rev. **14**, 357–412 (1975)
4.30 I.B. Bersuker, V.Z. Polinger: Phys. Lett. A **44**, 495–496 (1973)
4.31 I.B. Bersuker, V.Z. Polinger: Zh. Eksp. Teor. Fiz. **66**, 2078–2091 (1974) [English transl.: Sov. Phys.—JETP]
4.32 I.B. Bersuker: "Tunneling Effects in Electronically Degenerate E-term Systems", in *Physics of Impurity Centres in Crystals*, Proc. Int. Seminar Selected Probl. Theory Impurity Centres Cryst., Tallinn, Sept. 1970, ed. by G.S. Zavt (Acad. Sci. Est. SSR, Tallinn 1972) pp. 479–482
4.33 A.I. Baz', Ya.B. Zeldovich. A.M. Perelomov: *Scattering, Reactions and Decays in Non-Relativistic Quantum Mechanics*, 2nd ed. (Nauka, Moscow 1971) [Russian ed.]
4.34 B.R. Judd, E.E. Vogel: Phys. Rev. B **11**, 2427–2435 (1975)
4.35 V.Z. Polinger: Fiz. Tverd. Tela **16**, 2578–2583 (1974) [English transl.: Sov. Phys.—Solid State]
4.36 R. Englman, M. Caner, S. Toaff: J. Phys. Soc. Jpn. **29**, 306–310 (1970)
4.37 S.P. Karkach, V.J. Osherov: Mol. Phys. **36**, 1069–1084 (1978)
4.38 B.P. Martinenas, R.S. Dagis: Teor. Eksp. Khim. **5**, 123–125 (1969)
4.39 G. Herzberg: *Molecular Spectra and Molecular Structure*, Vol. 3: Electronic Spectra and Electronic Structure of Polyatomic Molecules (Van Nostrand, New York 1966) p. 422
4.40 R.L. Flurry, Jr.: *Symmetry Groups. Theory and Chemical Applications* (Prentice-Hall, Englewood Cliffs, NJ 1980)
4.41 G.F. Koster, J.O. Dimmock, R.G. Wheeler, H. Statz: *Properties of the Thirty-Two Point Groups* (M.I.T. Press, Cambridge, Mass. 1963)
4.42 A.F. Devonshire: Proc. R. Soc. London, Ser. A **153**, 601–621 (1936)
4.43 P. Sauer: Z. Phys. **194**, 360–372 (1966)
4.44 F. Bridges: CRC Crit. Rev. Solid State Sci. **5**, 1–88 (1975)
4.45 M.E. Bauer, W.R. Salzman: Phys. Rev. **151**, 710–720 (1966)
4.46 M. Gomez. S.P. Bowen, J.A. Krumhansl: Phys. Rev. **153**, 1009–1024 (1967)
4.47 B.G. Dick: Phys. Status Solidi B **60**, 567–577 (1973)
4.48 F. Bridges: Phys. Status Solidi B **65**, 743–750 (1974)
4.49 C.W. Struck, F. Herzfeld: J. Chem. Phys. **44**, 464–468 (1966)
4.50 H. Uehara: J. Chem. Phys. **45**, 4536–4542 (1966)
4.51 S. Muramatsu, N. Sakamoto: J. Phys. Soc. Jpn. **44**, 1640–1646 (1978)

4.52 J.R. Hoffman, T.L. Estle: Phys. Rev. B **27**, 2640–2651 (1983)
4.53 B.R. Judd: Can. J. Phys. **52**, 999–1044 (1974)
4.54 B.R. Judd: "Group Theoretical Approach", in *The Dynamical Jahn-Teller Effect in Localized Systems*, ed. by Yu.E. Perlin, M. Wagner (North-Holland, Amsterdam 1984) pp. 87–116
4.55 M. Caner, R. Englman: J. Chem. Phys. **44**, 4054–4055 (1966)
4.56 N. Sakamoto: J. Phys. C **17**, 4791–4798 (1984)
4.57 N. Sakamoto: J. Phys. C **15**, 6379–6388 (1982)
4.58 E.R. Bernstein, J.D. Webb: Mol. Phys. **36**, 1113–1118 (1978)
4.59 M. Perić, S.D. Peyerimhoff, R.J. Buenker: Chem. Phys. Lett. **105**, 44–48 (1984)
4.60 M. Perić, S.D. Peyerimhoff: Mol. Phys. **43**, 379–400 (1983)
4.61 G. Duxburg, Ch. Jungen, J. Rostas: Mol. Phys. **48**, 719–752 (1983)
4.62 W.H. Henneker, A.P. Penner, W. Siebrand, M.Z. Zgierski: J. Chem. Phys. **69**, 1884–1896 (1978)
4.63 R. Lacroix, J. Weber, E. Duval: J. Phys. C **12**, 2065–2080 (1979)
4.64 M. Zgierski, M. Pawlikowski: J. Chem. Phys. **70**, 3444–3452 (1979)
4.65 H. Köppel, L.S. Cederbaum, W. Domcke: Chem. Phys. **69**, 175–183 (1982)
4.66 B.N. Parlett: *The Symmetrical Eigenvalue Problem* (Prentice-Hall, Englewood Cliffs, NJ 1980) Sect. 13
4.67 D.K. Faddeev, V.N. Faddeeva: *Computational Methods of Linear Algebra* (Fizmatgiz, Moscow 1963) [Russian ed.]
4.68 J.H. Wilkinson, C. Reinsch: *Handbook for Automatic Computation. Linear Algebra*, Grundlehren der mathematischen Wissenschaften, Band 186, Vol. 2 (Springer, Berlin, Heidelberg 1971)
4.69 B.R. Judd: J. Chem. Phys. **67**, 1174–1179 (1977)
4.70 S. Muramatsu, N. Sakamoto: J. Phys. Soc. Jpn. **46**, 1273–1279 (1979)
4.71 M. Sakamoto, S. Muramatsu: Phys. Rev. B **17**, 868–875 (1978)
4.72 N. Sakamoto: J. Phys. Soc. Jpn. **48**, 527–533 (1980)
4.73 N. Sakamoto: J. Phys. Soc. Jpn. **51**, 1516–1519 (1982)
4.74 N. Sakamoto: Phys. Rev. B **26**, 6438–6443 (1982)
4.75 E. Haller, L.S. Cederbaum, W. Domcke, H. Köppel: Chem. Phys. Lett. **72**, 427–431 (1980)
4.76 E. Haller, L.S. Cederbaum, W. Domcke: Mol. Phys. **41**, 1291–1315 (1980)
4.77 M.C.M. O'Brien, S.N. Evangelou: J. Phys. C **13**, 611–623 (1980)
4.78 J.R. Fletcher, M.C.M. O'Brien, S.N. Evangelou: J. Phys. A **13**, 2035–2047 (1980)
4.79 S.N. Evangelou, M.C.M. O'Brien, R.S. Perkins: J. Phys. C **C13**, 4175–4198 (1980)
4.80 D.R. Pooler: "Numerical Diagonalization Techniques in the Jahn-Teller Effect", in *The Dynamical Jahn-Teller Effect in Localized Systems*, ed. by Yu.E. Perlin, M. Wagner (North-Holland, Amsterdam 1984) pp. 199–250
4.81 M. Escribe, A.E. Hughes: J. Phys. C **4**, 2537–2549 (1971)
4.82 M.C.M. O'Brien: J. Phys. C **9**, 3153–3164 (1976)
4.83 M.C.M. O'Brien: J. Phys. C **18**, 4963–4974 (1985)
4.84 S.I. Boldyrev, V.Z. Polinger, I.B. Bersuker: Fiz. Tverd. Tela **23**, 746–753 (1981) [English transl.: Sov. Phys.—Solid State]
4.85 W. Thorson, W. Moffitt: Phys. Rev. **168**, 362–365 (1968)
4.86 D.R. Pooler, M.C.M. O'Brien: J. Phys. C **10**, 3769–3791 (1977)
4.87 A. Abragam, B. Bleaney: *Electron Paramagnetic Resonance of Transition Ions* (Clarendon, Oxford 1970) Chap. 21
4.88 J.M. Ziman: *Elements of Advanced Quantum Theory* (Cambridge University Press, Cambridge 1969)
4.89 A. Barchielli, E. Mulazzi, G. Parravicini: Phys. Rev. B **24**, 3166–3185 (1981)
4.90 H.A.M. Van Eekelen, K.W.H. Stevens: Proc. Phys. Soc. London **90**, 199–205 (1967)
4.91 N.N. Bogolubov: "Quasiaverages in the Problems of Statistical Mechanics", Preprint D-781 (Joint Inst. Nucl. Res., Dubna 1961), see also N.N. Bogolubov: *Selected Works*, Vol. 3 (Naukova Dumka, Kiev 1971) p. 174 [Russian ed.]
4.92 A.A. Abrikosov: Physics (NY) **2**, 5–20 (1965)

4.93 N. Gauthier, M.B. Walker: Phys. Rev. Lett. **31**, 1211–1214 (1973)
4.94 N. Gauthier, M.B. Walker: Can. J. Phys. **54**, 9–25 (1976)
4.95 S.N. Payne, G.E. Stedman: J. Phys. C **16**, 2679–2703, 2705–2723, 2725–2748 (1983)
4.96 A.V. Vaisleib, V.P. Oleinikov, Yu.B. Rosenfeld: Phys. Status Solidi B **109**, K35–K38 (1982)
4.97 A.V. Vaisleib, Yu.B. Rosenfeld, V.P. Oleinikov: Phys. Lett. A **89**, 41–42 (1982)
4.98 A.V. Vaisleib, V.P. Oleinikov, Yu.B. Rozenfeld: Fiz. Tverd. Tela **24**, 1074–1080 (1982) [English transl.: Sov. Phys.—Solid State]
4.99 Yu.B. Rosenfeld, A.V. Vaisleib: Zh. Eksp. Teor. Phys. **86**, 1059–1065 (1984) [English transl.: Sov. Phys.—JETP]
4.100 Yu.B. Rosenfeld, A.V. Vaisleib: J. Phys. C **19**, 1721–1738, 1739–1752 (1986)
4.101 E.M. Henley, W. Thirring: *Elementary Quantum Field Theory* (McGraw-Hill, New York 1962), Chap. 18
4.102 Yu.B. Rosenfeld, V.Z. Polinger: Zh. Eksp. Teor. Fiz. **70**, 597–609 (1976) [English transl.: Sov. Phys.—JETP]
4.103 J.S. Griffith: *The Irreducible Tensor Method for Molecular Symmetry Groups* (Prentice-Hall Englewood Cliffs, NJ 1962)
4.104 H. Lehmann: Nuovo Cimento **11**, 342–357 (1954)
4.105 D.N. Zubarev: Usp. Fiz. Nauk **71**, 71–116 (1960) [English transl.: Sov. Phys.—Usp.]
4.106 H. Barentzen, O.E. Polansky: J. Chem. Phys. **68**, 4398–4405 (1978)
4.107 M. Wagner: Z. Phys. **256**, 291–308 (1972)
4.108 M. Wagner: "Unitary Transformation Methods in Vibronic Problems", in *The Dynamical Jahn-Teller Effect in Localized Systems*, ed. by Yu.E. Perlin, M. Wagner (North-Holland, Amsterdam 1984) pp. 155–198
4.109 C.C. Chancey: J. Phys. A **17**, 3183–3194 (1984)
4.110 M. Schmutz: Physica A **101**, 1–21 (1980)
4.111 H. Barentzen, G. Olbrich, M.C.M. O'Brien: J. Phys. A **14**, 111–124 (1981)
4.112 J. Singh: Int. J. Quantum Chem. **19**, 859–871 (1981)
4.113 H.G. Reik, H. Nusser, L.A. Amarante Ribeiro: J. Phys. A **15**, 3491–3507 (1982)
4.114 H.G. Reik, F. Kaspar: Chem. Phys. Lett. **97**, 253–255 (1983)
2.115 H.G. Reik: "Non-Adiabatic Systems: Analytical Approach and Exact Results", in *The Dynamical Jahn-Teller Effect in Localized Systems*, ed. by Yu.E. Perlin, M. Wagner (North-Holland, Amsterdam 1984) pp. 117–154
4.116 C.C. Chancey, B.R. Judd: J. Phys. A **16**, 875–890 (1983)
4.117 R. Rai: Physica B + C **115**, 247–253 (1983)
4.118 J.R. Fletcher: Solid Stable Commun. **11**, 601–603 (1972)
4.119 J. Maier, E. Sigmund: Solid State Commun. **51**, 961–965 (1984)
4.120 C.J. Ballhausen: Chem. Phys. Lett. **93**, 407–409 (1982)
4.121 K.W.H. Stewens: J. Phys. C **2**, 1934–1946 (1969)
4.122 P. Steggles: J. Phys. C **10**, 2817–2830 (1977)
4.123 H. Köppel, W. Domcke, L.S. Cederbaum: "Multimode Molecular Dynamics Beyond the Born-Oppenheimer Approximation", in Adv. Chem. Phys., Vol. 57 ed. by I. Prigogine, S.A. Rice (Wiley, New York 1984) pp. 59–246
4.124 C.S. Sloane, R. Silbey: J. Chem. Phys. **56**, 6031–6043 (1972)
4.125 M.C.M. O'Brien: J. Phys. C **16**, 6345–6357 (1983)
4.126 B.M. Kogan, R.A. Suris: Zh. Eksp. Teor. Fiz. **50**, 1279–1284 (1966) [English transl.: Sov. Phys.—JETP]
4.127 N.M. Aminov, B.I. Kochelaev: Fiz. Tverd. Tela **11**, 2906–2909 (1969) [English transl.: Sov. Phys.—Solid State]
4.128 Y.B. Levinson, E.I. Rashba: Rep. Prog. Phys. **36**, 1499–1565 (1973)
4.129 V.Z. Polinger, G.I. Bersuker: Phys. Status Solidi B **95**, 403–411 (1979)
4.130 V.Z. Polinger, G.I. Bersuker: Fiz. Tverd. Tela **22**, 2545–2553 (1980) [English transl.: Sov. Phys.—Solid State]
4.131 J.C. Slonczewski, W.L. Moruzzi: Physics **3**, 237–254 (1967)

4.132 Fan Li-Chin, Ou Chin: Kexue Tonbao **17**, 101–104 (1966)
4.133 Fan Li-Chin, Ou Chin: Acta Phys. Sin. **22**, 471–486 (1966)
4.134 G.I. Bersuker, V.Z. Polinger: Zh. Eksp. Teor. Fiz. **80**, 1788–1809 (1981) [English transl.: Sov. Phys.—JETP]
4.135 G.I. Bersuker, V.Z. Polinger: Solid State Commun. **38**, 795–797 (1981)
4.136 V.Z. Polinger: Zh. Eksp. Teor. Fiz. **77**, 1503–1508 (1979) [English transl.: Sov. Phys.—JETP]
4.137 J.R. Taylor: *Scattering Theory. The Quantum Theory of Nonrelativistic Collisions* (Wiley, New York 1972)
4.138 G.I. Bersuker, V.Z. Polinger: Phys. Status Solidi B **125**, 401–408 (1984)
4.139 E. Sigmund, K. Lassmann: Phys. Status Solidi B **111**, 631–637 (1982)
4.140 S.I. Pekar: Zh. Eksp. Teor. Fiz. **20**, 510–522 (1950) [English transl.: Sov. Phys.—JETP]
4.141 S.I. Pekar: Usp. Fiz. Nauk, **50**, 197–252 (1953) [English transl.: Sov. Phys.—Usp.]
4.142 K. Huang, A. Rhys: Proc. R. Soc. London, Ser. A **204**, 406–423 (1950)
4.143 N.N. Kristoffel: *Theory of Small Radius Impurity Centers in Ionic Crystals* (Nauka, Moscow 1974) [Russian ed.]
4.144 H. Bill: "Observation of the Jahn-Teller Effect with Electron Paramagnetic Resonance" in *The Dynamical Jahn-Teller Effect in Localized Systems*, ed. by Yu.E. Perlin, M. Wagner (North-Holland, Amsterdam 1984) pp. 709–818
4.145 I.B. Bersuker, V.Z. Polinger: Zh. Strukt. Khim. **24**, 62–73 (1983) [English transl.: J. Struct. Chem. (USSR)]
4.146 M.C.M. O'Brien: J. Phys. C **5**, 2045–2063 (1972)
4.147 R. Englman, B. Halperin: Ann. de Phys. **3**, 453–478 (1978)
4.148 S.N. Evangelou. A.C. Hewson: "Numerical Renormalization-Group Approach for Locally-Coupled Electron-Phonon System", Preprint Ref. 34/82 (University of Oxford 1982) p. 27
4.149 E. Haller, H. Köppel, L.S. Cederbaum: Chem. Phys. Lett. **101**, 215–220 (1983)
4.150 E. Haller, H. Köppel. L.S. Cederbaum: Phys. Rev. Lett. **52**, 1665–1668 (1984)
4.151 H.D. Heyer, E. Haller, H. Köppel. L.S. Cederbaum: J. Phys. A **17**, L831–L836 (1984)
4.152 F.S. Ham: "A Model for the Dynamical Jahn-Teller Effect in the $^4T_{2g}$ Excited State of V^{2+} in MgO", in *Optical Properties of Ions in Crystals*, ed. By H.M. Crosswhite, H.M. Moos, Proc. Conf. Opt. Properties Ions Cryst., Baltimore 1966 (Interscience, New York 1967) pp. 357–374
4.153 F.S. Ham: Phys. Rev. Lett. **58**, 725–728 (1987)
4.154 F.S. Ham: Phys. Rev. A **138**, 1727–1740 (1965)
4.155 C.H. Leung, W.H. Kleiner: Phys. Rev. B **10**, 4434–4436 (1974)
4.156 F.S. Ham: Phys. Rev. **166**, 307–321 (1968)
4.157 J.R. Fletcher: J. Phys. C **14**, L491–L492 (1981)
4.158 R. Romestain, J. Merle d'Aubigne: Phys. Rev. B **4**, 4611–4616 (1971)
4.159 F.S. Ham: "Jahn-Teller Effects in Electron Paramagnetic Resonance Spectra", in *Electron Paramagnetic Resonance*, ed. by S. Geschwind (Plenum, New York 1972) pp. 1–119
4.160 J.R. Fletcher, G.E. Stedman: J. Phys. C **17**, 3441–3447 (1984)
4.161 N. Sakamoto: Phys. Rev. B **31**, 785–790 (1985)
4.162 M.S. Child, H.C. Longuet-Higgins: Philos. Trans. R. Soc. London, Ser. A **254**, 259–294 (1962)
4.163 B. Halperin, R. Englman: Phys. Rev. Lett. **31**, 1052–1055 (1973)
4.164 J.C. Slonczewski: Solid State Commun. **7**, 519–520 (1969)
4.165 B. Halperin, R. Emglman: Solid State Commun. **7**, 1579–1580 (1969)
4.166 B.G. Vekhter: Fiz. Tverd. Tela **25**, 1537–1539 (1983) [English transl.: Sov. Phys.—Solid State]
4.167 I.B. Bersuker, B.G. Vekhter: Fiz. Tverd. Tela **5**, 2432–2440 (1963) [English transl.: Sov. Phys.—Solid State]
4.168 M.C.M. O'Brien, D.R. Pooler: J. Phys. C **12**, 311–320 (1979)
4.169 G.G. Setser. T.L. Estle: Phys. Rev. B **17**, 999–1014 (1978)
4.170 G. Davies, C. Foy: J. Phys. C **13**, 2203–2213 (1980)
4.171 B.G. Vekhter: Phys. Lett. A **45**, 133–134 (1973)

Chapter 5

5.1 G. Herzberg: *Spectra of Diatomic Molecules*, 2nd ed. (Van Nostrand, Princeton, NJ 1950)
5.2 Yu.E. Perlin: Usp. Fiz. Nauk **80**, 553–595 (1963) [English transl.: Sov. Phys.—Usp.]
5.3 K.K. Rebane: *Elementary Theory of Vibrational Structure of Impurity Spectra in Crystals* (Nauka, Moscow 1968) [Russian ed.]
5.4 A.M. Stoneham: *Theory of Defects in Solids. Electronic Structure of Defects in Insulators and Semiconductors* (Clarendon, Oxford 1975) Chap. 17
5.5 Yu.E. Perlin, B.S. Tsukerblat: *The Effects of Electron-Vibrational Interactions in the Optical Spectra of Paramagnetic Impurity Ions* (Shtiintsa, Kishinev 1974) [in Russian]
5.6 Y. Farge, M.P. Fontana: *Electronic and Vibrational Properties of Point Defects in Ionic Crystals* (North-Holland, Amsterdam 1979) Chaps. 3, 4
5.7 M.C.M. O'Brien: "Vibronic Spectra and Structure Associated with Jahn-Teller Interactions in the Solid State", in *Vibrational Spectra and Structure: A Series of Advances* ed. by J.R. Durig, Vol. 10 (Elsevier, Amsterdam 1981) pp. 321–394
5.8 Yu.E. Perlin, B.S. Tsukerblat: "Optical Bands and Polarization Dichroism of Hahn-Teller Centers" in *The Dynamical Jahn-Teller Effect in Localized Systems*, ed. by Yu.E. Perlin, M. Wagner (North-Holland, Amsterdam 1984) pp. 251–346
5.9 A.L. Natadze, A.I. Ryskin, B.G. Vekhter: "Jahn-Teller Effects in Optical Spectra of II-IV and III–V Impurity Crystals", in *The Dynamical Jahn-Teller Effect in Localized Systems*, ed. by Yu.E. Perlin, M. Wagner (North-Holland, Amsterdam 1984) pp. 347–382
5.10 V.V. Hizhnyakov, N.N. Kristoffel: "Jahn-Teller Mercury-Like Impurities in Lonic Crystals", in *The Dynamical Jahn-Teller Effect in Localized Systems*, ed. by Yu.E. Perlin, M. Wagner (North-Holland, Amsterdam 1984) pp. 383–438
5.11 W. Ulrici: "Manifestations of the Jahn-Teller Effect in the Optical Spectra of Transition Model Impurities in Crystals", in *The Dynamical Jahn-Teller Effect in Localized Systems*, ed. by Yu.E. Perlin, M. Wagner (North-Holland, Amsterdam 1984), pp. 439–494
5.12 H. Köppel, W. Domcke, L.S. Cederbaum: Adv. Chem. Phys. **57**, 59–246 (1984)
5.13 S.I. Pekar: *Studies on the Electronic Theory of Crystals* (Gostechizdat, Moscow 1951) [Russian ed.]
5.14 W. Moffitt, W. Thorson: Some Calculations Related to Jahn-Teller Effect", in *Calcul des fonctions d'onde moleculaire*, ed. by R.A. Daudel (CNRS, Paris 1958) pp. 141–148
5.15 H.C. Longuet-Higgins, U. Öpik, M.H.L. Pryce, R.A. Sack: Proc. R. Soc. London, Ser. A **244**, 1–16 (1958)
5.16 P. Habitz, W.H.E. Schwarz: Theor. Chim. Acta **28**, 267–282 (1973)
5.17 V. Loorits: Izv. Akad. Nauk Est. SSR, Fiz.—Mat. **29**, 208–212 (1980)
5.18 H. Köppel, E. Haller, L.S. Cederbaum, W. Domcke: Mol. Phys. **41**, 669–677 (1980)
5.19 M.C.M. O'Brien: Solid State Commun. **36**, 29–32 (1980)
5.20 N. Sakamoto: J. Phys. C **15**, 6379–6388 (1982)
5.21 W. Domcke, H. Köppel, L.S. Cederbaum: Mol. Phys. **43**, 851–875 (1981)
5.22 I.B. Bersuker, V.Z. Polinger: *Vibronic Interactions in Molecules and Crystals* (Nauka, Moscow 1983) Sect. 15 [Russian ed.]
5.23 S. Muramatsu, N. Sakamoto: J. Phys. Soc. Jpn. **36**, 839–842 (1974)
5.24 M.C.M. O'Brien: Proc. Phys. Soc. London **86**, Pt. 4, 847–856 (1965)
5.25 P.R. Moran: Phys. Rev. A **137**, 1016–1027 (1965)
5.26 Y. Toyozawa, M. Inoue: J. Phys. Soc. Jpn. **20**, 1289–1290 (1965); ibid. **21**, 1663–1679 (1966)
5.27 B.G. Vekhter, Yu.E. Perlin, V.Z. Polinger, Yu.B. Rosenfeld, B.S. Tsukerblat: Cryst. Lattice Defects **3**, 61–68, 69–76 (1972)
5.28 B.G. Vekhter, Yu.E. Perlin, V.Z. Polinger, Yu.B. Rosenfeld, B.S. Tsukerblat: "Optical Transitions in Systems with Electronic Degeneracy", in *Physics of Impurity Centres in Crystals*, Proc. Int. Seminar Selected Probl. Theory Impurity Centres Cryst., Tallinn, Sept. 1970, ed. by G.S. Zavt (Acad. Sci. Est. SSR, Tallinn 1972) pp. 463–478

5.29 Yu.E. Perlin: Izv. Akad. Nauk SSSR, Ser. Fiz. **46**, 6262–6273 (1982) [English transl.: Bull. Acad. Sci. USSR, Phys. Ser.]
5.30 S. Muramatsu, K. Nasu: Phys. Status Solidi B **68**, 761–766 (1975)
5.31 M. Rueff, E. Sigmund: Phys. Status Solidi B **80**, 215–223 (1977)
5.32 H. Barentzen, G. Olbrich, M.C.M. O'Brien: J. Phys. A **14**, 111–124 (1981)
5.33 A. Barchielli: Physica A **110**, 451–470 (1982)
5.34 C. Cossart-Magos, D. Cossart, S. Leach: J. Chem. Phys. **69**, 4313–4314 (1978)
5.35 C. Cossart-Magos, D. Cossart, S. Leach: Mol. Phys. **37**, 793–830 (1979)
5.36 C. Cossart-Magos, D. Cossart, S. Leach: Chem. Phys. **41**, 345–362, 363–372 (1979)
5.37 C. Cossart-Magos, S. Leach: Chem. Phys. **48**, 329–348, 349–358 (1980)
5.38 T.A. Miller, V.E. Bondybey: Chem. Phys. Lett. **58**, 454–456 (1978)
5.39 V.E. Bondybey, J.H. English, T.A. Miller: J. Am. Chem. Soc. **100**, 5251–5252 (1978)
5.40 V.E. Bondybey, T.A. Miller, J.H. English: J. Am. Chem. Soc. **101**, 1248–1253 (1979)
5.41 V.E. Bondybey, T.A. Miller, J.H. English: J. Chem. Phys. **70**, 138–146 (1979); ibid. **71**, 1088–1100 (1979)
5.42 T.A. Miller, V.E. Bondybey, J.H. English: J. Chem. Phys. **70**, 2919–2925 (1979)
5.43 T.J. Sears. T.A. Miller, V.E. Bondybey: J. Chem. Phys. **72**, 6070–6080 (1980); ibid. **74**, 3240–3248 (1981); Faraday Discuss. Chem. Soc. **71**, 175–180 (1981)
5.44 K. Raghavachari, R.C. Haddon, T.A. Miller, V.E. Bondybey: J. Chem. Phys. **79**, 1387–1395 (1983)
5.45 E. Haller, H. Köppel, L.S. Cederbaum, W. von Nissen, G. Bieri: J. Chem. Phys. **78**, 1359–1370 (1983)
3.46 E. Haller, H. Köppel, L.S. Cederbaum, G. Bieri, W. von Niessen: Chem. Phys. Lett. **85**, 12–16 (1982)
5.47 P.J. Bassett, D.R. Lloyd: Chem. Commun. N **1**, 36–37 (1970)
5.48 W. Duch, G.A. Segal: J. Chem. Phys. **79**, 2951–2963 (1983)
5.49 J.A. Davis: "R′ Center in Lithium Fluoride: An Optical Study of Jahn-Teller Distorted System"; Ph.D. Thesis, Cornell University (1970)
5.50 F.T. Chau, L. Karlsson: Phys. Scr. **16**, 248–257 (1977)
5.51 L.S. Cederbaum, W. Domcke, H. Köppel: Chem. Phys. **33**, 319–326 (1978)
5.52 D.W. Turner, C. Baker, A.D. Baker, C.R. Brundle: *Molecular Photoelectron Spectroscopy* (Wiley-Interscience, New York 1970)
5.53 B. Scharf: Chem. Phys. Lett. **96**, 89–92 (1983)
5.54 R. Englman, M. Caner, S. Toaff: J. Phys. Soc. Jpn. **29**, 306–310 (1970)
5.55 N. Sakamoto: J. Phys. Soc. Jpn. **48**, 527–533 (1980)
5.56 C.M. Weinert, F. Forstmann, R. Grinter, D.M. Kolb: Chem. Phys. **80**, 95–111 (1983)
5.57 K. Cho: J. Phys. Soc. Jpn. **25**, 1372–1387 (1968)
5.58 M.C.M. O'Brien: J. Phys. C **9**, 3153–3164 (1976)
5.59 J. Hormes, J. Schiller: Chem. Phys. **74**, 433–439 (1983)
5.60 A.J. Honma: Sci. Light (Tokyo) **18**, 33–38 (1969)
5.61 A. Matsushima, A. Fukuda: Phys. Rev. B **14**, 3664–3671 (1976)
5.62 A. Ranfagni, D. Mugnai, M. Bacci, G. Viliani, M.P. Fontana: Adv. Phys. **32**, 823–905 (1983)
5.63 M.F. Trinkler: Izv. Akad. Nauk SSSR, Ser. Fiz. **45**, 332–336 (1981) [English transl.: Bull. Acad. Sci. USSR, Phys. Ser.]
5.64 S. Asano, Y. Nakao: J. Phys. C **14**, 3897–3914 (1981)
5.65 V.S. Sivasankar, Y. Kamishina, P.W.M. Jacobs: J. Chem. Phys. **76**, 4681–4688 (1982)
5.66 F. Cussó, F. Jaque, J.L. Martinez, F. Agulló-López, M. Manfredi, J. García Solé: Phys. Rev. B **31**, 5437–5442 (1985)
5.67 T. Schonherr. R. Wernicke, H.-H. Schmidtke: Spectrochim. Acta A **38**, 679–685 (1982)
5.68 D.J. Simkin, J.P. Martin, M. Authier-Martin, K. Oyama-Gannon, P. Fabeni, G.P. Pazzi, A. Ranfagni: Phys. Rev. B **23**, 1999–2003 (1981)
5.69 M.C.M. O'Brien: J. Phys. C **4**, 2524–2536 (1971)
5.70 M.C.M. O'Brien, S.N. Evangelou: J. Phys. C **13**, 611–623 (1980)

5.71 M.C.M. O'Brien: J. Phys. C **18**, 4963–4974 (1985)
5.72 J. Duran: Semicond. Insulators **3**, 329–350 (1978)
5.73 K. Nasu, T. Kojima: Prog. Theor. Phys. **51**, 26–42 (1974)
5.74 K. Nasu: Z. Naturforsch. **30**, 1060–1070 (1975)
5.75 S.N. Evangelou: J. Phys. C **14**, 2117–2133 (1981)
5.76 Y. Kayanuma, T. Kojima: J. Phys. Soc. Jpn. **48**, 1990–1997 (1980)
5.77 M.J. Schultz, R. Silbey: J. Chem. Phys. **65**, 4375–4383 (1976)
5.78 M. Bacci, B.D. Bhattacharyya, A. Ranfagni, G. Viliani: Phys. Lett. A **55**, 489–490 (1976)
5.79 D.R. Pooler, M.C.M. O'Brien: J. Phys. C **10**, 3769–3791 (1977)
5.80 G. Herzberg, E. Teller: Z. Phys. Chem., Abt. B **B21**, 410–446 (1933)
5.81 J. Brickmann: Mol. Phys. **35**, 155–176 (1978)
5.82 B.G. Vekhter, B.S. Tsukerblat, Yu.B. Rosenfeld: Theor. Chim. Acta **27**, 49–54 (1972)
5.83 W.H. Henneker, A.P. Penner, W. Siebrand, M.Z. Zgierski: J. Chem. Phys. **69**, 1884–1896 (1978)
5.84 M. García Sucre, F. Gény, R. Lefebore: J. Chem. Phys. **49**, 458–464 (1968)
5.85 G. Fisher, G.J. Small: J. Chem. Phys. **56**, 5934–5944 (1972)
5.86 R.J. Shaw, J.E. Kent, M.F. O'Dwyer: J. Mol. Spectrosc. **82**, 1–26 (1980)
5.87 M.V. Priyutov: Opt. Spektrosk. **51**, 90–96 (1981) [English transl.: Opt. Spectrosc. (USSR)]
5.88 H. Köppel, L.S. Cederbaum, W. Domcke: J. Chem. Phys. **77**, 2014–2022 (1982)
5.89 F. Brogli, E. Heilbronner, E. Kloster-Jensen, A. Schmelzer, A.S. Manocha, J.A. Pople, L. Radom: Chem. Phys. **4**, 107–119, (1974)
5.90 L.S. Cederbaum, W. Domcke, H. Köppel, W. von Niessen: Chem. Phys. **26**, 169–177 (1977)
5.91 H. Köppel, L.S. Cederbaum, W. Domcke, S.S. Shaik: Angew. Chem., Int. Ed. Engl. **22**, 210–224 (1983)
5.92 A.D. Liehr: Z. Naturforsch. A **16**, 641–668 (1961)
5.93 L.S. Cederbaum, W. Domcke, J. Schirmer, H. Köppel: J. Chem. Phys. **72**, 1348–1358 (1980)
5.94 H. Köppel, L.S. Cederbaum, W. Domcke, W. von Niessen: Chem. Phys. **37**, 303–317 (1979)
5.95 C. Fridh, L. Asbrink: J. Electron Spectrosc. **7**, 119–138 (1975)
5.96 D.C. Frost, S.T. Lee, C.A. McDowell: Chem. Phys. Lett. **23**, 472–475 (1973)
5.97 Ch. Jungen, A.J. Merer: "The Renner-Teller Effect", in *Molecular Spectroscopy: Modern Research*, ed. by K.N. Rao, Vol. 2 (Academic, New York 1976), pp. 127–164
5.98 G. Duxbury: "Electronic Spectra of Triatomic Molecules and the Renner-Teller Effects", in *Molecular Spectroscopy, Specialist Periodical Report*, Vol. 3 (Chemical Society, London 1975) pp. 497–573
5.99 V.P. Khlopin, B.S. Tsukerblat, Yu.B. Rosenfeld, I.B. Bersuker: Fiz. Tverd. Tela **14**, 1060–1068 (1972) [English transl.: Sov. Phys.—Solid State]
5.100 V.P. Khlopin, B.S. Tsukerblat, Yu.B. Rosenfeld, I.B. Bersuker: Phys. Lett. A **38**, 437–438 (1972)
5.101 H. Köppel, W. Domcke, L.S. Cederbaum: J. Chem. Phys. **74**, 2945–2968 (1981)
5.102 A. Agresti, J.H. Ammeter, M. Bacci: J. Chem. Phys. **84**, 1861–1871 (1984)
5.103 G. Davies: Solid State Commun. **32**, 745–747 (1979)
5.104 S. Muramatsu: J. Phys. Soc. Jpn. **50**, 1645–1651 (1981)
5.105 G. Davies: J. Phys. C **5**, L141–L154 (1982)
5.106 S. Muramatsu: J. Phys. C **16**, 4849–4852 (1983)
5.107 G. Davies, C. Foy: J. Phys. C **13**, 2203–2213 (1980)
5.108 G. Davies: Rep. Prog. Phys. **44**, 787–830 (1981)
5.109 V.Z. Arutyunyan, A.K. Petrosyan, E.G. Sharoyan, R.M. Khachatryan: Izv. Akad. Nauk Arm. SSR. Ser. Fiz. **19**, 264–270 (1984)
5.110 B. Clerjaud, A. Gelineau, F. Gendron, C. Porte, J.M. Baranowski, Z. Liro: J. Phys. C **17**, 3837–3848 (1984)
5.111 N.N. Kristoffel: Tr. Inst. Fiz. Astron., Akad. Nauk Est. SSR **12**, 20–41 (1960)
5.112 K. Nasu: Prog. Theor. Phys. **57**, 361–379 (1977)
5.113 S.G. Zazubovich: Tr. Inst. Fiz. Astron., Akad. Nauk Est. SSR **36**, 109–153 (1969)
5.114 S.G. Zazubovich, V.P. Nagirnyi, T.A. Soovik: Opt. Spektrosk. **57**, 952–954 (1984) [English transl.: Opt. Spectrosc. (USSR)]

5.115 B.G. Vekhter, B.S. Tsukerblat: Fiz. Tverd. Tela **10**, 1574–1576 (1968) [English transl.: Sov. Phys.—Solid State]

5.116 B.S. Tsukerblat, B.G. Vekhter, I.B. Bersuker, A.V. Ablov: Zh. Strukt. Khim. **11**, 102–107 (1970) [English transl.: J. Struct. Chem. (USSR)]

5.117 V.V. Khizhnyakov, I. Tehver: "Theory of Polarized Luminescence of Impurity Centers", in *Physics of Impurity Centres in Crystals*, Proc. Int. Seminar Selected Probl. Theory Impurity Centres Cryst., Tallinn, Sept. 1970, ed. by G.S. Zavt (Acad. Sci. Est. SSR, Tallinn 1972) pp. 607–626

5.118 G.S. Zavt, V.S. Plekhanov, V.V. Hizhnyakov, V.V. Shepelev: J. Phys. C **17**, 2839–2858 (1984)

5.119 Yu.E. Perlin, V.M. Polyanovskii: Fiz. Tverd. Tela **26**, 3325–3333 (1984) [English transl.: Sov. Phys.—Solid State]

5.120 R. Englman: *Non-Radiative Decay of Ions and Molecules in Solids* (North-Holland, Amsterdam 1979)

5.121 R. Englman, A. Ranfagni: Physica B + C **98**, 151–160, 161–175 (1980)

5.122 D. Mugnai, A. Ranfagni: Physica B + C **98**, 282–288 (1980)

5.123 D. Mugnai, A. Ranfagni: Phys. Rev. B **25**, 4284–4287 (1982)

5.124 R. Englman, A. Ranfagni, A. Agresti, D. Mugnai: Phys. Rev. B **31**, 6766–6774 (1985)

5.125 A. Ranfagni, D. Mugnai, R. Englman: Phys. Rep. **108**, 165–216 (1984)

5.126 M. Wagner: Phys. Status Solidi B **115**, 427–435 (1983)

5.127 H. Köppel: Chem. Phys. **77**, 359–375 (1983)

5.128 A.E. Douglas: J. Chem. Phys. **45**, 1007–1015 (1966)

5.129 V.M. Donnelly, F. Kaufman: J. Chem. Phys. **66**, 4100–4110 (1977)

5.130 E. Haller, H. Köppel, L.S. Cederbaum: J. Mol. Spectrosc. **111**, 377–397 (1985)

5.131 J.R. Flectcher, M.C.M. O'Brien, S.N. Evangelou: J. Phys. A **13**, 2035–2047 (1980)

5.132 L.S. Cederbaum, E. Haller, W. Domcke: Solid State Commun. **35**, 879–881 (1980)

5.133 E. Haller, L.S. Cederbaum, W. Domcke: Mol. Phys. **41**, 1291–1315 (1980)

5.134 H. Köppel, W. Domcke, L.S. Cederbaum: "Multimode Molecular Dynamics Beyond the Born-Oppenheimer Approximation", in Adv. Chem. Phys., Vol. 57 ed. by I. Prigogine, S.A. Rice (Wiley, New York 1984) pp. 59–246

5.135 C. Cossart-Magos: "The Jahn-Teller Effect as a Case of Electronic Nonrigidity. Application to the Ground State of the Symmetric $C_6F_3H_3$ion", in *Symmetries and Properties of Non-Rigid Molecules. A Comprehensive Survey*, Proc. Int. Sympt., Paris, July 1982; Studies in Physical and Theoretical Chemistry, Vol. 23 (Elsevier, Amsterdam 1983) pp. 265–272

5.136 E. Mulazzi, A.A. Maradudin: Solid State Commun. **41**, 487–490 (1982)

5.137 Yu.B. Rosenfeld, V.Z. Polinger: Zh. Eksp. Teor. Fiz. **70**, 597–609 (1976) [English transl.: Sov. Phys.—JETP]

5.138 C.H. Henry, S.E. Schnatterly, C.P. Slichter: Phys. Rev. **137**, 583–602 (1965)

5.139 Yu.E. Perlin, B.S. Tsukerblat: "Piezoelectric Spectroscopy of Broad Optical Bands with a Dynamic Jahn-Teller Effect", in *Crystal Spectroscopy* (Nauka, Moscow 1975) pp. 61–72 [Russian ed.]

5.140 B.S. Tsukerblat, S.Kh. Brunstein: Fiz. Tverd. Tela **20**, 1220–1222 (1978) [English transl.: Sov. Phys.—Solid State]

5.141 S.Kh. Brunstein: "The Electron-Vibrational Bands of Jahn-Teller Complexes with Open d Shells"; C.Sc. Thesis, Kishinev State University (1978) [In Russian]

5.142 S. Sugano, J. Tanabe, H. Kamimura: *Multiplets of Transition Metal Ions in Crystals* (Academic, New York 1970)

5.143 L.M. Kushkuley, Yu.E. Perlin, B.S. Tsukerblat, G.R. Engelgardt: Zh. Eksp. Teor. Fiz. **70**, 2226–2235 (1976) [English transl.: Sov. Phys.—JETP]

5.144 Yu.E. Perlin, L.S. Kharchenko, B.S. Tsukerblat: Izv. Akad. Nauk SSSR, Ser. Fiz. **40**, 1770–1777 (1976) [Bull. Acad. Sci. USSR, Phys. Ser. (Engl. Transl.)]

5.145 L.M. Kushkuley, B.S. Tsukerblat: Fiz. Tverd. Tela **21**, 2254–2263 (1979) [English Transl.: Sov. Phys.—Solid State]

5.146 L.M. Kushkuley, Yu.E. Perlin, B.S. Tsukerblat: Fiz. Tverd. Tela **19**, 2178–2188 (1977) [English transl.: Sov. Phys.—Solid State]
5.147 Yu.E. Perlin, B.S. Tsukerblat, D.T. Singh: Fiz. Tverd. Tela **19**, 1569–1579 (1977) [English transl.: Sov. Phys.—Solid State]
5.148 Yu. E. Perlin, B.S. Tsukerblat, D.T. Singh: "Anisotropic Jahn-Teller Impurity Centers in Cublic Crystals", in *Crystal Spectroscopy* (Nauka, Leningrad 1978) pp. 11–27 [Russian ed.]
5.149 S. Sakoda, Y. Toyozawa: J. Phys. Soc. Jpn. **35**, 172–179 (1973)
5.150 S.U. Klokishner, Yu.E. Perlin, B.S. Tsukerblat: Fiz. Tverd. Tela **20**, 3201–3210 (1978) [English transl.: Sov. Phys.—Solid State]
5.151 R. Barrie, R.G. Rystephanick: Can. J. Phys. **44**, 109–138 (1966)
5.152 K.V. Korsak, M.A. Krivoglaz: Ukr. Fiz. Zh. **14**, 2019–2028 (1969) [English transl.: Ukr. Phys. J.]
5.153 E. Mulazzi, A.A. Maradudin: J. de Phys. Colloq. **37**, 114–116 (1976)
5.154 B.S. Tsukerblat, Yu.B. Rosenfeld, V.Z. Polinger, B.G. Vekhter: Zh. Eksp. Teor. Fiz. **68**, 1117–1126 (1975) [English transl.: Sov. Phys.—JETP]
5.155 M.A. Krivoglaz: Zh. Eksp. Teor. Fiz. **48**, 310–326 (1965) [English transl.: Sov. Phys.—JETP]
5.156 R.H. Silsbee: Phys. Rev. **128**, 1726–1733 (1962)
5.157 B. Scharf: Chem. Phys. Lett. **98**, 81–85 (1983)
5.158 M.D. Sturge: Phys. Rev. B **1**, 1005–1012 (1970)
5.159 A.S. Abhvani, C.A. Bates, B. Clerjaud, D.R. Pooler: J. Phys. C **15**, 1345–1351 (1982)
5.160 M.J. Ponnambalam, J.J. Markham: Solid State Commun. **43**, 555–559 (1982)
5.161 A. Hjortsberg, B. Nygren, J. Vallin, F.S. Ham: Phys. Rev. Lett. **39**, 1233–1236 (1977)
5.162 A.A. Kaplyanskiy, A.K. Przhevuskii: Opt. Spektrosk. **19**, 597–610 (1966); ibid. **20**, 1045–1057 (1966) [English transl.: Opt. Spectrosc. (USSR)]
5.163 L.L. Chase: Phys. Rev. Lett. **23**, 275–277 (1969); Phys. Rev. B **2**, 2308–2318 (1970)
5.164 J.E. Lowther: J. Phys. C **8**, 3448–3456 (1975)
5.165 A.M. Stoneham: Solid State Commun. **21**, 339–341 (1977)
5.166 S. Guha, L.L. Chase: Phys. Rev. B **12**, 1658–1675 (1975)
5.167 L.D. Landau, E.M. Lifshitz: *Quantum Mechanics: Non-Relativistic Theory*, 3rd ed., Course of Theoretical Physics, Vol. 3 (Pergamon, Oxford 1977)
5.168 W.R. Thorson: J. Chem. Phys. **29**, 938–944 (1958)
5.169 M.S. Child, H.C. Longuet-Higgins: Philos. Trans. R. Soc. London, Ser. A **254**, 259–294 (1962)
5.170 V.Z. Polinger, G.I. Bersuker: Phys. Status Solidi B **96**, 153–161 (1979)
5.171 I.B. Bersuker, B.G. Vekhter: "The Microwave "Inversion" Spectrum of the Transition Metal Complexes", in *Transactions of the Spectroscopy Commission of the Academy of Sciences of the USSR*. 1st issue: Proc. 15th Conf. Spectrosc., Minsk, 1963, Vol. 3 (Nauka, Moscow 1965) pp. 520–528 [Russian ed.]
5.172 E.M. Gyorgy, R.C. Le Graw, M.D. Sturge: J. Appl. Phys. **37**, 1303–1309 (1966)
5.173 A.I. Burstein, Yu.I. Naberukhin: Zh. Eksp. Teor. Fiz. **52**, 1202–1211 (1967) [English transl.: Sov. Phys.—JETP]
5.174 M.S. Child: J. Mol. Spectrosc. **10**, 357–365 (1963)
5.175 M.S. Child: Philos. Trans. R. Soc. London, Ser. A **255**, 31–53 (1963)
5.176 S.I. Boldyrev, V.Z. Polinger, I.B. Bersuker: Fiz. Tverd. Tela **23**, 746–753 (1981) [English transl.: Sov. Phys.—Solid State]
5.177 R.L. Fulton, M. Gouterman: J. Chem. Phys. **41**, 2280–2286 (1964)
5.178 V. Loorits: J. Phys. C **16**, L711–L715 (1983)
5.179 M.S. Child: Mol. Phys. **5**, 391–396 (1962)
5.180 J.T. Hougen: J. Mol. Spectrosc. **81**, 73–92 (1980)
5.181 I.B. Bersuker: Teor. Eksp. Khim. **5**, 293–299 (1969) [Theor. Exp. Chem. (Engl. Transl.)]
5.182 I.B. Bersuker, I.Ya. Ogurtsov, Yu.V. Shaparev: Teor. Eksp. Khim. **9**, 451–459 (1973)
5.183 I.B. Bersuker, I.Ya. Ogurtsov, Yu.V. Shaparev: Opt. Spektrosk. **36**, 315–321 (1974) [English transl.: Opt. Spectrosc. (USSR)]

5.184 G. Herzberg: *Molecular Spectra and Molecular Structure*, Vol. 2, Infrared and Raman Spectra (Van Nostrand, Princeton, NJ 1945)
5.185 I.Ya. Ogurtsov: Opt. Spektrosk. **56**, 60–65 (1984) [English transl.: Opt Spectrosc. (USSR)]
5.186 Ch. Jungen, A.J. Merer: Mol. Phys. **40**, 1–23 (1980)
5.187 Ch. Jungen, K.-E. Hallin, A.J. Merer: Mol. Phys. **40**, 25–63, 65–94 (1980)
5.188 Ch. Jungen, A.J. Merer: Mol. Phys. **40**, 95–114 (1980)
5.189 D. Gauyaco, Ch. Jungen: Mol. Phys. **41**, 383–407 (1980)
5.190 G. Duxburg, Ch. Jungen, J. Rostas: Mol. Phys. **48**, 719–752 (1983)
5.191 I.Ya. Ogurtsov: Opt. Spektrosk. **58**, 799–803 (1985) [English transl.: Opt. Spectrosc. (USSR)]
5.192 C.G. Gray: J. Phys. B **4**, 1661 (1971)
5.193 V.B. Berestetskiy, E.M. Lifshitz, L.P. Pitaevskiy: *Relativistic Quantum Theory*, Part 1 (Nauka, Moscow 1968) Sect. 60 [Russian ed.]
5.194 I.Ya. Ogurtsov, Yu.V. Shaparev, I.B. Bersuker: Opt. Spektrosk. **45**, 672–678 (1978) [English transl.: Opt. Spectrosc. (USSR)]
5.195 B. Scharf, Y.B. Band: J. Chem. Phys. **79**, 3175–3181 (1983)
5.196 Yu.B. Rosenfeld, A.V. Vaisleib: J. Phys. C **19**, 1721–1738, 1739–1752 (1986)
5.197 A.V. Vaisleib, V.P. Oleinikov, Yu.B. Rosenfeld: Fiz. Tverd. Tela **23**, 3486–3487 (1981) [English transl.: Sov. Phys.—Solid State]
5.198 S. Guha, L.L. Chase: Phys. Rev. Lett. **32**, 869–872 (1974)
5.199 M.M. Sushchinskiy: *Raman Spectra of Molecules and Crystals* (Nauka, Moscow 1969) [Russian ed.]
5.200 B.Weinstock, G.L. Goodman: Adv. Chem. Phys. **9**, 169–316 (1965)
5.201 H. Bill, O. Pilla: J. Phys. C **17**, 3263–3267 (1984)
5.202 M.C.M. O'Brien: J. Phys. C **16**, 85–106 (1983)
5.203 A.V. Vaisleib, V.P. Oleinikov, Yu.B. Rosenfeld: Phys. Status Solidi B **109**, K35–K38 (1982)
5.204 A.V. Vaisleib, Yu.B. Rosenfeld, V.P. Oleinikov: Phys. Lett. A **89**, 41–42 (1982)
5.205 A.V. Vaisleib, V.P. Oleinikov, Yu.B. Rosenfeld: Fiz. Tverd. Tela **24**, 1074–1080 (1982) [English transl.: Sov. Phys.—Solid State]
5.206 Yu.B. Rosenfeld, A.V. Vaisleib: Zh. Eksp. Teor. Phys. **86**, 1059–1065 (1984) [English transl.: Sov. Phys.—JETP]
5.207 E. Mulazzi, N. Terzi: Solid State Commun. **18**, 721–724 (1976); Phys. Rev. B **19**, 2332–2342 (1979)
5.208 I.Ya. Ogurtsov, V.L. Ostrovskiy, I.B. Bersuker: Opt. Spektrosk. **53**, 356–358 (1982) [English transl.: Opt. Spectrosc. (USSR)]
5.209 I.Ya. Ogurtsov, V.L. Ostrovski, I.B. Bersuker: Mol. Phys. **50**, 315–328 (1983)
5.210 V.L. Ostrovski, I.Ya. Ogurtsov, I.B. Bersuker: Mol. Phys. **48**, 13–24 (1983)
5.211 V.L. Ostrovski, I.Ya. Ogurtsov, I.B. Bersuker: Mol. Phys. **51**, 1205–1216 (1984)
5.212 I.B. Bersuker, I.Ya. Ogurtsov: Adv. Quantum Chem. **18**, 1–84 (1986)
5.213 Y. Fujumura, S.N. Lin: J. Chem. Phys. **70**, 247–262 (1979)
5.214 M. Pawlikowski, M.Z. Zgierski: Chem. Phys. Lett. **48**, 201–206 (1977)
5.215 M. Pawlikowski, M.Z. Zgierski: J. Raman Spectrosc. **7**, 106–110 (1978)
5.216 S. Muramatsu, K. Nasu, M. Takahashi, K. Kaya: Chem. Phys. Lett. **50**, 284–288 (1977)
5.217 S. Muramatsu, K. Nasu: J. Phys. Soc. Jpn. **46**, 189–197 (1979)
5.218 I.L. Fabelinskii: *Molecular Scattering of Light* (Nauka, Moscow 1966) [Russian ed.]
5.219 I.Ya. Ogurtsov, V.L. Ostrovskii, I.B. Bersuker: Opt. Spektrosk. **53**, 356–358 (1982) [English transl.: Opt. Spectrosc. (USSR)]
5.220 I.Ya. Ogurtsov, V.L. Ostrovskii, I.B. Bersuker: Khim. Fiz. **5**, 579–589 (1983)
5.221 A.D. Buchingham, B.J. Orr: Trans. Faraday Soc. **65**, 673–682 (1969)
5.222 I.Ya. Ogurtsov, L.A. Kazantseva: "Adiabatic Approximation for Density Matrices of Systems with the Jahn-Teller Effect", in *Cooperative Phenomena*, Proc. Int. Symp. Synergetics and Coooperative Phenomena in Solids and Macromolecules, Tallin, Sept. 1982 (Valgus, Tallin 1983) pp. 44–48
5.223 I.Ya. Ogurtsov, V.L. Ostrovski: Mol. Phys. **54**, 119–127 (1985)

5.224 I.Ya. Ogurtsov, V.L. Ostrovski, I.B. Bersuker: Mol. Phys. **50**, 315–328 (1983)
5.225 V.L. Ostrovski, I.Ya. Ogurtsov, I.B. Bersuker: Mol. Phys. **51**, 1205–1216 (1984)
5.226 B. Bleaney, D.J.E. Ingram: Proc. Phys. Soc. London A **63**, 408–409 (1950)
5.227 B. Bleaney, K.D. Bowers: Proc. Phys. Soc. London A **65**, 667–668 (1952)
5.228 A. Abragam, M.H.L. Pryce: Proc. Phys. Soc. London A **63**, 409–411 (1950)
5.229 F.S. Ham: "Jahn-Teller Effects in Electron Paramagnetic Resonance Spectra", in *Electron Paramagnetic Resonance*, ed. by S. Geschwind (Plenum, New York 1972) pp. 1–119
5.230 C.A. Bates: Phys. Rep. **35**, 187–304 (1978)
5.231 H. Bill: "Observation of the Jahn-Teller Effect with Electron Paramagnetic Resonance" in *The Dynamical Jahn-Teller Effect in Localized System*, ed. by Yu.E. Perlin, M. Wagner (North-Holland, Amsterdam 1984) pp. 709–818
5.232 A. Abragam, B. Bleaney: *Electron Paramagnetic Resonance of Transition Ions* (Clarendon, Oxford 1970) Chap. 21
5.233 R. Englman: *The Jahn-Teller Effect in Molecules and Crystals* (Wiley, New York 1972)
5.234 I.B. Bersuker (ed.): *The Jahn-Teller Effect. A Bibliographic Review* (IFI/Plenum, New York 1984)
5.235 P. Mehran, K.W.H. Stevens: Phys. Rep. **85**, 123–160 (1982)
5.236 R.E. Coffman: J. Chem. Phys. **48**, 609–618 (1968)
5.237 R.E. Coffman: Phys. Lett. **21**, 381–383 (1966)
5.238 M. Abou-Chantous, P.C. Jaussaud, C.A. Bates, J.R. Fletcher, W.S. Moore: Phys. Rev. Lett. **33**, 530–533 (1974)
5.239 B.G. Vekhter: Phys. Lett. A **45**, 133–134 (1973)
5.240 L.J. Challis, A.M. de Goer: "Phonon Spectroscopy of Jahn-Teller Ions", in *The Dynamical Jahn-Teller Effect in Localized Systems*, ed. by Yu.E. Perlin, M. Wagner (North-Holland, Amsterdam 1984) pp. 533–708
5.241 F.S. Ham: Phys. Rev. **166**, 307–321 (1968)
5.242 R.W. Reynolds, L.A. Boatner: Phys. Rev. B **12**, 4735–4754 (1975)
5.243 K. Blum: *Density Matrix. Theory and Applications* (Plenum, New York 1981) Chap. 7
5.244 L.D. Landau, E.M. Lifshitz: *Statistical Physics*, Course of Theoretical Physics, Vol. 5, 2nd ed. (Pergamon, Oxford 1969)
5.245 L.D. Landau, E.M. Lifshitz: *Electrodynamics of Continuous Media*, Course of Theoretical Physics, Vol. 8 (Pergamon, Oxford 1960)
5.246 N. Gauthier, M.B. Walker: Can. J. Phys. **54**, 9–25 (1976)
5.247 A. Abragam: *The Principles of Nuclear Magnetism* (Clarendon, Oxford 1961)
5.248 V. Z. Polinger, G.I. Bersuker: Fiz. Tverd. Tela **22**, 2545–2553 (1980) [English transl.: Sov. Phys.—Solid State]
5.249 G.I. Bersuker, V.Z. Polinger: Solid State Commun. **38**, 795–797 (1981)
5.250 I.B. Bersuker: Zh. Eksp. Teor. Fiz. **44**, 1239–1247 (1963) [English transl.: Sov. Phys—JETP]
5.251 I.B. Bersuker: "Investigations in the Field of Quantum Theory of Transition Metal Complexes"; D.Sc. Thesis, Leningrad State University (1964) [In Russian]
5.252 G.G. Setser, A.O. Barksdale, T.L. Estle: Phys. Rev. B **12**, 4720–4734 (1975)
5.253 R.W. Reynolds, L.A. Boatner, M.M. Abraham, Y. Chen: Phys. Rev. B **10**, 3802–3817 (1974)
5.254 G. Van Kalkeren, C.P. Keijzers, R. Srinivasan, E.De Boer, J.S. Wood: Mol. Phys. **48**, 1–11 (1983)
5.255 H. Bill, Y. Ravi Sekhar: J. Phys. C **16**, L889–L894 (1983)
5.256 M. Iwasaki, K. Toriyama, K. Nunome: J. Chem. Soc., Chem. Commun. No. 6, 320–322 (1983)
5.257 A.A. Bugai, V.S. Vikhnin, V.E. Kustov: Fiz. Tverd. Tela **25**, 1466–1472 (1983) [English transl.: Sov. Phys.—Solid State]
5.258 M.A. Ivanov, A.Ya. Fishman: Fiz. Tverd. Tela **26**, 3497–3499 (1984) [English transl.: Sov. Phys.—Solid State]
5.259 R.S. Rubins, N. Tello Lucio, D.K. De, T.D. Black: J. Chem. Phys. **81**, 4230–4233 (1984)
5.260 O.A. Anikeenok, R.M. Gumerov, M.V. Eremin, T.A. Ivanova, Yu.V. Yablokov: Fiz. Tverd. Tela, **26**, 2249–2253 (1984) [English transl.: Sov. Phys.—Solid State]

5.261 A.K. Petrosyan, R.M. Khachatryan, E.G. Sharoyan: Fiz. Tverd. Tela **26**, 22–28 (1984) [English transl.: Sov. Phys.—Solid State]
5.262 C.M. Srivastava, M.J. Patni, T.T. Srinivasan: J. Appl. Phys. **53**, 2107–2109 (1982)
5.263 V.N. Vasyukov, S.N. Lukin: Fiz. Tverd. Tela **27**, 1056–1061 (1985) [English transl.: Sov. Phys.—Solid State]
5.264 E. Kratochvílová, P. Novák, I. Veltruský, B.V. Mill: J. Phys. C **18**, 1671–1676 (1985)
5.265 L.A. Boatner, R.W. Reynolds, Y. Chen, M.M. Abraham: Phys. Rev. B **16**, 86–106 (1977)
5.266 A. Hudson: Mol. Phys. **10**, 575–581 (1966)
5.267 I.B. Bersuker: *The Jahn-Teller Effect and Vibronic Interactions in Modern Chemistry* (Plenum, New York 1984)
5.268 I.B. Bersuker: *Electronic Structure and Properties of Coordination Compounds: Introduction to the Theory*, 2nd ed. (Khimiya, Leningrad 1976) [Russian ed.]
5.269 I.B. Bersuker, V.Z. Polinger: Zh. Eksp. Teor. Fiz. **66**, 2078–2091 (1974) [English transl.: Sov. Phys.—JETP]
5.270 I.B. Bersuker, B.G. Vekhter: Fiz. Tverd. Tela **5**, 2432–2440 (1963) [English transl.: Sov. Phys.—Solid State]
5.271 I.B. Bersuker, S.S. Budnikov: Izv. Akad. Nauk Mold. SSR, No. 11, 14–25 (1965)
5.272 G.W. Ludwig, H.H. Woodburg: "Electron Spin Resonance in Semiconductors", in *Solid State Physics*, ed. by F. Seitz, D. Turnbull, H. Ehrenreich, Vol. 13 (Academic, New York 1962) pp. 223–304
5.273 L.N. Shen, T.L. Estle: J. Phys. C **12**, 2103–2118, 2119–2132 (1979)
5.274 A. Nahmani, R. Buisson, R. Romestain: J. de Phys. **41**, 59–65 (1980)
5.275 A.S. Abhvani, S.P. Austen, C.A. Bates: Solid State Commun. **37**, 777–778 (1981)
5.276 A.S. Abhvani, S.P. Austen, C.A. Bates, L.W. Parker, D.R. Pooler: J. Phys. C **15**, 2217–2231 (1982)
5.277 M.H. de A. Viccaro, S. Sundaram, R.R. Sharma: Phys. Rev. B, **25**, 7731–7740 (1982)
5.278 C.A. Bates: Postepy Fiz. **24**, 613–629 (1973)
5.279 M.D. Glinchuk: "Paraelectric Resonance of Off-Center Ions", in *The Dynamical Jahn-Teller Effect in Localized Systems*, ed. by Yu.E. Perlin, M. Wagner, Modern. Probl. Cond. Matter Sci., Vol. 7 (North-Holland, Amsterdam 1984) pp. 819–872
5.280 L.C. Goodfellow: "The Theory of Acoustic Paramagnetic Resonance for V^{3+} Ion in MgO and Al_2O_3"; Ph.D. Thesis, University of Nottingam, U.K. (1973)
5.281 J.K. Fletcher, K.W.H. Stevens: Solid State Phys. **2**, 444–456 (1969)
5.282 M. Abou-Ghantous, C.A. Bates, J.R. Fletcher, P.C. Jaussaud: J. Phys. C **8**, 3641–3652 (1975)
5.283 S. Guha, J. Lange: Phys. Rev. B **15**, 4157–4167 (1977)
5.284 G. Bemski, J.C. Fernandes: Phys. Lett. **6**, 10–11 (1963)
5.285 A. Gerard: Bull. Soc. Belge. Phys. No. 6, 379–388 (1965)
5.286 J.A. Tjon, A. Blume: Phys. Rev. **165**, 456–461 (1968)
5.287 F. Hartmann-Boutron: J. Phys. **29**, 47–56 (1968)
5.288 F.S. Ham: Phys. Rev. **160**, 328–333 (1967)
5.289 I.B. Bersuker, I.Ya. Ogurtsov: "The Hyperfine Interactions in Molecular Systems Taking into Account the Jahn-Teller Effect", in Conf. Republ. de Chimie Fizica Generală şi Aplicată: Resum. Lucr., Bucureşti (1968), p. 157
5.290 I.B. Bersuker, I.Ya. Ogurtsov: Fiz. Tverd. Tela **10**, 3651–3660 (1968) [English transl.: Sov. Phys.—Solid State]
5.291 I.B. Bersuker, I.Ya. Ogurtsov: "Jahn-Teller Effect and Inversion (Tunneling) Splitting in Mössbauer Spectra of Coordination Systems", in Abstr. Papers 22nd Int. Conf. Pure and Appl. Chem. Sydney (1969) p. 92
5.292 I.B. Bersuker, I.Ya. Ogurtsov, E.I. Polinkovskii: Izv. Akad. Nauk. Mold. SSR No. 4, 70–73 (1969)
5.293 I.B. Bersuker, S.A. Borshch, I.Ya. Ogurtsov: "Quadrupole Splitting in Nuclear Gamma Resonance Spectra of Orbital Degenerated *E*-Term Systems", in Abstr. 11th Eur. Congr. Mol. Spectrosc., Tallinn 1973, p. 273 (C7)

5.294 I.B. Bersuker, S.A. Borshch, I.Ya. Ogurtsov: Phys. Status Solidi B **59**, 707–714 (1973)
5.295 J.R. Regnard, J. Chappert, A. Ribeyron: Solid State Commun. **15**, 1539–1542 (1974)

Chapter 6

6.1 B.G. Vekhter, M.D. Kaplan: "Spectroscopy of Jahn-Teller Crystals", in *Crystal Spectroscopy* (Nauka, Leningrad 1978) pp. 149–159 [Russian ed.]
6.2 J. Sarfatt, A.M. Stoneham: Proc. Phys. Soc. London, Ser. A **91**, 214–221 (1967)
6.3 G.A. Gehring, K.A. Gehring: Rep. Prog. Phys. **38**, 1–89 (1975)
6.4 H. Thomas: "Theory of Jahn-Teller Transitions", in *Electron-Phonon Interactions and Phase Transitions*, ed. by T. Riste, NATO Adv. Study Inst. Ser., Ser. B, No. 29 (Plenum, New York 1977) pp. 245–270
6.5 R.L. Melcher: Phys. Acoust., No. 12, 1–77 (1976)
6.6 J.C.M. Tindemans van Eijndhoven, C.J. Kroese: J. Phys. C **8**, 3963–3974 (1975)
6.7 L.N. Bulaevskiy: Usp. Fiz. Nauk **115**, 263–300 (1973) [English transl.: Sov. Phys.—Usp.]
6.8 K.I. Kugel, D.I. Khomskiy: Zh. Eksp. Teor. Fiz. **64**, 1429–1439 (1973) [English transl.: Sov. Phys.—JETP]
6.9 KI. Kugel, D.I. Khomskiy: Fiz. Tverd. Tela **17**, 454–461 (1975) [English transl.: Sov. Phys.—Solid State]
6.10 V.G. Vaks: *Introduction to the Microscopic Theory of Ferroelectricity* (Nauka, Moscow 1973) [Russian ed.]
6.11 S.R.P. Smith: J. Phys. C **16**, 41–50 (1984)
6.12 J.H. Page, D.R. Taylor, S.R.P. Smith: J. Phys. C **17**, 51–71 (1984)
6.13 L.D. Landau, E.M. Lifshitz: *Statistical Physics*, Course of Theoretical Physics, Vol. 5, 2nd ed. (Pergamon, Oxford 1969)
6.14 P. J. Becker, M.J.M. Leask, R.N. Tyte: J. Phys. C **5**, 2027–2036 (1972)
6.15 B.S. Lee: J. Phys. C **12**, 855–863 (1979)
6.16 W.J.A. Maaskant, W.G. Haije: J. Phys. C **19**, 5295–5308 (1986)
6.17 W.G. Haije, J.A.L. Dobbelaar, M.H. Welter, W.J.A. Maaskant: J. Chem. Phys., in press
6.18 W.G. Haije, W.J.A. Maaskant: J. Phys. C **19**, 6943–6949 (1986)
6.19 W.G. Haije, W.J.A. Maaskant: J. Phys. C **20**, 2089–2096 (1987)
6.20 W.G. Haije: "The Stability of Helically Deformed Crystal Structures: Phase Transitions and the Phenomenon of Structural Resonance"; Thesis, Leiden State University (1988)
6.21 H. Tanaka, H. Dachs, K. Ito, K. Hagata: J. Phys. C **19**, 4861–4878, 4879–4896 (1986)
6.22 S.V. Vonsovskiy: *Magnetism* (Nauka, Moscow 1971) [Russian ed.]
6.23 J.B. Goodenough: *Magnetism and the Chemical Bond* (Wiley-Interscience, New York 1963) Chap. 1
6.24 R.B. Stinchcombe: J. Phys. C **6**, 2459–2483, 2484–2506, 2507–2524 (1973)
6.25 G. Schröder, H. Thomas: Z. Phys. B **25**, 369–380 (1976)
6.26 R. Englman, B. Halperin: Phys. Rev. B **2**, 75–94 (1970)
6.27 S. Kashida: J. Phys. Soc. Jpn. **45**, 414–421 (1978)
6.28 Y. Noda, M. Mori, Y. Yamada: J. Phys. Soc. Jpn. **45**, 945–966 (1978)
6.29 B. Halperin, R. Englman: Phys. Rev. B **3**, 1698–1708 (1971)
6.30 B.G. Vekhter, M.D. Kaplan: Phys. Lett. **43A**, 389–390 (1973)
6.31 R. Brout, K.A. Müller, H. Thomas: Solid State Commun. **4**, 507–511 (1966)
6.32 F.W. Sheard, G.A. Toombs: J. Phys. C **4**, 313–323 (1971)
6.33 F.W. Sheard, G.A. Toombs: Solid State Commun. **12**, 713–716 (1973)
6.34 K.M. Leung, D.L. Huber: Phys. Rev. B **19**, 5483–5494 (1979)
6.35 A.P. Young: J. Phys. C **8**, 3158–3170 (1975)
6.36 J. Feder, E. Pytte: Phys. Rev. B **8**, 3978–3981 (1973)
6.37 B.S. Lee: J. Phys. C **13**, 2651–2665 (1980)
6.38 S.V. Tyablikov: *Methods of the Quantum Theory of Magnetism*, 2nd ed. (Nauka, Moscow 1975) [Russian ed.]

6.39 J.E. Battison, A. Kasten, M.J.M. Leask, J.B. Lawry, K.J. Maxwell: J. Phys. C **9**, 1345–1350 (1976)
6.40 J. Kanamori: J. Appl. Phys. **31** suppl., 14S–23S (1960)
6.41 M. Kataoka, J. Kanamori: J. Phys. Soc. Jpn. **32**, 113–114 (1972)
6.42 Y. Kino, B. Luthi, M.E. Mulllen: J. Phys. Soc. Jpn. **33**, 687–697 (1972)
6.43 M.D. Kaplan, B.G. Vekhter: J. Phys. C **16**, L191–L194 (1983)
6.44 M.D. Kaplan: Phys. Status Solidi B **118**, 81–88 (1983)
6.45 M.D. Kaplan: Fiz. Tverd. Tela **26**, 89–95 (1984) [English transl.: Sov. Phys.—Solid State]
6.46 B.G. Vekhter, M.D. Kaplan: Zh. Eksp. Teor. Fiz. **87**, 1774–1783 (1984) [English transl.: Sov. Phys.—JETP]
6.47 M.D. Kaplan, B.G. Vekhter: Jpn. J. Appl. Phys. **24**, Suppl. 24–2, 24–26 (1985)
6.48 M.D. Kaplan: Pis'ma Zh. Eksp. Teor. Fiz. **41**, 92–95 (1985) [English transl.: JETP Lett.]
6.49 M.D. Kaplan: Pis'ma Zh. Eksp. Teor. Fiz. **35**, 89–91 (1982) [English transl.: JETP Lett.]
6.50 M.D. Kaplan: Phys. Status Solidi B **112**, 351–356 (1982)
6.51 V.L. Ginzburg: Zh. Eksp. Teor. Fiz. **19**, 36–41 (1949) [English transl.: Sov. Phys.—JETP]
6.52 A.F. Devonshire: Philos. Mag. **40**, 1040–1063 (1949)
6.53 A.F. Devonshire: Philos. Mag. **42**, 1055–1079 (1951)
6.54 V.L. Ginzburg: Fiz. Tverd. Tela **2**, 2031–2043 (1960) [English transl.: Sov. Phys.—Solid State]
6.55 P.W. Anderson: "Qualitative Considerations on the Statistics of the Phase Transitions in Ferroelectrics of the $BaTiO_3$ Type", in *Physics of Dielectrics*, ed. by G.I. Skanavi, K.V. Filippova, Proc. 2nd All-Union Conf. Phys. Dielectrics, November 1958 (Acad. Sci. USSR, Moscow 1960) pp. 290–296 [Russian ed.]
6.56 W. Cochran: Phys. Rev. Lett. **3**, 412–414 (1959)
6.57 W. Cochran: Adv. Phys. **9**, 387–423 (1960)
6.58 W. Cochran: Adv. Phys. **10**, 401–420 (1961)
6.59 I.B. Bersuker: Phys. Lett. **20**, 589–590 (1966)
6.60 K. Sinha, A. Sinha: Indian J. Pure Appl. Phys. **2**, 91–94 (1964)
6.61 R. Englman: "Microscopic Theory of Ionic Dielectrics", AEC Accession N14129, Rep. No. IA-994 (1964)
6.62 I.B. Bersuker, B.G. Vekhter: Ferroelectrics **19**, 137–150 (1978)
6.63 N.N. Kristoffel: "The Electron-Phonon Interaction and Ferroelectricity", Preprint F-3 (Inst. Phys., Dept. Phys. Mat. Techn. Sci., Acad. Sci. Est. SSR, 1977) p. 46 [In Russian]
6.64 I.B. Bersuker: "The Vibronic Theory of Ferroelectricity and Structural Phase Transitions in Dielectrics", in *Interband Model of Ferroelectrics*, ed. by E.V. Bursian et al. (Leningrad State Pedagogical Inst., Leningrad 1987) pp. 8–32 [In Russian]
6.65 I.B. Bersuker: Fiz. Tverd. Tela **30**, 1738–1744 (1988) [English transl.: Sov. Phys.—Solid State]
6.66 P.I. Konsin, N.N. Kristofel: "On the Vibronic Theory of Ferroelectrics", in *Interban Model of Ferroelectrics*, ed. by E.V. Bursian et al. (Leningrad State Pedagogical Inst., Leningrad 1987) pp. 32–68 [In Russian]
6.67 E.V. Bursian: "Polar and Coherent Effects in Ferroelectrics", in *Interband Model of Ferroelectrics*, ed. by E.V. Bursian et al. (Leningrad State Pedagogical Inst., Leningrad 1987) pp. 88–109 [In Russian]
6.68 Ya.G. Girshberg, V.I. Tamarchenko: Fiz. Tverd. Tela **18**, 1066–1072 (1976) [English transl.: Sov. Phys.—Solid State]
6.69 Ya.G. Girshberg, V.I. Tamarchenko: Fiz. Tverd. Tela **18**, 3340–3347 (1976) [English transl.: Sov. Phys.—Solid State]
6.70 N.M. Plakida, G.L. Mailyan: Fiz. Tverd. Tela **19**, 121–126 (1977) [English transl.: Sov. Phys.—Solid State]
6.71 G.L. Mailyan, N.M. Plakida: Phys. Status Solidi B **80**, 543–547 (1977)
6.72 I.B. Bersuker, B.G. Vekhter: Fiz. Tverd. Tela **9**, 2652–2660 (1967) [English transl.: Sov. Phys.—Solid State]
6.73 R. Comes, M. Lambert, A. Guiner: Solid State Commun. **6**, 715 (1968)
6.74 A.M. Quittet, M. Lambert: Solid State Commun. **12**, 1053–1055 (1973)

6.75 K.A. Müller: "Microscopic Probing of $BaTiO_3$ Ferroelectric Phase Transitions by EPR", in *Nonlinearity in Condensed Matter*, ed. by A.R. Bishop, D.K. Campbell, P. Kumar, S.E. Trullinger, Springer Ser. Solid-State Sci., Vol. 69 (Springer, Berlin, Heidelberg 1986) pp. 1–13; Helv. Phys. Acta **59**, 874–884 (1986)
6.76 V.A. Vaks: *Introduction to the Microscopic Theory of Ferroelectricity* (Nauka, Moscow 1973) [Russian ed.]
6.77 H. Bilz, A. Bussmann, G. Benedek, H. Büttner, D. Strauch: Ferroelectrics **25**, 339–342 (1980)
6.78 A. Bussman-Holder, H. Bilz, P. Vogl: Springer Tracts Mod. Phys. **99**, 51–98 (1983)
6.79 I.B. Bersuker, B.G. Vekhter, A.A. Muzalevskiy: Ferroelectrics **6**, 197–202 (1974)
6.80 Z.K. Petru, G.L. Mailyan: Teor. Fiz. **27**, 233–241 (1976) [Theor. Math. Phys. (Engl. Transl.)]
6.81 B.G. Vekhter, I.B. Bersuker: Ferroelectrics **6**, 13–14 (1973)
6.82 B.G. Vekhter, V.P. Zenchenko, I.B. Bersuker: Ferroelectrics **25**, 443–446 (1980)
6.83 V.P. Zenchenko: Fiz. Tverd. Tela **19**, 3345–3348 (1977) [English transl.: Sov. Phys.—Solid State]
6.84 N.N. Kristoffel, A.V. Gulbis: Fiz. Tverd. Tela **19**, 3071–3074 (1977) [English transl.: Sov. Phys.—Solid State]
6.85 B.G. Vekhter, V.P. Zenchenko, I.B. Bersuker: Fiz. Tverd. Tela **18**, 2325–2330 (1976) [English transl.: Sov. Phys.—Solid State]
6.86 V.P. Zenchenko, B.G. Vekhter, I.B. Bersuker: Zh. Eksp. Teor. Fiz. **82**, 1628–1639 (1982); ibid **55**, 943–949 (1982) [English transl.: Sov. Phys.—JETP]
6.87 S. Takaoka, K. Murase: Phys. Rev. B **20**, 2823–2833 (1979)
6.88 D. Wagner, D. Bäuerle: Phys. Lett. **83A**, 347–350 (1981)
6.89 V.M. Fridkin: Pis'ma Zh. Eksp. Teor. Fiz. **3**, 252–255 (1966) [English transl.: Sov. Phys.—JETP Lett.]
6.90 V.M. Fridkin: *Photoferroelectrics* (Springer, Berlin, 1979)
6.91 Ya.G. Girshberg, N.N. Trunov, E.V. Bursian. Izv. Akad. Nauk SSSR, Ser. Fiz. **47**, 541–547 (1983) [English transl.: Bull. Acad. Sci. USSR, Phys. Ser.]
6.92 N.N. Kristofel: Preprint F-21, Acad. Sci. Est. SSR, Tartu, 1984, pp. 15–26
6.93 Ya.G. Girshberg, R.Kh. Kalmullin, V.A. Egorov, E.V. Bursian: Solid State Commun. **53**, 633–636 (1985)
6.94 V.M. Fridkin, V.G. Lazarev, A.L. Shlenskii: "Magnetovoltaic Effect in Crystals without Inversion Symmetry", in *Interband Model of Ferroelectrics*, ed. by E.V. Bursian et al. (Leningrad State Pedagogical Inst., Leningrad 1987) pp. 68–88 [In Russian]
6.95 N.N. Kristoffel, P.I. Konsin: Fiz. Tverd. Tela **13**, 2513–2520; 3513–3516 (1971) [English transl.: Sov. Phys.—Solid State]
6.96 G. Chanussot: Ferroelectrics **8**, 671–683 (1974)
6.97 I.B. Bersuker, B.G. Vekhter, M.L. Rafalovich: Fiz. Tverd. Tela **15**, 946–948 (1973) [English transl.: Sov. Phys.—Solid State]
6.98 N.N. Kristoffel: Fiz. Tverd. Tela **19**, 775–780 (1977) [English transl.: Sov. Phys.—Solid State]
6.99 B.G. Vekhter: Fiz. Tverd. Tela **15**, 509–513 (1973) [English transl.: Sov. Phys.—Solid State]
6.100 B.G. Vekhter, M.D. Kaplan: Fiz. Tverd. Tela **18**, 784–788 (1976) [English transl.: Sov. Phys.—Solid State]
6.101 J. Gažo, I.B. Bersuker, J. Garaj, M. Kabesova, J. Kohout, H. Langfelderova, M. Melnik, M. Serator, F. Valach: Coord. Chem. Rev. **19**, 253–297 (1976)
6.102 J. Burdett: *Molecular Shapes. Theoretical Models of Inorganic Stereochemistry* (Wiley-Interscience, New York 1980)
6.103 I.B. Bersuker: *The Jahn-Teller Effect and Vibronic Interactions in Modern Chemistry* (Plenum, New York 1984)
6.104 I.B. Bersuker: Phase Transition **2**, 53–66 (1980)
6.105 I.B. Bersuker: Ferroelectrics **63**, 135–142 (1985)
6.106 T.K. Rebane: Teor. Eksp. Khim. **20**, 532–539 (1984) [Theor. Exp. Chem. (Engl. Transl.)]
6.107 H. Hellmann: *Einführung in die Quantenchemie* (Franz Deuticke, Leipzig 1937)
6.108 R.P. Feynman: Phys. Rev. **56**, 340–343 (1939)

6.109 I.B. Bersuker, N.N. Gorinchoi, V.Z. Polinger: Theor. Chim. Acta **66**, 161–172 (1984)
6.110 J. Birman: Phys. Rev. **125**, 1959–1961 (1962); ibid. **127**, 1093–1106 (1962)
6.111 R. Berenson, J.N. Kotzev, D.B. Litvin: Phys. Rev. B **25**, 7523–7543 (1982)
6.112 N.N. Kristofel: Fiz. Tverd. Tela **6**, 3266–3271 (1964) [English Transl.: Sov. Phys.—Solid State]
6.113 R.E. Peierls: *Quantum Theory of Solids* (Clarendon, Oxford 1955) p. 108
6.114 H. Fröhlich: Proc. R. London, Ser. A **223**, 296–305 (1954)
6.115 A.M. Afanasjev, Yu. Kagan: Zh. Eksp. Teor. Fiz. **43**, 1456–1463 (1962) [English transl.: Sov. Phys.—JETP]
6.116 P.W. Anderson, E.I. Blount: Phys. Rev. Lett. **14**, 217–219 (1965)
6.117 J. Friedel: "Phase Transitions and Electron-Phonon Couplings in Perfect Crystals. Modulated Structures", in *Electron-Phonon Interaction and Phase Transitions*, ed. by T. Riste (Plenum, New York 1977) pp. 1–49
6.118 R. Hoffman: *How Chemistry and Physics Meet in the Solid State and on Surfaces*, in press
6.119 D. Keszler, R. Hoffmann: J. Am. Chem. Soc. **109**, 118–124 (1987)
6.120 W. Tremel, R. Hoffmann: J. Am. Chem. Soc. **109**, 124–140 (1987)
6.121 W. Tremel, R. Hoffmann: Inorg. Chem. **26**, 118–127 (1987)
6.122 F. Hulliger, R. Schmelczer, D. Schwarzenbach: J. Solid State Chem. **21**, 374–374 (1977)
6.123 J.K. Burdett, T.J. McLarnan: J. Chem. **75**, 5764–5773 (1981)
6.124 J.K. Burdett, P. Haaland, T.J. McLarnan: J. Chem. Phys. **75**, 5774–5781 (1981)
6.125 Yu.A. Bychkov, L.P. Gor'kov, I.E. Dzyaloshinskii: Zh. Eksp. Teor. Fiz. **50**, 738–758 (1966) [English transl.: Sov. Phys.—JETP]
6.126 I.E. Dzyaloshinskii, A.I. Larkin: Zh. Eksp. Teor. Fiz. **61**, 791–800 (1971) [English transl.: Sov. Phys.—JETP]
6.127 A.W. Overhauser: Phys. Rev. Lett. **4**, 462–465 (1960)
6.128 G. Bilbro, W.L. McMillan: Phys. Rev. **14**, 1887–1892 (1976)
6.129 E. Matsushitz, T. Matsubara: Prog. Theo. Phys. **62**, 862–873 (1979)
6.130 A.A. Ovchinnikov, I.I. Ukrainskii, G.F. Kventsel: Usp. Fiz. Nauk **108**, 81–111 (1972) [English transl.: Sov. Phys.—Usp.]
6.131 H. Fukutome, M. Sasai: Prog. Theor. Phys. **67**, 41–67 (1982)
6.132 I.I. Ukrainskii: Zh. Eksp. Teor. Fiz. **76**, 760–768 (1979) [English transl.: Sov. Phys.—JETP]
6.133 P. Horsch: Phys. Rev. B **24**, 7351–7360 (1981)
6.134 D. Baeriswyl, K. Maki: Phys. Rev. B **31**, 6633 (1985)
6.135 S. Kivelson, D.E. Heim: Phys. Rev. B **26**, 4278–4292 (1982)
6.136 V.Ya. Krivnov, A.A. Ovchinnikov: Zh. Eksp. Teor. Fiz. **90**, 709–723 (1986) [English transl.: Sov. Phys.—JETP]
6.137 J.E. Hirsh: Phys. Rev. Lett. **51**, 296–299 (1983)
6.138 S. Mazumdar, S.N. Dixit: Phys. Rev. Lett. **51**, 292–295 (1983)
6.139 W.A. Little: Phys. Rev. A **134**, 1416–1424 (1966)
6.140 J. Labbé, J. Friedel: J. de Phys. **27**, 153–165, 303–308, 708–716 (1966)
6.141 J. Labbé: Phys. Rev. **158**, 647–654 (1967)
6.142 S. Barišić: Phys. Lett. **34A**, 188–189 (1971)
6.143 S. Barišić, S. Marcelja: Solid State Commun. **7**, 1395–1398 (1969)
6.144 J. Labbé: J. de Phys. **29**, 195–200 (1968)
6.145 J. Labbé, S. Barišić, J. Friedel: Phys. Rev. Lett. **19**, 1039–1041 (1967)
6.146 L.P. Gor'kov: Pis'ma Zh. Eksp. Teor. Fiz. **17**, 525–529 (1973) [English transl.: Sov. Phys.—JETP Lett.]
6.147 L.P. Gor'kov: Zh. Eksp. Teor. Fiz. **65**, 1658–1676 (1973) [English transl.: Sov. Phys.—JETP]
6.148 L.P. Gor'kov, O.N. Dorokhov: J. Low Temp. Phys. **22**, 1–26 (1976)
6.149 W.P. Su, J.R. Schrieffer, A.J. Heeger: Phys. Rev. B **22**, 2099–2111 (1980)
6.150 S.A. Brazovskii: Zh. Eksp. Teor. Fiz. **78**, 677–699 (1980) [English transl.: Sov. Phys.—JETP]
6.151 H. Takayama, Y.R. Lin-Liu, K. Maki: Phys. Rev. B **21**, 2388–2393 (1980)
6.152 J.-J. André, A. Bieber, F. Gautier: Ann. de Phys. **1**, 145–256 (1976)
6.153 G.A. Toombs: Phys. Rep. **40C**, 181–240 (1978)

6.154 R.H. Friend, D. Jerome: J. Phys. C **12**, 1441–1447 (1979)
6.155 A.J. Berlinsky: Rep. Prog. Phys. **42**, 1243–1283 (1979)
6.156 A.J. Heeger: Comments Solid State Phys. **10**, 133 (1982)
6.157 M. Kertész: Adv. Quantum Chem. **15**, 161–214 (1982)
6.158 A.-H. Whangbo: In *Crystal Chemistry and Properties of Materials with Quasi-One-Dimensional Structures: A Chemical and Physical Synthetic Approach*, ed. by. J. Rouxel (Springer, Berlin, Heidelberg 1986) p. 27
6.159 H. Böttger: *Principles of the Theory of Lattice Dynamics* (Akademie-Verlag, Berlin 1983) Sect. 4.4
6.160 M.E. Eberhart, K.H. Johnson, D. Adler, R.C. O'Handley: J. Non-Cryst. Solids **83**, 12–26 (1986)
6.161 J.F. Sadoc: J. Non-Cryst. Solids **44**, 1–16 (1981)
6.162 P.J. Steinhardt, D.R. Nelson, M. Renchetti: Phys. Rev. Lett. **47**, 1297–1300 (1981)
6.163 M.E. Eberhart: Solid State Commun. **54**, 187–191 (1985)
6.164 J.G. Bednorz, M. Takashide, K.A. Müller: Europhys. Lett. **3**, 379–385 (1987)
6.165 J.G. Bednorz, K.A. Müller: Z. Phys. B **64**, 189–193 (1986)
6.166 C.W. Chu, P.H. Hor, R.L. Meng, L. Gao, Z.J. Huang, Y.O. Wang: Phys. Rev. Lett. **58**, 405–407 (1987)
6.167 R.L. Gava, R.B. van Dover, B. Battlog, E.A. Rietman: Phys. Rev. Lett. **58**, 408–410 (1987)
6.168 R.K. Nesbet: Phys. Rev. **126**, 2014–2020 (1962)
6.169 K.H. Johnson, R.P. Messmer: Synth. Met. **5**, 151–204 (1983)
6.170 M.D. Kaplan, D.I. Khomskii: "Pseudo-Jahn-Teler Interactions in the High Temperature Superconductors", in *Programme and Abstracts IXth Jahn-Teller Symp.*, Nottingham, Sept. 14–18, 1987 (Physics Department, Nottingham University 1987) p. 24
6.171 M.D. Kaplan, D.I. Khomskii: Pis'ma Zh. Eksp. Teor. Fiz. **47**, 631–633 (1988) [English transl.: JETP Lett.
6.172 H. Aoki, H. Kamimura: Solid State Commun. **63**, 665–669 (1987)
6.173 M. Georgiev, M. Borissov: Physica C **153–155**, 208–209 (1988)
6.174 K.H. Johnson, M.E. McHenry, C. Counterman, A. Collins, M.M. Donovan, R.C. O'Handley, G. Kalonji: Physica C **153–155**, 1165–1166 (1988)
6.175 G. Ries: Physica C **153–155**, 235–236 (1988)
6.176 Xiong Shi-jie: J. Phys. C **21**, L459–L462 (1988)
6.177 H. Kuratsuji: Phys. Lett. A **128**, 286–288 (1988)

Appendix A

A.1 E.B. Wilson, Jr., J.C. Decius, P.C. Cross: *Molecular Vibrations* (McGraw-Hill, New York 1955)
A.2 H.A. Jahn, E. Teller: Proc. R. Soc. London, Ser. A **161**, 220–235 (1937)
A.3 E. Ruch, A. Schönhofer: Theor. Chim. Acta **3**, 291–304 (1965)
A.4 E.J. Blount: J. Math. Phys. (NY) **12**, 1890–1896 (1971)

Subject Index

Printed by Publishers' Graphics LLC